Vector & Tensor Analysis

LOUIS BRAND

Dover Publications
Garden City, New York

Bibliographical Note

This Dover edition, first published in 2020, is an unabridged republication of the work originally published in 1947 by John Wiley & Sons, Inc., New York.

Library of Congress Cataloging-in-Publication Data

Names: Brand, Louis, 1885-1971, author.
Title: Vector and tensor analysis / Louis Brand.
Description: Dover edition. | Garden City, New York : Dover Publications, 2020. | Originally published: New York : John Wiley & Sons, Inc., 1947. | Summary: "An outstanding introduction to tensor analysis for physics and engineering students, this text admirably covers the expected topics in a careful step-by-step manor. The treatment also delves into several related topics such as quaternions, dyadics, and the application of vectors to perfect fluids and to elementary differential geometry. A chapter on tensor analysis goes beyond covariant differentiation and includes surface geometry in tensor notation. 1947 edition"—Provided by publisher.
Identifiers: LCCN 2019045271 | ISBN 9780486842837 (trade paperback)
Subjects: LCSH: Vector analysis. | Calculus of tensors.
Classification: LCC QA261 .B63 2020 | DDC 515/.63—dc23
LC record available at https://lccn.loc.gov/2019045271

Manufactured in the United States of America
84283502 2023
www.doverpublications.com

To My Wife

PREFACE

The vector analysis of Gibbs and Heaviside and the more general tensor analysis of Ricci are now recognized as standard tools in mechanics, hydrodynamics, and electrodynamics. These disciplines have also proved their worth in pure mathematics, especially in differential geometry. Their use not only materially simplifies and condenses the exposition, but also makes mathematical concepts more tangible and easy to grasp. Moreover tensor analysis provides a simple automatic method for constructing invariants. Since a tensor equation has precisely the same form in all coordinate systems, the desirability of stating physical laws or geometrical properties in tensor form is manifest. The perfect adaptability of the tensor calculus to the theory of relativity was responsible for its original renown. It has since won a firm place in mathematical physics and engineering technology. Thus the British analyst E. T. Whittaker rates the discovery of the tensor calculus as one of the three principal mathematical advances in the last quarter of the 19th century.

The first volume of this work not only comprises the standard vector analysis of Gibbs, including dyadics or tensors of valence two, but also supplies an introduction to the algebra of *motors*, which is apparently destined to play an important role in mechanics as well as in line geometry. The entire theory is illustrated by many significant applications; and surface geometry and hydrodynamics * are treated at some length by vector methods in separate chapters.

For the sake of concreteness, tensor analysis is first developed in 3-space, then extended to space of n dimensions. As in the case of vectors and dyadics, I have distinguished the *invariant tensor* from its *components*. This leads to a straightforward treatment of the affine connection and of covariant differentiation; and also to a simple introduction of the curvature tensor. Applications of tensor analysis to relativity, electrodynamics and rotating elec-

* For a systematic development of mechanics in vector notation see the author's *Vectorial Mechanics*, John Wiley & Sons, New York, 1930.

tric machines are reserved for the second volume. The present volume concludes with a brief introduction to quaternions, the source of vector analysis, and their use in dealing with finite rotations.

Nearly all of the important results are formulated as theorems, in which the essential conditions are explicitly stated. In this connection the student should observe the distinction between *necessary* and *sufficient* conditions. If the assumption of a certain property P leads deductively to a condition C, the condition is *necessary*. But if the assumption of the condition C leads deductively to the property P, the condition C is *sufficient*. Thus we have symbolically

$$P \rightarrow C \text{ (necessary)}, \qquad C \text{ (sufficient)} \rightarrow P.$$

When $P \rightleftarrows C$, the condition C is *necessary and sufficient*.

The problems at the end of each chapter have been chosen not only to develop the student's technical skill, but also to introduce new and important applications. Some of the problems are *mathematical projects* which the student may carry through step by step and thus arrive at really important results.

As very full cross references are given in this book, an article as well as a page number is given at the top of each page. Equations are numbered serially (1), (2), . . . in each article. A reference to an equation in another article is made by giving article and number to the left and right of a point; thus (24.9) means article 24, equation 9. Figures are given the number of the article in which they appear followed by a serial letter; Fig. 6d, for example, is the fourth figure in article 6.

Bold-face type is used in the text to denote vectors or tensors of higher valence with their complement of base vectors. Scalar components of vectors and tensors are printed in italic type.

The rich and diverse field amenable to vector and tensor methods is one of the most fascinating in applied mathematics. It is hoped that the reasoning will not only appeal to the mind but also impinge on the reader's aesthetic sense. For mathematics, which Gauss esteemed as "the queen of the sciences" is also one of the great arts. For, in the eloquent words of Bertrand Russell:

"The true spirit of delight, the exaltation, the sense of being more than man, which is the touchstone of the highest excellence, is to be found in mathematics as surely as in poetry. What is best in mathematics deserves not merely to be learned as a task, but also to be assimilated as a part of daily thought, and brought again and again before the mind with ever-renewed encouragement. Real life is, to most men, a long second-best, a perpetual compromise between the real and the possible; but the world of pure reason knows no compromise, no practical limitations, no barrier to the creative activity embodying in splendid edifices the passionate aspiration after the perfect from which all great work springs."

The material in this book may be adapted to several short courses. Thus Chapters I, III, IV, V, and VI may serve as a course in vector analysis; and Chapters I (in part), IV, V, and IX as one in tensor analysis. But the prime purpose of the author was to cover the theory and simpler applications of vector and tensor analysis in ordinary space, and to weave into this fabric such concepts as dyadics, matrices, motors, and quaternions.

The author wishes, finally, to express his thanks to his colleagues, Professor J. W. Surbaugh and Mr. Louis Doty for their help with the figures. Mr. Doty also suggested the notation used in the problems dealing with air navigation and read the entire page proof.

LOUIS BRAND

University of Cincinnati
January 15, 1947

CONTENTS

CHAPTER V

Differential Invariants

CHAPTER VI

Integral Transformations

CHAPTER VII

Hydrodynamics

CHAPTER VIII

Geometry on a Surface

CHAPTER IX

Tensor Analysis

CHAPTER X

Quaternions

CHAPTER I

VECTOR ALGEBRA

1. Scalars and Vectors. There are certain physical quantities, such as length, time, mass, temperature, electric charge, that may be specified by a single real number. A mass, for example, may be specified by the positive number equal to the ratio of the given mass to the unit mass. Similarly an electric charge may be specified by a number, positive or negative, according as the charge is "positive" or "negative." Quantities of this sort are called *scalar quantities;* and the numbers that represent them are often called scalars.

On the other hand, some physical quantities require a *direction* as well as magnitude for their specification. Thus a rectilinear displacement can only be completely specified by its length and direction. A displacement may be represented graphically by a segment of a straight line having a definite length and direction. We shall call such a directed segment a *vector.* Any physical quantity that involves both magnitude and direction, so that it may be represented by a line segment of definite length and direction, and which moreover conforms to the parallelogram law of addition (§ 2), is called a *vector quantity.* Velocity, acceleration, force, electric and magnetic field intensities are examples of vector quantities. It is customary, however, in applied mathematics, to speak of *vector quantities* as *vectors.*

DEFINITIONS: *A vector is a segment of a straight line regarded as having a definite length and direction.* Thus we may represent a vector by an *arrow.* The vector directed from the point A to the point B is denoted by the symbol $\overrightarrow{AB}$. With this notation $\overrightarrow{AB}$ and $\overrightarrow{BA}$ denote different vectors; they have the same length but opposite directions.

Besides the *proper vectors* just defined, we extend the term vector to include the *zero vector*, an "arrow" of length zero but devoid of direction.

If the initial point of a vector may be chosen at pleasure, the vector is said to be *free*. If, however, its initial point is restricted to a certain set of points, the vector is said to be *localized* in this set. When the set consists of a single point (initial point fixed), the vector is said to be a *bound vector*. If the vector is restricted to the line of which it forms a part, it is called a *line vector*. For example, the forces acting upon rigid bodies must be regarded as line vectors; they may only be shifted along their *line of action*.

Two vectors are said to be equal when they have the same length and direction.

Vectors are said to be collinear when they are parallel to the same line. In this sense two parallel vectors are collinear.

Vectors are said to be coplanar when they are parallel to the same plane. In this sense any *two* vectors are coplanar.

A unit vector is a vector of unit length.

In addition to the foregoing notation, in which we denote a vector by giving its end points, we also shall employ single letters in heavy **(bold-face)** type to denote vectors. Thus, in Fig. 2b the vectors forming the opposite sides of the parallelogram are equal (they have the same length and direction) and may therefore be represented by the same symbol,

$$\overrightarrow{AB} = \overrightarrow{DC} = \mathbf{u}, \qquad \overrightarrow{AD} = \overrightarrow{BC} = \mathbf{v}.$$

A vector symbol between vertical bars, as $|\overrightarrow{AB}|$ or $|\mathbf{u}|$, denotes the length of the vector. We shall also, on occasion, denote the lengths of vectors, $\mathbf{u}$, $\mathbf{v}$, $\mathbf{F}$ by the corresponding letters u, v, F in italic type. Bars about real numbers denote their positive magnitudes: thus $|-3| = 3$.

The preceding definition of a vector is adequate for the elementary applications to Euclidean space of three dimensions. For purposes of generalization, however, it is far better to define a vector as a new type of number—a hypernumber—which is given by a set of real numbers written in a definite order. Thus, in our ordinary space it will be seen that, when a suitable system of reference has been adopted, a vector can be represented by a set of three real numbers, $[a, b, c]$, called the *components* of the vector. With this definition the zero vector is denoted by $[0, 0, 0]$. In order to complete this definition of a vector, a rule must be given to enable us to compute the components when the system of reference is changed.

2. Addition of Vectors. To obtain a rule for adding vectors, let us regard them, for the moment, as representing rectilinear displacements in space. If a particle is given two rectilinear displacements, one from A to B, and a second from B to C, the result is the same as if the particle were given a single displacement from A to C. This equivalence may be represented by the notation,

$$\overrightarrow{AB} + \overrightarrow{BC} = \overrightarrow{AC}. \tag{1}$$

We shall regard this equation as the *definition* of vector addition. The sum of two vectors, $\mathbf{u}$, $\mathbf{v}$, therefore is defined by the following *triangle construction* (Fig. 2a):

Draw $\mathbf{v}$ *from the end of* $\mathbf{u}$; *then the vector directed from the beginning of* $\mathbf{u}$ *to the end of* $\mathbf{v}$ *is the sum of u and v and is written* $\mathbf{u} + \mathbf{v}$.

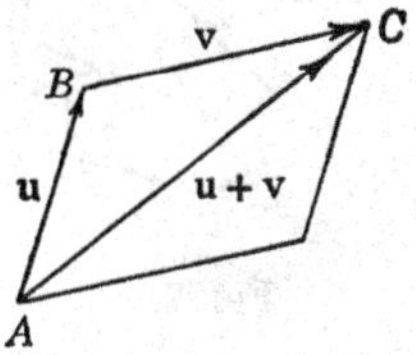

FIG. 2a

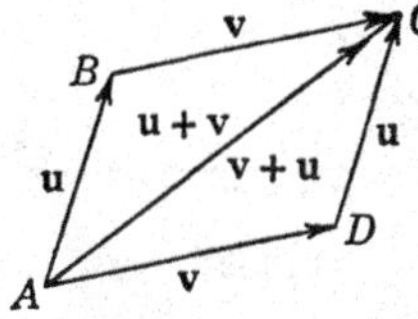

FIG. 2b

Since any side of a triangle is less than the sum of the other two sides,

$$|\mathbf{u} + \mathbf{v}| \leqq |\mathbf{u}| + |\mathbf{v}|,$$

the equal sign holding only when $\mathbf{u}$ and $\mathbf{v}$ have the same direction.

Vector addition obeys both the commutative and associative laws:

$$\mathbf{u} + \mathbf{v} = \mathbf{v} + \mathbf{u}, \tag{2}$$

$$(\mathbf{u} + \mathbf{v}) + \mathbf{w} = \mathbf{u} + (\mathbf{v} + \mathbf{w}). \tag{3}$$

In the parallelogram formed with $\mathbf{u}$ and $\mathbf{v}$ as sides (Fig. 2b),

$$\mathbf{u} + \mathbf{v} = \overrightarrow{AB} + \overrightarrow{BC} = \overrightarrow{AC}, \qquad \mathbf{v} + \mathbf{u} = \overrightarrow{AD} + \overrightarrow{DC} = \overrightarrow{AC}.$$

This proves (2). In view of this construction, the rule for vector addition is called the *parallelogram law*.

To find $\mathbf{u} + \mathbf{v}$ when $\mathbf{u}$ and $\mathbf{v}$ are *line vectors* whose lines intersect at A, shift the vectors along their lines so that both issue from A (Fig. 2b) and complete the parallelogram $ABCD$; then

$$\mathbf{u} + \mathbf{v} = \overrightarrow{AB} + \overrightarrow{AD} = \overrightarrow{AC}.$$

Since the diagonal of the parallelogram on **u**, **v** gives the line of action of **u** + **v**, the term "parallelogram law" is especially appropriate for the addition of intersecting line vectors. We shall speak of the addition of line vectors as *statical addition*.

The associative law (3) is evident from Fig. 2c:

$$(\mathbf{u} + \mathbf{v}) + \mathbf{w} = (\overrightarrow{AB} + \overrightarrow{BC}) + \overrightarrow{CD} = \overrightarrow{AC} + \overrightarrow{CD} = \overrightarrow{AD},$$

$$\mathbf{u} + (\mathbf{v} + \mathbf{w}) = \overrightarrow{AB} + (\overrightarrow{BC} + \overrightarrow{CD}) = \overrightarrow{AB} + \overrightarrow{BD} = \overrightarrow{AD}.$$

Since the grouping of the vectors is immaterial, the preceding sum is simply written **u** + **v** + **w**.

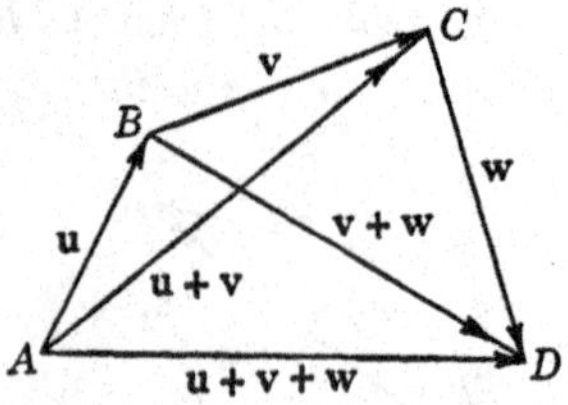

FIG. 2c

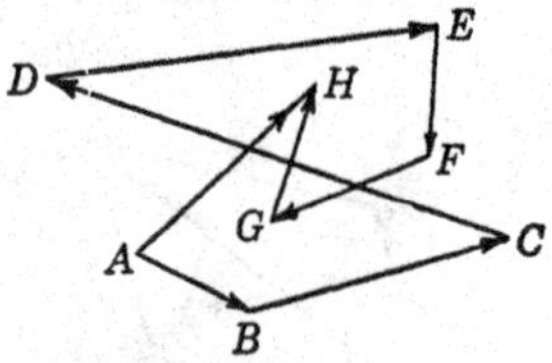

FIG. 2d

From the commutative and associative laws we may deduce the following general result: *The sum of any number of vectors is independent of the order in which they are added, and of their grouping to form partial sums.*

To construct the sum of any number of vectors, form a broken line whose segments, in length and direction, are these vectors taken in any order whatever; then the vector directed from the beginning to the end of the broken line will be the required sum. The figure formed by the vectors and their sum is called a *vector polygon*. If $A, B, C, \cdots, G, H$ are the successive vertices of a vector polygon (Fig. 2d), then

$$\overrightarrow{AB} + \overrightarrow{BC} + \cdots + \overrightarrow{GH} = \overrightarrow{AH}. \tag{4}$$

When the vectors to be added are all parallel, the vector "polygon" becomes a portion of a straight line described twice.

If, in the construction of a vector sum, the end point of the last vector coincides with the origin of the first, we say that the sum of the vectors is *zero*. Thus, if in (4) the point H coincides with A, we write

$$\overrightarrow{AB} + \overrightarrow{BC} + \cdots + \overrightarrow{GA} = 0. \tag{5}$$

This equation may be regarded as a special case of (4) if we agree that $\overrightarrow{AA} = 0$.

The zero vector $\overrightarrow{AA}$ (or $\overrightarrow{BB}$, etc.) is not a vector in the proper sense since it has no definite direction; it is an extension of our original vector concept. From

$$\overrightarrow{AB} + \overrightarrow{BB} = \overrightarrow{AB}, \qquad \overrightarrow{AA} + \overrightarrow{AB} = \overrightarrow{AB}$$

we have, on writing $AB = \mathbf{u}$,

$$\mathbf{u} + 0 = \mathbf{u}, \qquad 0 + \mathbf{u} = \mathbf{u}.$$

We shall refer to vectors which are not zero as *proper vectors.*

3. Subtraction of Vectors. The sum of two vectors is zero when, and only when, they have the same length and opposite directions: $\overrightarrow{AB} + \overrightarrow{BA} = 0$. If $\overrightarrow{AB} = \mathbf{u}$, it is natural to write $\overrightarrow{BA} = -\mathbf{u}$ in order that the characteristic equation for negatives,

(1) $$\mathbf{u} + (-\mathbf{u}) = 0,$$

will hold for vectors as well as for numbers. Hence by definition:

The negative of a vector is a vector of the same length but opposite direction.

Note also that $-(-\mathbf{u}) = \mathbf{u}$.

The difference $\mathbf{u} - \mathbf{v}$ of two vectors is defined by the equation,

(2) $$(\mathbf{u} - \mathbf{v}) + \mathbf{v} = \mathbf{u}.$$

Adding $-\mathbf{v}$ to both sides of (2), we have

(3) $$\mathbf{u} - \mathbf{v} = \mathbf{u} + (-\mathbf{v});$$

that is, *subtracting a vector is the same as adding its negative.* The construction of $\mathbf{u} - \mathbf{v}$ is shown in Fig. 3a.

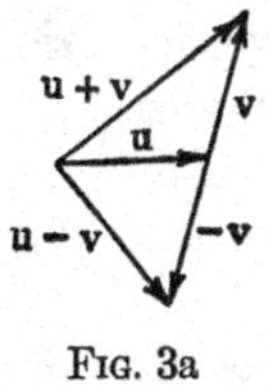

FIG. 3a

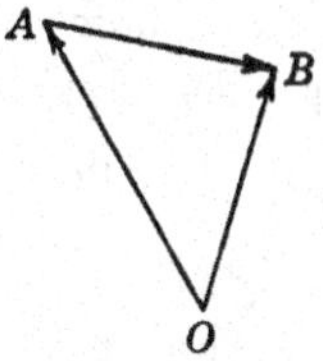

FIG. 3b

If O is chosen as a point of reference, any point P in space may be located by giving its *position vector* $\overrightarrow{OP}$. Any vector $\overrightarrow{AB}$ may be expressed in terms of the position vectors of its end points (Fig. 3b),

$$\overrightarrow{AB} = \overrightarrow{AO} + \overrightarrow{OB} = \overrightarrow{OB} + (-\overrightarrow{OA}),$$

and, from (3),

$$\overrightarrow{AB} = \overrightarrow{OB} - \overrightarrow{OA}. \tag{4}$$

4. Multiplication of Vectors by Numbers. The vector $\mathbf{u} + \mathbf{u}$ is naturally denoted by $2\mathbf{u}$; similarly, we write $-\mathbf{u} + (-\mathbf{u}) = -2\mathbf{u}$. Thus, both $2\mathbf{u}$ and $-2\mathbf{u}$ denote vectors twice as long as $\mathbf{u}$; the former has the same direction as $\mathbf{u}$, the latter the opposite direction. This definition is generalized as follows:

The product $\alpha\mathbf{u}$ *or* $\mathbf{u}\alpha$ *of a vector* $\mathbf{u}$ *and a real number* α *is defined as a vector* α *times as long as* $\mathbf{u}$, *and having the same direction as* $\mathbf{u}$, *or the opposite, according as* α *is positive or negative. If* $\alpha = 0$, $\alpha\mathbf{u} = 0$.

In accordance with this definition,

$$\alpha(-\mathbf{u}) = (-\alpha)\mathbf{u} = -\alpha\mathbf{u}, \qquad (-\alpha)(-\mathbf{u}) = \alpha\mathbf{u}.$$

These relations have the same form as the rules for multiplying numbers. Moreover, the multiplication of a vector by numbers is commutative (by definition), associative, and distributive:

$$\alpha\mathbf{u} = \mathbf{u}\alpha, \tag{1}$$

$$(\alpha\beta)\mathbf{u} = \alpha(\beta\mathbf{u}), \tag{2}$$

$$(\alpha + \beta)\mathbf{u} = \alpha\mathbf{u} + \beta\mathbf{u}. \tag{3}$$

The product of the sum of two vectors by a given number is also distributive:

$$\alpha(\mathbf{u} + \mathbf{v}) = \alpha\mathbf{u} + \alpha\mathbf{v}. \tag{4}$$

The proof of (4) follows immediately from the theorem that the corresponding sides of similar triangles are proportional. Figure 4 applies to the case when $\alpha > 0$.

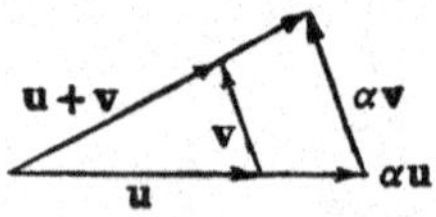

FIG. 4

The quotient $\mathbf{u}/\alpha$ of a vector by a number α (not zero) is defined as the product of $\mathbf{u}$ by $1/\alpha$.

The developments thus far show that: *As far as addition, subtraction, and multiplication by numbers are concerned, vectors may be treated formally in accordance with the rules of ordinary algebra.*

5. Linear Dependence. The n vectors $\mathbf{u}_1, \mathbf{u}_2, \cdots, \mathbf{u}_n$ are said to be *linearly dependent* if there exist n real numbers $\lambda_1, \lambda_2, \cdots, \lambda_n$, not all zero, such that

$$(1) \qquad \lambda_1\mathbf{u}_1 + \lambda_2\mathbf{u}_2 + \cdots + \lambda_n\mathbf{u}_n = 0.$$

If the vectors are not linearly dependent, they are said to be *linearly independent.* Consequently, if a relation (1) exists between n linearly independent vectors, all the constants must be zero.

If m vectors $\mathbf{u}_1, \mathbf{u}_2, \cdots, \mathbf{u}_m$ are linearly dependent, any greater number n of vectors including these are also linearly dependent. For if $\mathbf{u}_1, \mathbf{u}_2, \cdots, \mathbf{u}_m$ satisfy

$$\lambda_1\mathbf{u}_1 + \lambda_2\mathbf{u}_2 + \cdots + \lambda_m\mathbf{u}_m = 0,$$

we can give $\lambda_1, \lambda_2, \cdots, \lambda_m$ the preceding values (at least one of these is not zero) and take $\lambda_{m+1} = \lambda_{m+2} = \cdots = \lambda_n = 0$. Then (1) is satisfied, and the n vectors $\mathbf{u}_i$ are linearly dependent.

If $\lambda\mathbf{u} = 0$ and $\lambda \neq 0$, $\mathbf{u} = 0$; hence *one vector is linearly dependent only when it is the zero vector.* Hence the vectors of any set that includes the zero vector are linearly dependent. Consequently, we need only consider sets of *proper* vectors in the theorems following.

If $\lambda_1\mathbf{u}_1 + \lambda_2\mathbf{u}_2 = 0$ and $\lambda_1 \neq 0$, we can write $\mathbf{u}_1 = \alpha\mathbf{u}_2$; hence $\mathbf{u}_1$ and $\mathbf{u}_2$ are collinear. Conversely, if $\mathbf{u}_1$ and $\mathbf{u}_2$ are collinear, $\mathbf{u}_1 = \alpha\mathbf{u}_2$ $(\alpha \neq 0)$. Therefore:

A necessary and sufficient condition that two proper vectors be linearly dependent is that they be collinear.

If $\lambda_1\mathbf{u}_1 + \lambda_2\mathbf{u}_2 + \lambda_3\mathbf{u}_3 = 0$ and $\lambda_1 \neq 0$, we can write $\mathbf{u}_1 = \alpha\mathbf{u}_2 + \beta\mathbf{u}_3$; the parallelogram construction (Fig. 5a) now shows that $\mathbf{u}_1$ is parallel to the plane of $\mathbf{u}_2$ and $\mathbf{u}_3$. Conversely, if $\mathbf{u}_1, \mathbf{u}_2, \mathbf{u}_3$ are coplanar, they are linearly dependent. For (a) if two of the vectors are collinear, they are linearly dependent, and the same is true of all three; and (b) if no two vectors are collinear, we can construct a parallelogram on $\mathbf{u}_1$ as diagonal whose sides are parallel to $\mathbf{u}_2$ and $\mathbf{u}_3$ (Fig. 5a), so that

$$\mathbf{u}_1 = \overrightarrow{AC} = \overrightarrow{AB} + \overrightarrow{BC} = \alpha\mathbf{u}_2 + \beta\mathbf{u}_3.$$

Therefore:

A necessary and sufficient condition that three proper vectors be linearly dependent is that they be coplanar.

In space of three dimensions, four vectors $\mathbf{u}_1$, $\mathbf{u}_2$, $\mathbf{u}_3$, $\mathbf{u}_4$ are always linearly dependent. For (*a*) if three of the vectors are coplanar, they are linearly dependent, and the same is true of all four; and (*b*) if no three vectors are coplanar, we can construct a

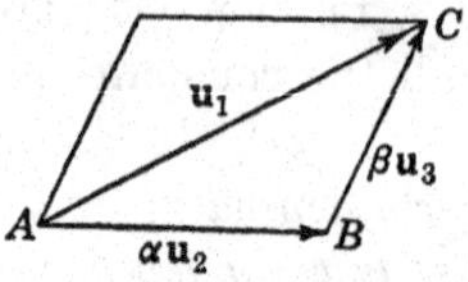

FIG. 5a

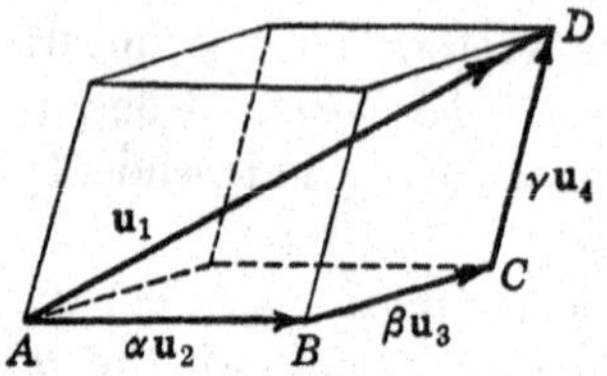

FIG. 5b

parallelepiped on $\mathbf{u}_1$ as a diagonal whose edges are parallel to $\mathbf{u}_2$, $\mathbf{u}_3$, $\mathbf{u}_4$ (Fig. 5b), so that

$$\mathbf{u}_1 = \overrightarrow{AD} = \overrightarrow{AB} + \overrightarrow{BC} + \overrightarrow{CD} = \alpha\mathbf{u}_2 + \beta\mathbf{u}_3 + \gamma\mathbf{u}_4.$$

Therefore: *Any four vectors are linearly dependent.*

6. Collinear Points. If A, B, P are points of a straight line, P is said to divide the segment AB in the ratio λ when

$$\overrightarrow{AP} = \lambda\,\overrightarrow{PB}. \tag{1}$$

As P passes from A to B (Fig. 6a), λ increases through all positive values from 0 to infinity. If P describes the line to the left of A, λ varies from 0 to -1; and, as P describes the line to the right of B, λ varies from $-\infty$ to -1. Thus $\lambda = 0$, $\lambda = \pm\infty$, $\lambda = -1$ correspond, respectively, to the points A, B, and the infinitely distant "point" of the line. The ratio λ is positive or negative, according as P lies within or without the segment AB.

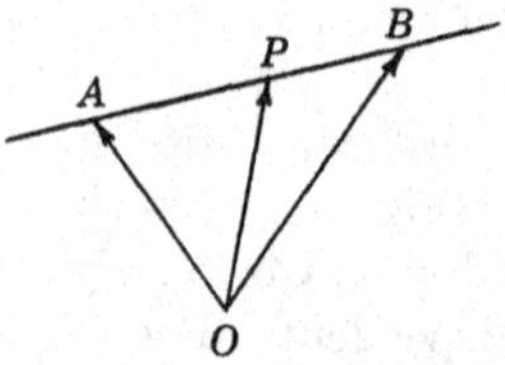

FIG. 6a

To find the position vector of P, relative to an origin O, write (1) in the form,

$$\overrightarrow{OP} - \overrightarrow{OA} = \lambda(\overrightarrow{OB} - \overrightarrow{OP}).$$

Then

$$\overrightarrow{OP} = \frac{\overrightarrow{OA} + \lambda\,\overrightarrow{OB}}{1 + \lambda}, \tag{2}$$

or, if we write $\lambda = \beta/\alpha$,

$$\overrightarrow{OP} = \frac{\alpha\,\overrightarrow{OA} + \beta\,\overrightarrow{OB}}{\alpha + \beta}. \tag{3}$$

In particular, if $\lambda = 1$, P is the mid-point of AB.

In the following we shall denote the position vectors of the points $A, B, C, \cdots, P$ by $\mathbf{a}, \mathbf{b}, \mathbf{c}, \cdots, \mathbf{p}$. Thus if C divides AB in the ratio β/α,

$$\mathbf{c} = \frac{\alpha\mathbf{a} + \beta\mathbf{b}}{\alpha + \beta}. \tag{4}$$

Thus the mid-point of AB has the position vector $\frac{1}{2}(\mathbf{a} + \mathbf{b})$.

When the points C, D divide a segment AB internally and externally in the same numerical ratios $\pm\lambda$, we have

$$\mathbf{c} = \frac{\mathbf{a} + \lambda\mathbf{b}}{1 + \lambda}, \qquad \mathbf{d} = \frac{\mathbf{a} - \lambda\mathbf{b}}{1 - \lambda}.$$

If we solve these equations for $\mathbf{a}$ and $\mathbf{b}$, we find that the points A, B also divide the segment CD in the same numerical ratios $\pm(1 - \lambda)/(1 + \lambda)$. Pairs of points A, B and C, D having this property are said to be *harmonic conjugates;* either pair is the harmonic conjugate of the other.

A useful test for collinearity is given by the following:

THEOREM. *Three distinct points A, B, C lie on a straight line when, and only when, there exist three numbers α, β, γ, different from zero, such that*

$$\alpha\mathbf{a} + \beta\mathbf{b} + \gamma\mathbf{c} = 0, \qquad \alpha + \beta + \gamma = 0. \tag{5}$$

Proof. If A, B, C are collinear, C divides AB in some ratio β/α; hence on putting $\gamma = -(\alpha + \beta)$ in (4) we obtain (5). Conversely, from (5) we can deduce (4) since $\alpha + \beta = -\gamma \neq 0$; hence C lies on the line AB.

From (5) we conclude that C, A, B divide AB, BC, CA, respectively, in the ratios β/α, γ/β, α/γ whose product is 1.

If an equation of the form (5) *subsists between three distinct non-collinear points, we must conclude that* $\alpha = \beta = \gamma = 0$. For at least one coefficient $\gamma = 0$; and from

$$\alpha\mathbf{a} + \beta\mathbf{b} = 0, \qquad \alpha + \beta = 0,$$

we have $\mathbf{a} = \mathbf{b}$ (A coincides with B) unless $\alpha = \beta = 0$.

Another criterion for collinearity may be based on the statical addition of line vectors (§ 2).

THEOREM. *The points A, B, C are collinear when the line vectors* $\overrightarrow{AB}, \overrightarrow{BC}, \overrightarrow{CA}$ *are statically equal to zero:*

$$(6) \qquad \overrightarrow{AB} + \overrightarrow{BC} + \overrightarrow{CA} \equiv 0.$$

Proof. If we use $\equiv$ to denote statical equivalence, $\overrightarrow{AB} + \overrightarrow{BC} \equiv \overrightarrow{BD}$, a vector through B; and, since $\overrightarrow{BD} + \overrightarrow{CA} \equiv 0$, B, C, A are collinear.

Example 1. In the parallelogram $ABCD$, E and F are the middle points of the sides AB, BC. Show that the lines DE, DF divide the diagonal AC into thirds and that AC cuts off a third of each line (Fig. 6b).

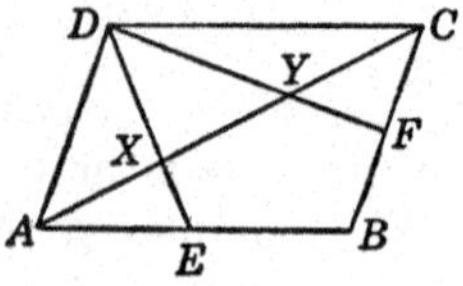

FIG. 6b

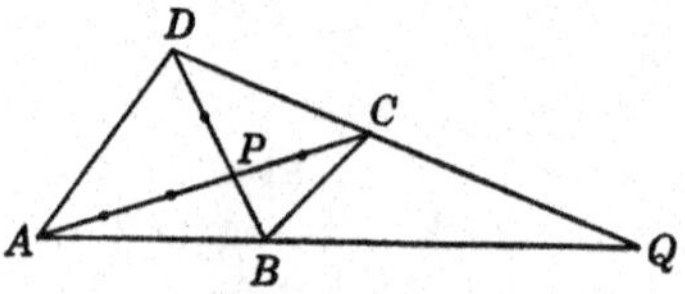

FIG. 6c

The hypotheses of our problem are expressed by the equations:

$$\mathbf{d} - \mathbf{a} = \mathbf{c} - \mathbf{b}, \qquad 2\mathbf{e} = \mathbf{a} + \mathbf{b}, \qquad 2\mathbf{f} = \mathbf{b} + \mathbf{c}.$$

Let DE cut AC at X. To find $\mathbf{x}$, eliminate $\mathbf{b}$ from the first and second equations. Thus

$$\mathbf{d} - \mathbf{a} + 2\mathbf{e} = \mathbf{a} + \mathbf{c} \quad \text{and} \quad \frac{\mathbf{d} + 2\mathbf{e}}{3} = \frac{2\mathbf{a} + \mathbf{c}}{3} = \mathbf{x};$$

for the first member represents a point on DE, the second member a point on AC, and, since the points are the same, the point is at the intersection X of these lines. Comparison with (4) now shows that X divides DE in the ratio 2/1, AC in the ratio 1/2.

Let DF cut AC at Y. To find $\mathbf{y}$, eliminate $\mathbf{b}$ from the first and third equations. Thus

$$\mathbf{d} - \mathbf{a} + 2\mathbf{f} = 2\mathbf{c} \quad \text{and} \quad \frac{\mathbf{d} + 2\mathbf{f}}{3} = \frac{\mathbf{a} + 2\mathbf{c}}{3} = \mathbf{y}.$$

Hence Y divides DF in the ratio 2/1, AC in the ratio 2/1.

Example 2. In a plane quadrilateral $ABCD$, the diagonals AC, BD intersect at P, the sides AB, CD intersect at Q (Fig. 6c). If P divides AC and BD in the ratios 3/2 and 1/2, respectively, in what ratios does Q divide the segments AB, CD?

By hypothesis

$$\mathbf{p} = \frac{2\mathbf{a} + 3\mathbf{c}}{5} = \frac{2\mathbf{b} + \mathbf{d}}{3};$$

hence

$$6\mathbf{a} + 9\mathbf{c} = 10\mathbf{b} + 5\mathbf{d} \quad \text{and} \quad \frac{9\mathbf{c} - 5\mathbf{d}}{4} = \frac{10\mathbf{b} - 6\mathbf{a}}{4} = \mathbf{q};$$

for the first fraction represents a point on CD, the second a point on AB, and both points are the same, that is, the point Q. Therefore Q divides CD in the ratio $-5/9$, AB in the ratio $-10/6$.

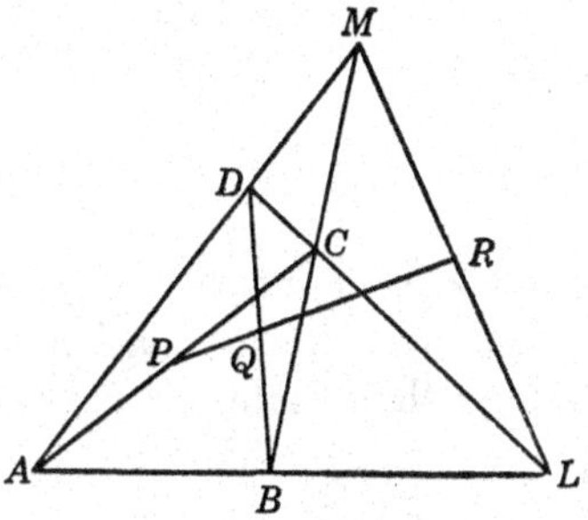

FIG. 6d

Example 3. Prove that the mid-points of the diagonals of a complete quadrilateral are collinear.

In the complete quadrilateral $ABCDLM$ (Fig. 6d) let P, Q, R be the mid-points of the diagonals AC, BD, LM. The sum of the line vectors,

$$\overrightarrow{AB} + \overrightarrow{AD} \equiv 2\,\overrightarrow{AQ}, \qquad \overrightarrow{CB} + \overrightarrow{CD} \equiv 2\,\overrightarrow{CQ}, \qquad \overrightarrow{QA} + \overrightarrow{QC} \equiv 2\,\overrightarrow{QP};$$

hence we have the statical equivalence,

$$\overrightarrow{AB} + \overrightarrow{AD} + \overrightarrow{CB} + \overrightarrow{CD} \equiv 4\,\overrightarrow{PQ},$$

for the quadrilateral $ABCD$ with diagonals AC, BD. Similarly, for the quadrilateral $BLDM$ with diagonals BD, LM,

$$\overrightarrow{BL} + \overrightarrow{BM} + \overrightarrow{DL} + \overrightarrow{DM} \equiv 4\,\overrightarrow{QR};$$

and, for the quadrilateral $LAMC$ with diagonals LM, AC,

$$\overrightarrow{LA} + \overrightarrow{LC} + \overrightarrow{MA} + \overrightarrow{MC} \equiv 4\,\overrightarrow{RP}.$$

On adding these three equations, we find that the entire left member is statically equal to zero; for

$$\overrightarrow{AB} + \overrightarrow{BL} + \overrightarrow{LA} \equiv 0, \qquad \overrightarrow{AD} + \overrightarrow{DM} + \overrightarrow{MA} \equiv 0,$$

$$\overrightarrow{CB} + \overrightarrow{BM} + \overrightarrow{MC} \equiv 0, \qquad \overrightarrow{CD} + \overrightarrow{DL} + \overrightarrow{LC} \equiv 0,$$

are statical equations, since the vectors in each are collinear. Hence

$$\overrightarrow{PQ} + \overrightarrow{QR} + \overrightarrow{RP} \equiv 0,$$

and P, Q, R are collinear.

7. Coplanar Points. THEOREM. *If no three of the points A, B, C, D are collinear, they will lie in a plane when, and only when, there exist four numbers α, β, γ, δ, different from zero, such that*

$$(1) \qquad \alpha\mathbf{a} + \beta\mathbf{b} + \gamma\mathbf{c} + \delta\mathbf{d} = 0, \qquad \alpha + \beta + \gamma + \delta = 0.$$

Proof. If A, B, C, D are coplanar, either AB is parallel to CD, or AB cuts CD in a point P (not A, B, C, or D). In the respective cases, we have

$$\mathbf{b} - \mathbf{a} = \kappa(\mathbf{d} - \mathbf{c}); \qquad \frac{\mathbf{a} + \lambda\mathbf{b}}{1 + \lambda} = \frac{\mathbf{c} + \lambda'\mathbf{d}}{1 + \lambda'} = \mathbf{p},$$

where λ, λ' are neither 0 nor -1. In both cases, $\mathbf{a}$, $\mathbf{b}$, $\mathbf{c}$, $\mathbf{d}$ are connected by a linear relation of the form (1). Conversely, let us assume that equations (1) hold good. If $\alpha + \beta = 0$ (and hence $\gamma + \delta = 0$), we have

$$\alpha(\mathbf{a} - \mathbf{b}) + \gamma(\mathbf{c} - \mathbf{d}) = 0$$

and the lines AB and CD are parallel. If $\alpha + \beta \neq 0$ (and hence $\gamma + \delta \neq 0$),

$$(2) \qquad \frac{\alpha\mathbf{a} + \beta\mathbf{b}}{\alpha + \beta} = \frac{\gamma\mathbf{c} + \delta\mathbf{d}}{\gamma + \delta} = \mathbf{p}$$

where P is a point common to the lines AB and CD. In both cases, A, B, C, D are coplanar.

Note that (2) states that the point P in which AB, CD intersect divides AB and CD in the ratios β/α, δ/γ. Similarly, if $\alpha + \gamma \neq 0$,

$$(3) \qquad \frac{\alpha\mathbf{a} + \gamma\mathbf{c}}{\alpha + \gamma} = \frac{\beta\mathbf{b} + \delta\mathbf{d}}{\beta + \delta} = \mathbf{q};$$

thus Q, the point in which AC and BD intersect, divides AC and BD in the ratios γ/α, δ/β.

What conclusion can be drawn if $\alpha + \delta \neq 0$?

If an equation of the form (1) *subsists between four distinct non-coplanar points, we must conclude that* $\alpha = \beta = \gamma = \delta = 0$. For at least one coefficient $\delta = 0$; then, from

$$\alpha\mathbf{a} + \beta\mathbf{b} + \gamma\mathbf{c} = 0, \qquad \alpha + \beta + \gamma = 0,$$

and the fact that A, B, C are not collinear, we deduce (§ 6) that $\alpha = \beta = \gamma = 0$.

Example 1. *The Trapezoid.* If $ABCD$ is a trapezoid with AB parallel to DC (Fig. 7a), then

$$\overrightarrow{AB} = \lambda\,\overrightarrow{DC} \quad\text{or}\quad \mathbf{b} - \mathbf{a} = \lambda(\mathbf{c} - \mathbf{d}).$$

Hence we have

$$\mathbf{b} + \lambda\mathbf{d} = \mathbf{a} + \lambda\mathbf{c} \quad\text{or}\quad \mathbf{b} - \lambda\mathbf{c} = \mathbf{a} - \lambda\mathbf{d}.$$

These equations may be written

$$\frac{\mathbf{b} + \lambda\mathbf{d}}{1 + \lambda} = \frac{\mathbf{a} + \lambda\mathbf{c}}{1 + \lambda} = \mathbf{p}, \qquad \frac{\mathbf{b} - \lambda\mathbf{c}}{1 - \lambda} = \frac{\mathbf{a} - \lambda\mathbf{d}}{1 - \lambda} = \mathbf{q};$$

for the former expressions represent the point P where the diagonals BD, AC meet, and the latter the point Q where the sides BC, AD meet. Evidently P divides both BD and AC in the same ratio λ; and Q divides both BC and AD in the same ratio $-\lambda$. In what ratio does the line PQ divide AB?

In particular, if $\lambda = 1$, the trapezoid becomes a parallelogram. The diagonals then bisect each other at P, while Q recedes to infinity.

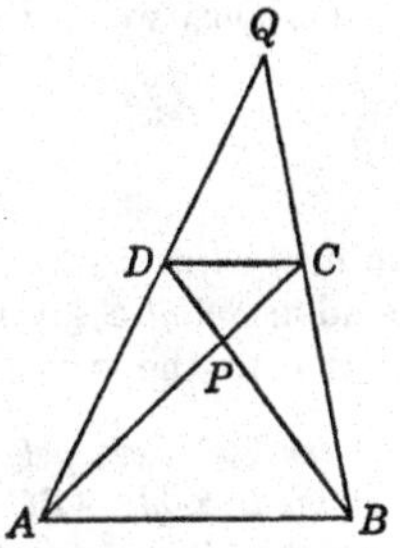

FIG. 7a

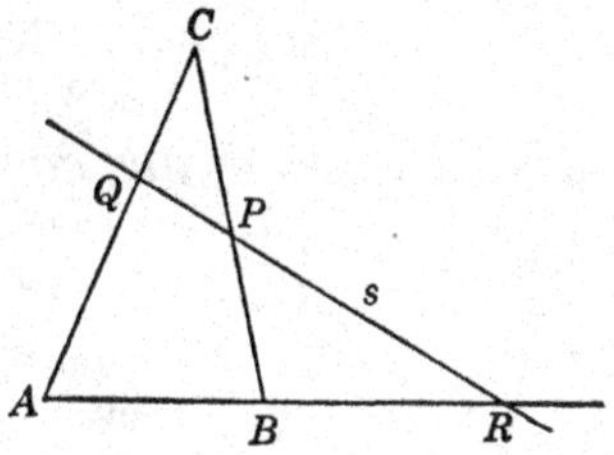

FIG. 7b

Example 2. *Theorem of Menelaus. If a line s cuts the sides BC, CA, AB of the triangle ABC in the points P, Q, R, respectively, the product of the ratios in which P, Q, R divide these sides equals* -1. *Conversely, if P, Q, R divide the sides of the triangle in ratios whose product is* -1, *the points are collinear.*

Proof. (Fig. 7b.) We lose no generality if we assume that P, Q divide BC, CA in the ratios $-\gamma/\beta$, $-\alpha/\gamma$:

(i) $$(\beta - \gamma)\mathbf{p} = \beta\mathbf{b} - \gamma\mathbf{c},$$

(ii) $$(\gamma - \alpha)\mathbf{q} = \gamma\mathbf{c} - \alpha\mathbf{a}.$$

In order to locate R, which lies on the lines PQ, AB, we seek a linear relation between $\mathbf{p}$, $\mathbf{q}$, $\mathbf{a}$, $\mathbf{b}$. Add (i) and (ii) to eliminate $\mathbf{c}$ and divide by $\beta - \alpha$; then

$$\frac{(\beta - \gamma)\mathbf{p} + (\gamma - \alpha)\mathbf{q}}{\beta - \alpha} = \frac{\beta\mathbf{b} - \alpha\mathbf{a}}{\beta - \alpha} = \mathbf{r}.$$

Thus R divides AB in the ratio $-\beta/\alpha$. The product of the division ratios $-\gamma/\beta$, $-\alpha/\gamma$, $-\beta/\alpha$ is -1.

Conversely, let us assume that P, Q, R divide BC, CA, AB in the ratios $-\gamma/\beta$, $-\alpha/\gamma$, $-\beta/\alpha$ whose product is -1. Then we have equations (i), (ii) and also

(iii) $$(\alpha - \beta)\mathbf{r} = \alpha\mathbf{a} - \beta\mathbf{b}.$$

From these we deduce the linear relation,

(iv) $$(\beta - \gamma)\mathbf{p} + (\gamma - \alpha)\mathbf{q} + (\alpha - \beta)\mathbf{r} = 0,$$

in which the sum of the coefficients is zero. The points P, Q, R are therefore collinear.*

Note. From (i), (ii), (iii) it is easily proved that the three pairs of lines BQ, CR; CR, AP; AP, BQ meet in the points A', B', C' given by

$$(-\alpha + \beta + \gamma)\mathbf{a}' = -\alpha\mathbf{a} + \beta\mathbf{b} + \gamma\mathbf{c},$$

$$(\alpha - \beta + \gamma)\mathbf{b}' = \alpha\mathbf{a} - \beta\mathbf{b} + \gamma\mathbf{c},$$

$$(\alpha + \beta - \gamma)\mathbf{c}' = \alpha\mathbf{a} + \beta\mathbf{b} - \gamma\mathbf{c}.$$

If we add $2\alpha\mathbf{a}$, $2\beta\mathbf{b}$, $2\gamma\mathbf{c}$, respectively, to these equations, we find that the point S given by

(v) $$\mathbf{s} = \frac{\alpha\mathbf{a} + \beta\mathbf{b} + \gamma\mathbf{c}}{\alpha + \beta + \gamma}$$

is common to the lines AA', BB', CC'. Thus to every *line* s given by (iv) we have a corresponding *point* S given by (v) —the *pole* of s relative to the triangle ABC.

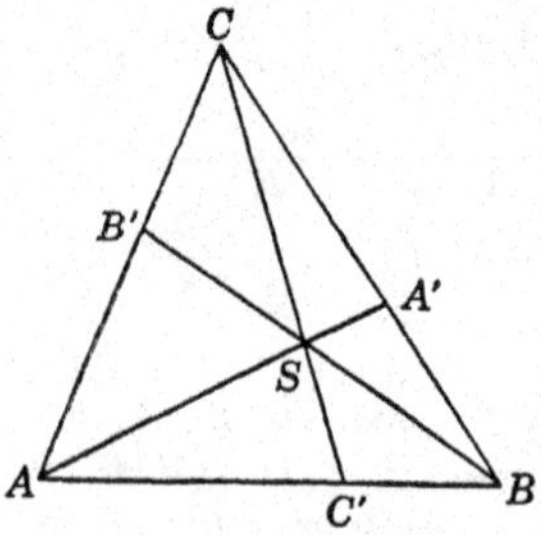

FIG. 7c

Example 3. Theorem of Ceva. If S is a point in the plane of the triangle ABC, and the lines SA, SB, SC cut the sides opposite in the points A', B', C', then the product of the ratios in which A', B', C' divide the sides BC, CA, AB equals 1. *Conversely, if A', B', C' divide the sides BC, CA, AB in ratios whose product is* 1, *the lines AA', BB', CC' meet in a point.*

Proof. (Fig. 7c.) Since A, B, C, S are coplanar,

$$\alpha\mathbf{a} + \beta\mathbf{b} + \gamma\mathbf{c} + \delta\mathbf{s} = 0, \qquad \alpha + \beta + \gamma + \delta = 0.$$

Hence

(i) $$\mathbf{a}' = \frac{\beta\mathbf{b} + \gamma\mathbf{c}}{\beta + \gamma} = \frac{\alpha\mathbf{a} + \delta\mathbf{s}}{\alpha + \delta},$$

(ii) $$\mathbf{b}' = \frac{\gamma\mathbf{c} + \alpha\mathbf{a}}{\gamma + \alpha} = \frac{\beta\mathbf{b} + \delta\mathbf{s}}{\beta + \delta},$$

(iii) $$\mathbf{c}' = \frac{\alpha\mathbf{a} + \beta\mathbf{b}}{\alpha + \beta} = \frac{\gamma\mathbf{c} + \delta\mathbf{s}}{\gamma + \delta}.$$

*This conclusion is obvious; for the line PQ must meet AB in the point R for which the product of the division ratios is -1.

These equations state that A', B', C' divide BC, CA, AB in the ratios γ/β, α/γ, β/α, whose product is 1. Incidentally, A', B', C' divide SA, SB, SC in the ratios α/δ, β/δ, γ/δ whose sum is -1.

Conversely, let us assume that A', B', C' divide BC, CA, AB in the ratios γ/β, α/γ, β/α whose product is 1. Then we have equations (i), (ii), (iii). From these we find that the vectors $\alpha\mathbf{a} + (\beta + \gamma)\mathbf{a}'$, $\beta\mathbf{b} + (\gamma + \alpha)\mathbf{b}'$, $\gamma\mathbf{c} + (\alpha + \beta)\mathbf{c}'$ are all equal to $\alpha\mathbf{a} + \beta\mathbf{b} + \gamma\mathbf{c}$; the point,

$$\text{(iv)} \qquad \mathbf{s} = \frac{\alpha\mathbf{a} + \beta\mathbf{b} + \gamma\mathbf{c}}{\alpha + \beta + \gamma},$$

is therefore common to the lines AA', BB', CC'.

Note. From (i), (ii), (iii) it is easily proved that the three pairs of lines BC, $B'C'$; CA, $C'A'$; AB, $A'B'$ meet in the points P, Q, R given by

$$\text{(v)} \qquad \begin{aligned} (\beta - \gamma)\mathbf{p} &= \beta\mathbf{b} - \gamma\mathbf{c}, \\ (\gamma - \alpha)\mathbf{q} &= \gamma\mathbf{c} - \alpha\mathbf{a}, \\ (\alpha - \beta)\mathbf{r} &= \alpha\mathbf{a} - \beta\mathbf{b}. \end{aligned}$$

From these equations we deduce the linear relation,

$$\text{(vi)} \qquad (\beta - \gamma)\mathbf{p} + (\gamma - \alpha)\mathbf{q} + (\alpha - \beta)\mathbf{r} = 0,$$

in which the sum of the coefficients is zero. The points P, Q, R therefore lie on a line s. Thus to every *point* S given by (iv) we have a corresponding *line* s whose points P, Q, R are given by (v)—the *polar* of S relative to the triangle ABC.

Example 4. Let ABC and $A'B'C'$ be two triangles, in the same or different planes, so that the vertices A, B, C correspond to A', B', C', and the sides AB, BC, CA correspond to $A'B'$, $B'C'$, $C'A'$. We then have (Fig. 7d)

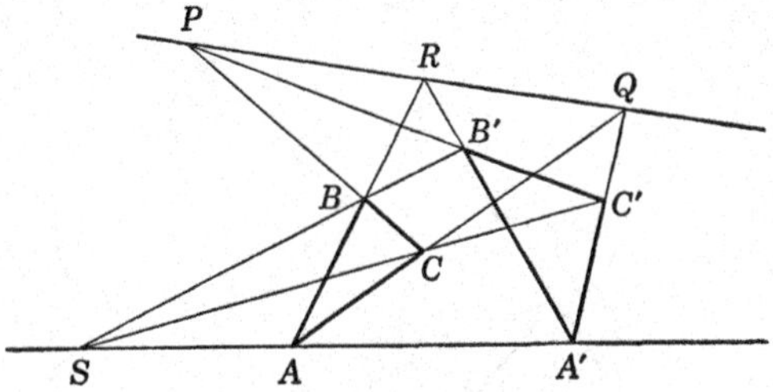

FIG. 7d

DESARGUES' THEOREM. *If the lines joining the corresponding vertices of two triangles are concurrent, the three pairs of corresponding sides intersect in collinear points, and conversely.*

Let the lines AA', BB', CC' intersect at S; then

$$\alpha\mathbf{a} + \alpha'\mathbf{a}' = \beta\mathbf{b} + \beta'\mathbf{b}' = \gamma\mathbf{c} + \gamma'\mathbf{c}' = \mathbf{s},$$
$$\alpha + \alpha' = \beta + \beta' = \gamma + \gamma' = 1.$$

From these equations we find in the usual manner the points P Q, R in which BC, $B'C'$; CA, $C'A'$; AB, $A'B'$ intersect:

(i) $$\mathbf{p} = \frac{\beta\mathbf{b} - \gamma\mathbf{c}}{\beta - \gamma} = \frac{\beta'\mathbf{b}' - \gamma'\mathbf{c}'}{\beta' - \gamma'},$$

(ii) $$\mathbf{q} = \frac{\gamma\mathbf{c} - \alpha\mathbf{a}}{\gamma - \alpha} = \frac{\gamma'\mathbf{c}' - \alpha'\mathbf{a}'}{\gamma' - \alpha'},$$

(iii) $$\mathbf{r} = \frac{\alpha\mathbf{a} - \beta\mathbf{b}}{\alpha - \beta} = \frac{\alpha'\mathbf{a}' - \beta'\mathbf{b}'}{\alpha' - \beta'}.$$

Hence

(iv) $$(\beta - \gamma)\mathbf{p} + (\gamma - \alpha)\mathbf{q} + (\alpha - \beta)\mathbf{r} = 0,$$

(v) $$(\beta' - \gamma')\mathbf{p} + (\gamma' - \alpha')\mathbf{q} + (\alpha' - \beta')\mathbf{r} = 0;$$

either equation shows that P, Q, R are collinear.

To prove the converse, we may start with the expressions (i), (ii), (iii) for $\mathbf{p}$, $\mathbf{q}$, $\mathbf{r}$. These ensure that P, Q, R are collinear; but, in order that (iv) and (v) determine the same division ratios for P, Q, R, we must have

(vi) $$\frac{\beta' - \gamma'}{\beta - \gamma} = \frac{\gamma' - \alpha'}{\gamma - \alpha} = \frac{\alpha' - \beta'}{\alpha - \beta} = h,$$

or

(vii) $$\alpha' - h\alpha = \beta' - h\beta = \gamma' - h\gamma = k,$$

h and k representing the values of the equal members of (vi) and (vii). In the usual way we now find from (i), (ii), (iii) that

$$\alpha'\mathbf{a}' - h\alpha\mathbf{a} = \beta'\mathbf{b}' - h\beta\mathbf{b} = \gamma'\mathbf{c}' - h\gamma\mathbf{c} = k\mathbf{s},$$

where S is a point common to AA', BB', CC'.

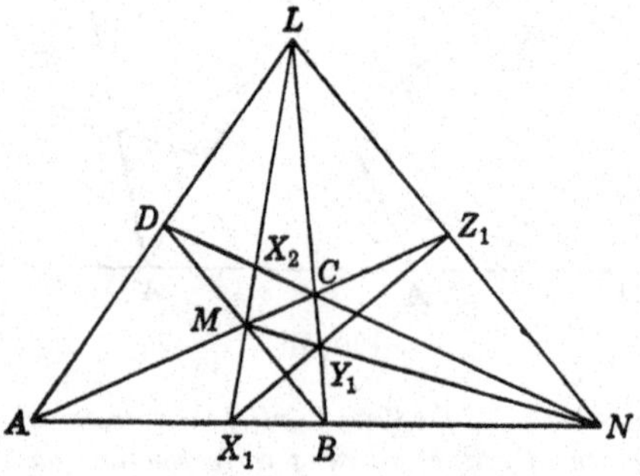

FIG. 7e

Example 5. *The Complete Quadrangle.* A *complete quadrangle* consists of four coplanar points, its *vertices*, no three collinear, and the six lines, its *sides*, which join them. The three pairs of sides which do not meet at a vertex are said to be *opposite;* and the three points in which they meet are called *diagonal points.* In the complete quadrangle $ABCD$ (Fig. 7e) the pairs of opposite sides (BC, AD), (CA, BD), (AB, CD), meet at the diagonal points L, M, N, respectively.

The properties of this configuration of points and lines must all be consequences of the fact that A, B, C, D are coplanar points, that is,

$$\text{(i)} \qquad \alpha\mathbf{a} + \beta\mathbf{b} + \gamma\mathbf{c} + \delta\mathbf{d} = 0, \qquad \alpha + \beta + \gamma + \delta = 0.$$

Since no three vertices are collinear, none of the scalars α, β, γ, δ are zero. From equations (i) we locate at once the diagonal points at the intersections of the opposite sides:

$$\text{(ii)} \qquad \mathbf{l} = \frac{\beta\mathbf{b} + \gamma\mathbf{c}}{\beta + \gamma} = \frac{\alpha\mathbf{a} + \delta\mathbf{d}}{\alpha + \delta},$$

$$\text{(iii)} \qquad \mathbf{m} = \frac{\gamma\mathbf{c} + \alpha\mathbf{a}}{\gamma + \alpha} = \frac{\beta\mathbf{b} + \delta\mathbf{d}}{\beta + \delta},$$

$$\text{(iv)} \qquad \mathbf{n} = \frac{\alpha\mathbf{a} + \beta\mathbf{b}}{\alpha + \beta} = \frac{\gamma\mathbf{c} + \delta\mathbf{d}}{\gamma + \delta}.$$

We here assume that no two of the scalars α, β, γ, δ have a zero sum; the diagonal points L, M, N are then all in the finite plane. But if for example, $\alpha + \beta = \gamma + \delta = 0$, AB and CD are parallel, and N is the point at infinity in their common direction.

To find the points X_1, X_2 where LM cuts AB and CD, we seek linear relations connecting $\mathbf{l}$, $\mathbf{m}$, $\mathbf{a}$, $\mathbf{b}$ and $\mathbf{l}$, $\mathbf{m}$, $\mathbf{c}$, $\mathbf{d}$, respectively.

Thus, from (ii) and (iii),

$$\text{(v)} \qquad \mathbf{x}_1 = \frac{(\gamma + \alpha)\mathbf{m} - (\beta + \gamma)\mathbf{l}}{\alpha - \beta} = \frac{\alpha\mathbf{a} - \beta\mathbf{b}}{\alpha - \beta},$$

$$\text{(vi)} \qquad \mathbf{x}_2 = \frac{(\beta + \gamma)\mathbf{l} - (\beta + \delta)\mathbf{m}}{\gamma - \delta} = \frac{\gamma\mathbf{c} - \delta\mathbf{d}}{\gamma - \delta}.$$

Equations (iv) and (v) show that N and X_1 divide AB in the ratios β/α and $-\beta/\alpha$; hence N, X_1 are harmonic conjugates of A, B.

Equations (iv) and (vi) show that N and X_2 divide CD in the ratios δ/γ and $-\delta/\gamma$; hence N, X_2 are harmonic conjugates of C, D.

Equation (v) shows that X_1 divides LM in the ratio $-(\gamma + \alpha)/(\beta + \gamma)$; and (vi) shows that X_2 divides LM in the ratio $-(\beta + \delta)/(\beta + \gamma) = (\gamma + \alpha)/(\beta + \gamma)$; hence X_1, X_2 are harmonic conjugates of L, M.

We may now state the following harmonic properties of the complete quadrangle in the

THEOREM. *Two vertices of a complete quadrangle are separated harmonically by the diagonal point on their side and by a point on the line joining the other two diagonal points.*

Two diagonal points are separated harmonically by points on the sides passing through the third diagonal point.

From (ii), (iii), (iv) we next find the points X_1, Y_1, Z_1 in which AB, BC, CA are cut by LM, MN, NL, respectively; thus

$$\mathbf{x}_1 = \frac{\alpha\mathbf{a} - \beta\mathbf{b}}{\alpha - \beta}, \qquad \mathbf{y}_1 = \frac{\beta\mathbf{b} - \gamma\mathbf{c}}{\beta - \gamma}, \qquad \mathbf{z}_1 = \frac{\gamma\mathbf{c} - \alpha\mathbf{a}}{\gamma - \alpha}.$$

Since

$$(\alpha - \beta)\mathbf{x}_1 + (\beta - \gamma)\mathbf{y}_1 + (\gamma - \alpha)\mathbf{z}_1 = 0, \qquad \alpha - \beta + \beta - \gamma + \gamma - \alpha = 0,$$

the points X_1, Y_1, Z_1 are collinear. This is also a consequence of Desargues' Theorem applied to the triangles ABC, LMN, which are in perspective from D.

Each vertex of the quadrangle is the center of perspective of the triangles formed by the other three vertices and by the three diagonal points. Thus, with careful regard to exact correspondence, the triangle LMN is in perspec-

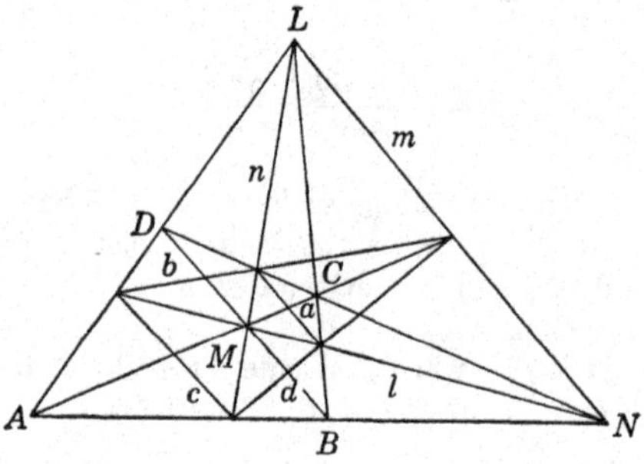

FIG. 7f

tive with DCB, CDA, BAD, ABC from A, B, C, D, respectively. Hence the corresponding sides of these triangles intersect in four lines a, b, c, d (only d, the line $X_1Y_1Z_1$, is shown in Fig. 7e). These lines are the polars of A, B, C, D with respect to the triangle LMN (ex. 3).

Corresponding to the complete *quadrangle* given by four points A, B, C, D, we now have a complete *quadrilateral* given by four lines a, b, c, d. In these configurations the roles of points and lines are interchanged (Fig. 7f):

The quadrangle has four *vertices* A, B, C, D and six *sides* consisting of three opposite pairs (BC, AD), (CA, BD), (AB, CD) which meet in the three *diagonal points* L, M, N.

The quadrilateral has four sides a, b, c, d and six vertices consisting of three opposite pairs (bc, ad), (ca, bd), (ab, cd) which lie on the three *diagonal lines* l, m, n.

The quadrangle and quadrilateral have the same *diagonal triangle* LMN or lmn; the sides l, m, n are opposite the vertices L, M, N.

8. Linear Relations Independent of the Origin. We have seen that the position vectors of collinear or coplanar points satisfy a linear equation in which the sum of the scalar coefficients is zero. The significance of such relations is given by the

THEOREM. *A linear relation of the form,*

$$(1) \qquad \lambda_1 \mathbf{p}_1 + \lambda_2 \mathbf{p}_2 + \cdots + \lambda_n \mathbf{p}_n = 0,$$

connecting the position vectors of the points $P_1, P_2, \cdots, P_n$ will be independent of the position of the origin O when, and only when, the sum of the scalar coefficients is zero:

$$(2) \qquad \lambda_1 + \lambda_2 + \cdots + \lambda_n = 0.$$

Proof. Change from O to a new origin O'. Writing $\overrightarrow{OP_i} = \mathbf{p}_i$, $\overrightarrow{O'P_i} = \mathbf{p}_i'$, $\overrightarrow{OO'} = \mathbf{d}$, we have $\mathbf{p}_i = \mathbf{d} + \mathbf{p}_i'$; hence (1) becomes

$$(\lambda_1 + \lambda_2 + \cdots + \lambda_n)\mathbf{d} + \lambda_1 \mathbf{p}_1' + \lambda_2 \mathbf{p}_2' + \cdots + \lambda_n \mathbf{p}_n' = 0.$$

This equation will have the same form as (1) when and only when (2) is satisfied.

For two, three, and four points relations independent of the origin have a simple geometric meaning; namely, the points are coincident, collinear, or coplanar, respectively. The question now arises: What geometric property relates five or more points whose position vectors satisfy a linear relation independent of the origin?

9. Centroid. We shall encounter problems in which each point of a given set is associated with a certain number. The points, for example, may represent particles of matter and the numbers, their masses or electric charges. In the latter case, the numbers may be positive or negative. We shall now define a point P^* called the *centroid* of a set of n points $P_1, P_2, \cdots, P_n$ associated with the numbers $m_1, m_2, \cdots, m_n$, respectively. Denote any one of these "weighted" points by the symbol m_iP_i; then, *if the sum of the numbers m_i is not zero, the centroid of the entire set is defined as the point for which the sum of all the vectors $m_i\overrightarrow{P^*P_i}$ is zero.* The defining equation for the centroid is thus

$$(1) \qquad \Sigma m_i \overrightarrow{P^*P_i} = 0, \quad \text{provided} \quad \Sigma m_i \neq 0.$$

As we wish P^* to be uniquely defined, the case $\Sigma m_i = 0$ must be excluded; for then (1) is a relation independent of the position of P^* (§ 8).

When $\Sigma m_i \neq 0$ there is always a unique point P^* which satisfies (1). For if we choose an origin O at pleasure, (1) can be written

$$\Sigma m_i(\overrightarrow{OP_i} - \overrightarrow{OP^*}) = 0,$$

or

$$(\Sigma m_i)\overrightarrow{OP^*} = \Sigma m_i \overrightarrow{OP_i}. \tag{2}$$

This relation, independent of the origin, fixes the position of P^* relative to O. The point P^* thus determined is the only point that satisfies (1); for if Q is a second point for which

$$\Sigma m_i \overrightarrow{QP_i} = 0,$$

we have, on subtraction from (1),

$$\Sigma m_i(\overrightarrow{P^*P_i} - \overrightarrow{QP_i}) = \Sigma m_i(\overrightarrow{P^*P_i} + \overrightarrow{P_iQ}) = (\Sigma m_i)\overrightarrow{P^*Q} = 0;$$

hence $\overrightarrow{P^*Q} = 0$, and Q coincides with P^*.

If the position vectors of P^*, P_i are written $\mathbf{p}^*$ and $\mathbf{p}_i$, (2) becomes

$$(\Sigma m_i)\mathbf{p}^* = \Sigma m_i \mathbf{p}_i. \tag{3}$$

The centroid of the points m_iP_i is not altered when the numbers m_i are replaced by any set of numbers cm_i proportional to them; for, in (2), the constant c may be canceled from numerator and denominator. In particular, if the numbers m_i are all equal, we may replace them all by unity; the centroid of the n points then is called their *mean center* and is given by the equation:

$$\overrightarrow{OP^*} = \frac{1}{n}\Sigma\overrightarrow{OP_i}, \quad \text{or} \quad \mathbf{p}^* = \frac{1}{n}\Sigma \mathbf{p}_i. \tag{4}$$

In finding the centroid P^* of any set of weighted points m_iP_i, we may replace any subset of points for which the sum of the weights is a number $m' \neq 0$ by their centroid P' with the weight m'. For, if Σ' and Σ'', respectively, denote summations extended over the points of the subset and over all the remaining points, we may write (3) in the form,

$$(\Sigma' m_i + \Sigma'' m_i)\mathbf{p}^* = \Sigma' m_i \mathbf{p}_i + \Sigma'' m_i \mathbf{p}_i,$$

or

$$(m' + \Sigma'' m_i)\mathbf{p}^* = m'\mathbf{p}' + \Sigma'' m_i \mathbf{p}_i.$$

This equation shows that P^* is also the centroid of the point $m'P'$ and the points m_iP_i not included in the subset.

Finally, let us consider the nature of a set of n points m_iP_i $(m_i \neq 0)$ for which $\Sigma m_i = 0$. If we attempt to find a point P^* which satisfies

$$(5) \qquad \Sigma m_i \overrightarrow{P^*P_i} = 0 \quad \text{when} \quad \Sigma m_i = 0,$$

we find that there are two possibilities.

(*a*) For any arbitrary choice of the points m_iP_i, subject only to the condition $\Sigma m_i = 0$, there will in general be no point P^* that will satisfy (5). Thus in the cases $n = 2, 3, 4$, equations (5) imply, respectively, that P_1 and P_2 coincide; P_1, P_2, P_3 are collinear; P_1, P_2, P_3, P_4 are coplanar. Hence, if these conditions are *not* fulfilled, P^* does not exist.

(*b*) If, however, a point P^* can be found which satisfies (5), any point whatever will serve (§ 8). In fact, we see, from (2), that the existence of P^* implies that

$$(6) \qquad \Sigma m_i \overrightarrow{OP_i} = 0 \qquad \Sigma m_i = 0,$$

for any choice of O. In particular, we may take any one of the given points P_i as origin. Thus, if we take O at P_1, (6) becomes

$$\sum_{i=2}^{n} m_i \overrightarrow{P_1P_i} = 0, \qquad \sum_{i=2}^{n} m_i = -m_1 \neq 0.$$

Hence, from the defining equation (1), we see that P_1 is the centroid of the points $m_2P_2, m_3P_3, \cdots, m_nP_n$. Precisely the same conclusion may be drawn for $P_2, P_3, \cdots, P_n$. We now can answer the question raised at the end of § 8. *Any set of weighted points* m_iP_i $(m_i \neq 0)$ *whose position vectors satisfy a linear relation* (6) *independent of the origin has the intrinsic property that any point of the set is the centroid of all the remaining weighted points.*

A set of n points having this property may be readily constructed. Take any set of $n - 1$ weighted points m_iP_i, such that $m_1 + m_2 + \cdots + m_{n-1} \neq 0$, and adjoin to the set their centroid $P^* = P_n$ with the weight $m_n = -(m_1 + m_2 + \cdots + m_{n-1})$. Then, since

$$-m_n\mathbf{p}_n = m_1\mathbf{p}_1 + m_2\mathbf{p}_2 + \cdots + m_{n-1}\mathbf{p}_{n-1},$$

we have

$$\sum_1^n m_i\mathbf{p}_i = 0, \qquad \sum_1^n m_i = 0.$$

Thus the relation,

$$\text{(7)} \qquad \alpha\mathbf{a} + \beta\mathbf{b} + \gamma\mathbf{c} = 0, \qquad \alpha + \beta + \gamma = 0,$$

between three collinear points shows that each point of the set αA, βB, γC is the centroid of the other two. Any point C on the line of A and B satisfies a relation of the type (7) and hence is the centroid of these points when suitably weighted.

The relation,

$$\text{(8)} \qquad \alpha\mathbf{a} + \beta\mathbf{b} + \gamma\mathbf{c} + \delta\mathbf{d} = 0, \qquad \alpha + \beta + \gamma + \delta = 0,$$

between four coplanar points shows that each point of the set αA, βB, γC, δD is the centroid of the other three. Any point D in the plane of A, B, C satisfies a relation of the type (8) and hence is the centroid of these points when suitably weighted.

Example 1. *Centroid of Two Points.* The centroid of αA and βB is given by

$$\mathbf{p}^* = \frac{\alpha\mathbf{a} + \beta\mathbf{b}}{\alpha + \beta}.$$

Hence P^* divides AB in the inverse ratio β/α of their weights. In particular, if $\alpha = \beta$, P^* divides AB in half. *The mean center of two points lies midway between them.*

Example 2. *Mean Center of Three Points A, B, C.* Let L, M, N be the mid-points of BC, CA, AB (Fig. 9a). Then the mean center P^* of A, B, C is the centroid of A and $2L$, B and $2M$, C and $2N$. Hence (ex. 1) the segments AL, BM, CN are all divided by P^* in the ratio of 2/1.

If A, B, C are not collinear, they are the vertices of a triangle. Its medians AL, BM, CN intersect at P^* and are there divided in the ratio of 2/1.

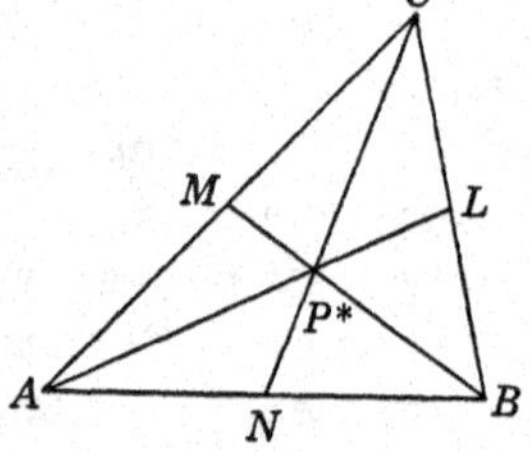

FIG. 9a

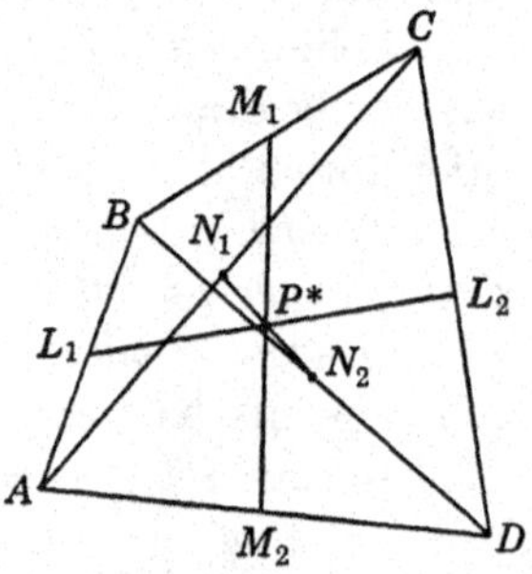

FIG. 9b

Example 3. *Mean Center of Four Points A, B, C, D.* Let L_1, L_2; M_1, M_2; N_1, N_2 be the mid-points of AB, CD; BC, DA; AC, BD, respectively (Fig. 9b). To find P^* we may replace A, B, C, D by $2L_1$ and $2L_2$, or by $2M_1$ and $2M_2$, or by $2N_1$ and $2N_2$. Therefore P^* is the mid-point of the segments L_1L_2, M_1M_2, N_1N_2. The bisectors of the three pairs of opposite sides of the quadrangle $ABCD$ (which may be plane or skew) intersect at P^* and are there divided in half.

If A, B, C, D are not coplanar, they determine a tetrahedron. Let A', B', C', D' be the mean centers of the triads BCD, CDA, DAB, ABC. Then P^* is the centroid of A and $3A'$, B and $3B'$, C and $3C'$, D and $3D'$, and hence divides each of the segments AA', BB', CC', DD' in the ratio of 3/1. The preceding result also shows that the bisectors of the three pairs of opposite sides of the tetrahedron meet at P^*.

Example 4. The sum of n vectors $\overrightarrow{A_iB_i}$ ($i = 1, 2, \cdots, n$) is given by

$$\Sigma(\mathbf{b}_i - \mathbf{a}_i) = n\mathbf{b}^* - n\mathbf{a}^* = n(\mathbf{b}^* - \mathbf{a}^*)$$

where A^* and B^* are the mean centers of the initial points A_i and the terminal points B_i, respectively; hence

$$\Sigma\overrightarrow{A_iB_i} = n\,\overrightarrow{A^*B^*}.$$

In particular,

$$\overrightarrow{A_1B_1} + \overrightarrow{A_2B_2} = 2\,\overrightarrow{A^*B^*},$$

where A^*, B^* are the mid-points of A_1A_2 and B_1B_2, respectively.

10. Barycentric Coordinates. If P is any point in the plane of the reference triangle ABC, the vectors $\overrightarrow{AP}$, $\overrightarrow{BP}$, $\overrightarrow{CP}$ are linearly dependent (§ 5); hence

$$\alpha(\mathbf{p} - \mathbf{a}) + \beta(\mathbf{p} - \mathbf{b}) + \gamma(\mathbf{p} - \mathbf{c}) = 0,$$

$$(1) \qquad \mathbf{p} = \frac{\alpha\mathbf{a} + \beta\mathbf{b} + \gamma\mathbf{c}}{\alpha + \beta + \gamma}.$$

The denominator is not zero; for, if

$$\alpha + \beta + \gamma = 0, \quad \text{then} \quad \alpha\mathbf{a} + \beta\mathbf{b} + \gamma\mathbf{c} = 0,$$

and A, B, C would be collinear, contrary to hypothesis. Thus P is the centroid of the weighted points αA, βB, γC. The three numbers α, β, γ are called the *barycentric coordinates* of P; as they all may be multiplied by the same number without altering $\mathbf{p}$, their ratios determine P. The vertices of the reference triangle have the coordinates $A(1, 0, 0)$, $B(0, 1, 0)$, $C(0, 0, 1)$; the *unit point* $(1, 1, 1)$ is the mean center of A, B, C.

If P is a point in space and $ABCD$ a reference tetrahedron, the vectors $\overrightarrow{AP}$, $\overrightarrow{BP}$, $\overrightarrow{CP}$, $\overrightarrow{DP}$ are linearly dependent (§ 5); hence

$$\alpha(\mathbf{p} - \mathbf{a}) + \beta(\mathbf{p} - \mathbf{b}) + \gamma(\mathbf{p} - \mathbf{c}) + \delta(\mathbf{p} - \mathbf{d}) = 0,$$

$$(2) \qquad \mathbf{p} = \frac{\alpha\mathbf{a} + \beta\mathbf{b} + \gamma\mathbf{c} + \delta\mathbf{d}}{\alpha + \beta + \gamma + \delta}.$$

The denominator is not zero; for, if

$$\alpha + \beta + \gamma + \delta = 0, \quad \text{then} \quad \alpha\mathbf{a} + \beta\mathbf{b} + \gamma\mathbf{c} + \delta\mathbf{d} = 0$$

and the points A, B, C, D would be coplanar, contrary to hypothesis. Thus P is the centroid of the weighted points αA, βB, γC, δD. The four numbers α, β, γ, δ are called *barycentric coordinates* of P; as they all may be multiplied by the same number without altering $\mathbf{p}$, their ratios determine P. The vertices of the reference tetrahedron have the barycentric coordinates $A(1, 0, 0, 0)$, $B(0, 1, 0, 0)$, $C(0, 0, 1, 0)$, $D(0, 0, 0, 1)$; the *unit point* $(1, 1, 1, 1)$ is the mean center of A, B, C, D.

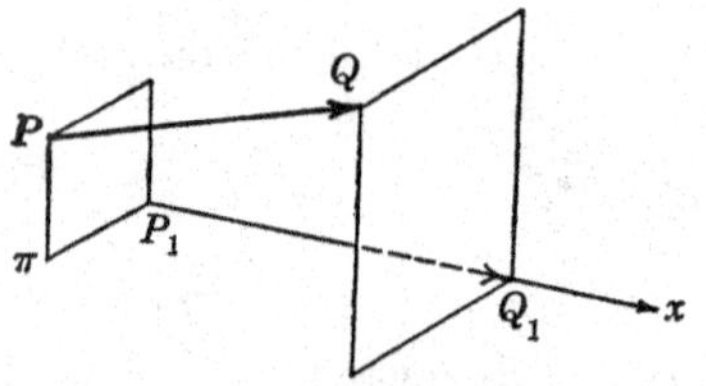

FIG. 11

11. Projection of a Vector. To find the *projection* of a vector $\overrightarrow{PQ}$ upon a line x, a director plane π must be specified. Pass planes through P and Q parallel to π and let them cut x in the points P_1, Q_1 (Fig. 11). Then the vector $\overrightarrow{P_1Q_1}$ is the π-projection of $\overrightarrow{PQ}$ upon x.

If no director plane π is specified, we tacitly assume that π is perpendicular to x; the projection is then *orthogonal.*

Let the vectors $\overrightarrow{PQ}$, $\overrightarrow{QR}$ and their sum $\overrightarrow{PR}$ have the projections $\overrightarrow{P_1Q_1}$, $\overrightarrow{Q_1R_1}$ and $\overrightarrow{P_1R_1}$ on the line x; then, since

$$\overrightarrow{P_1Q_1} + \overrightarrow{Q_1R_1} = \overrightarrow{P_1R_1},$$

the projection of the sum of two vectors on a line is equal to the sum of their projection on this line.

12. Base Vectors. Let $\mathbf{e}_1$, $\mathbf{e}_2$, $\mathbf{e}_3$ be three *linearly independent* vectors; they are then non-coplanar. If the vectors are drawn from a common origin O (Fig. 12a), we may pass a closed plane curve through their end points E_1, E_2, E_3. If this curve is viewed from the side of its plane opposite to that on which O lies, the order $E_1E_2E_3$ defines a sense of circulation. If this sense is counterclockwise, the set $\mathbf{e}_1$, $\mathbf{e}_2$, $\mathbf{e}_3$ is said to be *right-handed* or *dextral;* for it is then possible to extend the thumb, index, and middle fingers of the right hand so that they have the directions of $\mathbf{e}_1$, $\mathbf{e}_2$, $\mathbf{e}_3$, respectively. If the sense defined by $E_1E_2E_3$ is clockwise the set is said to be *left-handed* or *sinistral.*

Any vector $\mathbf{u} = \overrightarrow{PQ}$ may be expressed as the sum of three vectors parallel to $\mathbf{e}_1$, $\mathbf{e}_2$, $\mathbf{e}_3$, respectively. For, if we construct a parallelepiped on PQ as diagonal by passing planes through P and

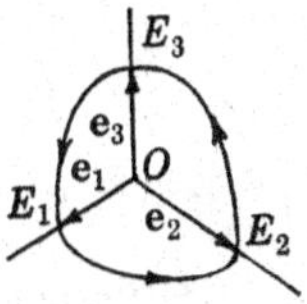

FIG. 12a

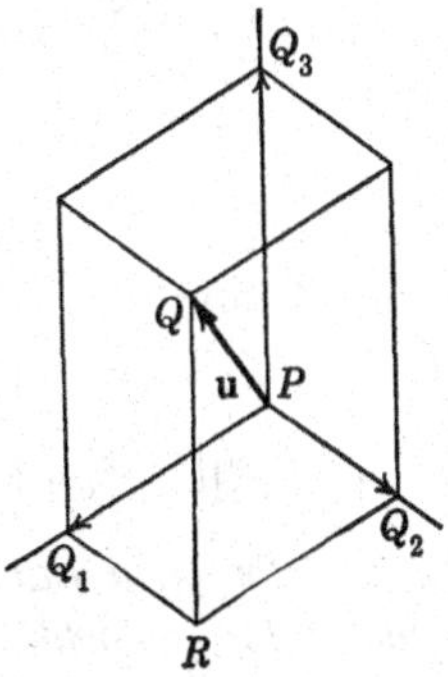

FIG 12b

Q parallel to $\mathbf{e}_2$ and $\mathbf{e}_3$, $\mathbf{e}_3$ and $\mathbf{e}_1$, $\mathbf{e}_1$ and $\mathbf{e}_2$ (Fig. 12b), its edges will be parallel to $\mathbf{e}_1$, $\mathbf{e}_2$, $\mathbf{e}_3$, and

$$\overrightarrow{PQ} = \overrightarrow{PQ_1} + \overrightarrow{Q_1R} + \overrightarrow{RQ} = \overrightarrow{PQ_1} + \overrightarrow{PQ_2} + \overrightarrow{PQ_3}.$$

Since $\overrightarrow{PQ_i}$ is a scalar multiple of $\mathbf{e}_i$, say $u^i\mathbf{e}_i$,

$$\mathbf{u} = u^1\mathbf{e}_1 + u^2\mathbf{e}_2 + u^3\mathbf{e}_3, \tag{1}$$

where u^1, u^2, u^3 are numbers called the *components* of $\mathbf{u}$ with respect to the *basis* $\mathbf{e}_1$, $\mathbf{e}_2$, $\mathbf{e}_3$. Their indices are *not exponents but mere identification tags;* they are written as superscripts for reasons given in Chapter IX. The vector $\mathbf{u}$ is often written $[u^1, u^2, u^3]$, with brackets to enclose its components.

If $\mathbf{u}$ is the position vector $\overrightarrow{OP}$, the components u^i are called the *Cartesian coordinates* of the point P with respect to the basis $\mathbf{e}_1$, $\mathbf{e}_2$, $\mathbf{e}_3$.

A straight line upon which two directions are distinguished is called an *axis.* One direction is called *positive,* the other *negative.* In a figure the positive direction is marked with an arrowhead. With a given basis, lines drawn through the origin O parallel to

$\mathbf{e}_1$, $\mathbf{e}_2$, $\mathbf{e}_3$ and with their directions positive are called the *coordinate axes.* The numbers u^1, u^2, u^3 often are called the components of $\mathbf{u}$ on these axes.

The components of a zero vector are all zero; for, since $\mathbf{e}_1$, $\mathbf{e}_2$, $\mathbf{e}_3$ are linearly independent, $\mathbf{u} = 0$ implies $u^1 = u^2 = u^3 = 0$.

If λ is any scalar,

$$(2) \qquad \lambda\mathbf{u} = \lambda u^1\mathbf{e}_1 + \lambda u^2\mathbf{e}_2 + \lambda u^3\mathbf{e}_3.$$

If $\mathbf{v} = v^1\mathbf{e}_1 + v^2\mathbf{e}_2 + v^3\mathbf{e}_3$,

$$(3) \qquad \mathbf{u} + \mathbf{v} = (u^1 + v^1)\mathbf{e}_1 + (u^2 + v^2)\mathbf{e}_2 + (u^3 + v^3)\mathbf{e}_3.$$

With the bracket notation, (2) and (3) become

$$(4) \qquad \lambda[u^1, u^2, u^3] = [\lambda u^1, \lambda u^2, \lambda u^3],$$

$$(5) \qquad [u^1, u^2, u^3] + [v^1, v^2, v^3] = [u^1 + v^1, u^2 + v^2, u^3 + v^3].$$

To multiply a vector by a number, multiply its components by that number; to add vectors, add their corresponding components. In particular,

$$(6) \qquad -[u^1, u^2, u^3] = [-u^1, -u^2, -u^3],$$

$$(7) \qquad [u^1, u^2, u^3] - [v^1, v^2, v^3] = [u^1 - v^1, u^2 - v^2, u^3 - v^3].$$

If $\mathbf{u} = \mathbf{v}$, then $\mathbf{u} - \mathbf{v} = 0$; hence

$$(8) \qquad \mathbf{u} = \mathbf{v} \quad \text{implies} \quad u^1 = v^1, \quad u^2 = v^2, \quad u^3 = v^3.$$

When referred to the same basis, equal vectors have their corresponding components equal.

13. Rectangular Components. Let $\mathbf{i}$, $\mathbf{j}$, $\mathbf{k}$ denote a dextral system of mutually perpendicular unit vectors. From an origin O draw the coordinate axes x, y, z with positive directions given by $\mathbf{i}$, $\mathbf{j}$, $\mathbf{k}$ (Fig. 13). Any vector $\mathbf{u}$ now is determined by giving its components u_1, u_2, u_3 on these rectangular axes:

$$(1) \qquad \mathbf{u} = u_1\mathbf{i} + u_2\mathbf{j} + u_3\mathbf{k}.$$

Draw $\overrightarrow{OP} = \mathbf{u}$ from the origin, and let $\overrightarrow{OP_1}$, $\overrightarrow{OP_2}$, $\overrightarrow{OP_3}$ be its orthogonal projections on the axes. Then u_1 is the length of $\overrightarrow{OP_1}$, taken positive or negative according as $\overrightarrow{OP_1}$ has the direction of $\mathbf{i}$ or $-\mathbf{i}$; hence

$$u_1 = |\mathbf{u}| \cos(\mathbf{i}, \mathbf{u}),$$

where $|\mathbf{u}|$ denotes the length of $\mathbf{u}$ and $(\mathbf{i}, \mathbf{u})$ the angle between $\mathbf{i}$ and $\mathbf{u}$. The *rectangular components* of $\mathbf{u}$ thus are obtained by multiplying its length by the corresponding direction cosines:

$$(2) \qquad u_1 = |\mathbf{u}| \cos(\mathbf{i}, \mathbf{u}), \qquad u_2 = |\mathbf{u}| \cos(\mathbf{j}, \mathbf{u}),$$

$$u_3 = |\mathbf{u}| \cos(\mathbf{k}, \mathbf{u}).$$

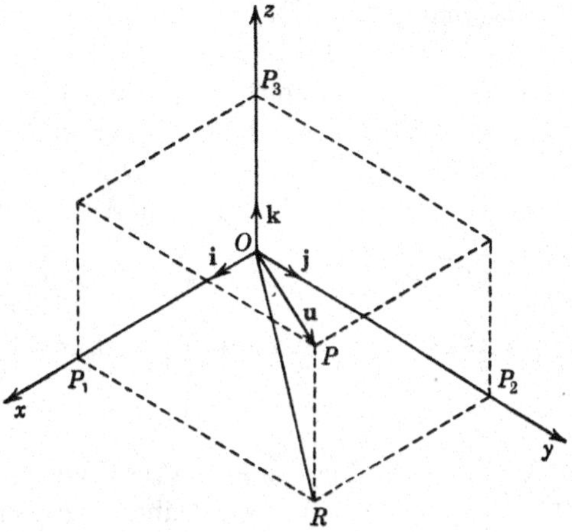

FIG. 13

The Pythagorean Theorem gives the length of $\mathbf{u}$ in terms of its components; from

$$OP^2 = (OR)^2 + (RP)^2 = (OP_1)^2 + (OP_2)^2 + (OP_3)^2,$$

$$(3) \qquad |\mathbf{u}|^2 = (u_1)^2 + (u_2)^2 + (u_3)^2.$$

If we substitute from (2) in (3), we obtain the relation,

$$\cos^2(\mathbf{i}, \mathbf{u}) + \cos^2(\mathbf{j}, \mathbf{u}) + \cos^2(\mathbf{k}, \mathbf{u}) = 1,$$

satisfied by the direction cosines of any vector.

The *rectangular coordinates* of any point P are defined as the components of its position vector $\overrightarrow{OP}$ on the x-, y-, and z- axes. Thus, if

$$(4) \qquad \overrightarrow{OP} = x\mathbf{i} + y\mathbf{j} + z\mathbf{k},$$

P has the rectangular coordinates (x, y, z). Since $\overrightarrow{P_1P_2} = \overrightarrow{OP_2}$

$-\overrightarrow{OP_1}$, this gives

$$\overrightarrow{P_1P_2} = (x_2 - x_1)\mathbf{i} + (y_2 - y_1)\mathbf{j} + (z_2 - z_1)\mathbf{k} \tag{5}$$

if (x_i, y_i, z_i) are the coordinates of P_i. *The components of* $\overrightarrow{P_1P_2}$ *are found by subtracting the coordinates of* P_1 *from the corresponding coordinates of* P_2.

With an orthogonal basis **i**, **j**, **k**, the components may be written with subscripts (as previously) or superscripts. Separately, they are called the x-, y-, and z- components and often are written u_1, u_2, u_3. Just as with a general basis, vectors are specified by giving their components: $\mathbf{u} = [u_1, u_2, u_3]$. Thus the point (x, y, z) has the position vector $[x, y, z]$; and (5) may be written

$$\overrightarrow{P_1P_2} = [x_2 - x_1, y_2 - y_1, z_2 - z_1].$$

The unit vectors **i**, **j** form a basis for all vectors **u** in their plane:

$$\mathbf{u} = u_1\mathbf{i} + u_2\mathbf{j}. \tag{6}$$

If **e** is a unit vector in the plane and θ is the angle $(\mathbf{i}, \mathbf{e})$, reckoned positive in the sense from **i** to **j**, we define $\cos\theta$ and $\sin\theta$ as the components of **e**:

$$\mathbf{e} = \mathbf{i}\cos\theta + \mathbf{j}\sin\theta. \tag{7}$$

Any vector in the plane may be written

$$\mathbf{u} = |\mathbf{u}|\,\mathbf{e} = |\mathbf{u}|\,\{\mathbf{i}\cos(\mathbf{i}, \mathbf{u}) + \mathbf{j}\sin(\mathbf{i}, \mathbf{u})\}; \tag{8}$$

its rectangular components are

$$u_1 = |\mathbf{u}|\cos(\mathbf{i}, \mathbf{u}), \qquad u_2 = |\mathbf{u}|\sin(\mathbf{i}, \mathbf{u}), \qquad u_3 = 0. \tag{9}$$

Example. Addition Theorems for the Sine and Cosine. Let **a** and **b** be two unit vectors such that the angles $(\mathbf{i}, \mathbf{a}) = \alpha$, $(\mathbf{a}, \mathbf{b}) = \beta$; then the angle $(\mathbf{i}, \mathbf{b}) = \alpha + \beta$, and, from (7),

$$\mathbf{a} = \mathbf{i}\cos\alpha + \mathbf{j}\sin\alpha,$$

$$\mathbf{b} = \mathbf{i}\cos(\alpha + \beta) + \mathbf{j}\sin(\alpha + \beta).$$

If we refer **b** to the new basis $\bar{\mathbf{i}} = \mathbf{a}$, $\bar{\mathbf{j}}$ (a unit vector 90° ahead of **a**), we have

$$\begin{aligned}\mathbf{b} &= \bar{\mathbf{i}}\cos\beta + \bar{\mathbf{j}}\sin\beta\\ &= \{\mathbf{i}\cos\alpha + \mathbf{j}\sin\alpha\}\cos\beta + \left\{\mathbf{i}\cos\left(\alpha + \frac{\pi}{2}\right) + \mathbf{j}\sin\left(\alpha + \frac{\pi}{2}\right)\right\}\sin\beta.\end{aligned}$$

Comparing the components of **b** in the two preceding expressions, we find

$$\cos(\alpha + \beta) = \cos\alpha\cos\beta + \cos\left(\alpha + \frac{\pi}{2}\right)\sin\beta,$$

$$\sin(\alpha + \beta) = \sin\alpha\cos\beta + \sin\left(\alpha + \frac{\pi}{2}\right)\sin\beta.$$

With $\alpha = \pi/2$, these equations give

$$\cos\left(\beta + \frac{\pi}{2}\right) = -\sin\beta, \qquad \sin\left(\beta + \frac{\pi}{2}\right) = \cos\beta;$$

hence

$$\cos(\alpha + \beta) = \cos\alpha\cos\beta - \sin\alpha\sin\beta,$$

$$\sin(\alpha + \beta) = \sin\alpha\cos\beta + \cos\alpha\sin\beta.$$

These addition theorems hold for all values of α and β, positive or negative.

14. Products of Two Vectors. Hitherto we have considered only the products of vectors by numbers. Next we shall define two operations between vectors, which are known as "products," because they have some properties in common with the products of numbers. These products of vectors, however, will also prove to have properties in striking disagreement with those of numbers.

Since one of these products is a scalar and the other a vector, they are called the *scalar product* and *vector product*, respectively. The definitions of these new products may seem rather arbitrary to one unfamiliar with the history of vector algebra. We present this algebra in the form and notation due to the American mathematical physicist, J. Willard Gibbs (1839–1903).† It is an offshoot of the *algebra of quaternions*, adapted to the uses of geometry and physics. In Chapter X quaternion algebra is developed briefly, and the origin of the foregoing products is revealed.

15. Scalar Product. *The scalar product of two vectors* **u** *and* **v**, *written* $\mathbf{u} \cdot \mathbf{v}$, *is defined as the product of their lengths and the cosine of their included angle:*

(1) $$\mathbf{u} \cdot \mathbf{v} = |\mathbf{u}|\,|\mathbf{v}|\cos(\mathbf{u}, \mathbf{v}).$$

The scalar product is therefore a *number* which for *proper* (non-zero) vectors is positive, zero, or negative, according as the angle $(\mathbf{u}, \mathbf{v})$ is acute, right, or obtuse. Hence, for *proper* vectors,

(2) $$\mathbf{u} \cdot \mathbf{v} = 0 \quad \text{means} \quad \mathbf{u} \perp \mathbf{v}.$$

† Professor of mathematical physics at Yale University. His pamphlet on the *Elements of Vector Analysis* was privately printed in 1881. A more complete treatise on *Vector Analysis* (New Haven, Yale University Press, 1901) based on Gibbs's lectures, was written by Professor E. B. Wilson.

When $\mathbf{u}$ and $\mathbf{v}$ are parallel,

$$\mathbf{u} \cdot \mathbf{v} = |\mathbf{u}||\mathbf{v}|, \quad \text{or} \quad -|\mathbf{u}||\mathbf{v}|,$$

according as the vectors have the same or opposite directions; thus $\mathbf{u} \cdot \mathbf{u} = |\mathbf{u}|^2$.

From (1) we see that

$$(-\mathbf{u}) \cdot \mathbf{v} = \mathbf{u} \cdot (-\mathbf{v}) = -\mathbf{u} \cdot \mathbf{v}, \qquad (-\mathbf{u}) \cdot (-\mathbf{v}) = \mathbf{u} \cdot \mathbf{v},$$

$$(3) \qquad (\alpha \mathbf{u}) \cdot (\beta \mathbf{v}) = \alpha\beta\, \mathbf{u} \cdot \mathbf{v}.$$

The last result is obvious when α and β are positive numbers; the other cases then follow from the equations preceding.

Besides the components of a vector on the coordinate axes (§§ 12, 13) we shall also use the *orthogonal* projection and component of a vector on an arbitrary directed line l. If the positive direction of l is given by the unit vector $\mathbf{e}$, the projection of $\mathbf{u}$ upon l ($\text{proj}_l\, \mathbf{u}$) is a scalar multiple of $\mathbf{e}$. This scalar is called the component of $\mathbf{u}$ upon l ($\text{comp}_l\, \mathbf{u}$); its defining equation is therefore

$$(4) \qquad \mathbf{e}\, \text{comp}_l\, \mathbf{u} = \text{proj}_l\, \mathbf{u}.$$

As in § 13, we compute $\text{comp}_l\, \mathbf{u}$ as

$$(5) \qquad \text{comp}_l\, \mathbf{u} = |\mathbf{u}| \cos(\mathbf{e}, \mathbf{u});$$

since $|\mathbf{e}| = 1$, this may also be written

$$(6) \qquad \text{comp}_l\, \mathbf{u} = \mathbf{e} \cdot \mathbf{u}.$$

From the projection theorem of § 11,

$$(7) \qquad \text{proj}_l\, (\mathbf{u} + \mathbf{v}) = \text{proj}_l\, \mathbf{u} + \text{proj}_l\, \mathbf{v};$$

hence, from (3),

$$(8) \qquad \text{comp}_l\, (\mathbf{u} + \mathbf{v}) = \text{comp}_l\, \mathbf{u} + \text{comp}_l\, \mathbf{v}.$$

The operations expressed by proj_l *and* comp_l *are distributive with respect to addition.*

The definition (1) of $\mathbf{u} \cdot \mathbf{v}$ now may be written

$$(9) \qquad \mathbf{u} \cdot \mathbf{v} = |\mathbf{u}|\, \text{comp}_u\, \mathbf{v} = |\mathbf{v}|\, \text{comp}_v\, \mathbf{u}.$$

Scalar or "dot" multiplication is commutative and distributive:

$$(10) \qquad \mathbf{v} \cdot \mathbf{u} = \mathbf{u} \cdot \mathbf{v},$$

$$(11) \qquad \mathbf{w} \cdot (\mathbf{u} + \mathbf{v}) = \mathbf{w} \cdot \mathbf{u} + \mathbf{w} \cdot \mathbf{v}.$$

The last equation is nothing more than (8) multiplied by $|\mathbf{w}|$ when l is taken along $\mathbf{w}$.

If $\mathbf{c} \neq 0$ in the equation,

$$\mathbf{a} \cdot \mathbf{c} = \mathbf{b} \cdot \mathbf{c}, \quad \text{or} \quad (\mathbf{a} - \mathbf{b}) \cdot \mathbf{c} = 0,$$

we can conclude *either* that $\mathbf{a} - \mathbf{b} = 0$ *or* that $\mathbf{a} - \mathbf{b}$ and $\mathbf{c}$ are perpendicular. We cannot "cancel" $\mathbf{c}$ to obtain $\mathbf{a} = \mathbf{b}$ unless $\mathbf{a} - \mathbf{b}$ and $\mathbf{c}$ are not perpendicular.

It is obvious that, in general, $(\mathbf{u} \cdot \mathbf{v})\mathbf{w} \neq \mathbf{u}(\mathbf{v} \cdot \mathbf{w})$.

Since $\mathbf{i}, \mathbf{j}, \mathbf{k}$ are mutually perpendicular unit vectors,

$$(12) \qquad \mathbf{i} \cdot \mathbf{i} = \mathbf{j} \cdot \mathbf{j} = \mathbf{k} \cdot \mathbf{k} = 1, \qquad \mathbf{i} \cdot \mathbf{j} = \mathbf{j} \cdot \mathbf{k} = \mathbf{k} \cdot \mathbf{i} = 0.$$

Hence, if we expand the product,

$$\mathbf{u} \cdot \mathbf{v} = (u_1\mathbf{i} + u_2\mathbf{j} + u_3\mathbf{k}) \cdot (v_1\mathbf{i} + v_2\mathbf{j} + v_3\mathbf{k})$$

we obtain

$$(13) \qquad \mathbf{u} \cdot \mathbf{v} = u_1v_1 + u_2v_2 + u_3v_3.$$

The scalar product of two vectors is equal to the sum of the products of their corresponding rectangular components.

Example 1. If $\mathbf{u} = [2, -1, 3]$, $\mathbf{v} = [0, 2, 4]$, $\mathbf{u} \cdot \mathbf{v} = 0 - 2 + 12 = 10$. Moreover,

$$\text{comp}_v\,\mathbf{u} = \frac{\mathbf{u} \cdot \mathbf{v}}{|\mathbf{v}|} = \frac{10}{\sqrt{20}} = 2.236, \qquad \text{comp}_u\,\mathbf{v} = \frac{\mathbf{u} \cdot \mathbf{v}}{|\mathbf{u}|} = \frac{10}{\sqrt{14}} = 2.673;$$

$$\cos(\mathbf{u}, \mathbf{v}) = \frac{\mathbf{u} \cdot \mathbf{v}}{|\mathbf{u}||\mathbf{v}|} = \frac{10}{\sqrt{280}} = 0.5979, \qquad \text{angle}(\mathbf{u}, \mathbf{v}) = 53^\circ\, 18'.$$

Example 2. Identities involving scalar products may be given a geometric interpretation. Thus (Fig. 15a)

$$(\mathbf{a} + \mathbf{b}) \cdot (\mathbf{a} - \mathbf{b}) = \mathbf{a} \cdot \mathbf{a} - \mathbf{b} \cdot \mathbf{b} \quad \text{gives} \quad cd \cos \varphi = a^2 - b^2,$$

and, if we write $\mathbf{c} = \mathbf{a} + \mathbf{b}$, $\mathbf{d} = \mathbf{a} - \mathbf{b}$, φ = angle $(\mathbf{c}, \mathbf{d})$. If $a = b$, $\mathbf{c} \cdot \mathbf{d} = 0$; then $PQRS$ is a rhombus and the angle PRT may be inscribed in a semicircle about Q. We thus have two geometric theorems:

1. The diagonals of a rhombus cut at right angles.
2. An angle inscribed in a semicircle is a right angle.

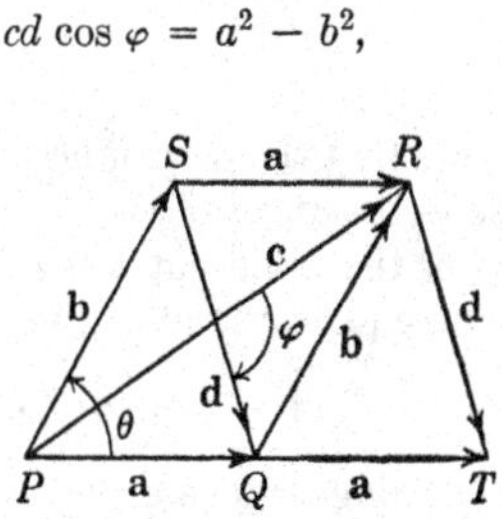

FIG. 15a

Moreover, from

$$(\mathbf{a} - \mathbf{b}) \cdot (\mathbf{a} - \mathbf{b}) = \mathbf{a} \cdot \mathbf{a} + \mathbf{b} \cdot \mathbf{b} - 2\mathbf{a} \cdot \mathbf{b},$$

we have the *cosine law:* $d^2 = a^2 + b^2 - 2ab \cos \theta$.

Example 3. We also may interpret identities involving scalar products by regarding the vectors $\mathbf{a}$, $\mathbf{b}$, $\cdots$ as position vectors $\overrightarrow{OA}$, $\overrightarrow{OB}$, $\cdots$ from an arbitrary origin O. Then $\mathbf{a} - \mathbf{b} = \overrightarrow{BA}$ and $\mathbf{a} + \mathbf{b} = 2\,\overrightarrow{OM}$ where M is the mid-point of AB.

Consider, for example, the identity,

$$(\mathbf{b} - \mathbf{a})^2 + (\mathbf{c} - \mathbf{b})^2 + (\mathbf{d} - \mathbf{c})^2 + (\mathbf{a} - \mathbf{d})^2$$
$$= (\mathbf{c} - \mathbf{a})^2 + (\mathbf{d} - \mathbf{b})^2 + (\mathbf{a} + \mathbf{c} - \mathbf{b} - \mathbf{d})^2,$$

which can be verified on expansion; $\mathbf{u}^2$ means $\mathbf{u} \cdot \mathbf{u}$. If we regard $\mathbf{a}$, $\mathbf{b}$, $\mathbf{c}$, $\mathbf{d}$ as the position vectors of the vertices of a space quadrilateral $ABCD$, $\mathbf{p} = \frac{1}{2}(\mathbf{a} + \mathbf{c})$, $\mathbf{q} = \frac{1}{2}(\mathbf{b} + \mathbf{d})$ locate the mid-points P, Q of the diagonals AC, BD; hence

$$(AB)^2 + (BC)^2 + (CD)^2 + (DA)^2 = (AC)^2 + (BD)^2 + 4(QP)^2.$$

The sum of the squares of the sides of any space quadrilateral equals the sum of the squares of its diagonals plus four times the square of the segment joining their middle points.

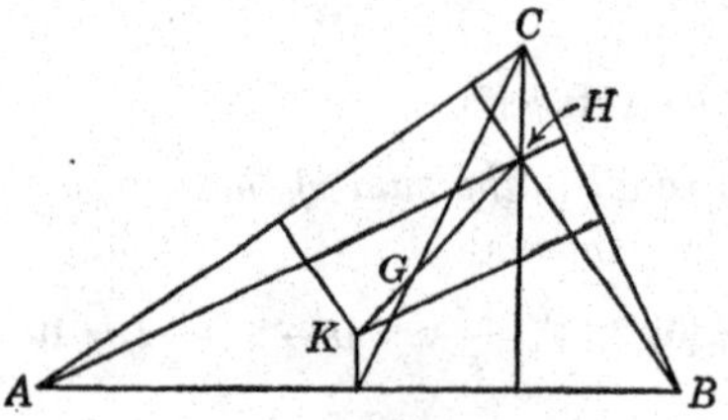

FIG. 15b

Example 4. The identity,

$$(\mathbf{a} - \mathbf{b}) \cdot (\mathbf{h} - \mathbf{c}) + (\mathbf{b} - \mathbf{c}) \cdot (\mathbf{h} - \mathbf{a})$$
$$+ (\mathbf{c} - \mathbf{a}) \cdot (\mathbf{h} - \mathbf{b}) = 0,$$

shows that the altitudes of a triangle ABC meet in a point H, the *orthocenter* of the triangle (Fig. 15b); for, if two terms of this equation are zero, the third is likewise.

Similarly the identity,

$$(\mathbf{a} - \mathbf{b}) \cdot \left(\mathbf{k} - \frac{\mathbf{a} + \mathbf{b}}{2}\right) + (\mathbf{b} - \mathbf{c}) \cdot \left(\mathbf{k} - \frac{\mathbf{b} + \mathbf{c}}{2}\right)$$
$$+ (\mathbf{c} - \mathbf{a}) \cdot \left(\mathbf{k} - \frac{\mathbf{c} + \mathbf{a}}{2}\right) = 0,$$

shows that the perpendicular bisectors of the triangle ABC meet in a point K, the *circumcenter* of the triangle.

For the orthocenter H and circumcenter K the individual terms of the foregoing equations vanish; for example,

$$(\mathbf{a} - \mathbf{b}) \cdot (\mathbf{h} - \mathbf{c}) = 0, \qquad (\mathbf{a} - \mathbf{b}) \cdot (2\mathbf{k} - \mathbf{a} - \mathbf{b}) = 0.$$

On adding these, we obtain

$$(\mathbf{a} - \mathbf{b}) \cdot (\mathbf{h} + 2\mathbf{k} - \mathbf{a} - \mathbf{b} - \mathbf{c}) = 0;$$

or on writing $\mathbf{g} = \frac{1}{3}(\mathbf{a} + \mathbf{b} + \mathbf{c})$ for the position vector of the mean center G of the triangle ABC.

$$(\mathbf{a} - \mathbf{b}) \cdot (\mathbf{h} + 2\mathbf{k} - 3\mathbf{g}) = 0.$$

Since this equation also holds when $\mathbf{a} - \mathbf{b}$ is replaced by $\mathbf{b} - \mathbf{c}$ and $\mathbf{c} - \mathbf{a}$, we conclude that

$$\mathbf{h} + 2\mathbf{k} - 3\mathbf{g} = 0.$$

Therefore *the mean center of a triangle lies on the line joining the orthocenter to the circumcenter and divides it in the ratio of* 2/1. This line is called the *Euler line* of the triangle.

Example 5. In order to interpret the identity,

$$(\mathbf{a} + \mathbf{b} - \mathbf{c} - \mathbf{d})^2 - (\mathbf{a} - \mathbf{b} - \mathbf{c} + \mathbf{d})^2 = 4(\mathbf{a} - \mathbf{c}) \cdot (\mathbf{b} - \mathbf{d}),$$

with reference to the plane quadrilateral $ABCD$, let P, Q, R, S denote the mid-points of AB, BC, CD, DA; then (Fig. 15c) we have

$$(PR)^2 - (QS)^2 = \overrightarrow{CA} \cdot \overrightarrow{DB}.$$

Hence, if the diagonals of a quadrilateral cut at right angles, the lines joining the mid-points of opposite sides are equal. Since the lines PR, QS intersect

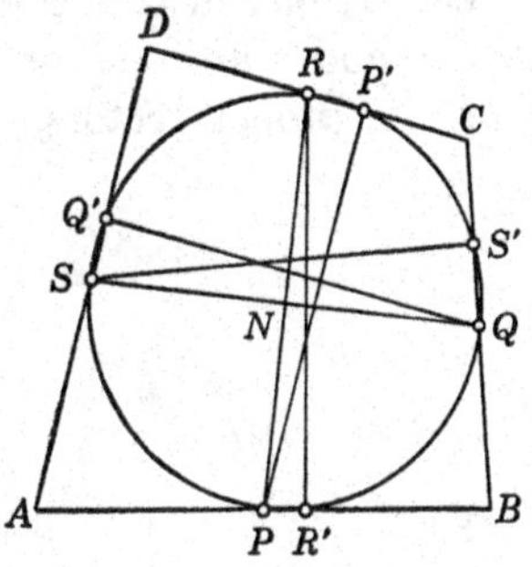

FIG. 15c

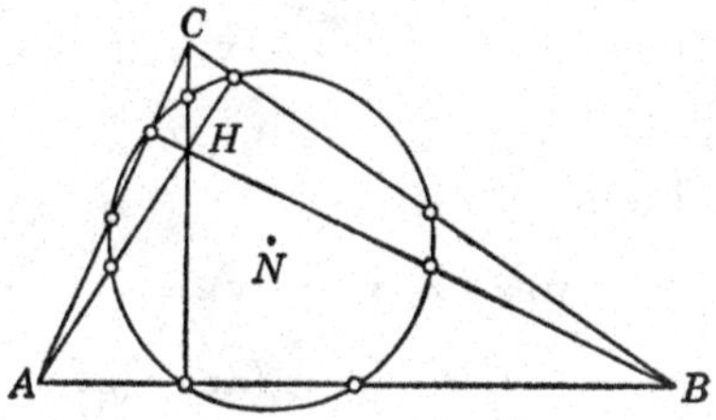

FIG. 15d

in the mean center N of the points A, B, C, D and are bisected there (§ 9, ex. 3), a circle with N as center will pass through P, Q, R, S. If perpendiculars from P, Q, R, S are dropped upon the sides opposite, their feet, P', Q', R', S also will lie on this circle. Thus we have proved the

THEOREM. *When the diagonals of a quadrilateral are perpendicular, the mid-points of its sides and the feet of the perpendiculars dropped from them on the opposite sides all lie on a circle described about the mean center of the vertices.*

In the figure formed by a triangle ABC and its three altitudes meeting at the orthocenter H (Fig. 15d), three quadrilaterals, $ABCH$, $BCAH$, $CABH$, all have perpendicular diagonals. Their three eight-point circles are all the same. This circle, whose center is at the mean center N of A, B, C, H, is the famous *nine-point circle* of the triangle.

THEOREM. *For any triangle ABC whose orthocenter is H, a circle whose center is the mean center of A, B, C, H, passes through the nine points: the mid-points of the sides, the feet of its altitudes, and the mid-points of the segments joining H to the vertices.*

The center N of the nine-point circle has the position vector n given by

$$4\mathbf{n} = \mathbf{a} + \mathbf{b} + \mathbf{c} + \mathbf{h} = 3\mathbf{g} + \mathbf{h} = 2\mathbf{h} + 2\mathbf{k},$$

in view of ex. 4. Therefore the center N of the nine-point circle is collinear with the mean center G of the triangle, its orthocenter H, and circumcenter K. Moreover N bisects the segment HK.

16. Vector Product. *The vector product of two vectors* $\mathbf{u}$ *and* $\mathbf{v}$, *written* $\mathbf{u} \times \mathbf{v}$, *is defined as the vector,*

$$\mathbf{u} \times \mathbf{v} = |\mathbf{u}|\,|\mathbf{v}| \sin(\mathbf{u}, \mathbf{v})\mathbf{e}, \tag{1}$$

where $\mathbf{e}$ *is a unit vector perpendicular to both* $\mathbf{u}$ *and* $\mathbf{v}$ *and forming with them a dextral set* $\mathbf{u}, \mathbf{v}, \mathbf{e}$. If $\mathbf{u}$ and $\mathbf{v}$ are not parallel, a right-handed screw revolved from $\mathbf{u}$ towards $\mathbf{v}$ will advance in its nut towards $\mathbf{u} \times \mathbf{v}$.

When $\mathbf{u}$ and $\mathbf{v}$ are parallel, $\mathbf{e}$ is not defined; but in this case $\sin(\mathbf{u}, \mathbf{v}) = 0$ and $\mathbf{u} \times \mathbf{v} = 0$. Moreover, if $\mathbf{u}$ and $\mathbf{v}$ are not zero, $\mathbf{u} \times \mathbf{v} = 0$ only when $\sin(\mathbf{u}, \mathbf{v}) = 0$. Hence, for proper vectors,

$$\mathbf{u} \times \mathbf{v} = 0 \quad \text{means} \quad \mathbf{u} \parallel \mathbf{v}. \tag{2}$$

In particular $\mathbf{u} \times \mathbf{u} = 0$.

From (1) we see that

$$(-\mathbf{u}) \times \mathbf{v} = \mathbf{u} \times (-\mathbf{v}) = -\mathbf{u} \times \mathbf{v}, \qquad (-\mathbf{u}) \times (-\mathbf{v}) = \mathbf{u} \times \mathbf{v},$$

$$(\alpha\mathbf{u}) \times (\beta\mathbf{v}) = \alpha\beta\, \mathbf{u} \times \mathbf{v}. \tag{3}$$

The last result is obvious when α and β are positive numbers; the other cases then follow from the equations preceding.

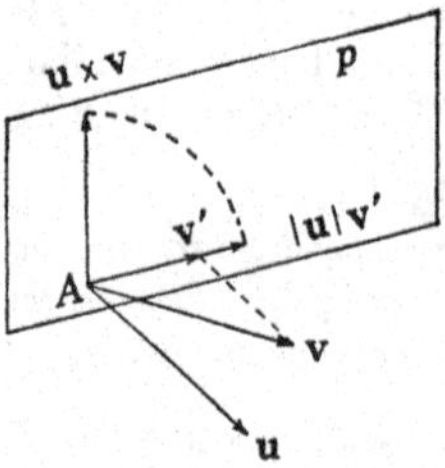

FIG. 16

If $\mathbf{u}$ and $\mathbf{v}$ are interchanged in (1), the scalar factor is not altered, but $\mathbf{e}$ is reversed; hence

$$\mathbf{v} \times \mathbf{u} = -\mathbf{u} \times \mathbf{v}. \tag{4}$$

Vector multiplication is not commutative.

Draw $\mathbf{u}$ and $\mathbf{v}$ from the point A, and let p be a plane perpendicular to $\mathbf{u}$ at A (Fig. 16). Then $\mathbf{u} \times \mathbf{v}$ may be formed by a sequence of three operations:

(P) Project $\mathbf{v}$ on p, and obtain $\mathbf{v}'$;

(M) Multiply $\mathbf{v}'$ by $|\mathbf{u}|$, and obtain $|\mathbf{u}|\,\mathbf{v}'$;

(R) Revolve $|\mathbf{u}|\,\mathbf{v}'$ about $\mathbf{u}$ through $+90°$.

The resulting vector agrees with $\mathbf{u} \times \mathbf{v}$ in magnitude, for $|\mathbf{v}'| = |\mathbf{v}| \sin(\mathbf{u}, \mathbf{v})$, and also in direction (upward in the figure). We indicate this method of forming $\mathbf{u} \times \mathbf{v}$ by the notation,

$$\mathbf{u} \times \mathbf{v} = RMP\mathbf{v}. \tag{5}$$

This means that $\mathbf{v}$ is *projected*, and the projection *multiplied*, and finally *revolved* as previously described. Now each of these operators is *distributive:* operating on the sum of two vectors is the same as operating on the vectors separately and adding the results; hence

$$RMP(\mathbf{v} + \mathbf{w}) = RM(P\mathbf{v} + P\mathbf{w})$$
$$= R(MP\mathbf{v} + MP\mathbf{w}) = RMP\mathbf{v} + RMP\mathbf{w}.$$

Thus, from (5),

$$\mathbf{u} \times (\mathbf{v} + \mathbf{w}) = \mathbf{u} \times \mathbf{v} + \mathbf{u} \times \mathbf{w}, \qquad (\mathbf{v} + \mathbf{w}) \times \mathbf{u} = \mathbf{v} \times \mathbf{u} + \mathbf{w} \times \mathbf{u}. \tag{6}$$

Vector or "cross" multiplication is distributive. By repeated applications of (6) we may expand the vector product of two vector sums just as in ordinary algebra, *provided that the order of the factors is not altered.* For example,

$$(\mathbf{a} + \mathbf{b}) \times (\mathbf{c} + \mathbf{d}) = \mathbf{a} \times \mathbf{c} + \mathbf{a} \times \mathbf{d} + \mathbf{b} \times \mathbf{c} + \mathbf{b} \times \mathbf{d}.$$

If $\mathbf{c} \neq 0$ in the equation,

$$\mathbf{a} \times \mathbf{c} = \mathbf{b} \times \mathbf{c}, \quad \text{or} \quad (\mathbf{a} - \mathbf{b}) \times \mathbf{c} = 0,$$

we can conclude *either* that $\mathbf{a} - \mathbf{b} = 0$ *or* that $\mathbf{a} - \mathbf{b}$ and $\mathbf{c}$ are parallel. We cannot "cancel" $\mathbf{c}$ to obtain $\mathbf{a} = \mathbf{b}$ unless $\mathbf{a} - \mathbf{b}$ and $\mathbf{c}$ are not parallel.

We shall see in § 18 that in general, $(\mathbf{u} \times \mathbf{v}) \times \mathbf{w} \neq \mathbf{u} \times (\mathbf{v} \times \mathbf{w})$.

Since the unit vectors $\mathbf{i}, \mathbf{j}, \mathbf{k}$ form a *dextral* orthogonal set, we have the cyclic relations.

$$\mathbf{i} \times \mathbf{j} = \mathbf{k}, \quad \mathbf{j} \times \mathbf{k} = \mathbf{i}, \quad \mathbf{k} \times \mathbf{i} = \mathbf{j}; \qquad \mathbf{i} \times \mathbf{i} = \mathbf{j} \times \mathbf{j} = \mathbf{k} \times \mathbf{k} = 0. \tag{7}$$

Hence, if we expand the product,

$$\mathbf{u} \times \mathbf{v} = (u_1\mathbf{i} + u_2\mathbf{j} + u_3\mathbf{k}) \times (v_1\mathbf{i} + v_2\mathbf{j} + v_3\mathbf{k}),$$

we obtain

$$\mathbf{u} \times \mathbf{v} = (u_2v_3 - u_3v_2)\mathbf{i} + (u_3v_1 - u_1v_3)\mathbf{j} + (u_1v_2 - u_2v_1)\mathbf{k}. \tag{8}$$

The components of $\mathbf{u} \times \mathbf{v}$ are the determinants formed by columns *2* and *3*, *3* and *1* (not *1* and *3*), *1* and *2* of the array $\begin{pmatrix} u_1 & u_2 & u_3 \\ v_1 & v_2 & v_3 \end{pmatrix}$; hence we may write

$$\mathbf{u} \times \mathbf{v} = \begin{vmatrix} \mathbf{i} & \mathbf{j} & \mathbf{k} \\ u_1 & u_2 & u_3 \\ v_1 & v_2 & v_3 \end{vmatrix}. \tag{9}$$

For example, if $\mathbf{u} = [2, -3, 5]$, $\mathbf{v} = [-1, 4, 2]$, we compute the components of $\mathbf{u} \times \mathbf{v}$ from the array,

$$\begin{pmatrix} 2 & -3 & 5 \\ -1 & 4 & 2 \end{pmatrix}; \quad \begin{vmatrix} -3 & 5 \\ 4 & 2 \end{vmatrix} = -26, \quad \begin{vmatrix} 5 & 2 \\ 2 & -1 \end{vmatrix} = -9, \quad \begin{vmatrix} 2 & -3 \\ -1 & 4 \end{vmatrix} = 5.$$

Thus $\mathbf{u} \times \mathbf{v} = [-26, -9, 5]$. As a check, we verify that $\mathbf{u} \times \mathbf{v}$ is perpendicular to both $\mathbf{u}$ and $\mathbf{v}$: $-52 + 27 + 25 = 0$, $26 - 36 + 10 = 0$.

Example 1. To find the shortest distance d from a point A to the line BC.

Method. Let $\mathbf{e}$ be the unit vector along BC. Then, if $\mathbf{u}$ is any vector from A to the line (as $\overrightarrow{AB}$ or $\overrightarrow{AC}$),

$$d = |\mathbf{u}| \sin(\mathbf{u}, \mathbf{e}) = |\mathbf{u} \times \mathbf{e}|.$$

Computation. If the points are $A(3, 1, -1)$, $B(2, 3, 0)$, $C(-1, 2, 4)$,

$$\mathbf{u} = \overrightarrow{AB} = [-1, 2, 1], \qquad \overrightarrow{BC} = [-3, -1, 4], \qquad \mathbf{e} = \frac{[-3, -1, 4]}{\sqrt{26}};$$

$$\mathbf{u} \times \mathbf{e} = \frac{[9, 1, 7]}{\sqrt{26}}, \qquad d = |\mathbf{u} \times \mathbf{e}| = \sqrt{\tfrac{131}{26}} = 2.245$$

Check. If we take $\mathbf{u} = \overrightarrow{AC} = [-4, 1, 5]$, $\mathbf{u} \times \mathbf{e}$ again has the preceding value.

Example 2. To find the shortest distance d from a point A to the plane BCD.

Method. Find a vector normal to the plane, such as $\overrightarrow{BC} \times \overrightarrow{BD}$, and let $\mathbf{n}$ be the unit vector in its direction. Then, if $\mathbf{u}$ is any vector from A to the plane (as $\overrightarrow{AB}$ or $\overrightarrow{AC}$), $d = |\mathbf{n} \cdot \mathbf{u}|$.

Computation. If the points are $A(1, -2, 1)$, $B(2, 4, 1)$, $C(-1, 0, 1)$, $D(-1, 4, 2)$,

$$\overrightarrow{BC} = [-3, -4, 0], \qquad \overrightarrow{BD} = [-3, 0, 1], \qquad \overrightarrow{BC} \times \overrightarrow{BD} = [-4, 3, -12];$$

$$\mathbf{u} = \overrightarrow{AB} = [1, 6, 0]; \qquad \mathbf{n} = \frac{[-4, 3, -12]}{13}; \qquad d = \mathbf{n} \cdot \mathbf{u} = \tfrac{14}{13}.$$

Check. If we take $\mathbf{u} = \overrightarrow{AC} = [-2, 2, 0]$, $\mathbf{n} \cdot \mathbf{u} = \tfrac{14}{13}$.

Example 3. To find the shortest distance d between two non-parallel lines AB, CD; and to locate the shortest vector $\overrightarrow{PQ}$ from AB to CD.

Method. Find the vector $\overrightarrow{AB} \times \overrightarrow{CD}$ which is perpendicular to both lines, and let $\mathbf{n}$ be a unit vector in its direction. Then, if $\mathbf{u}$ is any vector from AB to CD (as $\overrightarrow{AC}$ or $\overrightarrow{BD}$), $d = |\mathbf{n} \cdot \mathbf{u}|$.

To find P and Q, write $\overrightarrow{AP} = \alpha\, \overrightarrow{AB}$, $\overrightarrow{CQ} = \gamma\, \overrightarrow{CD}$, and find the scalars α, γ from the condition that $\overrightarrow{PQ} = \overrightarrow{PA} + \overrightarrow{AC} + \overrightarrow{CQ}$ is parallel to $\overrightarrow{AB} \times \overrightarrow{CD}$. The length $PQ = d$.

Computation. If the lines AB, CD are given by the points $A(1, -2, -1)$, $B(4, 0, -3)$; $C(1, 2, -1)$, $D(2, -4, -5)$;

$$\overrightarrow{AB} = [3, 2, -2], \qquad \overrightarrow{CD} = [1, -6, -4], \qquad \overrightarrow{AB} \times \overrightarrow{CD} = 10[-2, 1, -2];$$

$$\mathbf{n} = \tfrac{1}{3}[-2, 1, -2], \qquad \overrightarrow{AC} = [0, 4, 0], \qquad d = \mathbf{n} \cdot \overrightarrow{AC} = \tfrac{4}{3}.$$

To find P and Q, we have

$$\begin{aligned} \overrightarrow{PQ} &= -\alpha\, \overrightarrow{AB} + \overrightarrow{AC} + \gamma\, \overrightarrow{CD} \\ &= -\alpha[3, 2, -2] + [0, 4, 0] + \gamma[1, -6, -4] \\ &= [-3\alpha + \gamma, -2\alpha + 4 - 6\gamma, 2\alpha - 4\gamma]; \end{aligned}$$

and, since $\overrightarrow{PQ}$ is parallel to $[-2, 1, -2]$,

$$\frac{-3\alpha + \gamma}{-2} = \frac{-2\alpha + 4 - 6\gamma}{1} = \frac{2\alpha - 4\gamma}{-2}.$$

These equations give $\alpha = \gamma = \frac{4}{9}$; hence

$$\overrightarrow{OP} = \overrightarrow{OA} + \overrightarrow{AP} = [1, -2, -1] + \tfrac{4}{9}[3, 2, -2] = \tfrac{1}{9}[21, -10, -17],$$

$$\overrightarrow{OQ} = \overrightarrow{OC} + \overrightarrow{CQ} = [1, 2, -1] + \tfrac{4}{9}[1, -6, -4] = \tfrac{1}{9}[13, -6, -25].$$

Check. The distance $PQ = \frac{4}{3}$.

17. Vector Areas. Consider a plane area A whose boundary is traced in a definite sense—shown by arrows in Fig. 17a. *We shall*

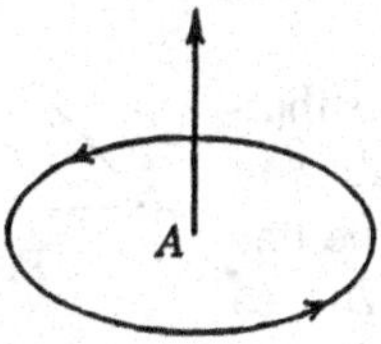

FIG. 17a

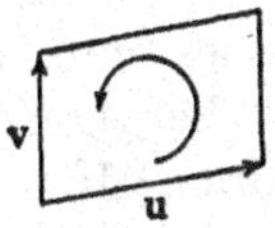

FIG. 17b

associate such an area with a vector of magnitude A, normal to its plane, and pointing in the direction a right-handed screw would move if turned in the given sense. Thus the parallelogram whose sides are the vectors **u**, **v** (Fig. 17b) and whose sense agrees with the rotation that carries **u** into **v** is associated with the vector $\mathbf{u} \times \mathbf{v}$; for its area is

$$A = |\mathbf{u}| \, |\mathbf{v}| \sin(\mathbf{u}, \mathbf{v}) = |\mathbf{u} \times \mathbf{v}|.$$

All plane areas associated with the same vector will be regarded as equal. The sum of two plane areas associated with the vectors, **u**, **v**, is defined as the plane area associated with $\mathbf{u} + \mathbf{v}$.

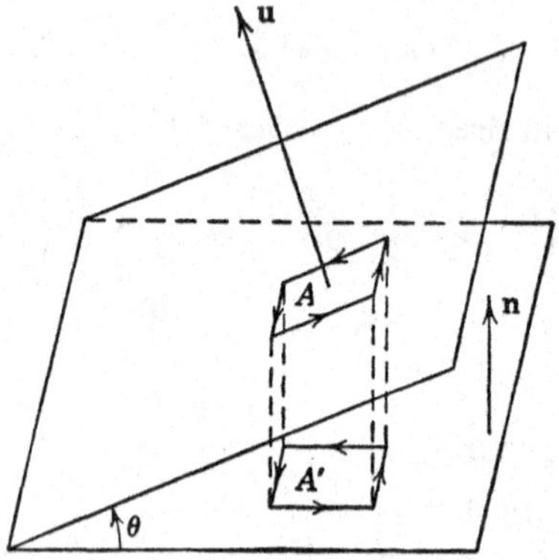

FIG. 17c

A *directed plane* is a plane associated with a definite normal vector, say the unit normal **n**. A circuit in a directed plane is positive when it is counterclockwise relative to **n**; a clockwise circuit is negative.

Consider now a plane area A associated with the vector **u**. The projection of A on a directed plane p is $A' = A \cos \theta$ (Fig. 17c). The circuital sense of A is projected on A'. We shall give A' the sign corresponding to its circuit on p, and call this signed area the *component* of A on p. Since $|\mathbf{u}| = A$ and $(\mathbf{n}, \mathbf{u}) = \theta$ or $\pi - \theta$,

$$\mathbf{n} \cdot \mathbf{u} = |\mathbf{u}| \cos(\mathbf{n}, \mathbf{u}) = A \cos \theta \quad \text{or} \quad -A \cos \theta$$

according as $(\mathbf{n}, \mathbf{u})$ is acute or obtuse. In both cases $\mathbf{n} \cdot \mathbf{u}$ gives the component of A on p:

$$\text{(1)} \qquad \text{comp}_p A = \mathbf{n} \cdot \mathbf{u}.$$

THEOREM. *When the vector areas of the faces of a closed polyhedron are all drawn in the direction of the outward normals, their sum is zero.*

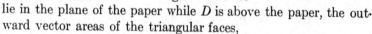

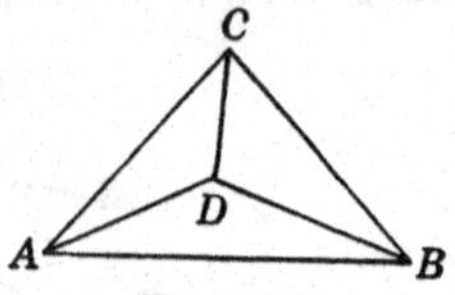

FIG. 17d

Proof. Any polyhedron may be subdivided by planes into a finite number of tetrahedrons. Let $ABCD$ (Fig. 17d) be one such tetrahedron. If we imagine ABC to lie in the plane of the paper while D is above the paper, the outward vector areas of the triangular faces,

$$DAB, \qquad DBC, \qquad DCA, \qquad ABC,$$

are given by one half of the respective vectors,

$$\overrightarrow{DA} \times \overrightarrow{DB}, \qquad \overrightarrow{DB} \times \overrightarrow{DC}, \qquad \overrightarrow{DC} \times \overrightarrow{DA}, \qquad -\overrightarrow{AB} \times \overrightarrow{AC}.$$

If we choose D as origin, the sum of these vectors is given by

$$\mathbf{a} \times \mathbf{b} + \mathbf{b} \times \mathbf{c} + \mathbf{c} \times \mathbf{a} - (\mathbf{b} - \mathbf{a}) \times (\mathbf{c} - \mathbf{a}) = 0.$$

Thus the theorem is true for a tetrahedron.

Now write such an equation for all the tetrahedrons that make up the polyhedron, and add the results. The vector areas over all inner faces cancel, for each appears twice but with opposed directions. The net result on the left is double the sum of the outward vector areas of the polyhedron's faces. Since this sum is zero, the theorem follows.

Consider now a polygon $P_1P_2 \cdots P_h$ of area A lying in a directed plane of unit normal $\mathbf{n}$ (Fig. 17e), and suppose that the circuit $P_1P_2 \cdots P_h$ is positive. In the figure $\mathbf{n}$ points *up* from the plane of the paper, so that a positive circuit is counterclockwise. Choose an origin O at pleasure above the plane (towards the reader), and let $\mathbf{r}_1, \mathbf{r}_2, \cdots, \mathbf{r}_h$ denote the position vectors of the vertices. These vectors are the edges of a pyramid having O as vertex and the given polygon as base. The triangular faces of this pyramid have as their outward vector areas $\frac{1}{2}\mathbf{r}_1 \times \mathbf{r}_2$, $\frac{1}{2}\mathbf{r}_2 \times \mathbf{r}_3, \cdots, \frac{1}{2}\mathbf{r}_h \times \mathbf{r}_1$, while the outward vector area of the base is $-A\mathbf{n}$. From the preceding theorem the sum of these vector areas is zero; hence

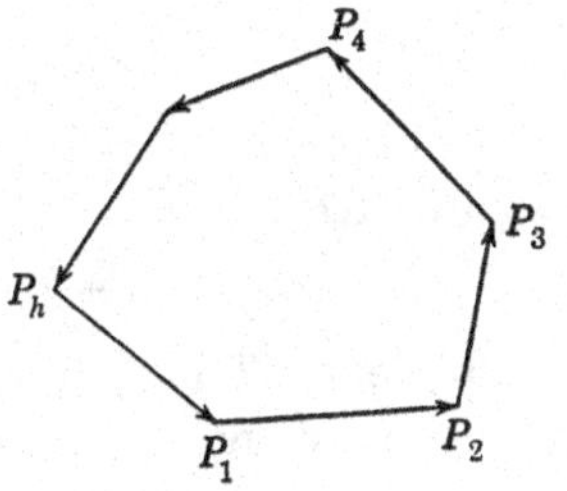

FIG. 17e

$$(2) \qquad A\mathbf{n} = \tfrac{1}{2}(\mathbf{r}_1 \times \mathbf{r}_2 + \mathbf{r}_2 \times \mathbf{r}_3 + \cdots + \mathbf{r}_h \times \mathbf{r}_1).$$

As the vertex O approaches the plane of the base, the terms on the right of (2) vary continuously, but their sum is always $A\mathbf{n}$. Hence when O is in the plane of the polygon, (2) remains valid. If O is any origin of rectangular coordinates in the plane, let the vertices of the polygon be $(x_1, y_1), (x_2, y_2), \cdots, (x_h, y_h)$ taken in counterclockwise order. Now $\mathbf{n} = \mathbf{k}$, and

$$\mathbf{r}_1 \times \mathbf{r}_2 = (x_1\mathbf{i} + y_1\mathbf{j}) \times (x_2\mathbf{i} + y_2\mathbf{j}) = \begin{vmatrix} x_1 & x_2 \\ y_1 & y_2 \end{vmatrix} \mathbf{k},$$

so that (2) gives

$$(3)\qquad 2A = \begin{vmatrix} x_1 & x_2 \\ y_1 & y_2 \end{vmatrix} + \begin{vmatrix} x_2 & x_3 \\ y_2 & y_3 \end{vmatrix} + \cdots + \begin{vmatrix} x_h & x_1 \\ y_h & y_1 \end{vmatrix}.$$

Example 1. The sum of the outward vector areas of the triangular prism (Fig. 17f) is

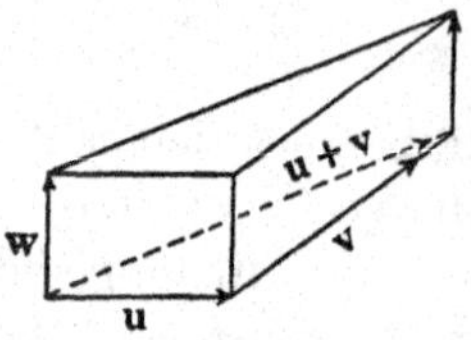

FIG. 17f

$$\mathbf{u} \times \mathbf{w} + \mathbf{v} \times \mathbf{w} - (\mathbf{u} + \mathbf{v}) \times \mathbf{w} = 0.$$

The vector areas of the triangular bases cancel, for they are equal in magnitude but opposite in direction.

Example 2. To find the area A of a triangle whose vertices are $(a, 0, 0)$, $(0, b, 0)$, $(0, 0, c)$, put $\mathbf{r}_1 = a\mathbf{i}$, $\mathbf{r}_2 = b\mathbf{j}$, $\mathbf{r}_3 = c\mathbf{k}$ in (2); then

$$A\mathbf{n} = \tfrac{1}{2}(ab\mathbf{k} + bc\mathbf{i} + ca\mathbf{j}), \qquad A = \tfrac{1}{2}\sqrt{a^2b^2 + b^2c^2 + c^2a^2}.$$

Example 3. To find the area of the polygon whose vertices in counterclockwise order are (1, 2), (5, 4), (−3, 7), (−5, 5), (−1, −3), we form the array,

$$\begin{array}{cccccccc} (x) & 1 & 5 & -3 & -5 & -1 & 1 \\ (y) & 2 & 4 & 7 & 5 & -3 & 2, \end{array}$$

repeating the first column. Then, from (3),

$$2A = -6 + 47 + 20 + 20 + 1 = 82, \qquad A = 41.$$

18. Vector Triple Product. The vector $(\mathbf{u} \times \mathbf{v}) \times \mathbf{w}$ is perpendicular to $\mathbf{u} \times \mathbf{v}$ and therefore coplanar with $\mathbf{u}$ and $\mathbf{v}$; hence (§ 5),

$$(\mathbf{u} \times \mathbf{v}) \times \mathbf{w} = \alpha\mathbf{u} + \beta\mathbf{v}.$$

But, since $(\mathbf{u} \times \mathbf{v}) \times \mathbf{w}$ is also perpendicular to $\mathbf{w}$,

$$\alpha\, \mathbf{u} \cdot \mathbf{w} + \beta\, \mathbf{v} \cdot \mathbf{w} = 0.$$

All numbers α, β that satisfy this equation must be of the form $\alpha = -\lambda\, \mathbf{v} \cdot \mathbf{w}$, $\beta = \lambda\, \mathbf{u} \cdot \mathbf{w}$, where λ is arbitrary. Thus we have

$$(\mathbf{u} \times \mathbf{v}) \times \mathbf{w} = \lambda(\mathbf{u} \cdot \mathbf{w}\, \mathbf{v} - \mathbf{v} \cdot \mathbf{w}\, \mathbf{u}).$$

In order to determine λ, we use a special basis in which $\mathbf{i}$ is collinear with $\mathbf{u}$, $\mathbf{j}$ coplanar with $\mathbf{u}$, $\mathbf{v}$; then

$$\mathbf{u} = u_1\mathbf{i}, \qquad \mathbf{v} = v_1\mathbf{i} + v_2\mathbf{j}, \qquad \mathbf{w} = w_1\mathbf{i} + w_2\mathbf{j} + w_3\mathbf{k}.$$

Substituting these values gives, after a simple calculation, $\lambda = 1$ We therefore have the important expansion formulas,

$$(1)\qquad (\mathbf{u} \times \mathbf{v}) \times \mathbf{w} = \mathbf{u} \cdot \mathbf{w}\, \mathbf{v} - \mathbf{v} \cdot \mathbf{w}\, \mathbf{u},$$

$$\mathbf{w} \times (\mathbf{u} \times \mathbf{v}) = \mathbf{w} \cdot \mathbf{v}\, \mathbf{u} - \mathbf{w} \cdot \mathbf{u}\, \mathbf{v}.$$

In the left-hand members of (1), one of the vectors in parenthesis is *adjacent* to the vector outside, the other *remote* from it. The right-hand members may be remembered as

(Outer dot Remote) Adjacent − (Outer dot Adjacent) Remote.

In general $(\mathbf{u} \times \mathbf{v}) \times \mathbf{w} \neq \mathbf{u} \times (\mathbf{v} \times \mathbf{w})$; for the former is coplanar with $\mathbf{u}$ and $\mathbf{v}$, the latter with $\mathbf{v}$ and $\mathbf{w}$. *Cross multiplication of vectors is not associative.*

From (1) we see that the sum of a vector triple product and its two cyclical permutations is zero:

$$(\mathbf{a} \times \mathbf{b}) \times \mathbf{c} + (\mathbf{b} \times \mathbf{c}) \times \mathbf{a} + (\mathbf{c} \times \mathbf{a}) \times \mathbf{b} = 0. \tag{2}$$

If l is a directed line carrying the unit vector $\mathbf{e}$, we may express any vector $\mathbf{u}$ as the sum of its orthogonal projections on l and on a plane p perpendicular to l:

$$\text{(3)} \qquad \text{proj}_l\, \mathbf{u} = \mathbf{e}\, \text{comp}_l\, \mathbf{u} = (\mathbf{e} \cdot \mathbf{u})\, \mathbf{e} \qquad (15.6),$$

$$\text{(4)} \qquad \text{proj}_p\, \mathbf{u} = \mathbf{u} - (\mathbf{e} \cdot \mathbf{u})\, \mathbf{e} = \mathbf{e} \times (\mathbf{u} \times \mathbf{e}).$$

19. Scalar Triple Product. The scalar product of $\mathbf{u} \times \mathbf{v}$ and $\mathbf{w}$ is written $\mathbf{u} \times \mathbf{v} \cdot \mathbf{w}$ or $[\mathbf{uvw}]$. No ambiguity can arise from the parentheses being omitted, since $\mathbf{u} \times (\mathbf{v} \cdot \mathbf{w})$ is meaningless.

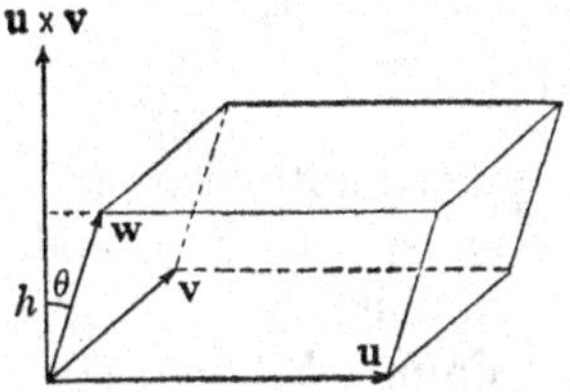

FIG. 19

THEOREM. *The product* $\mathbf{u} \times \mathbf{v} \cdot \mathbf{w}$ *is numerically equal to the volume* V *of a parallelepiped having* $\mathbf{u}$, $\mathbf{v}$, $\mathbf{w}$ *as concurrent edges. Its sign is positive or negative according as* $\mathbf{u}$, $\mathbf{v}$, $\mathbf{w}$ *form a right-handed or left-handed set.*

Proof. The volume V (Fig. 19) may be computed by multiplying the area of a face parallel to $\mathbf{u}$ and $\mathbf{v}$,

$$A = |\mathbf{u}|\,|\mathbf{v}| \sin(\mathbf{u}, \mathbf{v}) = |\mathbf{u} \times \mathbf{v}|,$$

by the corresponding altitude $h = |\mathbf{w}| \cos\theta$:

$$V = |\mathbf{u} \times \mathbf{v}|\,|\mathbf{w}| \cos\theta.$$

Now the angle between $\mathbf{u} \times \mathbf{v}$ and $\mathbf{w}$ is θ or $\pi - \theta$ according as $\mathbf{u}$, $\mathbf{v}$, $\mathbf{w}$ form a dextral (as in the figure) or a sinistral set (§ 12). The definition of a scalar product now shows that

$$\text{(1)} \qquad \mathbf{u} \times \mathbf{v} \cdot \mathbf{w} = \left\{\begin{matrix} V \\ -V \end{matrix}\right\} \text{ when the set } \mathbf{u}, \mathbf{v}, \mathbf{w} \text{ is } \left\{\begin{matrix} \text{dextral.} \\ \text{sinistral.} \end{matrix}\right.$$

The dextral or sinistral character of a set $\mathbf{u}$, $\mathbf{v}$, $\mathbf{w}$ is not altered by a cyclical change in their order. Hence, from (1),

$$\mathbf{u} \times \mathbf{v} \cdot \mathbf{w} = \mathbf{v} \times \mathbf{w} \cdot \mathbf{u} = \mathbf{w} \times \mathbf{u} \cdot \mathbf{v}. \tag{2}$$

But a dextral set becomes sinistral and vice versa, when the cyclical order is changed:

$$\mathbf{u} \times \mathbf{w} \cdot \mathbf{v} = -\mathbf{u} \times \mathbf{v} \cdot \mathbf{w}. \tag{3}$$

Thus, if the set $\mathbf{u}$, $\mathbf{v}$, $\mathbf{w}$ is dextral, the products in (2) all equal V, while the products,

$$\mathbf{u} \times \mathbf{w} \cdot \mathbf{v} = \mathbf{w} \times \mathbf{v} \cdot \mathbf{u} = \mathbf{v} \times \mathbf{u} \cdot \mathbf{w},$$

all equal $-V$.

On account of the geometric meaning of the scalar triple product, we shall call it the "*box product.*" ‡

If $\mathbf{u}$, $\mathbf{v}$, $\mathbf{w}$ are *proper* vectors, $V = 0$ only when the vectors are coplanar. Therefore: *three proper vectors are coplanar (parallel to the same plane) when and only when their box product is zero.* In particular, a box product containing two parallel vectors is zero; for example $\mathbf{u} \times \mathbf{v} \cdot \mathbf{u} = 0$.

The value of a box product is not altered by an interchange of the dot and cross. For

$$\mathbf{u} \times \mathbf{v} \cdot \mathbf{w} = \mathbf{v} \times \mathbf{w} \cdot \mathbf{u} = \mathbf{u} \cdot \mathbf{v} \times \mathbf{w} \tag{4}$$

as dot multiplication is commutative. The notation $[\mathbf{uvw}]$ often is used for the box product, as the omission of dot and cross causes no ambiguity.

From (15.3) and (16.3), we have

$$(\alpha\mathbf{u}) \times (\beta\mathbf{v}) \cdot (\gamma\mathbf{w}) = \alpha\beta\gamma\, \mathbf{u} \times \mathbf{v} \cdot \mathbf{w}. \tag{5}$$

Finally, the distributive law for scalar and vector products shows that a box product of vector sums may be expanded just as in ordinary algebra, provided that the order of the vector factors is not altered. Thus, if we expand the product $\mathbf{u} \times \mathbf{v} \cdot \mathbf{w}$ when the vectors are referred to the basis $\mathbf{i}$, $\mathbf{j}$, $\mathbf{k}$, we obtain 27 terms of which all but six vanish as they contain box products with two or three equal vectors. The remaining six terms are those containing

$$[\mathbf{ijk}] = [\mathbf{jki}] = [\mathbf{kij}] = 1, \qquad [\mathbf{ikj}] = [\mathbf{kji}] = [\mathbf{jik}] = -1,$$

‡ The name proposed by J. H. Taylor, *Vector Analysis*, New York, 1939 p. 46.

and constitute the expansion of the determinant,

$$(6)\qquad [\mathbf{uvw}] = \begin{vmatrix} u_1 & u_2 & u_3 \\ v_1 & v_2 & v_3 \\ w_1 & w_2 & w_3 \end{vmatrix}.$$

This result also follows at once from (16.8).

Example. To find the point P where the line AB pierces the plane CDE.

Method. $\overrightarrow{AP} = \lambda\,\overrightarrow{AB}$; the scalar λ is then determined by the equation,

$$\text{(i)}\qquad \overrightarrow{CP}\cdot\overrightarrow{CD}\times\overrightarrow{CE} = (\overrightarrow{CA} + \lambda\,\overrightarrow{AB})\cdot CD \times CE = 0$$

which expresses that P is coplanar with C, D, E.

Computation. With the points,

$$A(1, 2, 0),\quad B(2, 3, 1);\quad C(2, 0, 3),\quad D(0, 4, 2),\quad E(-1, 2, -2);$$

$$\overrightarrow{CP} = \overrightarrow{CA} + \lambda\,\overrightarrow{AB} = [-1, 2, -3] + \lambda[1, 1, 1] = [\lambda - 1, \lambda + 2, \lambda - 3];$$

$$\overrightarrow{CD}\times\overrightarrow{CE} = [-2, 4, -1]\times[-3, 2, -5] = [-18, -7, 8];$$

hence, from (i),

$$-18(\lambda - 1) - 7(\lambda + 2) + 8(\lambda - 3) = -17\lambda - 20 = 0,\qquad \lambda = -\tfrac{20}{17};$$

$$\overrightarrow{OP} = \overrightarrow{OA} + \lambda\,\overrightarrow{AB} = [1, 2, 0] - \tfrac{20}{17}[1, 1, 1] = \tfrac{1}{17}[-3, 14, -20].$$

Check. $\overrightarrow{DP}\cdot\overrightarrow{CD}\times\overrightarrow{CE} = 0$; for

$$\overrightarrow{DP} = -\tfrac{3}{17}[1, 18, 18],\qquad [1, 18, 18]\cdot[-18, -7, 8] = 0.$$

20. Products of Four Vectors. Since

$$(\mathbf{a}\times\mathbf{b})\cdot(\mathbf{c}\times\mathbf{d}) = \mathbf{a}\cdot\mathbf{b}\times(\mathbf{c}\times\mathbf{d}) = \mathbf{a}\cdot(\mathbf{b}\cdot\mathbf{d}\,\mathbf{c} - \mathbf{b}\cdot\mathbf{c}\,\mathbf{d}),$$

$$(1)\qquad (\mathbf{a}\times\mathbf{b})\cdot(\mathbf{c}\times\mathbf{d}) = \mathbf{a}\cdot\mathbf{c}\,\mathbf{b}\cdot\mathbf{d} - \mathbf{a}\cdot\mathbf{d}\,\mathbf{b}\cdot\mathbf{c} = \begin{vmatrix} \mathbf{a}\cdot\mathbf{c} & \mathbf{a}\cdot\mathbf{d} \\ \mathbf{b}\cdot\mathbf{c} & \mathbf{b}\cdot\mathbf{d} \end{vmatrix}$$

The product $(\mathbf{a}\times\mathbf{b})\times(\mathbf{c}\times\mathbf{d})$ may be regarded as a triple product of $\mathbf{a}\times\mathbf{b}$, $\mathbf{c}$, $\mathbf{d}$ or of $\mathbf{a}$, $\mathbf{b}$, $\mathbf{c}\times\mathbf{d}$. We thus are led to the two expansions,

$$(2)\qquad (\mathbf{a}\times\mathbf{b})\times(\mathbf{c}\times\mathbf{d}) = [\mathbf{acd}]\mathbf{b} - [\mathbf{bcd}]\mathbf{a} = [\mathbf{abd}]\mathbf{c} - [\mathbf{abc}]\mathbf{d},$$

which give, in turn, the following equation connecting any four vectors:

$$(3)\qquad \mathbf{a}[\mathbf{bcd}] - \mathbf{b}[\mathbf{cda}] + \mathbf{c}[\mathbf{dab}] - \mathbf{d}[\mathbf{abc}] = 0.$$

When $[\mathbf{abc}] \neq 0$,

$$[\mathbf{abc}]\mathbf{d} = [\mathbf{dbc}]\mathbf{a} + [\mathbf{adc}]\mathbf{b} + [\mathbf{abd}]\mathbf{c} \tag{4}$$

gives **d** explicitly in terms of the basis **a**, **b**, **c**.

21. Plane Trigonometry. If the vectors **a**, **b**, **c** form a closed triangle when placed end to end (Fig. 21),

$$\mathbf{a} + \mathbf{b} + \mathbf{c} = 0. \tag{1}$$

We shall denote the lengths of the sides by a, b, c and the interior angles opposite them by A, B, C; then the angles $(\mathbf{b}, \mathbf{c})$, $(\mathbf{c}, \mathbf{a})$, $(\mathbf{a}, \mathbf{b})$ are equal, respectively, to $\pi - A$, $\pi - B$, $\pi - C$.

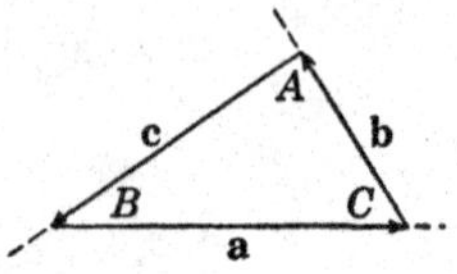

FIG. 21

From the identity,

$$(\mathbf{a} + \mathbf{b}) \cdot (\mathbf{a} + \mathbf{b}) = \mathbf{a} \cdot \mathbf{a} + \mathbf{b} \cdot \mathbf{b} + 2\,\mathbf{a} \cdot \mathbf{b},$$

we obtain the *cosine law* for plane triangles:

$$c^2 = a^2 + b^2 - 2ab \cos C. \tag{2}$$

Cyclical interchanges of the letters in (2) give two other forms of this law.

On multiplying (1) first by $\mathbf{a} \times$, then by $\mathbf{b} \times$, we find

$$\mathbf{b} \times \mathbf{c} = \mathbf{c} \times \mathbf{a} = \mathbf{a} \times \mathbf{b}. \tag{3}$$

Division by abc gives the *sine law* for plane triangles:

$$\frac{\sin A}{a} = \frac{\sin B}{b} = \frac{\sin C}{c}. \tag{4}$$

Since each product in (3) is double the vector area of the triangle, its

$$\text{(5)}\quad \text{Area} = \tfrac{1}{2}bc \sin A$$
$$= \tfrac{1}{2}ca \sin B = \tfrac{1}{2}ab \sin C.$$

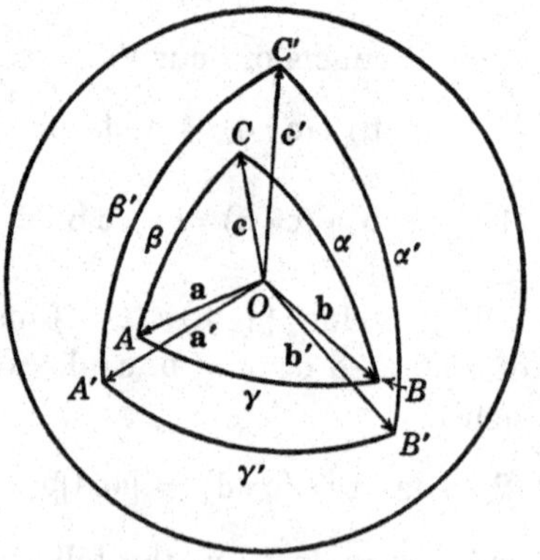

FIG. 22

22. Spherical Trigonometry. Consider a spherical triangle ABC on a sphere of unit radius, and let **a**, **b**, **c**, denote the position vectors of the vertices referred to its center (Fig. 22). The notation is so chosen that **a**, **b**, **c**, form a dextral set: then $[\mathbf{abc}] > 0$.

Let α, β, γ denote the *sides* (arcs of great circles) opposite the vertices, A, B, C. The *interior* dihedral angles at these vertices also are denoted by A, B, C. We shall consider only spherical triangles in which the sides and angles are each less than 180°. Now

$$(1) \quad \mathbf{b} \cdot \mathbf{c} = \cos\alpha, \qquad \mathbf{c} \cdot \mathbf{a} = \cos\beta, \qquad \mathbf{a} \cdot \mathbf{b} = \cos\gamma;$$

$$(2) \quad \mathbf{b} \times \mathbf{c} = \sin\alpha\, \mathbf{a}', \qquad \mathbf{c} \times \mathbf{a} = \sin\beta\, \mathbf{b}', \qquad \mathbf{a} \times \mathbf{b} = \sin\gamma\, \mathbf{c}',$$

where $\mathbf{a}'$, $\mathbf{b}'$, $\mathbf{c}'$ are unit vectors. The vectors $\mathbf{b}'$ and $\mathbf{c}'$ are perpendicular to the planes of $\mathbf{c}$, $\mathbf{a}$ and $\mathbf{a}$, $\mathbf{b}$, respectively, and include an angle $\alpha' = \pi - A$, the *exterior* dihedral angle at A. Moreover, from (2),

$$\mathbf{b}' \times \mathbf{c}' = \frac{(\mathbf{c} \times \mathbf{a}) \times (\mathbf{a} \times \mathbf{b})}{\sin\beta \sin\gamma} = \frac{[\mathbf{abc}]}{\sin\beta\sin\gamma}\, \mathbf{a},$$

a *positive* multiple of $\mathbf{a}$. Hence if α', β', γ' denote the exterior dihedral angles at A, B, C, we have

$$(3) \quad \mathbf{b}' \cdot \mathbf{c}' = \cos\alpha', \qquad \mathbf{c}' \cdot \mathbf{a}' = \cos\beta', \qquad \mathbf{a}' \cdot \mathbf{b}' = \cos\gamma'$$

$$(4) \quad \mathbf{b}' \times \mathbf{c}' = \sin\alpha'\, \mathbf{a}, \qquad \mathbf{c}' \times \mathbf{a}' = \sin\beta'\, \mathbf{b}, \qquad \mathbf{a}' \times \mathbf{b}' = \sin\gamma'\, \mathbf{c}.$$

Thus there is complete reciprocity between the vector sets $\mathbf{a}$, $\mathbf{b}$, $\mathbf{c}$, and $\mathbf{a}'$, $\mathbf{b}'$, $\mathbf{c}'$, and also between the spherical triangles they determine. While ABC has α, β, γ for sides and α', β', γ' for exterior angles, $A'B'C'$ has α', β', γ' for sides and α, β, γ for exterior angles. Since the vertices of one triangle are poles of the corresponding sides of the other, the triangles are said to be ***polar.***

From (2) and (4), we have

$$[\mathbf{abc}] = \sin\alpha\, \mathbf{a} \cdot \mathbf{a}' = \sin\beta\, \mathbf{b} \cdot \mathbf{b}' = \sin\gamma\, \mathbf{c} \cdot \mathbf{c}',$$

$$[\mathbf{a}'\mathbf{b}'\mathbf{c}'] = \sin\alpha'\, \mathbf{a} \cdot \mathbf{a}' = \sin\beta'\, \mathbf{b} \cdot \mathbf{b}' = \sin\gamma'\, \mathbf{c} \cdot \mathbf{c}';$$

hence, on division,

$$(5) \qquad \frac{[\mathbf{abc}]}{[\mathbf{a}'\mathbf{b}'\mathbf{c}']} = \frac{\sin\alpha}{\sin\alpha'} = \frac{\sin\beta}{\sin\beta'} = \frac{\sin\gamma}{\sin\gamma'}.$$

This is the *sine law* for spherical triangles.

Again, from (2), we have

$$\sin\beta\sin\gamma\, \mathbf{b}' \cdot \mathbf{c}' = (\mathbf{c} \times \mathbf{a}) \cdot (\mathbf{a} \times \mathbf{b}) = (\mathbf{c} \cdot \mathbf{a})(\mathbf{a} \cdot \mathbf{b}) - \mathbf{b} \cdot \mathbf{c};$$

hence, from (1) and (3),

$$\text{(6)} \qquad \cos\alpha = \cos\beta\cos\gamma - \sin\beta\sin\gamma\cos\alpha'.$$

Similarly, from (4), we deduce

$$\text{(7)} \qquad \cos\alpha' = \cos\beta'\cos\gamma' - \sin\beta'\sin\gamma'\cos\alpha.$$

Equations (6) and (7) and the four others derived from them by cyclical permutation constitute the *cosine laws* for spherical triangles.

By replacing α', β', γ' in (5), (6), and (7) by $\pi - A$, $\pi - B$, $\pi - C$ we may express the sine and cosine laws in terms of the sides and *interior* angles. Thus, we find, for the sine law,

$$\frac{\sin\alpha}{\sin A} = \frac{\sin\beta}{\sin B} = \frac{\sin\gamma}{\sin C};$$

and, for the cosine laws,

$$\cos\alpha = \cos\beta\cos\gamma + \sin\beta\sin\gamma\cos A,$$

$$\cos A = -\cos B\cos C + \sin B\sin C\cos\alpha.$$

In this version of the cosine laws, the structural similarity exhibited by (6) and (7) is lost.

23. Reciprocal Bases. Two bases, $\mathbf{e}_1$, $\mathbf{e}_2$, $\mathbf{e}_3$ and $\mathbf{e}^1$, $\mathbf{e}^2$, $\mathbf{e}^3$, are said to be *reciprocal* when they satisfy the nine equations:

$$\text{(1)} \qquad \begin{array}{lll} \mathbf{e}_1\cdot\mathbf{e}^1 = 1, & \mathbf{e}_1\cdot\mathbf{e}^2 = 0, & \mathbf{e}_1\cdot\mathbf{e}^3 = 0, \\ \mathbf{e}_2\cdot\mathbf{e}^1 = 0, & \mathbf{e}_2\cdot\mathbf{e}^2 = 1, & \mathbf{e}_2\cdot\mathbf{e}^3 = 0, \\ \mathbf{e}_3\cdot\mathbf{e}^1 = 0, & \mathbf{e}_3\cdot\mathbf{e}^2 = 0, & \mathbf{e}_3\cdot\mathbf{e}^3 = 1. \end{array}$$

The superscripts applied to the base vectors are not exponents, but mere identification tags. By use of the *Kronecker delta* δ_i^j, defined as

$$\text{(2)} \qquad \delta_i^j = \begin{cases} 1 & \text{when} \quad i = j \\ 0 & \text{when} \quad i \neq j, \end{cases}$$

equations (1) condense to

$$\text{(3)} \qquad \mathbf{e}_i\cdot\mathbf{e}^j = \delta_i^j \qquad (i, j = 1, 2, 3).$$

Consider the three equations in the first column of (1). The second and third state that $\mathbf{e}^1$ is perpendicular to both $\mathbf{e}_2$ and $\mathbf{e}_3$, that is, parallel to $\mathbf{e}_2 \times \mathbf{e}_3$. Hence $\mathbf{e}^1 = \lambda\, \mathbf{e}_2 \times \mathbf{e}_3$; and, from the first equation, $1 = \lambda\, \mathbf{e}_1 \cdot \mathbf{e}_2 \times \mathbf{e}_3$. We thus obtain

$$(4) \qquad \mathbf{e}^1 = \frac{\mathbf{e}_2 \times \mathbf{e}_3}{[\mathbf{e}_1\mathbf{e}_2\mathbf{e}_3]}, \qquad \mathbf{e}^2 = \frac{\mathbf{e}_3 \times \mathbf{e}_1}{[\mathbf{e}_1\mathbf{e}_2\mathbf{e}_3]}, \qquad \mathbf{e}^3 = \frac{\mathbf{e}_1 \times \mathbf{e}_2}{[\mathbf{e}_1\mathbf{e}_2\mathbf{e}_3]},$$

$\mathbf{e}^2$ and $\mathbf{e}^3$ being derived from $\mathbf{e}^1$ by cyclical permutation. From the symmetry of equations (1) in the two sets $\mathbf{e}_i$, $\mathbf{e}^i$, we have also

$$(5) \qquad \mathbf{e}_1 = \frac{\mathbf{e}^2 \times \mathbf{e}^3}{[\mathbf{e}^1\mathbf{e}^2\mathbf{e}^3]}, \qquad \mathbf{e}_2 = \frac{\mathbf{e}^3 \times \mathbf{e}^1}{[\mathbf{e}^1\mathbf{e}^2\mathbf{e}^3]}, \qquad \mathbf{e}_3 = \frac{\mathbf{e}^1 \times \mathbf{e}^2}{[\mathbf{e}^1\mathbf{e}^2\mathbf{e}^3]}.$$

Thus, either basis is expressed in terms of the other by precisely the same formulas.

From (4) and (5), we have

$$\mathbf{e}^1 \cdot \mathbf{e}_1 = \frac{(\mathbf{e}_2 \times \mathbf{e}_3) \cdot (\mathbf{e}^2 \times \mathbf{e}^3)}{[\mathbf{e}_1\mathbf{e}_2\mathbf{e}_3][\mathbf{e}^1\mathbf{e}^2\mathbf{e}^3]} = \frac{1}{[\mathbf{e}_1\mathbf{e}_2\mathbf{e}_3][\mathbf{e}^1\mathbf{e}^2\mathbf{e}^3]},$$

on making use of (20.1) and equations (1); hence

$$(6) \qquad [\mathbf{e}_1\mathbf{e}_2\mathbf{e}_3][\mathbf{e}^1\mathbf{e}^2\mathbf{e}^3] = 1,$$

an equation which gives further justification for the name *reciprocal* applied to the *sets*. Since the box products in (6) must have the same sign, a basis and its reciprocal are both right-handed or both left-handed.

As to the orientation of reciprocal sets, we see from (1) that $\mathbf{e}^1$ is perpendicular to the plane of $\mathbf{e}_2$ and $\mathbf{e}_3$ in the direction which makes an acute angle with $\mathbf{e}_1$. Similar statements apply to $\mathbf{e}^2$ and $\mathbf{e}^3$. If the vectors $\mathbf{e}_i$, $\mathbf{e}^i$ are all drawn from the same point O and cut by a sphere s about O in the points E_i, E^i, respectively, the three planes OE_iE_j cut a spherical triangle $E_1E_2E_3$ from s; and the three planes OE^iE^j cut out a second spherical triangle $E^1E^2E^3$ (Fig. 23). Either triangle is the *polar* of the other; for E^1, E^2, E^3 are poles of the great-circle arcs, E_2E_3, E_3E_1, E_1E_2; and similarly for E_1, E_2, E_3.

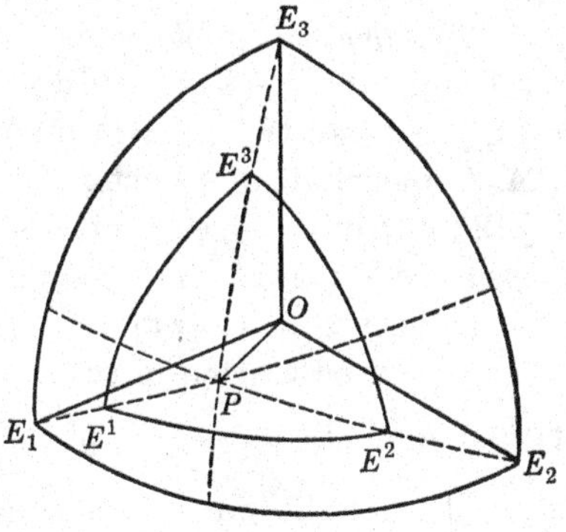

FIG. 23

On dividing the identity,

$$\mathbf{e}_1 \times (\mathbf{e}_2 \times \mathbf{e}_3) + \mathbf{e}_2 \times (\mathbf{e}_3 \times \mathbf{e}_1) + \mathbf{e}_3 \times (\mathbf{e}_1 \times \mathbf{e}_2) = 0 \quad (18.2),$$

by $[\mathbf{e}_1\mathbf{e}_2\mathbf{e}_3]$, we obtain the relation,

$$(7) \qquad \mathbf{e}_1 \times \mathbf{e}^1 + \mathbf{e}_2 \times \mathbf{e}^2 + \mathbf{e}_3 \times \mathbf{e}^3 = 0.$$

This has an interesting geometric interpretation. The great circle E_1E^1 is perpendicular to both great circles E_2E_3 and E^2E^3; it thus contains the altitudes of both triangles through the vertex labeled 1. Similarly, the great circles E_2E^2, E_3E^3 contain the altitudes of both triangles through the vertices labeled 2 and 3. The planes of the great circles E_1E^1, E_2E^2, E_3E^3 are perpendicular to $\mathbf{e}_1 \times \mathbf{e}^1$, $\mathbf{e}_2 \times \mathbf{e}^2$, $\mathbf{e}_3 \times \mathbf{e}^3$, respectively; and these vectors, in view of (7), lie in a plane p. The poles of this plane (the ends of the diameter of s perpendicular to p) are points common to all of the great circles E_iE^i. Therefore, *the three altitudes E_iE^i of the polar triangles $E_1E_2E_3$, $E^1E^2E^3$ meet in a pole P of the plane p.*

When a basis and its reciprocal are identical, the basis is called *self-reciprocal.* The equations (3) then become

$$(8) \qquad \mathbf{e}_i \cdot \mathbf{e}_j = \delta_{ij}.\dagger$$

These equations characterize an orthogonal triple of unit vectors. Hence *a basis is self-reciprocal when and only when it consists of a mutually orthogonal triple of unit vectors.* The triple will be dextral if $[\mathbf{e}_1\mathbf{e}_2\mathbf{e}_3] = 1$ (§ 19), sinistral if $[\mathbf{e}_1\mathbf{e}_2\mathbf{e}_3] = -1$. Thus $\mathbf{i}$, $\mathbf{j}$, $\mathbf{k}$ and $\mathbf{i}$, $\mathbf{j}$, $-\mathbf{k}$ are typical dextral and sinistral orthogonal bases.

24. Components of a Vector. We now supplement the notation of § 12 by writing the components of a vector $\mathbf{u}$ u^i or u_i according as the basis is $\mathbf{e}_i$ or $\mathbf{e}^i$. The components u^i are called *contravariant,* the components u_i *covariant,* for reasons given in § 148. Any vector now may be written in two forms:

$$(1)\ (2) \qquad \mathbf{u} = u^1\mathbf{e}_1 + u^2\mathbf{e}_2 + u^3\mathbf{e}_3, \qquad \mathbf{u} = u_1\mathbf{e}^1 + u_2\mathbf{e}^2 + u_3\mathbf{e}^3.$$

From (1) and (2), we obtain equations of the type $\mathbf{u} \cdot \mathbf{e}^1 = u^1$, $\mathbf{u} \cdot \mathbf{e}_1 = u_1$; all six are included in

$$(3)\ (4) \qquad u^i = \mathbf{u} \cdot \mathbf{e}^i, \qquad u_i = \mathbf{u} \cdot \mathbf{e}_i \qquad (i = 1, 2, 3).$$

When a basis is given, a vector $\mathbf{u}$ is completely specified by giving its components written in order. With the notation of

† With a self-reciprocal basis there is no need for superscripts; we therefore write δ_i^j as δ_{ij}.

§ 12, we write a vector **u** as a number triple, $[u^1, u^2, u^3]$ or $[u_1, u_2, u_3]$ according as it is referred to the basis $\mathbf{e}_i$ or $\mathbf{e}^i$. We denote the (non-zero) box products of the base vectors by

$$(5) \qquad E = [\mathbf{e}_1\mathbf{e}_2\mathbf{e}_3], \qquad 1/E = [\mathbf{e}^1\mathbf{e}^2\mathbf{e}^3].$$

Addition and multiplication by scalars follow the formulas of § 12; thus, with covariant components,

$$(6) \qquad \lambda[u_1, u_2, u_3] = [\lambda u_1, \lambda u_2, \lambda u_3].$$

$$(7) \qquad [u_1, u_2, u_3] + [v_1, v_2, v_3] = [u_1 + v_1, u_2 + v_2, u_3 + v_3].$$

A simple expression for the scalar product $\mathbf{u} \cdot \mathbf{v}$ is obtained when **u** and **v** are referred to different, but reciprocal, bases. Thus, if we compute

$$\mathbf{u} \cdot \mathbf{v} = \begin{cases} (u^1\mathbf{e}_1 + u^2\mathbf{e}_2 + u^3\mathbf{e}_3) \cdot (v_1\mathbf{e}^1 + v_2\mathbf{e}^2 + v_3\mathbf{e}^3) \\ (u_1\mathbf{e}^1 + u_2\mathbf{e}^2 + u_3\mathbf{e}^3) \cdot (v^1\mathbf{e}_1 + v^2\mathbf{e}_2 + v^3\mathbf{e}_3) \end{cases}$$

we obtain, by virtue of equations (23.1),

$$(8) \qquad \mathbf{u} \cdot \mathbf{v} = u^1v_1 + u^2v_2 + u^3v_3 = u_1v^1 + u_2v^2 + u_3v^3.$$

We next compute the components of $\mathbf{u} \times \mathbf{v}$, relative to bases $\mathbf{e}_i$ and $\mathbf{e}^i$. Using covariant components gives

$$\mathbf{u} \times \mathbf{v} = (u_1\mathbf{e}^1 + u_2\mathbf{e}^2 + u_3\mathbf{e}^3) \times (v_1\mathbf{e}^1 + v_2\mathbf{e}^2 + v_3\mathbf{e}^3)$$

$$= \begin{vmatrix} u_2 & u_3 \\ v_2 & v_3 \end{vmatrix} \mathbf{e}^2 \times \mathbf{e}^3 + \begin{vmatrix} u_3 & u_1 \\ v_3 & v_1 \end{vmatrix} \mathbf{e}^3 \times \mathbf{e}^1 + \begin{vmatrix} u_1 & u_2 \\ v_1 & v_2 \end{vmatrix} \mathbf{e}^1 \times \mathbf{e}^2$$

$$= \begin{vmatrix} u_2 & u_3 \\ v_2 & v_3 \end{vmatrix} \frac{\mathbf{e}_1}{E} + \begin{vmatrix} u_3 & u_1 \\ v_3 & v_1 \end{vmatrix} \frac{\mathbf{e}_2}{E} + \begin{vmatrix} u_1 & u_2 \\ v_1 & v_2 \end{vmatrix} \frac{\mathbf{e}_3}{E};$$

or, more compactly,

$$(9) \qquad \mathbf{u} \times \mathbf{v} = E^{-1} \begin{vmatrix} \mathbf{e}_1 & \mathbf{e}_2 & \mathbf{e}_3 \\ u_1 & u_2 & u_3 \\ v_1 & v_2 & v_3 \end{vmatrix}.$$

Using contravariant components, we find, in similar fashion,

$$(10) \qquad \mathbf{u} \times \mathbf{v} = E \begin{vmatrix} \mathbf{e}^1 & \mathbf{e}^2 & \mathbf{e}^3 \\ u^1 & u^2 & u^3 \\ v^1 & v^2 & v^3 \end{vmatrix}.$$

From (9) and (10), we see that the components of $\mathbf{q} = \mathbf{u} \times \mathbf{v}$ are

(11) $$q^i = E^{-1} \begin{vmatrix} u_j & u_k \\ v_j & v_k \end{vmatrix}, \qquad q_i = E \begin{vmatrix} u^j & u^k \\ v^j & v^k \end{vmatrix},$$

where the indices ijk form a cyclical permutation of 123.

From (9) and (10), we next obtain two expressions for the box product $\mathbf{u} \times \mathbf{v} \cdot \mathbf{w}$:

(12) $$[\mathbf{uvw}] = E^{-1} \begin{vmatrix} u_1 & u_2 & u_3 \\ v_1 & v_2 & v_3 \\ w_1 & w_2 & w_3 \end{vmatrix} = E \begin{vmatrix} u^1 & u^2 & u^3 \\ v^1 & v^2 & v^3 \\ w^1 & w^2 & w^3 \end{vmatrix}.$$

The first formula of (12) may be written

(13) $$[\mathbf{uvw}][\mathbf{e}_1\mathbf{e}_2\mathbf{e}_3] = \begin{vmatrix} \mathbf{u} \cdot \mathbf{e}_1 & \mathbf{u} \cdot \mathbf{e}_2 & \mathbf{u} \cdot \mathbf{e}_3 \\ \mathbf{v} \cdot \mathbf{e}_1 & \mathbf{v} \cdot \mathbf{e}_2 & \mathbf{v} \cdot \mathbf{e}_3 \\ \mathbf{w} \cdot \mathbf{e}_1 & \mathbf{w} \cdot \mathbf{e}_2 & \mathbf{w} \cdot \mathbf{e}_3 \end{vmatrix},$$

whereas the second gives the analogous equation obtained by replacing $\mathbf{e}_i$ by $\mathbf{e}^i$. Since $\mathbf{e}_1$, $\mathbf{e}_2$, $\mathbf{e}_3$ may be any linearly independent set of vectors $\mathbf{a}$, $\mathbf{b}$, $\mathbf{c}$, we also have

(14) $$[\mathbf{uvw}][\mathbf{abc}] = \begin{vmatrix} \mathbf{u} \cdot \mathbf{a} & \mathbf{u} \cdot \mathbf{b} & \mathbf{u} \cdot \mathbf{c} \\ \mathbf{v} \cdot \mathbf{a} & \mathbf{v} \cdot \mathbf{b} & \mathbf{v} \cdot \mathbf{c} \\ \mathbf{w} \cdot \mathbf{a} & \mathbf{w} \cdot \mathbf{b} & \mathbf{w} \cdot \mathbf{c} \end{vmatrix}.$$

When the basis $\mathbf{e}_i$ is self-reciprocal (an orthogonal triple of unit vectors) $\mathbf{e}^i = \mathbf{e}_i$ and $E = \pm 1$. The two sets of components of $\mathbf{u}$ then coalesce into a single set which we arbitrarily write u_i. If the self-reciprocal basis is *dextral*, we can write $\mathbf{e}_1 = \mathbf{i}$, $\mathbf{e}_2 = \mathbf{j}$, $\mathbf{e}_3 = \mathbf{k}$, $E = 1$; then

$$\mathbf{u} \cdot \mathbf{v} = u_1v_1 + u_2v_2 + u_3v_3 ,$$

$$\mathbf{u} \times \mathbf{v} = \begin{vmatrix} \mathbf{i} & \mathbf{j} & \mathbf{k} \\ u_1 & u_2 & u_3 \\ v_1 & v_2 & v_3 \end{vmatrix}, \qquad [\mathbf{uvw}] = \begin{vmatrix} u_1 & u_2 & u_3 \\ v_1 & v_2 & v_3 \\ w_1 & w_2 & w_3 \end{vmatrix},$$

in agreement with our previous results.

25. Vector Equations. A vector is uniquely determined when its scalar and vector products with two known *non-perpendicular* vectors are given. Thus let

(1) $$\mathbf{u} \cdot \mathbf{a} = \alpha, \qquad \mathbf{u} \times \mathbf{b} = \mathbf{c}, \qquad (\mathbf{a} \cdot \mathbf{b} \neq 0);$$

the scalar α and the vectors **a**, **b**, **c** are regarded as known, and $\mathbf{b} \cdot \mathbf{c} = 0$. We have

$$\mathbf{a} \times (\mathbf{u} \times \mathbf{b}) = \mathbf{a} \times \mathbf{c}, \quad \text{or} \quad \mathbf{a} \cdot \mathbf{b}\,\mathbf{u} - \alpha\mathbf{b} = \mathbf{a} \times \mathbf{c};$$

$$\mathbf{u} = \frac{\mathbf{a} \times \mathbf{c} + \alpha\mathbf{b}}{\mathbf{a} \cdot \mathbf{b}}. \tag{2}$$

By direct substitution, (2) is seen to be a solution of equations (1).

A vector is also determined when its scalar products with three known non-coplanar vectors are given. For example, if

$$\mathbf{u} \cdot \mathbf{e}_1 = u_1, \qquad \mathbf{u} \cdot \mathbf{e}_2 = u_2, \qquad \mathbf{u} \cdot \mathbf{e}_3 = u_3, \tag{3}$$

and the sets $\mathbf{e}_i$, $\mathbf{e}^i$ are reciprocal, we have, from (24.2),

$$\mathbf{u} = u_1\mathbf{e}^1 + u_2\mathbf{e}^2 + u_3\mathbf{e}^3. \tag{4}$$

By direct substitution, (4) is seen to be a solution of equations (3).

26. Homogeneous Coordinates. Coordinates of a point, line or plane are called *homogeneous* if the entity they determine is not altered when the coordinates are multiplied by the same scalar. A coordinate that depends upon the choice of origin O bears the subscript O; such coordinates are written *after* those independent of the origin. In equations **r** denotes any position vector from O to the entity in question, whereas $\mathbf{r}_1$, $\mathbf{r}_2$ denote position vectors to points R_1, R_2 given in advance. In this article we use Latin letters for vectors, Greek for scalars.

Point Coordinates. If $\mathbf{r} = \overrightarrow{OA}$, we write $\mathbf{r} = \mathbf{a}_O/\alpha$. The "equation" of the point A is then

$$\mathbf{r}\alpha = \mathbf{a}_O, \tag{1}$$

and its homogeneous coordinates are $(\alpha, \mathbf{a}_O)$. Note that $(\lambda\alpha, \lambda\mathbf{a}_O)$ determine the same point; hence the coordinates $(\alpha, \mathbf{a}_O)$ depend on three independent scalars: "there are ∞^3 points in space."

Plane Coordinates. The equation of a plane through R_1 and perpendicular to the vector **a** is $(\mathbf{r} - \mathbf{r}_1) \cdot \mathbf{a} = 0$; or, on writing $\alpha_O = \mathbf{r}_1 \cdot \mathbf{a}$,

$$\mathbf{r} \cdot \mathbf{a} = \alpha_O. \tag{2}$$

The homogeneous coordinates of the plane are $(\mathbf{a}, \alpha_O)$. Since $(\lambda\mathbf{a}, \lambda\alpha_O)$ determine the same plane, the coordinates $(\mathbf{a}, \alpha_O)$ depend on three independent scalars: "there are ∞^3 planes in space."

Line Coordinates. The equation of a line through R_1 and parallel to the vector $\mathbf{a}$ is $(\mathbf{r} - \mathbf{r}_1) \times \mathbf{a} = 0$; or, on writing $\mathbf{a}_O = \mathbf{r}_1 \times \mathbf{a}$,

$$\mathbf{r} \times \mathbf{a} = \mathbf{a}_O. \tag{3}$$

The homogeneous (or Plücker) coordinates of the line are $(\mathbf{a}, \mathbf{a}_O)$ and are connected by the relation,

$$\mathbf{a} \cdot \mathbf{a}_O = 0. \tag{4}$$

In view of (4) and the fact that $(\lambda\mathbf{a}, \lambda\mathbf{a}_O)$ determine the same line, the coordinates $(\mathbf{a}, \mathbf{a}_O)$ depend on four independent scalars: "there are ∞^4 lines in space."

We thus have the homogeneous coordinates:

$$\text{Point } (\alpha, \mathbf{a}_O), \quad \text{Plane } (\mathbf{a}, \alpha_O), \quad \text{Line } (\mathbf{a}, \mathbf{a}_O).$$

The first coordinate cannot vanish. When the second coordinate (subscript O) is zero, the origin O is on the point, plane, or line, respectively.

The distances of the point, plane, or line from the origin are, respectively,

$$\frac{|\mathbf{a}_O|}{|\alpha|}, \quad \frac{|\alpha_O|}{|\mathbf{a}|}, \quad \frac{|\mathbf{a}_O|}{|\mathbf{a}|}. \tag{5}$$

The first result is obvious. If $\mathbf{p}$ is the vector from O perpendicular to the plane $\mathbf{r} \cdot \mathbf{a} = \alpha_O$,

$$\mathbf{p} \cdot \mathbf{a} = \alpha_O, \qquad \mathbf{p} \times \mathbf{a} = 0; \qquad \text{hence} \quad \mathbf{p} = \frac{\alpha_O \mathbf{a}}{\mathbf{a} \cdot \mathbf{a}}$$

from (25.2), and $|\mathbf{p}| = |\alpha_O|/|\mathbf{a}|$. Again if $\mathbf{p}$ is the vector from O perpendicular to the line $\mathbf{r} \times \mathbf{a} = \mathbf{a}_O$,

$$\mathbf{p} \times \mathbf{a} = \mathbf{a}_O, \qquad \mathbf{p} \cdot \mathbf{a} = 0; \qquad \text{hence} \quad \mathbf{p} = \frac{\mathbf{a} \times \mathbf{a}_O}{\mathbf{a} \cdot \mathbf{a}}$$

from (25.2), and $|\mathbf{p}| = |\mathbf{a}_O|/|\mathbf{a}|$.

If the origin is shifted from O to P, we must replace $\mathbf{r}$ in (1), (2), (3) by $\overrightarrow{PO} + \mathbf{r}$ to obtain the second coordinate referred to P. We thus obtain the *shift formulas:*

$$\mathbf{a}_P = \mathbf{a}_O + \overrightarrow{PO}\,\alpha, \tag{6}$$

$$\alpha_P = \alpha_O + \overrightarrow{PO} \cdot \mathbf{a}, \tag{7}$$

$$\mathbf{a}_P = \mathbf{a}_O + \overrightarrow{PO} \times \mathbf{a}, \tag{8}$$

in the respective cases.

Two Elements Determine a Third. We have the following cases: Two distinct points determine a line:

$$(9) \qquad (\alpha, \mathbf{a}_0), \quad (\beta, \mathbf{b}_0) \to (\alpha\mathbf{b}_0 - \beta\mathbf{a}_0, \mathbf{a}_0 \times \mathbf{b}_0).$$

Two non-parallel planes determine a line:

$$(10) \qquad (\mathbf{a}, \alpha_0), \quad (\mathbf{b}, \beta_0) \to (\mathbf{a} \times \mathbf{b}, \beta_0\mathbf{a} - \alpha_0\mathbf{b}).$$

A point and line determine a plane:

$$(11) \qquad (\alpha, \mathbf{a}_0), \quad (\mathbf{b}, \mathbf{b}_0) \to (\alpha\mathbf{b}_0 - \mathbf{a}_0 \times \mathbf{b}, \mathbf{a}_0 \cdot \mathbf{b}_0),$$

unless $\alpha\mathbf{b}_0 - \mathbf{a}_0 \times \mathbf{b} = 0$; then also $\mathbf{a}_0 \cdot \mathbf{b}_0 = 0$, and the point lies on the line.

A line and plane determine a point:

$$(12) \qquad (\mathbf{a}, \mathbf{a}_0), \quad (\mathbf{b}, \beta_0) \to (\mathbf{a} \cdot \mathbf{b}, \beta_0\mathbf{a} - \mathbf{a}_0 \times \mathbf{b}),$$

unless $\mathbf{a} \cdot \mathbf{b} = 0$. If $\beta_0\mathbf{a} - \mathbf{a}_0 \times \mathbf{b} = 0$ (and hence $\mathbf{a} \cdot \mathbf{b} = 0$), the line lies in the plane.

Proofs in outline.

For (9): The line AB has the equation,

$$\left(\mathbf{r} - \frac{\mathbf{a}_0}{\alpha}\right) \times \left(\frac{\mathbf{b}_0}{\beta} - \frac{\mathbf{a}_0}{\alpha}\right) = 0, \quad \text{or} \quad \mathbf{r} \times (\alpha\mathbf{b}_0 - \beta\mathbf{a}_0) = \mathbf{a}_0 \times \mathbf{b}_0.$$

For (10): The line is parallel to $\mathbf{a} \times \mathbf{b}$; and

$$\mathbf{r} \times (\mathbf{a} \times \mathbf{b}) = \mathbf{r} \cdot \mathbf{b}\,\mathbf{a} - \mathbf{r} \cdot \mathbf{a}\,\mathbf{b} = \beta_0\mathbf{a} - \alpha_0\mathbf{b}.$$

For (11): Refer the line $(\mathbf{b}, \mathbf{b}_0)$ to the point $A(\alpha, \mathbf{a}_0)$; then, from (8),

$$\mathbf{b}_A = \mathbf{b}_0 - \overrightarrow{OA} \times \mathbf{b} = \mathbf{b}_0 - \frac{\mathbf{a}_0 \times \mathbf{b}}{\alpha}.$$

The plane through A normal to $\mathbf{b}_A$ has the equation

$$\left(\mathbf{r} - \frac{\mathbf{a}_0}{\alpha}\right) \cdot (\alpha\mathbf{b}_0 - \mathbf{a}_0 \times \mathbf{b}) = 0, \quad \text{or} \quad \mathbf{r} \cdot (\alpha\mathbf{b}_0 - \mathbf{a}_0 \times \mathbf{b}) = \mathbf{a}_0 \cdot \mathbf{b}_0.$$

For (12): The equations of line and plane, $\mathbf{r} \times \mathbf{a} = \mathbf{a}_0$ and $\mathbf{r} \cdot \mathbf{b} = \beta_0$, have the solution (§ 25):

$$\mathbf{r} = \frac{\beta_0\mathbf{a} - \mathbf{a}_0 \times \mathbf{b}}{\mathbf{a} \cdot \mathbf{b}}.$$

Two Lines. If R_1 and R_2 are points on the lines $(\mathbf{a}, \mathbf{a}_O)$, $(\mathbf{b}, \mathbf{b}_O)$, the lines are coplanar when and only when the vectors $\mathbf{r}_1 - \mathbf{r}_2$, $\mathbf{a}$, $\mathbf{b}$ are coplanar. Since

$$(13)\quad (\mathbf{r}_1 - \mathbf{r}_2) \cdot \mathbf{a} \times \mathbf{b} = \mathbf{r}_1 \times \mathbf{a} \cdot \mathbf{b} + \mathbf{r}_2 \times \mathbf{b} \cdot \mathbf{a} = \mathbf{a}_O \cdot \mathbf{b} + \mathbf{b}_O \cdot \mathbf{a},$$

a necessary and sufficient condition that the lines be coplanar is

$$(14)\qquad \mathbf{a} \cdot \mathbf{b}_O + \mathbf{b} \cdot \mathbf{a}_O = 0.$$

If the lines are not coplanar, let $\mathbf{e}$ be a unit vector in the direction $\mathbf{a} \times \mathbf{b}$ of their common normal; then, from (13),

$$(15)\qquad (\mathbf{r}_1 - \mathbf{r}_2) \cdot \mathbf{e} = \frac{\mathbf{a} \cdot \mathbf{b}_O + \mathbf{b} \cdot \mathbf{a}_O}{|\mathbf{a} \times \mathbf{b}|}.$$

This is the component of $\overrightarrow{R_2R_1}$ in the direction of $\mathbf{e}$; its numerical value gives the shortest distance between the lines.

Equations in Point and Plane Coordinates. If the point $(\sigma, \mathbf{s}_O)$ lies on the plane $(\mathbf{t}, \tau_O)$ we have

$$(16)\qquad \mathbf{s}_O \cdot \mathbf{t} - \sigma\tau_O = 0,$$

on eliminating $\mathbf{r}$ from $\mathbf{r}\sigma = \mathbf{s}_O$ and $\mathbf{r} \cdot \mathbf{t} = \tau_O$. This may be regarded as the equation of the point $(\sigma, \mathbf{s}_O)$ in plane coordinates; or as the equation of the plane $(\mathbf{t}, \tau_O)$ in point coordinates.

If the line $(\mathbf{p}, \mathbf{p}_O)$ contains the point $(\sigma, \mathbf{s}_O)$ we have

$$(17)\qquad \mathbf{s}_O \times \mathbf{p} - \sigma\mathbf{p}_O = 0,$$

on eliminating $\mathbf{r}$ from $\mathbf{r}\sigma = \mathbf{s}_O$, $\mathbf{r} \times \mathbf{p} = \mathbf{p}_O$. This is the equation of the line $(\mathbf{p}, \mathbf{p}_O)$ in point coordinates.

If the line $(\mathbf{p}, \mathbf{p}_O)$ lies on the plane $(\mathbf{t}, \tau_O)$ we have $\mathbf{p} \cdot \mathbf{t} = 0$ and hence

$$(18)\qquad \tau_O\mathbf{p} - \mathbf{p}_O \times \mathbf{t} = 0,$$

on eliminating $\mathbf{r}$ from $\mathbf{r} \times \mathbf{p} = \mathbf{p}_O$ and $\mathbf{r} \cdot \mathbf{t} = \tau_O$. This is the equation of the line $(\mathbf{p}, \mathbf{p}_O)$ in plane coordinates.

Example. If three points $(\alpha, \mathbf{a}_O)$, $(\beta, \mathbf{b}_O)$, $(\gamma, \mathbf{c}_O)$ lie on the line $(\mathbf{p}, \mathbf{p}_O)$, we have, from (17),

$$\frac{\mathbf{a}_O}{\alpha} \times \mathbf{p} = \frac{\mathbf{b}_O}{\beta} \times \mathbf{p} = \frac{\mathbf{c}_O}{\gamma} \times \mathbf{p} = \mathbf{p}_O;$$

hence

$$(\mathrm{i})\qquad \lambda\left(\frac{\mathbf{a}_O}{\alpha} - \frac{\mathbf{c}_O}{\gamma}\right) + \mu\left(\frac{\mathbf{b}_O}{\beta} - \frac{\mathbf{c}_O}{\gamma}\right) = 0,$$

since both vectors in parenthesis are parallel to $\mathbf{p}$. This is a linear relation between $\mathbf{a}_O/\alpha$, $\mathbf{b}_O/\beta$, $\mathbf{c}_O/\gamma$ in which the sum of the coefficients is zero (§ 6).

If the three planes $(\mathbf{a}, \alpha_O)$, $(\mathbf{b}, \beta_O)$, $(\mathbf{c}, \gamma_O)$ pass through the line $(\mathbf{p}, \mathbf{p}_O)$, we have, from (18),

$$\mathbf{p}_O \times \frac{\mathbf{a}}{\alpha_O} = \mathbf{p}_O \times \frac{\mathbf{b}}{\beta_O} = \mathbf{p}_O \times \frac{\mathbf{c}}{\gamma_O} = \mathbf{p},$$

provided α_O, β_O, γ_O are not zero; hence

(ii) $$\lambda\left(\frac{\mathbf{a}}{\alpha_O} - \frac{\mathbf{c}}{\gamma_O}\right) + \mu\left(\frac{\mathbf{b}}{\beta_O} - \frac{\mathbf{c}}{\gamma_O}\right) = 0,$$

since both vectors in parenthesis are parallel to $\mathbf{p}_O$. This is a linear relation connecting $\mathbf{a}/\alpha_O$, $\mathbf{b}/\beta_O$, $\mathbf{c}/\gamma_O$ in which the sum of the coefficients is zero.

A shift in origin does not alter the coefficients in (i). In (ii), however, the coefficients are changed but their sum still remains zero; if we write $\nu = -\lambda - \mu$ and shift the origin to P,

$$\lambda\frac{\mathbf{a}}{\alpha_O} + \mu\frac{\mathbf{b}}{\beta_O} + \nu\frac{\mathbf{c}}{\gamma_O} = 0 \quad \text{becomes} \quad \lambda'\frac{\mathbf{a}}{\alpha_P} + \mu'\frac{\mathbf{b}}{\beta_P} + \nu'\frac{\mathbf{c}}{\gamma_P} = 0,$$

where $\lambda' = \lambda\alpha_P/\alpha_O$, $\mu' = \mu\beta_P/\beta_O$, $\nu' = \nu\gamma_P/\gamma_O$. Let the student prove that $\lambda' + \mu' + \nu' = 0$.

27. Line Vectors and Moments. A vector which is restricted to lie in a definite line is called a *line vector*. If $\mathbf{f}$ is the vector, and

(1) $$\mathbf{r} \times \mathbf{f} = \mathbf{f}_O, \qquad (\mathbf{f} \cdot \mathbf{f}_O = 0)$$

the equation of its line of action, the line vector is completely specified by the Plücker coordinates $\mathbf{f}$, $\mathbf{f}_O$.

The vector $\mathbf{f}_O$ is called the *moment of* $\mathbf{f}$ *about the point* O. If $\mathbf{f} = \overrightarrow{AB}$,

(2) $$\mathbf{f}_O = \overrightarrow{OA} \times \overrightarrow{AB}$$

is twice the vector area of the triangle OAB (§ 17); $\mathbf{f}_O$ remains constant as $\mathbf{f}$ is shifted along its line of action.

If the origin is shifted from O to P, the moment of $\mathbf{f}$ about P is $\overrightarrow{PA} \times \overrightarrow{AB} = (\overrightarrow{PO} + \overrightarrow{OA}) \times \overrightarrow{AB}$; hence

(3) $$\mathbf{f}_P = \mathbf{f}_O + \overrightarrow{PO} \times \mathbf{f}.$$

If s is any axis through P with the unit vector $\mathbf{e}$, the component of $\mathbf{f}_P$ on s, namely, $\mathbf{e} \cdot \mathbf{f}_P$ (15.6), is called the *moment of* $\mathbf{f}$ *about the axis* s. We speak of moment about an *axis* because $\mathbf{e} \cdot \mathbf{f}_P$ is independent of the position of P on this axis; for, if the axis is given by the *unit* line vector $(\mathbf{e}, \mathbf{e}_O)$,

$$\mathbf{e} \cdot \mathbf{f}_P = \mathbf{e} \cdot (\mathbf{f}_O + \overrightarrow{PO} \times \mathbf{f}) = \mathbf{e} \cdot \mathbf{f}_O + \overrightarrow{OP} \times \mathbf{e} \cdot \mathbf{f},$$

(4) $$\mathbf{e} \cdot \mathbf{f}_P = \mathbf{e} \cdot \mathbf{f}_O + \mathbf{e}_O \cdot \mathbf{f}.$$

We also may regard $\mathbf{e} \cdot \mathbf{f}_P$ as the component of twice the vector area PAB on a directed plane of unit normal $\mathbf{e}$ (§ 17). In Fig. 27, A_1B_1 is the projection of AB on the plane; the moment of $\mathbf{f}$ about s is numerically equal to twice the area PA_1B_1, that is, to the product of the length A_1B_1 and the perpendicular distance h of A_1B_1 from P; its sign is plus or minus according as a turn in the sense PA_1B_1 would advance a right-handed screw in the direction of s or $-s$.

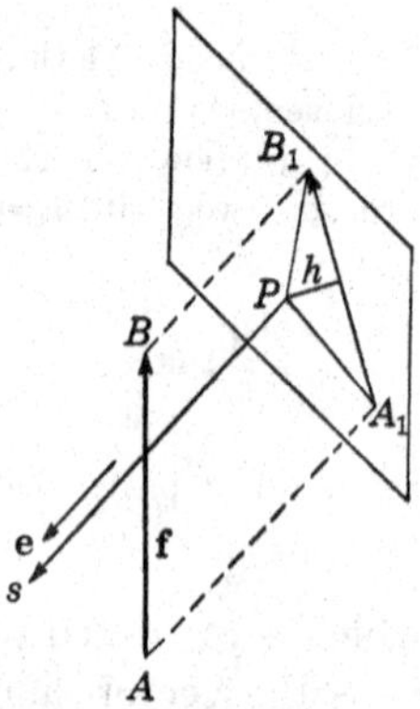

FIG. 27

We repeat these important definitions.

The moment of the line vector $\mathbf{f} = \overrightarrow{AB}$ about a *point* P is

(5) $$\mathbf{f}_P = \overrightarrow{PA} \times \overrightarrow{AB} = \mathbf{r} \times \mathbf{f},$$

where $\mathbf{r}$ is a vector from P to any point on $\mathbf{f}$'s line of action. The moment of $\mathbf{f}$ about an *axis* is the component of $\mathbf{r} \times \mathbf{f}$ on the axis, where $\mathbf{r}$ is a vector from any point on the axis to any point on the line of action. The moment about a *point* is a *vector*; about an *axis, a scalar.*

Example. The line of action of the force $\mathbf{f} = [1, -1, 2]$ passes through the point $A(2, 4, -1)$. Find its moment about an axis through the point $P(3, -1, 2)$ and having the direction of the vector $[2, -1, 2]$.

Since $\overrightarrow{PA} = [-1, 5, -3]$ and $\mathbf{e} = \frac{1}{3}[2, -1, 2]$ is a unit vector along the axis,

$$\mathbf{f}_P = \overrightarrow{PA} \times \mathbf{f} = \begin{vmatrix} \mathbf{i} & \mathbf{j} & \mathbf{k} \\ -1 & 5 & -3 \\ 1 & -1 & 2 \end{vmatrix} = [7, -1, -4]$$

is the moment of $\mathbf{f}$ about P; and

$$\mathbf{e} \cdot \mathbf{f}_P = \tfrac{1}{3}(14 + 1 - 8) = \tfrac{7}{3}$$

is the moment of $\mathbf{f}$ about the axis.

28. Summary: Vector Algebra. Equal vectors have the same length and direction. Free vectors are added by the triangle construction. The negative of a vector is the vector reversed. To subtract a vector, add its negative. The product $\lambda\mathbf{u}$ is a vector $|\lambda|$ times as long as $\mathbf{u}$; its direction is the same as $\mathbf{u}$ if $\lambda > 0$, the reverse if $\lambda < 0$.

Vectors may be added, subtracted, and multiplied by real numbers in conformity with the laws of ordinary algebra.

The n vectors $\mathbf{u}_i$ are *linearly dependent* if there exist n real numbers λ_i, not all zero, such that $\Sigma\lambda_i\mathbf{u}_i = 0$. When $n = 2, 3$, linear dependence implies that the vectors are collinear or coplanar, respectively; and conversely. In space of three dimensions any four vectors are linearly dependent.

A point P divides a segment AB in the ratio $\lambda = \beta/\alpha$ when $\overrightarrow{AP} = \lambda\overrightarrow{PB}$; then

$$(\alpha + \beta)\overrightarrow{OP} = \alpha\overrightarrow{OA} + \beta\overrightarrow{OB}.$$

A linear relation $\Sigma\lambda_i\overrightarrow{OP_i} = 0$ connecting n position vectors will hold for any origin when and only when $\Sigma\lambda_i = 0$. When $n = 2, 3, 4$, the points P_i are coincident, collinear, or coplanar, respectively.

A set of weighted points m_iP_i has a unique *centroid* P if $\Sigma m_i \neq 0$; its defining equation is

$$\Sigma m_i\overrightarrow{P^*P_i} = 0; \quad \text{and} \quad \overrightarrow{OP^*} = \frac{\Sigma m_i\overrightarrow{OP_i}}{\Sigma m_i}.$$

If all $m_i = 1$, P^* is called the *mean center*.

The *scalar product* $\mathbf{u}\cdot\mathbf{v}$ is defined as

$$\mathbf{u}\cdot\mathbf{v} = |\mathbf{u}|\,|\mathbf{v}|\cos(\mathbf{u}, \mathbf{v}).$$

Laws:

$$\mathbf{u}\cdot\mathbf{v} = \mathbf{v}\cdot\mathbf{u}, \qquad \mathbf{u}\cdot(\mathbf{v}+\mathbf{w}) = \mathbf{u}\cdot\mathbf{v} + \mathbf{u}\cdot\mathbf{w}.$$

If $\mathbf{u}, \mathbf{v} \neq 0$, $\mathbf{u}\cdot\mathbf{v} = 0$ implies $\mathbf{u} \perp \mathbf{v}$, and conversely.

The *vector product* $\mathbf{u}\times\mathbf{v}$ is defined as

$$\mathbf{u}\times\mathbf{v} = |\mathbf{u}|\,|\mathbf{v}|\sin(\mathbf{u}, \mathbf{v})\mathbf{e}.$$

where $\mathbf{e}$ is a unit vector perpendicular to both $\mathbf{u}$ and $\mathbf{v}$ so that $\mathbf{u}, \mathbf{v}, \mathbf{e}$ form a dextral set.

Laws:

$$\mathbf{u}\times\mathbf{v} = -\mathbf{v}\times\mathbf{u}, \qquad \mathbf{u}\times(\mathbf{v}+\mathbf{w}) = \mathbf{u}\times\mathbf{v}+\mathbf{u}\times\mathbf{w}.$$

If $\mathbf{u}, \mathbf{v} \neq 0$, $\mathbf{u}\times\mathbf{v} = 0$ implies that $\mathbf{u} \parallel \mathbf{v}$ and conversely.

Expansion rule:

$$\mathbf{u}\times(\mathbf{v}\times\mathbf{w}) = \mathbf{u}\cdot\mathbf{w}\,\mathbf{v} - \mathbf{u}\cdot\mathbf{v}\,\mathbf{w}.$$

Cross multiplication is not associative.

The *box product*, $\mathbf{u}\times\mathbf{v}\cdot\mathbf{w}$ or $[\mathbf{uvw}]$, is numerically equal to the volume of a "box" having $\mathbf{u}, \mathbf{v}, \mathbf{w}$ as concurrent edges; its sign is $+$ or $-$ according as $\mathbf{u}, \mathbf{v}, \mathbf{w}$ form a dextral or sinistral set. The value of $\mathbf{u}\times\mathbf{v}\cdot\mathbf{w}$ is not affected by a change in cyclical order of the vectors, or by an interchange of dot and cross. If $\mathbf{u}, \mathbf{v}, \mathbf{w} \neq 0$, $[\mathbf{uvw}] = 0$ implies that $\mathbf{u}, \mathbf{v}, \mathbf{w}$ are coplanar, and conversely.

A set of three vectors $\mathbf{e}_i$ forms a *basis* if $E = [\mathbf{e}_1\mathbf{e}_2\mathbf{e}_3] \neq 0$. To every basis there corresponds a unique basis $\mathbf{e}^i$ such that $\mathbf{e}_i\cdot\mathbf{e}^j = \delta_i^j$; such bases are called *reciprocal*. If the indices i, j, k form a cyclical permutation of 1 2 3,

$$\mathbf{e}^i = \frac{\mathbf{e}_j\times\mathbf{e}_k}{[\mathbf{e}_1\mathbf{e}_2\mathbf{e}_3]}, \qquad \mathbf{e}_i = \frac{\mathbf{e}^j\times\mathbf{e}^k}{[\mathbf{e}^1\mathbf{e}^2\mathbf{e}^3]}; \qquad [\mathbf{e}_1\mathbf{e}_2\mathbf{e}_3][\mathbf{e}^1\mathbf{e}^2\mathbf{e}^3] = 1.$$

Both bases are dextral if $E > 0$; sinistral, if $E < 0$.

Given a basis $\mathbf{e}_i$, a vector $\mathbf{u}$ may be written as $\Sigma u^i\mathbf{e}_i$ or $\Sigma u_i\mathbf{e}^i$; the numbers u^i are *contravariant* components of $\mathbf{u}$, u_i *covariant components*.

$$\mathbf{u}+\mathbf{v} = \Sigma(u^i + v^i)\mathbf{e}_i = \Sigma(u_i + v_i)\mathbf{e}^i;$$

$$\lambda\mathbf{u} = \Sigma\lambda u^i\mathbf{e}_i = \Sigma\lambda u_i\mathbf{e}^i;$$

$$\mathbf{u}\cdot\mathbf{v} = u^1v_1 + u^2v_2 + u^3v_3 = u_1v^1 + u_2v^2 + u_3v^3;$$

$$\mathbf{u}\times\mathbf{v} = E^{-1}\begin{vmatrix} \mathbf{e}_1 & \mathbf{e}_2 & \mathbf{e}_3 \\ u_1 & u_2 & u_3 \\ v_1 & v_2 & v_3 \end{vmatrix} = E\begin{vmatrix} \mathbf{e}^1 & \mathbf{e}^2 & \mathbf{e}^3 \\ u^1 & u^2 & u^3 \\ v^1 & v^2 & v^3 \end{vmatrix};$$

$$[\mathbf{uvw}] = E^{-1}\begin{vmatrix} u_1 & u_2 & u_3 \\ v_1 & v_2 & v_3 \\ w_1 & w_2 & w_3 \end{vmatrix} = E\begin{vmatrix} u^1 & u^2 & u^3 \\ v^1 & v^2 & v^3 \\ w^1 & w^2 & w^3 \end{vmatrix}.$$

When the basis $\mathbf{e}_i$ is self-reciprocal ($\mathbf{e}_i = \mathbf{e}^i$), its vectors form a mutually orthogonal set of unit vectors; $E = 1$, if the basis is dextral, $E = -1$, if sinistral. A dextral self-reciprocal basis is written

$\mathbf{i}, \mathbf{j}, \mathbf{k}$. For such a basis $u^i = u_i$; these rectangular components are given by

$$u_i = |\mathbf{u}| \cos(\mathbf{e}_i, \mathbf{u}), \quad \text{and} \quad |\mathbf{u}|^2 = (u_1)^2 + (u_2)^2 + (u_3)^2.$$

The two formulas previously given for $\mathbf{u} + \mathbf{v}$, $\lambda\mathbf{u}$, $\mathbf{u} \cdot \mathbf{v}$, $\mathbf{u} \times \mathbf{v}$, $\mathbf{u} \times \mathbf{v} \cdot \mathbf{w}$ in each case become identical.

The equations of a point, line, and plane, in terms of their homogeneous coordinates $(\alpha, \mathbf{a}_O)$, $(\mathbf{b}, \mathbf{b}_O)$, $(\mathbf{c}, \gamma_O)$, are

$$\mathbf{r}\alpha = \mathbf{a}_O, \qquad \mathbf{r} \times \mathbf{b} = \mathbf{b}_O, \qquad \mathbf{r} \cdot \mathbf{c} = \gamma_O,$$

respectively. When the origin is shifted to P,

$$\mathbf{a}_P = \mathbf{a}_O + \overrightarrow{PO}\alpha, \qquad \mathbf{b}_P = \mathbf{b}_O + \overrightarrow{PO} \times \mathbf{b}, \qquad \gamma_P = \gamma_O + \overrightarrow{PO} \cdot \mathbf{c}.$$

The moment of the line vector $(\mathbf{f}, \mathbf{f}_O)$ about the *point* P is $\mathbf{f}_P = \mathbf{r} \times \mathbf{f}$, where $\mathbf{r}$ is any vector from P to its line of action. The moment of $(\mathbf{f}, \mathbf{f}_O)$ about an *axis* through P is the component of $\mathbf{f}_P$ on this axis; if the axis is given by the *unit* line vector $(\mathbf{e}, \mathbf{e}_O)$, this axial moment

$$\mathbf{e} \cdot \mathbf{f}_P = \mathbf{e} \cdot \mathbf{f}_O + \mathbf{e}_O \cdot \mathbf{f}.$$

PROBLEMS

1. If ABC is any triangle and L, M, N are the mid-points of its sides, show that, for any choice of O,

$$\overrightarrow{OA} + \overrightarrow{OB} + \overrightarrow{OC} = \overrightarrow{OL} + \overrightarrow{OM} + \overrightarrow{ON}.$$

2. If $\overrightarrow{OA'} = 3\,\overrightarrow{OA}$, $\overrightarrow{OB'} = 2\,\overrightarrow{OB}$, in what ratio does the point P in which AB and $A'B'$ intersect divide these segments?

3. Show that the mid-points of the four sides of any quadrilateral (plane or skew) form the vertices of a parallelogram.

4. P and Q divide the sides CA, CB of the triangle ABC in the ratios $x/(1 - x)$, $y/(1 - y)$. If $\overrightarrow{PQ} = \lambda\,\overrightarrow{AB}$, show that $x = y = \lambda$.

5. E, F are the mid-points of the sides AB, BC of the parallelogram $ABCD$. Show that the lines DE, DF divide the diagonal AC into thirds and that AC cuts off a third of each line.

6. OAA', OBB', OCC', ODD' are four rays of a pencil of lines through O cut by two straight lines $ABCD$ and $A'B'C'D'$. If C and D divide AB in the ratios r and s, C' and D' divide $A'B'$ in the ratios r' and s', prove that $r/s = r'/s'$.

7. The points A, B, C and A', B', C' lie, respectively, on two intersecting lines. Show that BC', CB'; CA', AC'; AB', BA' intersect in the collinear points P, Q, R (Pascal's Theorem).

8. Lines drawn through a point P and the vertices A, B, C, D of a tetrahedron cut the planes of the opposite faces at A', B', C', D'. Show that the sum of the ratios in which these points divide the segments PA, PB, PC, PD is -1. [Equation (10.2), with $\epsilon = -(\alpha + \beta + \gamma + \delta)$ may be written

$$\alpha\mathbf{a} + \beta\mathbf{b} + \gamma\mathbf{c} + \delta\mathbf{d} + \epsilon\mathbf{p} = 0, \qquad \alpha + \beta + \gamma + \delta + \epsilon = 0.$$

From this we conclude, as in § 7, ex. 3, that $\mathbf{a}' = (\alpha\mathbf{a} + \epsilon\mathbf{p})/(\alpha + \epsilon)$, etc.]

9. The line DE is drawn parallel to the base AB of the triangle ABC and is included between its sides. If the lines AE, BD meet at P, show that the line CP bisects AB.

10. The points P, Q, R divide the sides BC, CA, AB of the triangle ABC in the ratios $\alpha/(1 - \alpha)$, $\beta/(1 - \beta)$, $\gamma/(1 - \gamma)$. If P, Q, R are collinear,

$$x\mathbf{p} + y\mathbf{q} + z\mathbf{r} = 0, \qquad x + y + z = 0;$$

putting $\mathbf{p} = (1 - \alpha)\mathbf{b} + \alpha\mathbf{c}$, etc. in this equation, we obtain a linear relation in $\mathbf{a}$, $\mathbf{b}$, $\mathbf{c}$ *whose coefficients have a zero sum.* Show that this implies that the separate coefficients vanish; hence deduce the *Theorem of Menelaus* (§ 7, ex. 2):

$$\alpha\beta\gamma/(1 - \alpha)(1 - \beta)(1 - \gamma) = -1.$$

11. Using an argument patterned after that in Problem 10, prove *Carnot's Theorem:*

If a plane cuts the sides AB, BC, CD, DA of a *skew* quadrilateral $ABCD$ in the points P, Q, R, S, respectively, the product of the ratios in which P, Q, R, S divide these sides equals 1.

12. In a plane quadrilateral $ABCD$, the point P in which the diagonals AC, BD intersect divides these segments in the ratios 4/3 and 2/3, respectively. In what ratio does the point Q, in which the sides AB, CD meet, divide these segments?

13. If $\mathbf{e}_1 = \overrightarrow{OE_1}$, $\mathbf{e}_2 = \overrightarrow{OE_2}$, $\mathbf{e}_3 = \overrightarrow{OE_3}$ form a basis, the reciprocal set $\mathbf{e}^1$, $\mathbf{e}^2$, $\mathbf{e}^3$ are vectors perpendicular to the planes OE_2E_3, OE_3E_1, OE_1E_2 and having lengths equal to the reciprocals of the distances of E_1, E_2, E_3 from these planes, respectively. Prove this theorem.

14. Prove that a necessary and sufficient condition that four points A, B, C, D be coplanar is that

$$[\mathbf{dbc}] + [\mathbf{adc}] + [\mathbf{abd}] - [\mathbf{abc}] = 0.$$

15. Show that the shortest distance from the point A to the line BC is $|\mathbf{a} \times \mathbf{b} + \mathbf{b} \times \mathbf{c} + \mathbf{c} \times \mathbf{a}|/|\mathbf{b} - \mathbf{c}|$.

16. If the vectors $\mathbf{e}_1$, $\mathbf{e}_2$, $\mathbf{e}_3$; $\mathbf{e}^1$, $\mathbf{e}^2$, $\mathbf{e}^3$ form reciprocal sets, show that the vectors $\mathbf{e}_2 \times \mathbf{e}_3$, $\mathbf{e}_3 \times \mathbf{e}_1$, $\mathbf{e}_1 \times \mathbf{e}_2$; $\mathbf{e}^2 \times \mathbf{e}^3$, $\mathbf{e}^3 \times \mathbf{e}^1$, $\mathbf{e}^1 \times \mathbf{e}^2$ do likewise.

17. Prove the formulas:

(*a*) $[\mathbf{a} \times \mathbf{b}, \mathbf{b} \times \mathbf{c}, \mathbf{c} \times \mathbf{a}] = [\mathbf{abc}]^2$,

(*b*) $(\mathbf{b} \times \mathbf{c}) \cdot (\mathbf{a} \times \mathbf{d}) + (\mathbf{c} \times \mathbf{a}) \cdot (\mathbf{b} \times \mathbf{d}) + (\mathbf{a} \times \mathbf{b}) \cdot (\mathbf{c} \times \mathbf{d}) = 0$,

(*c*) $(\mathbf{b} \times \mathbf{c}) \times (\mathbf{a} \times \mathbf{d}) + (\mathbf{c} \times \mathbf{a}) \times (\mathbf{b} \times \mathbf{d}) + (\mathbf{a} \times \mathbf{b}) \times (\mathbf{c} \times \mathbf{d}) = -2[\mathbf{abc}]\mathbf{d}$,

(*d*) $(\mathbf{d} - \mathbf{a}) \cdot (\mathbf{b} - \mathbf{c}) + (\mathbf{d} - \mathbf{b}) \cdot (\mathbf{c} - \mathbf{a}) + (\mathbf{d} - \mathbf{c}) \cdot (\mathbf{a} - \mathbf{b}) = 0$,

(*e*) $(\mathbf{a} - \mathbf{d}) \times (\mathbf{b} - \mathbf{c}) + (\mathbf{b} - \mathbf{d}) \times (\mathbf{c} - \mathbf{a}) + (\mathbf{c} - \mathbf{d}) \times (\mathbf{a} - \mathbf{b}) = 2(\mathbf{a} \times \mathbf{b} + \mathbf{b} \times \mathbf{c} + \mathbf{c} \times \mathbf{a})$

18. Find the shortest distance between the straight lines AB and CD when

(a) $A(-2, 4, 3)$, $B(2, -8, 0)$; $C(1, -3, 5)$, $D(4, 1, -7)$,

(b) $A(2, 3, 1)$, $B(0, -1, 2)$; $C(1, 2, 5)$, $D(-3, 1, 0)$.

19. Show that the lines AB, CD are coplanar, and find the point P in which they meet:

$$A(-2, -3, 4),\quad B(2, 3, 0);\qquad C(-2, 3, 2),\quad D(2, 0, 1).$$

[$\overrightarrow{CD}$ is parallel to $\overrightarrow{CP} = \overrightarrow{AP} - \overrightarrow{AC} = \lambda\overrightarrow{AB} - \overrightarrow{AC}$; find λ and then P from $\overrightarrow{OP} = \overrightarrow{OA} + \lambda\overrightarrow{AB}$.]

20. If the points P, Q, R divide the sides BC, CA, AB of the triangle ABC in the same ratio, show that A, B, C and P, Q, R have the same mean center.

21. If the vectors $\overrightarrow{OA}$, $\overrightarrow{OB}$, $\overrightarrow{OC}$ lie in a plane and are equal in length, prove that

$$\overrightarrow{OA} + \overrightarrow{OB} + \overrightarrow{OC} = \overrightarrow{OH},$$

where H is the orthocenter of the triangle ABC. [§ 15, ex. 4.]

22. The *power* of a point P with respect to a sphere of center C and radius r is defined as $(CP)^2 - r^2$. Prove that the power of P with respect to a sphere having AB as diameter is $\overrightarrow{PA} \cdot \overrightarrow{PB}$.

23. Prove that the sum of the n^2 powers of n given points, $P_1, P_2, \cdots, P_n$, with respect to the n spheres having for diameters the n segments joining the given points to a variable point P in space, is constant. [*Am. Math. Monthly*, vol. 51, p. 96.]

24. The lines $(\mathbf{a}, \mathbf{a}_O)$, $(\mathbf{b}, \mathbf{b}_O)$ are coplanar; then $\mathbf{a} \cdot \mathbf{b}_O + \mathbf{b} \cdot \mathbf{a}_O = 0$ (26.14). Show that they determine the plane $(\mathbf{a} \times \mathbf{b}, \mathbf{a}_O \cdot \mathbf{b})$ if $\mathbf{a} \times \mathbf{b} \neq 0$; and the point $(\mathbf{a}_O \cdot \mathbf{b}, \mathbf{a}_O \times \mathbf{b}_O)$ if $\mathbf{a}_O \cdot \mathbf{b} \neq 0$.

Solve Problem 19 using these results.

25. Show that the three planes $(\mathbf{a}, \alpha_O)$, $(\mathbf{b}, \beta_O)$, $(\mathbf{c}, \gamma_O)$ meet in the point,

$$([\mathbf{abc}],\quad \alpha_O\, \mathbf{b} \times \mathbf{c} + \beta_O\, \mathbf{c} \times \mathbf{a} + \gamma_O\, \mathbf{a} \times \mathbf{b}),$$

provided $[\mathbf{abc}] \neq 0$.

26. Show that the three points $(\alpha, \mathbf{a}_O)$, $(\beta, \mathbf{b}_O)$, $(\gamma, \mathbf{c}_O)$ determine the plane,

$$(\alpha\, \mathbf{b}_O \times \mathbf{c}_O + \beta\, \mathbf{c}_O \times \mathbf{a}_O + \gamma\, \mathbf{a}_O \times \mathbf{b}_O,\quad [\mathbf{a}_O\mathbf{b}_O\mathbf{c}_O]),$$

provided $\alpha\, \mathbf{b}_O \times \mathbf{c}_O + \beta\, \mathbf{c}_O \times \mathbf{a}_O + \gamma\, \mathbf{a}_O \times \mathbf{b}_O \neq 0$.

27. The points P, Q, R divide the sides BC, CA, AB of the triangle ABC in the ratio of $1/2$. The pairs of lines (AP, BQ), (BQ, CR), (CR, AP) intersect at X, Y, Z, respectively. Show that the area of the triangle XYZ is $1/7$ of the area of ABC. [Twice the vector area of XYZ is $\mathbf{y} \times \mathbf{z} + \mathbf{z} \times \mathbf{x} + \mathbf{x} \times \mathbf{y}$.]

When P, Q, R divide the sides in the ratio $t/1$, the area of XYZ is $(1 - t)^2/(1 + t + t^2)$ times the area of ABC.

28. Using (26.16), prove that

(*a*) If four points $(\alpha, \mathbf{a}_O)$, $(\beta, \mathbf{b}_O)$, $(\gamma, \mathbf{c}_O)$, $(\delta, \mathbf{d}_O)$ lie on a plane, the vectors $\mathbf{a}_O/\alpha, \cdots, \mathbf{d}_O/\delta$ are connected by a linear relation in which the sum of the coefficients is zero;

(*b*) If four planes $(\mathbf{a}, \alpha_O)$, $(\mathbf{b}, \beta_O)$, $(\mathbf{c}, \gamma_O)$, $(\mathbf{d}, \delta_O)$ pass through a point, the vectors $\mathbf{a}/\alpha_O, \cdots, \mathbf{d}/\delta_O$ are connected by a linear relation in which the sum of the coefficients is zero, provided $\alpha_O, \beta_O, \gamma_O, \delta_O \neq 0$.

29. If a sphere S with a fixed center cuts two concentric spheres S_1, S_2, prove that the distance between the planes of intersection SS_1 and SS_2 is independent of the radius of S. Extend this theorem to cover the case when S fails to cut one or both of the concentric spheres. [Consider the radical planes SS_1, SS_2.]

30. Prove that the radical planes of the three spheres $(\mathbf{r} - \mathbf{c}_i)^2 = \rho_i^2$ $(i = 1, 2, 3)$ meet in the line

$$\{\mathbf{c}_1 \times \mathbf{c}_2 + \mathbf{c}_2 \times \mathbf{c}_3 + \mathbf{c}_3 \times \mathbf{c}_1, \quad \tfrac{1}{2}\mathbf{c}_1(\mathbf{c}_2^2 - \mathbf{c}_3^2 - \rho_2^2 + \rho_3^2) + \text{cycl.}\},$$

their *radical axis*. Consider the case when $\mathbf{c}_1 \times \mathbf{c}_2 + \mathbf{c}_2 \times \mathbf{c}_3 + \mathbf{c}_3 \times \mathbf{c}_1 = 0$.

31. A plane system of forces $\mathbf{F}_i$ acting through the points $\mathbf{r}_i$ is equivalent to a single force $\mathbf{F} = \Sigma\mathbf{F}_i$. When all the forces $\mathbf{F}_i$ are revolved about these points through an angle θ, show that their resultant $\mathbf{F}$ also revolves through θ about the point $\bar{\mathbf{r}}$ (the *astatic center*) given by

$$\bar{\mathbf{r}} = \frac{\mathbf{F} \times \mathbf{M} - (\Sigma\mathbf{r}_i \cdot \mathbf{F}_i)\mathbf{F}}{\mathbf{F}^2}, \quad \text{where} \quad \mathbf{M} = \Sigma\mathbf{r}_i \times \mathbf{F}_i.$$

When the forces are parallel $(\mathbf{F}_i = \lambda_i\mathbf{e})$, show that the astatic center is the centroid of the weighted points $\lambda_i\mathbf{r}_i$.

32. P^* is the centroid of a set of n weighted points m_iP_i for which $\Sigma m_i = m$. Prove the *Theorems of Lagrange:*

(*a*) $$\Sigma m_i(OP_i)^2 = m(OP^*)^2 + \Sigma m_i(P^*P_i)^2;$$

(*b*) $$\Sigma m_im_j(P_iP_j)^2 = m\Sigma m_i(P^*P_i)^2,$$

where ij ranges over $\frac{1}{2}n(n-1)$ *combinations.*

$$[\overrightarrow{OP_i} = \overrightarrow{OP^*} + \overrightarrow{P^*P_i}; \qquad \overrightarrow{P_iP_j} = \overrightarrow{P^*P_j} - \overrightarrow{P^*P_i}.]$$

33. Deduce the following results from Prob. 32.

(*a*) If $ABCD$ are the vertices of a square in circuital order, $(OA)^2 + (OC)^2 = (OB)^2 + (OD)^2$ for any O. Generalize.

(*b*) If r is the radius of a sphere circumscribed about a regular tetrahedron of side a, $r^2 = 3a^2/8$.

(*c*) The mean square of the mutual distances of all the points within a sphere of radius r is $6r^2/5$.

34. If P^* and Q^* are the mean centers of p points P_i and q points Q_j, respectively, prove that

$$\text{Mean } (P_iQ_j)^2 = \text{Mean } (P^*P_i)^2 + \text{Mean } (Q^*Q_j)^2 + (P^*Q^*)^2.$$

CHAPTER II

MOTOR ALGEBRA

29. Dual Vectors. The vector $\mathbf{f}$, bound to the line whose equation, referred to the origin O, is

$$(1) \qquad \mathbf{r} \times \mathbf{f} = \mathbf{f}_O,$$

is completely determined by the two vectors $\mathbf{f}$, $\mathbf{f}_O$, its *Plücker coordinates.* These obviously satisfy the relation,

$$(2) \qquad \mathbf{f} \cdot \mathbf{f}_O = 0.$$

The vector $\mathbf{f}$ does not depend upon O; but $\mathbf{f}_O$, the *moment* of $\mathbf{f}$ about O, becomes

$$(3) \qquad \mathbf{f}_P = \mathbf{f}_O + \overrightarrow{PO} \times \mathbf{f},$$

when the origin is shifted from O to P.

We now amalgamate $\mathbf{f}$ and $\mathbf{f}_O$ into a *dual vector.*

$$(4) \qquad \mathbf{F} = \mathbf{f} + \epsilon \mathbf{f}_O$$

where ϵ is an algebraic unit having the property $\epsilon^2 = 0$. If $|\mathbf{f}| = 1$, $\mathbf{F}$ is called a *unit* line vector. The unit line vectors,

$$\mathbf{A} = \mathbf{a} + \epsilon \mathbf{a}_O, \qquad (\mathbf{a} \cdot \mathbf{a} = 1, \mathbf{a} \cdot \mathbf{a}_O = 0)$$

stand in one-to-one correspondence with the ∞^4 lines of space. A line vector $\mathbf{F}$ of length λ always may be written $\mathbf{F} = \lambda \mathbf{A}$, where $\mathbf{A}$ is a unit line vector. Unit line vectors depend upon *four* independent scalars, general line vectors upon *five.*

Finally, for applications in mechanics, we consider dual vectors $\mathbf{F}$ without the restriction (2). The vectors $\mathbf{f}$, $\mathbf{f}_O$ then involve *six* independent scalars. The dual vector (or the entity it represents) then is called a *motor*, provided its *resultant vector* $\mathbf{f}$ is independent of the choice of O, while its *moment vector* $\mathbf{f}_O$ changes in accordance with (3) when the origin is shifted to P.

Line vectors are thus special motors for which $\mathbf{f} \cdot \mathbf{f}_O = 0$; for unit line vectors also $\mathbf{f} \cdot \mathbf{f} = 1$.

30. Dual Numbers. In analogy with the complex numbers $x + ix'$, W. K. Clifford introduced *dual numbers* $x + \epsilon x'$, in which x, x' are real and ϵ is a unit with the property $\epsilon^2 = 0$.

In $x + \epsilon x'$, x is called the real part and x' the dual part. We write

$$x + \epsilon x' = y + \epsilon y' \quad \text{when} \quad x = y,\ x' = y';$$

$$x + \epsilon x' = 0 \quad \text{when} \quad x = 0,\ x' = 0.$$

Addition and multiplication of dual numbers are defined by the equations:

$$(1) \qquad (x + \epsilon x') + (y + \epsilon y') = x + y + \epsilon(x' + y'),$$

$$(2) \qquad (x + \epsilon x')(y + \epsilon y') = xy + \epsilon(xy' + x'y).$$

Observe that (2) may be obtained by distributing the product on the left and putting $\epsilon^2 = 0$. From these definitions we see that addition and multiplication are commutative and associative and that multiplication is distributive with respect to addition. In fact, the formal operations are precisely those of ordinary algebra followed by setting $\epsilon^2 = \epsilon^3 = \cdots = 0$.

The *negative* of $x + \epsilon x'$ is defined as $-x - \epsilon x'$.

If $A = a + \epsilon a'$, $B = b + \epsilon b'$, the difference $X = A - B$ and quotient $Y = A/B$ satisfy, by definition, the equations $B + X = A$, $BY = A$. We find that

$$(3) \qquad A - B = a - b + \epsilon(a' - b');$$

and that $BY = A$ has the unique solution,

$$(4) \qquad \frac{A}{B} = \frac{a}{b} + \epsilon \frac{a'b - ab'}{b^2},$$

when and only when $b \neq 0$. *Division by a pure dual number* $\epsilon b'$ *is not defined.* The quotient (4) may be remembered by means of the device $(a + \epsilon a')(b - \epsilon b')/(b + \epsilon b')(b - \epsilon b')$ used in complex algebra.

A dual number $a + \epsilon a'$ in which $a \neq 0$ is said to be *proper;* the product and quotient of proper dual numbers are also proper.

If the dual product (2) is zero, there are three alternatives:

$$x = x' = 0; \qquad y = y' = 0; \qquad x = y = 0.$$

The last shows that a dual product can vanish when neither factor is zero; for any two pure dual numbers have zero as their product.

If the function $f(x)$ has the derivative $f'(x)$, we define its value for the dual argument $X = x + \epsilon x'$ by writing down its formal Taylor expansion and setting $\epsilon^2 = \epsilon^3 = \cdots = 0$; thus

$$(5) \qquad f(x + \epsilon x') = f(x) + \epsilon x' f'(x).$$

In particular,

$$(6) \qquad \sin (x + \epsilon x') = \sin x + \epsilon x' \cos x,$$

$$(7) \qquad \cos (x + \epsilon x') = \cos x - \epsilon x' \sin x.$$

When $x = 0$, we have $\sin \epsilon x' = \epsilon x'$, $\cos \epsilon x' = 1$; consequently (6) and (7) have the form of the usual addition theorems of the sine and cosine. Note also that $\sin^2 X + \cos^2 X = 1$.

31. Motors. We now can characterize a *motor* as a *dual multiple of a unit line vector.* Thus, on multiplying the unit line vector $\mathbf{A} = \mathbf{a} + \epsilon \mathbf{a}_O$ ($\mathbf{a} \cdot \mathbf{a} = 1$, $\mathbf{a} \cdot \mathbf{a}_O = 0$) by the dual number $\lambda + \epsilon\lambda'$, we obtain the dual vector,

$$(1) \qquad \mathbf{M} = \mathbf{m} + \epsilon \mathbf{m}_O = (\lambda + \epsilon\lambda')(\mathbf{a} + \epsilon \mathbf{a}_O).$$

On equating the real and dual parts of both members, we obtain

$$(2) \qquad \mathbf{m} = \lambda \mathbf{a}, \qquad \mathbf{m}_O = \lambda \mathbf{a}_O + \lambda' \mathbf{a}.$$

To show that $\mathbf{M}$ is a motor we need only verify that $\mathbf{m}_O$ transforms in accordance with (29.3):

$$(3) \qquad \mathbf{m}_P = \mathbf{m}_O + \overrightarrow{PO} \times \mathbf{m}.$$

In view of (2), this result follows from

$$\mathbf{m}_P = \lambda \mathbf{a}_P + \lambda' \mathbf{a} = \lambda(\mathbf{a}_O + \overrightarrow{PO} \times \mathbf{a}) + \lambda' \mathbf{a}$$
$$= (\lambda \mathbf{a}_O + \lambda' \mathbf{a}) + \overrightarrow{PO} \times (\lambda \mathbf{a}).$$

From (3), $\mathbf{m} \cdot \mathbf{m}_P = \mathbf{m} \cdot \mathbf{m}_O$; hence the scalars $\mathbf{m} \cdot \mathbf{m}$ and $\mathbf{m} \cdot \mathbf{m}_O$ are invariants in the sense that they are not altered by a shift of origin. From (2),

$$(4) \qquad \mathbf{m} \cdot \mathbf{m} = \lambda^2, \qquad \mathbf{m} \cdot \mathbf{m}_O = \lambda\lambda';$$

and, when $\mathbf{m} \neq 0$, we call the invariant,

$$(5) \qquad \mu = \frac{\lambda'}{\lambda} = \frac{\mathbf{m} \cdot \mathbf{m}_O}{\mathbf{m} \cdot \mathbf{m}}$$

the *pitch* of the motor. Choosing $\lambda > 0$, we have the unique solution,

$$\lambda = |\mathbf{m}|, \qquad \lambda' = \mu|\mathbf{m}|,$$

$$\lambda\mathbf{a} = \mathbf{m}, \qquad \lambda\mathbf{a}_O = \mathbf{m}_O - \mu\mathbf{m}.$$

When $\mathbf{m} \neq 0$, $\mathbf{M}$ has the *dual length*,

$$\lambda + \epsilon\lambda' = |\mathbf{m}|(1 + \epsilon\mu) \tag{6}$$

and its *axis* is along the line vector,

$$|\mathbf{m}|\mathbf{A} = \mathbf{m} + \epsilon(\mathbf{m}_O - \mu\mathbf{m}). \tag{7}$$

The equation of the axis is therefore

$$\mathbf{r} \times \mathbf{m} = \mathbf{m}_O - \mu\mathbf{m} = \frac{(\mathbf{m} \times \mathbf{m}_O) \times \mathbf{m}}{\mathbf{m} \cdot \mathbf{m}}; \tag{8}$$

this shows that the axis passes through the point Q given by

$$\mathbf{r} = \overrightarrow{OQ} = \frac{\mathbf{m} \times \mathbf{m}_O}{\mathbf{m} \cdot \mathbf{m}}; \tag{9}$$

and, from (3),

$$\mathbf{m}_Q = \mathbf{m}_O - \frac{(\mathbf{m} \times \mathbf{m}_O) \times \mathbf{m}}{\mathbf{m} \cdot \mathbf{m}} = \mu\mathbf{m}. \tag{10}$$

At all points on its axis $\mathbf{M}$ has the same moment $\mu\mathbf{m}$, a fact also apparent from (7). *Only for points P on the axis is the moment* $\mathbf{m}_P$ *parallel to* $\mathbf{m}$.

Motors for which $\mathbf{m} \neq 0$ $(\lambda \neq 0)$ are called *proper*. Proper motors are *screws* if $\mathbf{m} \cdot \mathbf{m}_O \neq 0$ $(\lambda, \lambda' \neq 0)$; *line vectors* if $\mathbf{m} \cdot \mathbf{m}_O = 0$ $(\lambda \neq 0, \lambda' = 0)$. Only proper motors have a definite axis.

If $\mathbf{m} = 0$, $\mathbf{m}_O \neq 0$ $(\lambda = 0, \lambda' \neq 0)$, $\mathbf{M} = \epsilon\mathbf{m}_O$ is pure dual; then $\mathbf{m}_O$ is not altered by a change of origin. In this case $\mathbf{M}$ is called a *couple* of moment $\mathbf{m}_O$. A couple may be regarded as a screw of infinite pitch with an axis of given direction but arbitrary position in space.

Finally if $\mathbf{m} = 0$, $\mathbf{m}_O = 0$, the motor $\mathbf{M} = 0$, then $\lambda = \lambda' = 0$ from (2). Hence $\mathbf{M} = 0$ only when its dual length $\lambda + \epsilon\lambda' = 0$.

Our classification of motors is therefore as follows:

Screw	$(\mathbf{m} \neq 0,\ \mathbf{m} \cdot \mathbf{m}_O \neq 0)$:	$\lambda \neq 0,\ \lambda' \neq 0;$	Proper
Line vector	$(\mathbf{m} \neq 0,\ \mathbf{m} \cdot \mathbf{m}_O = 0)$:	$\lambda \neq 0,\ \lambda' = 0;$	Proper
Couple	$(\mathbf{m} = 0,\ \mathbf{m}_O \neq 0)$:	$\lambda = 0,\ \lambda' \neq 0;$	
Zero	$(\mathbf{m} = 0,\ \mathbf{m}_O = 0)$:	$\lambda = 0,\ \lambda' = 0.$	

If two proper motors **M**, **N** are connected by a linear relation with proper coefficients,

$$(\alpha + \epsilon\alpha')\mathbf{M} + (\beta + \epsilon\beta')\mathbf{N} = 0, \qquad \alpha\beta \neq 0,$$

either may be expressed as a dual multiple of the other; hence both **M** and **N** are multiples of the same unit line vector and are therefore coaxial. Conversely, if **M** and **N** are coaxial, they satisfy a linear relation with proper coefficients.

32. Motor Sum. The sum of the motors $\mathbf{M} = \mathbf{m} + \epsilon\mathbf{m}_O$, $\mathbf{N} = \mathbf{n} + \epsilon\mathbf{n}_O$ is defined as the motor,

$$(1) \qquad \mathbf{M} + \mathbf{N} = \mathbf{m} + \mathbf{n} + \epsilon(\mathbf{m}_O + \mathbf{n}_O).$$

That $\mathbf{M} + \mathbf{N}$ is a motor follows from (29.3):

$$\mathbf{m}_P + \mathbf{n}_P = \mathbf{m}_O + \mathbf{n}_O + \overrightarrow{PO} \times (\mathbf{m} + \mathbf{n}).$$

THEOREM 1. *The sum of two line vectors is a line vector only when their axes are coplanar and their vectors have a non-zero sum.*

Proof. If **M** and **N** are line vectors, $\mathbf{M} + \mathbf{N}$ is a line vector when and only when

$$\mathbf{m} + \mathbf{n} \neq 0, \quad \text{and} \quad (\mathbf{m} + \mathbf{n}) \cdot (\mathbf{m}_O + \mathbf{n}_O) = 0.$$

Since $\mathbf{m} \cdot \mathbf{m}_O = 0$, $\mathbf{n} \cdot \mathbf{n}_O = 0$, the latter condition reduces to

$$(2) \qquad \mathbf{m} \cdot \mathbf{n}_O + \mathbf{n} \cdot \mathbf{m}_O = 0,$$

which is precisely the condition (26.14) that the axes be coplanar.

THEOREM 2. *If two line vectors intersect, their sum is a line vector through the point of intersection.*

Proof. If $\mathbf{r}_1$ is the position vector of the point P_1 in which the line vectors **M** and **N** intersect, we may take $\mathbf{m}_O = \mathbf{r}_1 \times \mathbf{m}$, $\mathbf{n}_O = \mathbf{r}_1 \times \mathbf{n}$. Since the axis of $\mathbf{M} + \mathbf{N}$ has the equation,

$$\mathbf{r} \times (\mathbf{m} + \mathbf{n}) = \mathbf{m}_O + \mathbf{n}_O = \mathbf{r}_1 \times (\mathbf{m} + \mathbf{n}),$$

it passes through the point P_1.

THEOREM 3. *If the line vectors* **M** *and* **N** *are parallel and* $\mathbf{n} = \lambda\mathbf{m}$ $(\lambda \neq -1)$, *the axis of their sum divides any segment from* **M** *to* **N** *in the ratio* $\lambda/1$.

Proof. If P_1, P_2 are points on the axes of **M** and **N**, we may take $\mathbf{m}_O = \mathbf{r}_1 \times \mathbf{m}$, $\mathbf{n}_O = \mathbf{r}_2 \times \mathbf{n}$. The line vector,

$$\mathbf{M} + \mathbf{N} = (1 + \lambda)\mathbf{m} + \epsilon(\mathbf{r}_1 + \lambda\mathbf{r}_2) \times \mathbf{m};$$

and the equation of its axis,

$$\mathbf{r} \times (1 + \lambda)\mathbf{m} = (\mathbf{r}_1 + \lambda\mathbf{r}_2) \times \mathbf{m},$$

is satisfied by the point $\mathbf{r} = (\mathbf{r}_1 + \lambda\mathbf{r}_2)/(1 + \lambda)$ which divides P_1P_2 in the ratio $\lambda/1$. The division is internal when **m** and **n** have the same direction ($\lambda > 0$), external when they have opposite directions ($\lambda < 0$).

THEOREM 4. *The sum of two couples, if not zero, is another couple.*

Proof. If **M** and **N** are pure dual, $\mathbf{M} + \mathbf{N}$, if not zero, is also pure dual.

THEOREM 5. *If* **M** *and* **N** *are line vectors such that* $\mathbf{n} = -\mathbf{m}$ *and* P_1, P_2 *are points on their respective axes,* $\mathbf{M} + \mathbf{N}$ *is a couple of moment* $(\mathbf{r}_1 - \mathbf{r}_2) \times \mathbf{m}$.

Proof. Since $\mathbf{M} = \mathbf{m} + \epsilon\mathbf{r}_1 \times \mathbf{m}$, $\mathbf{N} = -\mathbf{m} + \epsilon\mathbf{r}_2 \times (-\mathbf{m})$,

$$\mathbf{M} + \mathbf{N} = \epsilon(\mathbf{r}_1 - \mathbf{r}_2) \times \mathbf{m}.$$

33. Scalar Product. The scalar product of the motors $\mathbf{M} = \mathbf{m} + \epsilon\mathbf{m}_O$, $\mathbf{N} = \mathbf{n} + \epsilon\mathbf{n}_O$ is defined as the result of distributing the product, $(\mathbf{m} + \epsilon\mathbf{m}_O) \cdot (\mathbf{n} + \epsilon\mathbf{n}_O)$, namely,

$$\mathbf{M} \cdot \mathbf{N} = \mathbf{m} \cdot \mathbf{n} + \epsilon(\mathbf{m} \cdot \mathbf{n}_O + \mathbf{n} \cdot \mathbf{m}_O). \tag{1}$$

This dual number is independent of the choice of origin; for on computing $\mathbf{m}_P$, $\mathbf{n}_P$ from (29.3), we find that

$$\mathbf{m} \cdot \mathbf{n}_P + \mathbf{n} \cdot \mathbf{m}_P = \mathbf{m} \cdot \mathbf{n}_O + \mathbf{n} \cdot \mathbf{m}_O.\text{*} \tag{2}$$

The definition (1) shows that

$$\mathbf{M} \cdot \mathbf{N} = \mathbf{N} \cdot \mathbf{M}, \qquad \mathbf{L} \cdot (\mathbf{M} + \mathbf{N}) = \mathbf{L} \cdot \mathbf{M} + \mathbf{L} \cdot \mathbf{N}. \tag{3}$$

* The definition (1) for $\mathbf{M} \cdot \mathbf{N}$ differs from that given by R. von Mises ("Motorrechnung, ein neues Hilfsmittel der Mechanik," *Z. angew. Math. Mech.*, vol. 4, 1924, p. 163) who defines $\mathbf{M} \cdot \mathbf{N}$ as the invariant real scalar (2). The definition (2) is suggested by applications in mechanics; its consequences are (*a*) the familiar rules of vector algebra do not all carry over into motor algebra, and (*b*) the elegant generalization of the scalar product of unit vectors, given in (5), is lost.

In order to compute the scalar product of two *unit* line vectors $\mathbf{A} = \mathbf{a} + \epsilon\mathbf{a}_O$, $\mathbf{B} = \mathbf{b} + \epsilon\mathbf{b}_O$, we shall suppose first that they are not parallel and write

$$\mathbf{a} \cdot \mathbf{b} = \cos\varphi, \qquad \mathbf{a} \times \mathbf{b} = \mathbf{e} \sin\varphi,$$

where $\mathbf{e}$ is a unit vector. There is, then, a definite unit line vector $\mathbf{E}$ in the direction of $\mathbf{e}$ which cuts both $\mathbf{A}$ and $\mathbf{B}$ at right angles. If we choose the origin at the point of intersection of $\mathbf{A}$ and $\mathbf{E}$, and let φ' denote the perpendicular distance between $\mathbf{A}$ and $\mathbf{B}$, taken positive if $\mathbf{a} \times \mathbf{b}$ points from $\mathbf{A}$ to $\mathbf{B}$, we may write

$$\mathbf{A} = \mathbf{a}, \qquad \mathbf{B} = \mathbf{b} + \epsilon\varphi' \, \mathbf{e} \times \mathbf{b}, \qquad \mathbf{E} = \mathbf{e}.$$

Then we have

$$\begin{aligned} \mathbf{A} \cdot \mathbf{B} &= \mathbf{a} \cdot \mathbf{b} + \epsilon\varphi' \, \mathbf{a} \cdot \mathbf{e} \times \mathbf{b} \\ &= \mathbf{a} \cdot \mathbf{b} - \epsilon\varphi' \, \mathbf{e} \cdot \mathbf{a} \times \mathbf{b} \\ &= \cos\varphi - \epsilon\varphi' \sin\varphi. \end{aligned}$$

If we now define the *dual angle* † between $\mathbf{A}$ and $\mathbf{B}$ as

$$\Phi = \varphi + \epsilon\varphi', \tag{4}$$

a reference to (30.7) shows that

$$\mathbf{A} \cdot \mathbf{B} = \cos\Phi, \tag{5}$$

in complete analogy with $\mathbf{a} \cdot \mathbf{b} = \cos\varphi$.

If $\mathbf{A}$ and $\mathbf{B}$ are parallel, let $\mathbf{e}$ be any unit vector perpendicular to both $\mathbf{A}$ and $\mathbf{B}$. If the origin is chosen on $\mathbf{A}$, the foregoing computation still applies; since $\sin\varphi = 0$, we find $\mathbf{A} \cdot \mathbf{B} = \cos\varphi = \pm 1$. If $\mathbf{A}$ and $\mathbf{B}$ are collinear, we choose the origin on their common line; then $\mathbf{A} = \mathbf{a}$, $\mathbf{B} = \mathbf{b}$, and $\mathbf{A} \cdot \mathbf{B} = \mathbf{a} \cdot \mathbf{b} = \pm 1$ as before. Formula (5) covers these cases; in both $\cos\Phi = \cos\varphi = \pm 1$ according as $\mathbf{a}$ and $\mathbf{b}$ have the same or opposite directions.

We note that

$$\mathbf{A} \cdot \mathbf{B} = \cos\Phi = \cos\varphi - \epsilon\varphi' \sin\varphi = 0,$$

when and only when

$$\cos\varphi = 0, \quad \varphi' \sin\varphi = 0, \quad \text{or} \quad \varphi = \frac{\pi}{2}, \quad \varphi' = 0;$$

that is, when the line vectors cut at right angles.

† The concept of dual angle is due to Study, *Geometrie der Dynamen*, Leipzig, 1903, p. 205.

THEOREM. *In order that the axes of two proper motors cut at right angles, it is necessary and sufficient that their scalar product vanish.*

Proof. We express the motors in the form,

$$\mathbf{M} = (\alpha + \epsilon\alpha')\mathbf{A}, \qquad \mathbf{N} = (\beta + \epsilon\beta')\mathbf{B},$$

where **A**, **B** are unit line vectors along their axes and $\alpha\beta \neq 0$. Then

$$\mathbf{M} \cdot \mathbf{N} = [\alpha\beta + \epsilon(\alpha\beta' + \alpha'\beta)]\mathbf{A} \cdot \mathbf{B},$$

and, since the first factor is neither zero nor pure dual, $\mathbf{M} \cdot \mathbf{N} = 0$ implies $\mathbf{A} \cdot \mathbf{B} = 0$, and conversely.

We denote the dual part of $\mathbf{M} \cdot \mathbf{N}$ by

$$(6) \qquad \mathbf{M} \circ \mathbf{N} = \mathbf{m} \cdot \mathbf{n}_O + \mathbf{n} \cdot \mathbf{m}_O.$$

As previously noted, this is von Mises' definition of the scalar product. In common with $\mathbf{M} \cdot \mathbf{N}$, the product $\mathbf{M} \circ \mathbf{N}$ is commutative and distributive with respect to addition. The preceding calculation shows that

$$\mathbf{A} \circ \mathbf{B} = -\varphi' \sin \varphi;$$

hence $\mathbf{A} \circ \mathbf{B} = 0$ implies $\varphi' = 0$ or $\sin \varphi = 0$, that is, the line vectors intersect or are parallel, and conversely. The same criterion applies to line vectors $\alpha\mathbf{A}$, $\beta\mathbf{B}$ of arbitrary length. Consequently *a necessary and sufficient condition that two line vectors* **M**, **N** *be coplanar is that* $\mathbf{M} \circ \mathbf{N} = 0$. This is the condition (26.14).

Writing the motor $\mathbf{M} = (\alpha + \epsilon\alpha')\mathbf{A}$ gives

$$\mathbf{M} \cdot \mathbf{M} = (\alpha + \epsilon\alpha')^2 = \alpha^2 + 2\epsilon\alpha\alpha'$$

as the square of its dual length; and $\mathbf{M} \circ \mathbf{M} = 2\alpha\alpha'$. Evidently $\mathbf{M} \circ \mathbf{M} = 0$ implies that **M** is a line vector ($\alpha' = 0$) or a couple ($\alpha = 0$).

34. Motor Product. The motor product of the motors $\mathbf{M} = \mathbf{m} + \epsilon\mathbf{m}_O$, $\mathbf{N} = \mathbf{n} + \epsilon\mathbf{n}_O$ is defined as the dual vector obtained by distributing the product $(\mathbf{m} + \epsilon\mathbf{m}_O) \times (\mathbf{n} + \epsilon\mathbf{n}_O)$, namely,

$$(1) \qquad \mathbf{M} \times \mathbf{N} = \mathbf{m} \times \mathbf{n} + \epsilon(\mathbf{m} \times \mathbf{n}_O + \mathbf{m}_O \times \mathbf{n}).$$

When the origin is shifted to P, the dual part of $\mathbf{M} \times \mathbf{N}$ becomes

$$\begin{aligned}\mathbf{m} \times \mathbf{n}_P + \mathbf{m}_P \times \mathbf{n} &= \mathbf{m} \times (\mathbf{n}_O + \overrightarrow{PO} \times \mathbf{n}) + (\mathbf{m}_O + \overrightarrow{PO} \times \mathbf{m}) \times \mathbf{n} \\ &= \mathbf{m} \times \mathbf{n}_O + \mathbf{m}_O \times \mathbf{n} + \overrightarrow{PO} \times (\mathbf{m} \times \mathbf{n}),\end{aligned}$$

in view of the identity,

$$\overrightarrow{PO} \times (\mathbf{m} \times \mathbf{n}) + \mathbf{m} \times (\mathbf{n} \times \overrightarrow{PO}) + \mathbf{n} \times (\overrightarrow{PO} \times \mathbf{m}) = 0.$$

This transformation conforms with (29.3) and shows that $\mathbf{M} \times \mathbf{N}$ is a motor.

The definition (1) shows that

$$\mathbf{M} \times \mathbf{N} = -\mathbf{N} \times \mathbf{M}, \qquad \mathbf{L} \times (\mathbf{M} + \mathbf{N}) = \mathbf{L} \times \mathbf{M} + \mathbf{L} \times \mathbf{N}. \tag{2}$$

In order to compute the motor product of two unit line vectors $\mathbf{A} = \mathbf{a} + \epsilon\mathbf{a}_O$, $\mathbf{B} = \mathbf{b} + \epsilon\mathbf{b}_O$, we shall suppose, first, that they are not parallel. With the same notation and choice of origin as in § 33, we have

$$\begin{aligned}\mathbf{A} \times \mathbf{B} &= \mathbf{a} \times \mathbf{b} + \epsilon\varphi' \, \mathbf{a} \times (\mathbf{e} \times \mathbf{b}) \\ &= \mathbf{e}(\sin\varphi + \epsilon\varphi' \cos\varphi)\end{aligned}$$

or, in terms of the dual angle $\Phi = \varphi + \epsilon\varphi'$,

$$\mathbf{A} \times \mathbf{B} = \mathbf{E} \sin\Phi \tag{3}$$

in complete analogy with $\mathbf{a} \times \mathbf{b} = \mathbf{e} \sin\varphi$.

When $\mathbf{A}$ and $\mathbf{B}$ are parallel, let $\mathbf{e}$ be any unit vector perpendicular to both $\mathbf{A}$ and $\mathbf{B}$. If the origin is chosen on $\mathbf{A}$, the foregoing computation still applies, and

$$\mathbf{A} \times \mathbf{B} = \epsilon\varphi' \cos\varphi \, \mathbf{e} = \pm\epsilon\varphi' \, \mathbf{E}$$

where $\mathbf{E} = \mathbf{e} + \epsilon\mathbf{e}_O$, an *arbitrary* line vector in the direction of $\mathbf{e}$. This conforms with the fact that $\mathbf{A} \times \mathbf{B}$, being pure dual (a couple), has its axis determined in direction but not as to position in space. When $\mathbf{A}$ and $\mathbf{B}$ are collinear, we choose an origin on their common line; then $\mathbf{A} = \mathbf{a}$, $\mathbf{B} = \mathbf{b}$ and $\mathbf{A} \times \mathbf{B} = \mathbf{a} \times \mathbf{b} = 0$ and $\mathbf{E}$ is entirely arbitrary. Formula (3) covers these cases in which $\sin\Phi = \pm\epsilon\varphi'$ and 0, respectively.

We note that

$$\mathbf{A} \times \mathbf{B} = \sin\Phi \, \mathbf{E} = (\sin\varphi + \epsilon\varphi' \cos\varphi) \, \mathbf{E} = 0,$$

(or $\sin\Phi = 0$) when and only when

$$\sin\varphi = 0, \quad \varphi' \cos\varphi = 0, \qquad \text{or } \varphi = 0, \pi, \quad \varphi' = 0,$$

that is, when the line vectors are collinear.

THEOREM. *In order that two proper motors be coaxial, it is necessary and sufficient that their motor product vanish.*

Proof. With the notation of § 33,

$$\mathbf{M} \times \mathbf{N} = [\alpha\beta + \epsilon(\alpha\beta' + \alpha'\beta)]\mathbf{A} \times \mathbf{B}, \qquad (\alpha\beta \neq 0).$$

Since the first factor is neither zero nor pure dual, $\mathbf{M} \times \mathbf{N} = 0$ implies $\sin \Phi = 0$, and conversely.

Finally let us consider the motor product,

$$\text{(4)} \qquad \mathbf{M} \times \mathbf{N} = [\alpha\beta + \epsilon(\alpha\beta' + \alpha'\beta)][\sin \varphi + \epsilon\varphi' \cos \varphi]\,\mathbf{E},$$

for any proper motors **M, N.** If their axes *coincide*, $\mathbf{M} \times \mathbf{N} = 0$. If their axes are *parallel*, $\sin \varphi = 0$, the coefficient of **E** is pure dual, and the axis of the couple $\mathbf{M} \times \mathbf{N}$ is any line parallel to the common normal to the axes of **M** and **N.** If their axes are *non-parallel*, $\sin \varphi \neq 0$, and $\mathbf{M} \times \mathbf{N}$ is a proper motor whose axis (along **E**) is the common normal to the axes of **M** and **N.**

35. Dual Triple Product. The dual triple product of three motors $\mathbf{L} = \mathbf{l} + \epsilon\mathbf{l}_O$, $\mathbf{M} = \mathbf{m} + \epsilon\mathbf{m}_O$, $\mathbf{N} = \mathbf{n} + \epsilon\mathbf{n}_O$ is defined as the dual number obtained by distributing the product $\mathbf{L} \times \mathbf{M} \cdot \mathbf{N}$; thus

$$\text{(1)} \quad \mathbf{L} \times \mathbf{M} \cdot \mathbf{N} = \mathbf{l} \times \mathbf{m} \cdot \mathbf{n} + \epsilon(\mathbf{l} \times \mathbf{m} \cdot \mathbf{n}_O + \mathbf{l} \times \mathbf{m}_O \cdot \mathbf{n} + \mathbf{l}_O \times \mathbf{m} \cdot \mathbf{n}).$$

This definition shows that

$$\text{(2)} \qquad \mathbf{L} \times \mathbf{M} \cdot \mathbf{N} = \mathbf{M} \times \mathbf{N} \cdot \mathbf{L} = \mathbf{N} \times \mathbf{L} \cdot \mathbf{M},$$

$$\text{(3)} \qquad \mathbf{L} \times \mathbf{M} \cdot \mathbf{N} = \mathbf{L} \cdot \mathbf{M} \times \mathbf{N},$$

$$\text{(4)} \qquad \mathbf{L} \times \mathbf{M} \cdot \mathbf{N} = -\mathbf{L} \times \mathbf{N} \cdot \mathbf{M}, \qquad \mathbf{L} \times \mathbf{M} \cdot \mathbf{M} = 0,$$

in exact analogy with vector products.

With von Mises' definition of a scalar product, the preceding triple product becomes

$$\text{(5)} \qquad \mathbf{L} \times \mathbf{M} \circ \mathbf{N} = (\mathbf{l} \times \mathbf{m}) \cdot \mathbf{n}_O + (\mathbf{l} \times \mathbf{m}_O + \mathbf{l}_O \times \mathbf{m}) \cdot \mathbf{n};$$

this is the dual part of $\mathbf{L} \times \mathbf{M} \cdot \mathbf{N}$. Equations (2), (3), (4) also apply to $\mathbf{L} \times \mathbf{M} \circ \mathbf{N}$.

We now propose to find under what conditions $\mathbf{L} \times \mathbf{M} \cdot \mathbf{N} = 0$ when **L, M, N** are proper motors. If the axes of **L, M, N** are all parallel, $\mathbf{l} \times \mathbf{m} = \mathbf{m} \times \mathbf{n} = \mathbf{n} \times \mathbf{l} = 0$ and $\mathbf{L} \times \mathbf{M} \cdot \mathbf{N} = 0$. If their axes are not all parallel, choose the notation so that $\mathbf{l} \times \mathbf{m} \neq 0$. Then $\mathbf{L} \times \mathbf{M}$ is a proper motor whose axis cuts the axes of **L** and **M** at right angles. The condition $\mathbf{L} \times \mathbf{M} \cdot \mathbf{N} = 0$ then holds when and only when the axis of **N** cuts the axis of $\mathbf{L} \times \mathbf{M}$ at right angles.

In this case the axes of **L**, **M**, **N** have a common normal (the axis of $\mathbf{L} \times \mathbf{M}$). We therefore may state the

THEOREM. *The dual triple product of three proper motors will vanish only when* (*a*) *their axes are all parallel, or* (*b*) *their axes have a common normal.*

If the proper motors **L**, **M**, **N** satisfy a linear relation with proper coefficients,

$$(6) \quad (\alpha + \epsilon\alpha')\mathbf{L} + (\beta + \epsilon\beta')\mathbf{M} + (\gamma + \epsilon\gamma')\mathbf{N} = 0, \qquad \alpha\beta\gamma \neq 0,$$

$\mathbf{L} \times \mathbf{M} \cdot \mathbf{N} = 0$, and we can apply the preceding theorem. More precise information, however, is given by the equations,

$$\frac{\mathbf{M} \times \mathbf{N}}{\alpha + \epsilon\alpha'} = \frac{\mathbf{N} \times \mathbf{L}}{\beta + \epsilon\beta'} = \frac{\mathbf{L} \times \mathbf{M}}{\gamma + \epsilon\gamma'}$$

which follow from (6). These show that the three motors $\mathbf{M} \times \mathbf{N}$, $\mathbf{N} \times \mathbf{L}$, $\mathbf{L} \times \mathbf{M}$ are all proper or all couples. When proper, these motors are all coaxial; when couples, their moments are all parallel. Hence from (6) we conclude that either (*a*) the axes of **L**, **M**, **N**, no two of which are parallel, have a common normal; or (*b*) their axes are all parallel and coplanar. *Thus the linear relation* (6) *implies that the axes of* **L**, **M**, **N** *have a common normal.* Conversely, if the axes of **L**, **M**, **N** have a common normal, we can infer a linear relation (6), provided the case in which two axes are parallel to each other but not to the third is excluded. We omit the proof.

36. Motor Identities. All the identities of vector algebra are still valid when the vectors are replaced by motors. For the motor products are defined as the results of applying the distributive law to dual vectors and subsequently equating $\epsilon^2, \epsilon^3, \cdots$ to zero. Now the identities apply to the ordinary vectors forming the dual vectors, so that the results before equating $\epsilon^2, \epsilon^3, \cdots$ to zero are true for any value of the scalar ϵ. In particular, the identities hold when $\epsilon^2, \epsilon^3, \cdots$ are replaced by zero. Thus we have the motor identities:

$$(1) \qquad \mathbf{L} \times (\mathbf{M} \times \mathbf{N}) = \mathbf{L} \cdot \mathbf{N}\,\mathbf{M} - \mathbf{L} \cdot \mathbf{M}\,\mathbf{N},$$

$$(2) \qquad \mathbf{L} \times (\mathbf{M} \times \mathbf{N}) + \mathbf{M} \times (\mathbf{N} \times \mathbf{L}) + \mathbf{N} \times (\mathbf{L} \times \mathbf{M}) = 0,$$

$$(3) \qquad (\mathbf{L} \times \mathbf{M}) \cdot (\mathbf{N} \times \mathbf{P}) = \mathbf{L} \cdot \mathbf{N}\,\mathbf{M} \cdot \mathbf{P} - \mathbf{L} \cdot \mathbf{P}\,\mathbf{M} \cdot \mathbf{N}.$$

The identity (2) has an interesting geometric interpretation when all nine motors,

$$\mathbf{L}, \mathbf{M}, \mathbf{N}; \mathbf{M} \times \mathbf{N}, \mathbf{N} \times \mathbf{L}, \mathbf{L} \times \mathbf{M}; \mathbf{L} \times (\mathbf{M} \times \mathbf{N}), \mathbf{M} \times (\mathbf{N} \times \mathbf{L}), \mathbf{N} \times (\mathbf{L} \times \mathbf{M}),$$

are proper. In this case, *the axes of the three last motors have a common normal* (§ 35). This common normal and the axes of the preceding nine motors form a configuration of ten lines, each one of which is normal to three others. This is essentially the theorem of Peterson and Morley.‡ When $\mathbf{L}$, $\mathbf{M}$, $\mathbf{N}$ are line vectors along the sides of a plane triangle, this theorem implies the concurrence of its altitudes. The theorem of Peterson and Morley therefore may be regarded as a generalization of this result.

37. Reciprocal Sets of Motors. Consider two sets of proper motors $\mathbf{M}_1$, $\mathbf{M}_2$, $\mathbf{M}_3$; $\mathbf{M}^1$, $\mathbf{M}^2$, $\mathbf{M}^3$ for which $\mathbf{m}_1 \cdot \mathbf{m}_2 \times \mathbf{m}_3 \neq 0$, $\mathbf{m}^1 \cdot \mathbf{m}^2 \times \mathbf{m}^3 \neq 0$. The sets are said to be reciprocal when the nine equations,

$$(1) \qquad \mathbf{M}_i \cdot \mathbf{M}^j = \delta_i^j \qquad (i, j = 1, 2, 3),$$

are satisfied. The indices are merely labels, and δ_i^j is the Kronecker delta.

Let the set $\mathbf{M}_i$ be given; we then may compute the reciprocal set $\mathbf{M}^i$ uniquely. Since $\mathbf{M}_2 \cdot \mathbf{M}^1 = 0$, $\mathbf{M}_3 \cdot \mathbf{M}^1 = 0$, the axis of $\mathbf{M}^1$ is the common normal to the axes of $\mathbf{M}_2$ and $\mathbf{M}_3$; hence $\mathbf{M}^1 = (\lambda + \epsilon\lambda')\mathbf{M}_2 \times \mathbf{M}_3$, and, since $\mathbf{M}_1 \cdot \mathbf{M}^1 = 1$, $\lambda + \epsilon\lambda' = 1/\mathbf{M}_1 \cdot \mathbf{M}_2 \times \mathbf{M}_3$; the condition $\mathbf{m}_1 \cdot \mathbf{m}_2 \times \mathbf{m}_3 \neq 0$ ensures that $\mathbf{M}_1 \cdot \mathbf{M}_2 \times \mathbf{M}_3$ is not pure dual. Thus we find

$$(2) \qquad \mathbf{M}^i = \frac{\mathbf{M}_j \times \mathbf{M}_k}{\mathbf{M}_1 \cdot \mathbf{M}_2 \times \mathbf{M}_3}, \qquad \mathbf{M}_i = \frac{\mathbf{M}^j \times \mathbf{M}^k}{\mathbf{M}^1 \cdot \mathbf{M}^2 \times \mathbf{M}^3},$$

where ijk represent any cyclical permutation of 123. That the motors (2) satisfy equations (1) is shown by direct substitution. Equations (2) have the same form as the corresponding equations in vector algebra. Moreover, in

$$\mathbf{M}^1 \cdot \mathbf{M}_1 = \frac{(\mathbf{M}_2 \times \mathbf{M}_3) \cdot (\mathbf{M}^2 \times \mathbf{M}^3)}{(\mathbf{M}_1 \cdot \mathbf{M}_2 \times \mathbf{M}_3)(\mathbf{M}^1 \cdot \mathbf{M}^2 \times \mathbf{M}^3)},$$

the left member is 1, and, from (36.3) the numerator on the right is 1; hence

$$(3) \qquad (\mathbf{M}_1 \cdot \mathbf{M}_2 \times \mathbf{M}_3)(\mathbf{M}^1 \cdot \mathbf{M}^2 \times \mathbf{M}^3) = 1.$$

‡ E. A. Weiss, *Einführung in die Linien-geometrie und Kinematik*, Leipzig 1935, p. 85.

The axes of two reciprocal sets taken in the order $\mathbf{M}_1\mathbf{M}^3\mathbf{M}_2\mathbf{M}^1\mathbf{M}_3\mathbf{M}^2$ form a skew hexagon with six right angles. Since

$$\mathbf{M}_1 \times \mathbf{M}^1 + \mathbf{M}_2 \times \mathbf{M}^2 + \mathbf{M}_3 \times \mathbf{M}^3 = 0, \tag{4}$$

the common normals of its opposite sides themselves admit a common normal. The six sides, the three common normals of the opposite sides, and *their* common normal, thus form a Peterson–Morley configuration (§ 36).

If we put $\mathbf{M}_i = \mathbf{M}^i$ in (1), we find that the only self-reciprocal motor sets are formed by three mutually orthogonal unit line vectors through a point.

Any motor $\mathbf{M}$ may be expressed linearly with dual coefficients in terms of $\mathbf{M}_1$, $\mathbf{M}_2$, $\mathbf{M}_3$, provided $\mathbf{m}_1 \cdot \mathbf{m}_2 \times \mathbf{m}_3 \neq 0$. To show that the representation is possible, we note that

$$\mathbf{M} = (\alpha + \epsilon\alpha')\mathbf{M}_1 + (\beta + \epsilon\beta')\mathbf{M}_2 + (\gamma + \epsilon\gamma')\mathbf{M}_3 \tag{5}$$

is equivalent to the two vector equations,

$$\mathbf{m} = \alpha\mathbf{m}_1 + \beta\mathbf{m}_2 + \gamma\mathbf{m}_3, \tag{6}$$

$$\mathbf{m}_O = \alpha\mathbf{m}_{1O} + \beta\mathbf{m}_{2O} + \gamma\mathbf{m}_{3O} + \alpha'\mathbf{m}_1 + \beta'\mathbf{m}_2 + \gamma'\mathbf{m}_3. \tag{7}$$

Since $\mathbf{m}_1 \cdot \mathbf{m}_2 \times \mathbf{m}_3 \neq 0$, the vectors $\mathbf{m}_1$, $\mathbf{m}_2$, $\mathbf{m}_3$ have a reciprocal set $\mathbf{m}^1$, $\mathbf{m}^2$, $\mathbf{m}^3$. From (6),

$$\alpha = \mathbf{m} \cdot \mathbf{m}^1, \qquad \beta = \mathbf{m} \cdot \mathbf{m}^2, \qquad \gamma = \mathbf{m} \cdot \mathbf{m}^3;$$

with α, β, γ known, we may determine α', β', γ' in the same way from (7). The actual computation, however, can be effected most simply from (5) by using the reciprocal set $\mathbf{M}^1$, $\mathbf{M}^2$, $\mathbf{M}^3$. Thus we find

$$\mathbf{M} = \mathbf{M} \cdot \mathbf{M}^1\, \mathbf{M}_1 + \mathbf{M} \cdot \mathbf{M}^2\, \mathbf{M}_2 + \mathbf{M} \cdot \mathbf{M}^3\, \mathbf{M}_3. \tag{8}$$

38. Statics. The statics of rigid bodies may be developed independently of dynamics from four fundamental principles: †

Principle I (*Vector Addition of Forces*). Two forces acting on the same particle may be replaced by a single force, acting on the particle, equal to their vector sum.

Principle II (*Transmissibility of a Force*). A force acting on a rigid body may be shifted along its line of action so as to act on any particle of that line.

† See Brand, L., *Vectorial Mechanics*, New York, 1930, Chapter II.

Principle III (*Static Equilibrium*). If the forces acting on a rigid body, initially at rest, can be reduced to zero by means of principles I and II, the body will remain at rest.

Principle IV (*Action and Reaction*). This need not concern us here.

Principle II states in effect that *a force acting on a rigid body is a line vector*. Such a force is determined by the vector $\mathbf{f}$ giving its magnitude and direction and by its moment $\mathbf{f}_O$ about O (§ 27). Thus the dual vector,

$$\text{(1)} \qquad \mathbf{F} = \mathbf{f} + \epsilon \mathbf{f}_O \qquad (\mathbf{f} \cdot \mathbf{f}_O = 0),$$

represents a force $\mathbf{f}$ acting along the line whose equation is $\mathbf{r} \times \mathbf{f} = \mathbf{f}_O$.

If two forces $\mathbf{P}$, $\mathbf{Q}$ act on a particle with position vector $\mathbf{r}$,

$$\mathbf{P} = \mathbf{p} + \epsilon\, \mathbf{r} \times \mathbf{p}, \qquad \mathbf{Q} = \mathbf{q} + \epsilon\, \mathbf{r} \times \mathbf{q};$$

principle I asserts that both forces may be replaced by a single force, their *resultant*, whose vector and moment are $\mathbf{p} + \mathbf{q}$, $\mathbf{r} \times (\mathbf{p} + \mathbf{q})$; that is, the intersecting line vectors $\mathbf{P}$ and $\mathbf{Q}$ are equivalent to the line vector $\mathbf{P} + \mathbf{Q}$, their motor sum (§ 32).

Two parallel forces $\mathbf{P}$, $\mathbf{Q}$ also have a resultant $\mathbf{P} + \mathbf{Q}$ provided $\mathbf{p} + \mathbf{q} \neq 0$. To see this, we need only introduce a pair of opposed forces $\mathbf{F}$, $-\mathbf{F}$ (equivalent to zero) acting along any line cutting the lines of $\mathbf{P}$ and $\mathbf{Q}$. We may now apply principle I to find the resultants of $\mathbf{P} + \mathbf{F}$ and $\mathbf{Q} - \mathbf{F}$. These coplanar forces intersect, since

$$(\mathbf{p} + \mathbf{f}) \times (\mathbf{q} - \mathbf{f}) = \mathbf{f} \times (\mathbf{p} + \mathbf{q}) \neq 0,$$

and therefore have the resultant,

$$(\mathbf{P} + \mathbf{F}) + (\mathbf{Q} - \mathbf{F}) = \mathbf{P} + \mathbf{Q}.$$

If $\mathbf{Q} = \lambda \mathbf{P}$ $(\lambda \neq -1)$, the line of $\mathbf{P} + \mathbf{Q}$ divides all segments from $\mathbf{P}$ to $\mathbf{Q}$ in the ratio $\lambda/1$ (§ 32, theorem 3).

If $\mathbf{p} + \mathbf{q} = 0$, $\mathbf{p}_O + \mathbf{q}_O \neq 0$, the parallel forces $\mathbf{P}$, $\mathbf{Q}$ are equal in magnitude and opposite in direction and are said to form a couple of moment $\mathbf{p}_O + \mathbf{q}_O$.

In any case, when a system of forces $\mathbf{F}_i$ acting on a rigid body is changed into an equivalent system $\mathbf{G}_j$ by the application of principles I and II, the force-sum and moment-sum remain unaltered:

$$\Sigma \mathbf{f}_i = \Sigma \mathbf{g}_j, \qquad \Sigma \mathbf{f}_{iO} = \Sigma \mathbf{g}_{jO}.$$

For each application of principle I leaves these sums unaltered; and shifting a force in accordance with principle II does not alter its vector or moment.

Now it can be shown that any system of forces $\mathbf{F}_i$ acting on a rigid body can be reduced by means of principles I and II to *two* forces,* say $\mathbf{P} = \mathbf{p} + \epsilon\mathbf{p}_O$, $\mathbf{Q} = \mathbf{q} + \epsilon\mathbf{q}_O$. Hence, if $\Sigma\mathbf{F}_i = \mathbf{m} + \epsilon\mathbf{m}_O$, we have

$$\mathbf{p} + \mathbf{q} = \mathbf{m}, \qquad \mathbf{p}_O + \mathbf{q}_O = \mathbf{m}_O.$$

If $\mathbf{m} \neq 0$, $\mathbf{m} \cdot \mathbf{m}_O \neq 0$, the system $\mathbf{F}_i$ is equivalent to a *screw* $\mathbf{M} = \mathbf{m} + \epsilon\mathbf{m}_O$ (§ 31) which may be expressed in many ways as two non-coplanar forces. A screw cannot be reduced to a single force. The screw $\mathbf{M}$ may be regarded as a force $\mathbf{m}$ acting through the origin and a couple of moment $\mathbf{m}_O$.

If $\mathbf{m} \neq 0$, $\mathbf{m} \cdot \mathbf{m}_O = 0$, the forces $\mathbf{P}$ and $\mathbf{Q}$ are coplanar; for $\mathbf{m} \cdot \mathbf{m}_O = 0$ implies $\mathbf{p} \cdot \mathbf{q}_O + \mathbf{q} \cdot \mathbf{p}_O = 0$ (26.14). The forces $\mathbf{P}, \mathbf{Q}$ now may be reduced to the *single force* $\mathbf{M} = \mathbf{m} + \epsilon\mathbf{m}_O$, their resultant.

If $\mathbf{m} = 0$, $\mathbf{m}_O \neq 0$, the system $\mathbf{F}_i$ is equivalent to the *couple* $\mathbf{P}, \mathbf{Q}$ of moment $\mathbf{m}_O$.

Finally, if $\mathbf{m} = 0$, $\mathbf{m}_O = 0$, the system $\mathbf{F}_i$ reduces to zero, and, in accordance with principle III, the rigid body is in *equilibrium.*

In brief, *the system of forces* $\mathbf{F}_i$ *acting on a rigid body is equivalent to the motor* $\mathbf{M} = \Sigma\mathbf{F}_i$; *this may be a screw, a single force, a couple, or zero; only in the last case is the rigid body in equilibrium.*

The moment of a force $\mathbf{F} = \mathbf{f} + \epsilon\mathbf{f}_O$ about an axis along the unit line vector $\mathbf{A} = \mathbf{a} + \epsilon\mathbf{a}_O$ was defined (27.4) as $\mathbf{a} \cdot \mathbf{f}_P$, P being a point on the axis of $\mathbf{A}$. The moment of a system of forces, equivalent to the motor $\mathbf{M} = \mathbf{m} + \epsilon\mathbf{m}_O$, about this axis is therefore

$$\Sigma\mathbf{a} \cdot \mathbf{f}_P = \mathbf{a} \cdot \Sigma\mathbf{f}_P = \mathbf{a} \cdot \mathbf{m}_P.$$

Now

$$\mathbf{a} \cdot \mathbf{m}_P = \mathbf{a} \cdot (\mathbf{m}_O + \overrightarrow{PO} \times \mathbf{m}) = \mathbf{a} \cdot \mathbf{m}_O + \mathbf{m} \cdot \overrightarrow{OP} \times \mathbf{a},$$

and, since $\overrightarrow{OP} \times \mathbf{a} = \mathbf{a}_O$,

$$(2) \qquad \mathbf{a} \cdot \mathbf{m}_P = \mathbf{a} \cdot \mathbf{m}_O + \mathbf{m} \cdot \mathbf{a}_O = \mathbf{A} \circ \mathbf{M}.$$

We call this the *moment of the motor* $\mathbf{M}$ *about the axis* $\mathbf{A}$. This moment is independent of the choice of P on the axis.

* Brand, L., *Vectorial Mechanics*, p. 143.

39. Null System. A line is called a *null line* with respect to a motor $\mathbf{M}$ if the moment of $\mathbf{M}$ about this line is zero. In order that a line, given by the unit line vector $\mathbf{A} = \mathbf{a} + \epsilon\mathbf{a}_O$, be a null line with respect to $\mathbf{M}$, it is necessary and sufficient that

$$(1) \qquad \mathbf{A} \circ \mathbf{M} = \mathbf{a} \cdot \mathbf{m}_O + \mathbf{a}_O \cdot \mathbf{m} = 0.$$

If $\mathbf{L} = \lambda\mathbf{A}$, $\mathbf{L} \circ \mathbf{M} = \lambda\mathbf{A} \circ \mathbf{M}$; hence the axis of any line vector $\mathbf{L}$ will be a null line when and only when $\mathbf{L} \circ \mathbf{M} = 0$. The totality of such lines constitutes the *null system* of the motor $\mathbf{M}$.

A motor $\mathbf{M}$ can be replaced in many ways by two line vectors $\mathbf{F}$, $\mathbf{G}$. When $\mathbf{M}$ is a screw ($\mathbf{m} \cdot \mathbf{m}_O \neq 0$), we shall prove that the line of one may be chosen at pleasure, *provided it is not a null line or parallel to the axis of* $\mathbf{M}$; after this choice $\mathbf{F}$ and $\mathbf{G}$ are *uniquely determined.*

Write $\mathbf{F} = \alpha\mathbf{A}$, $\mathbf{G} = \beta\mathbf{B}$ as multiples of unit line vectors, and choose $\mathbf{A}$ as any line not in the null system. Now $\mathbf{M} = \mathbf{F} + \mathbf{G}$ is equivalent to

$$\mathbf{m} = \alpha\mathbf{a} + \beta\mathbf{b}, \qquad \mathbf{m}_O = \alpha\mathbf{a}_O + \beta\mathbf{b}_O.$$

Since $\mathbf{a} \cdot \mathbf{a}_O = \mathbf{b} \cdot \mathbf{b}_O = 0$,

$$(\mathbf{m} - \alpha\mathbf{a}) \cdot (\mathbf{m}_O - \alpha\mathbf{a}_O) = \beta^2\, \mathbf{b} \cdot \mathbf{b}_O = 0,$$

$$(2) \qquad \alpha = \frac{\mathbf{m} \cdot \mathbf{m}_O}{\mathbf{a} \cdot \mathbf{m}_O + \mathbf{m} \cdot \mathbf{a}_O};$$

thus $\mathbf{F} = \alpha\mathbf{A}$ is known, and $\mathbf{G} = \mathbf{M} - \mathbf{F}$. Since $\mathbf{A}$ is not a null line, $\mathbf{a} \cdot \mathbf{m}_O + \mathbf{m} \cdot \mathbf{a}_O \neq 0$; and, since $\mathbf{A}$ is not parallel to $\mathbf{m}$, $\beta \neq 0$, and $\mathbf{G}$ is actually a line vector.

All null lines through a point P are perpendicular to $\mathbf{m}_P$ and therefore lie in a plane p, the *null plane* of P. Moreover all the null lines on any plane p pass through a point P, the *null point* of p, where $\mathbf{m}_P$ is perpendicular to p.

Analytic Proof. Consider a null line through the points $(\alpha, \mathbf{a}_O)$, $(\beta, \mathbf{b}_O)$; the coordinates are $(\alpha\mathbf{b}_O - \mathbf{a}_O\beta, \mathbf{a}_O \times \mathbf{b}_O)$ from (26.9). Since it is a null line,

$$(3) \qquad \mathbf{m} \cdot \mathbf{a}_O \times \mathbf{b}_O + \mathbf{m}_O \cdot (\alpha\mathbf{b}_O - \mathbf{a}_O\beta) = 0.$$

This may be written

$$(4) \qquad (\alpha\mathbf{m}_O - \mathbf{a}_O \times \mathbf{m}) \cdot \mathbf{b}_O = \mathbf{a}_O \cdot \mathbf{m}_O\beta;$$

hence, if the point $(\alpha, \mathbf{a}_O)$ is fixed and $(\beta, \mathbf{b}_O)$ varies over all null lines through $(\alpha, \mathbf{a}_O)$, $(\beta, \mathbf{b}_O)$ always will lie in the plane whose coordinates are $(\alpha\mathbf{m}_O - \mathbf{a}_O \times \mathbf{m}, \mathbf{a}_O \cdot \mathbf{m}_O)$.

Next, let the null line be common to the planes $(\mathbf{a}, \alpha_O)$, $(\mathbf{b}, \beta_O)$; its coordinates are $(\mathbf{a} \times \mathbf{b}, \mathbf{a}\beta_O - \alpha_O\mathbf{b})$ from (26.10). Since it is a null line,

$$(5) \qquad \mathbf{m} \cdot (\mathbf{a}\beta_O - \alpha_O\mathbf{b}) + \mathbf{m}_O \cdot \mathbf{a} \times \mathbf{b} = 0.$$

This may be written

$$(6) \qquad (\mathbf{m}\alpha_O - \mathbf{m}_O \times \mathbf{a}) \cdot \mathbf{b} = \mathbf{m} \cdot \mathbf{a}\, \beta_O;$$

hence, if the plane $(\mathbf{a}, \alpha_O)$ is fixed and $(\mathbf{b}, \beta_O)$ turns about all null lines in $(\mathbf{a}, \alpha_O)$, $(\mathbf{b}, \beta_O)$ always will pass through the point whose coordinates are $(\mathbf{m} \cdot \mathbf{a}, \mathbf{m}\alpha_O - \mathbf{m}_O \times \mathbf{a})$.

Equations (3) and (5) are not altered by an interchange of the two points or two planes; hence the same is true of (4) and (6). Therefore:

If the null plane of point A passes through B, the null plane of B passes through A.

If the null point of plane a lies on b, the null point of b lies on a.

From the preceding results:

$$(7) \quad \text{Point } (\alpha, \mathbf{a}_O) \sim \text{null plane} \qquad (\alpha\mathbf{m}_O - \mathbf{a}_O \times \mathbf{m}, \quad \mathbf{a}_O \cdot \mathbf{m}_O);$$

$$(8) \quad \text{Plane } (\mathbf{b}, \beta_O) \sim \text{null point} \qquad (\mathbf{m} \cdot \mathbf{b}, \quad \mathbf{m}\beta_O - \mathbf{m}_O \times \mathbf{b}).$$

Indeed, if we solve

$$\mathbf{b} = \alpha\mathbf{m}_O - \mathbf{a}_O \times \mathbf{m}, \qquad \beta_O = \mathbf{a}_O \cdot \mathbf{m}_O,$$

for $(\alpha, \mathbf{a}_O)$, we obtain the null point in (8). Note also that $\mathbf{a}_O \cdot \mathbf{b} = \alpha\beta_O$; let the student interpret this relation in the light of (26.16).

When $\mathbf{M}$ is a line vector, the null lines are lines that intersect its axis. Then (7) gives the plane through the point and $\mathbf{M}$ (26.11); and (8) the point on the plane and $\mathbf{M}$ (26.12).

If $\mathbf{M}$ is a screw ($\mathbf{m} \cdot \mathbf{m}_O \neq 0$), not only are points associated with planes, but also lines with lines in general. For if $\mathbf{A}$ is any unit line vector not in the null system of $\mathbf{M}$ or parallel to its axis, we can determine uniquely a unit line vector $\mathbf{B}$, its *conjugate*, such that

$$(9) \qquad \mathbf{M} = \alpha\mathbf{A} + \beta\mathbf{B},$$

where $\alpha = \mathbf{m} \cdot \mathbf{m}_O / \mathbf{A} \circ \mathbf{M}$ from (2). *Any line vector* $\mathbf{C}$ *that cuts two conjugate lines is a null line;* for

$$\mathbf{M} \circ \mathbf{C} = \alpha \mathbf{A} \circ \mathbf{C} + \beta \mathbf{B} \circ \mathbf{C} = 0.$$

In view of this property we may characterize conjugate lines as follows: The conjugate of a line $\mathbf{A}$ is the line $\mathbf{B}$: (i) which is common to the null planes of all points on $\mathbf{A}$, or (ii), which contains the null points of all planes through $\mathbf{A}$. With this point of view, *null lines are self-conjugate.* Along a line parallel to the axis of $\mathbf{M}$, the moment of $\mathbf{M}$ is constant; since the null planes of its points are all parallel, we may say that its conjugate is at infinity.

40. Summary: Motor Algebra. The number $X = x + \epsilon x'$ is called *dual* if x and x' are real and ϵ is a unit with the property $\epsilon^2 = 0$. If $x \neq 0$, the dual number is *proper*. Operations with dual numbers are carried out as in ordinary algebra and then $\epsilon^2, \epsilon^3, \cdots$ set equal to zero. Division, however, is only defined for proper dual numbers. The function $f(x + \epsilon x')$ is defined by means of its formal Taylor expansion; only the first two terms, $f(x) + \epsilon x' f'(x)$, appear since $\epsilon^2 = \epsilon^3 = \cdots 0$.

A *line vector* with the Plücker coordinates $\mathbf{a}, \mathbf{a}_O$ ($\mathbf{a} \cdot \mathbf{a}_O = 0$) is written as a *dual vector*, $\mathbf{A} = \mathbf{a} + \epsilon \mathbf{a}_O$; for a *unit* line vector $|\mathbf{a}| = 1$. When the origin is shifted to P, $\mathbf{a}_O$ becomes

$$\mathbf{a}_P = \mathbf{a}_O + \overrightarrow{PO} \times \mathbf{a}.$$

A *motor* $\mathbf{M} = \mathbf{m} + \epsilon \mathbf{m}_O$ is a dual multiple of a *unit* line vector,

$$\mathbf{M} = (\lambda + \epsilon\lambda')\mathbf{A} = (\lambda + \epsilon\lambda')(\mathbf{a} + \epsilon \mathbf{a}_O);$$

now $\mathbf{m} \cdot \mathbf{m}_O = 0$ only when $\lambda' = 0$. A shift of origin to P changes the moment vector as before:

$$\mathbf{m}_P = \mathbf{m}_O + \overrightarrow{PO} \times \mathbf{m},$$

but $\mathbf{m} \cdot \mathbf{m}_P = \mathbf{m} \cdot \mathbf{m}_O$ is invariant.

Motors for which $\lambda \neq 0$ are said to be *proper*. Proper motors are *screws* when $\lambda' \neq 0$, *line vectors* when $\lambda' = 0$. Proper motors have a definite *axis* given by the unit line vector $\mathbf{A}$. When $\lambda = 0$, $\lambda' \neq 0$ we have a pure dual motor $\epsilon \mathbf{m}_O$; it is called a *couple* of moment $\mathbf{m}_O$. The moment of a couple is a *free vector*, determined in direction but not in position.

When $\lambda = \lambda' = 0$ ($\lambda + \epsilon\lambda' = 0$), the motor reduces to zero.

The *sum* of the motors $\mathbf{M} = \mathbf{m} + \epsilon\mathbf{m}_O$, $\mathbf{N} = \mathbf{n} + \epsilon\mathbf{n}_O$ is defined as

$$\mathbf{M} + \mathbf{N} = \mathbf{m} + \mathbf{n} + \epsilon(\mathbf{m}_O + \mathbf{n}_O).$$

If $\mathbf{M}$ and $\mathbf{N}$ are line vectors, $\mathbf{M} + \mathbf{N}$ is a line vector only when $\mathbf{m} \cdot \mathbf{n}_O + \mathbf{n} \cdot \mathbf{m}_O = 0$.

The *scalar* and *motor* products, $\mathbf{M} \cdot \mathbf{N}$ and $\mathbf{M} \times \mathbf{N}$ are obtained by distributing the products as in ordinary vector analysis and then setting $\epsilon^2 = 0$:

$$\mathbf{M} \cdot \mathbf{N} = \mathbf{m} \cdot \mathbf{n} + \epsilon(\mathbf{m} \cdot \mathbf{n}_O + \mathbf{m}_O \cdot \mathbf{n}),$$

$$\mathbf{M} \times \mathbf{N} = \mathbf{m} \times \mathbf{n} + \epsilon(\mathbf{m} \times \mathbf{n}_O + \mathbf{m}_O \times \mathbf{n}).$$

If $\mathbf{A}$ and $\mathbf{B}$ are line vectors making an angle φ and at a normal distance φ' apart (reckoned positive if $\mathbf{a} \times \mathbf{b}$ points from $\mathbf{A}$ to $\mathbf{B}$), the *dual angle* between $\mathbf{A}$ and $\mathbf{B}$ is defined as $\Phi = \varphi + \epsilon\varphi'$. For *unit* line vectors,

$$\mathbf{A} \cdot \mathbf{B} = \cos \Phi, \qquad \mathbf{A} \times \mathbf{B} = \sin \Phi\, \mathbf{E}$$

where $\mathbf{E}$ is a unit line vector in the direction $\mathbf{a} \times \mathbf{b}$ and cutting $\mathbf{A}, \mathbf{B}$ at right angles. These results are perfect analogues of $\mathbf{a} \cdot \mathbf{b} = \cos \varphi$, $\mathbf{a} \times \mathbf{b} = \sin \varphi\, \mathbf{e}$.

If $\mathbf{M}$ and $\mathbf{N}$ are non-parallel proper motors, the common normal to their axes is the axis of $\mathbf{M} \times \mathbf{N}$. When $\mathbf{M}$ and $\mathbf{N}$ are parallel, $\mathbf{M} \times \mathbf{N}$ is a couple whose moment is perpendicular to their axes.

If $\mathbf{M}$ and $\mathbf{N}$ are *proper* motors, $\mathbf{M} \cdot \mathbf{N} = 0$ implies that their axes cut at right angles; $\mathbf{M} \times \mathbf{N} = 0$ that their axes coincide; and conversely.

For three motors, the dual number $\mathbf{L} \times \mathbf{M} \cdot \mathbf{N}$ is obtained by distributing the product $(\mathbf{l} + \epsilon\mathbf{l}_O) \times (\mathbf{m} + \epsilon\mathbf{m}_O) \cdot (\mathbf{n} + \epsilon\mathbf{n}_O)$:

$$\mathbf{L} \times \mathbf{M} \cdot \mathbf{N} = \mathbf{l} \times \mathbf{m} \cdot \mathbf{n} + \epsilon(\mathbf{l} \times \mathbf{m} \cdot \mathbf{n}_O + \mathbf{l} \times \mathbf{m}_O \cdot \mathbf{n} + \mathbf{l}_O \times \mathbf{m} \cdot \mathbf{n}).$$

If $\mathbf{L}, \mathbf{M}, \mathbf{N}$ are proper, $\mathbf{L} \times \mathbf{M} \cdot \mathbf{N} = 0$ implies that their axes are all parallel or have a common normal, and conversely.

Two sets of three motors $\mathbf{M}_i$, $\mathbf{M}^j$ whose triple products are proper are said to be *reciprocal* when $\mathbf{M}_i \cdot \mathbf{M}^j = \delta_i^j$ $(i, j = 1, 2, 3)$. All formulas for reciprocal vectors have exact analogues for reciprocal motors.

The forces acting on a rigid body are line vectors. A system $\mathbf{F}_i$ of such forces is equivalent to their motor sum $\mathbf{M} = \Sigma\mathbf{F}_i$, which may be a screw, a single-force, a couple, or zero; only when $\mathbf{M} = 0$ is the body in equilibrium.

All the rules of calculation in "real" vector algebra have exact analogues in motor algebra.

PROBLEMS

1. If $\mathbf{A}$ is a unit line vector, show that the screw $\mathbf{M} = (\lambda + \epsilon\lambda')\mathbf{A}$ may be expressed as the sum of a line vector and a couple whose moment is parallel to the line vector, namely,

$$\mathbf{M} = \lambda\mathbf{A} + \epsilon\mu\mathbf{m},$$

where $\mu = \lambda'/\lambda$ is the pitch of the screw.

2. Prove that the axes of the motors $\mathbf{M}$, $\mathbf{N}$ and $\mathbf{M} + \mathbf{N}$ either are parallel or have a common normal.

3. Express the screw $\mathbf{M} = 2\mathbf{i} + \epsilon(\mathbf{i} + \mathbf{j} + \mathbf{k})$

(*a*) As the dual multiple of a unit line vector.

(*b*) As the sum of a line vector and a couple whose moment is parallel to the line vector.

(*c*) Find the equation of the axis of $\mathbf{M}$.

4. In the tetrahedron $OABC$, $\mathbf{e}_1 = \overrightarrow{OA}$, $\mathbf{e}_2 = \overrightarrow{OB}$, $\mathbf{e}_3 = \overrightarrow{OC}$. If the forces

$$P\mathbf{e}_1,\ Q\mathbf{e}_2,\ R\mathbf{e}_3;\quad P'(\mathbf{e}_3 - \mathbf{e}_2),\ Q'(\mathbf{e}_1 - \mathbf{e}_3),\ R'(\mathbf{e}_2 - \mathbf{e}_1)$$

acting along its six edges are equivalent to the motor $\mathbf{m} + \epsilon\mathbf{m}_O$, show that they are uniquely determined by the six equations:

$$P' = \frac{\mathbf{m}_O \cdot \mathbf{e}_1}{[\mathbf{e}_1\mathbf{e}_2\mathbf{e}_3]}, \cdots; \qquad P + Q' - R' = \mathbf{m} \cdot \mathbf{e}^1, \cdots.$$

5. Show that the n forces $\overrightarrow{P_1Q_1}$, $\overrightarrow{P_2Q_2}$, $\cdots$, $\overrightarrow{P_nQ_n}$ are equivalent to the motor,

$$n(\mathbf{q}^* - \mathbf{p}^*) + \epsilon\Sigma\, \mathbf{p}_i \times \mathbf{q}_i\,,$$

where $\mathbf{p}^*$, $\mathbf{q}^*$ give the mean centers of the points $\mathbf{p}_i$ and $\mathbf{q}_i$, respectively. [Cf. (9.4)].

6. In order that the axes of the motors, $\mathbf{M} = \mathbf{m} + \epsilon\mathbf{m}_O$, $\mathbf{N} = \mathbf{n} + \epsilon\mathbf{n}_O$, be coplanar, it is necessary and sufficient that

$$\mathbf{m} \cdot \mathbf{n}_O + \mathbf{n} \cdot \mathbf{m}_O = (\mu + \nu)\mathbf{m} \cdot \mathbf{n},$$

where μ and ν denote the pitch of $\mathbf{M}$ and $\mathbf{N}$.

7. If three line vectors, $\mathbf{a} + \epsilon\mathbf{a}_O$, $\mathbf{b} + \epsilon\mathbf{b}_O$, $\mathbf{c} + \epsilon\mathbf{c}_O$, are parallel to a plane, $\alpha\mathbf{a} + \beta\mathbf{b} + \gamma\mathbf{c} = 0$ (§ 5). Prove that any line vector $\mathbf{d} + \epsilon\mathbf{d}_O$ meeting them is parallel to a fixed plane normal to $\alpha\mathbf{a}_O + \beta\mathbf{b}_O + \gamma\mathbf{c}_O$.

8. The forces $X\mathbf{i}$, $Y\mathbf{j}$, $-Z\mathbf{k}$ act through the points $(0, 0, 0)$, (a, b, c), $(0, b, 0)$, respectively. Show that they reduce to a single force if

$$X/a + Y/b + Z/c = 0.$$

9. If $ABCD$ are the vertices of a skew quadrilateral and $PQRS$ are the midpoints of its sides taken in order, show that the motor equivalent to the forces $\overrightarrow{AB}$, $\overrightarrow{BC}$, $\overrightarrow{CD}$, $\overrightarrow{DA}$ is also equivalent to four times the couple formed by $\overrightarrow{PQ}$ and $\overrightarrow{RS}$.

10. If the moment of the motor **M** about each of the six edges of a tetrahedron is zero, show that $\mathbf{M} = 0$. [Cf. (38.2)].

11. If the origin O is taken on the axis of a screw **M** of pitch μ show that the equation of the null plane of the point $\mathbf{r}_1$ is

$$(\mu(\mathbf{r} - \mathbf{r}_1) - \mathbf{r} \times \mathbf{r}_1) \cdot \mathbf{m} = 0.$$

If the axis of **M** is chosen as z-axis, this equation becomes $xy_1 - yx_1 = \mu(z - z_1)$.

CHAPTER III

VECTOR FUNCTIONS OF ONE VARIABLE

41. Derivative of a Vector. Let $\mathbf{u}(t)$ denote a vector function of a scalar variable t over the interval $a \leqq t \leqq b$; that is, when t is given, $\mathbf{u}(t)$ is uniquely determined.* The function $\mathbf{u}(t)$ is said to be *continuous* for the value t if $\mathbf{u}(t + \Delta t) \to \mathbf{u}(t)$ as $\Delta t \to 0$.

To obtain a clear idea of the way in which $\mathbf{u}$ varies with t, we may regard $\mathbf{u}(t) = \overrightarrow{OP}$ as a position vector issuing from the origin. Then as t varies, the point P traces a certain curve Γ in space. For the values t and $t + \Delta t$, let $\mathbf{u}(t) = \overrightarrow{OP}$, $\mathbf{u}(t + \Delta t) = \overrightarrow{OQ}$; the change in $\mathbf{u}$ for the increment Δt is

$$\Delta \mathbf{u} = \mathbf{u}(t + \Delta t) - \mathbf{u}(t) = \overrightarrow{OQ} - \overrightarrow{OP} = \overrightarrow{PQ}.$$

The average change per unit of t is $\Delta \mathbf{u}/\Delta t$. When $\Delta t > 0$, $\Delta \mathbf{u}/\Delta t$ is a vector ($\overrightarrow{PA'}$ in Fig. 41) in the direction of $\Delta \mathbf{u}$ and $1/\Delta t$ times as long; but if $\Delta t < 0$, $\Delta \mathbf{u}/\Delta t$ has the direction of $-\Delta \mathbf{u}$ (then $\overrightarrow{PA'}$ in Fig. 41 must be reversed). If $\overrightarrow{PA'}$ approaches a limiting vector $\overrightarrow{PA}$ as $\Delta t \to 0$, we call $\overrightarrow{PA}$ the *derivative* of $\mathbf{u}(t)$ with respect to t and denote it by $d\mathbf{u}/dt$ or $\mathbf{u}'(t)$. The equation defining the derivative is therefore

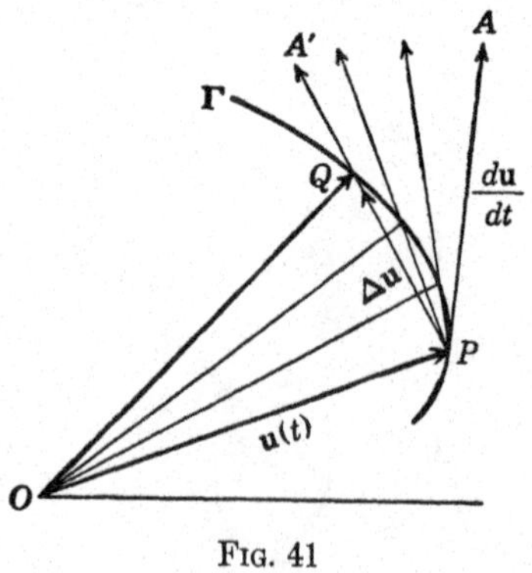

FIG. 41

$$\frac{d\mathbf{u}}{dt} = \lim_{\Delta t \to 0} \frac{\mathbf{u}(t + \Delta t) - \mathbf{u}(t)}{\Delta t} = \lim_{\Delta t \to 0} \frac{\Delta \mathbf{u}}{\Delta t}. \tag{1}$$

* The word function implies a *single-valued* function.

If the limit $d\mathbf{u}/dt$ exists for the value t, $\mathbf{u}(t)$ is said to be *differentiable* at t. Then we may write $\Delta\mathbf{u}/\Delta t = d\mathbf{u}/dt + \mathbf{h}$ where $\mathbf{h} \to 0$ as $\Delta t \to 0$; hence

$$\Delta\mathbf{u} = \Delta t\left(\frac{d\mathbf{u}}{dt} + \mathbf{h}\right) \to 0, \quad \text{as} \quad \Delta t \to 0,$$

and $\mathbf{u}(t + \Delta t) \to \mathbf{u}(t)$. Thus *if* $\mathbf{u}(t)$ *is differentiable at* t, *it is also continuous there.*

If $\mathbf{u}(t)$ is differentiable at P, Q describes the arc QP of the curve Γ as $\Delta t \to 0$, and the limiting direction of the chord PQ, and hence of the vector PA', is along the tangent at P. The limiting vector $PA = d\mathbf{u}/dt$ is therefore tangent to Γ at P. Since $\Delta\mathbf{u}/\Delta t$ has the same direction as $\Delta\mathbf{u}$ when $\Delta t > 0$, the opposite when $\Delta t < 0$, $d\mathbf{u}/dt$ is a vector tangent to Γ in the direction of increasing t. We restate this important result as follows:

If the vector $\mathbf{u}(t) = \overrightarrow{OP}$ *varies with* t, *so that* P *describes the curve* Γ *when* O *is held fast, the derivative* $d\mathbf{u}/dt$, *for any value of* t, *is a vector tangent to* Γ *at* P *in the direction of increasing* t.

Example 1. If $\mathbf{u} = \overrightarrow{OP}$ is a variable vector of constant direction, P will move on a straight line when O is held fast; hence $d\mathbf{u}/dt$, being tangent to this line, will be parallel to $\mathbf{u}$.

Example 2. If $\mathbf{u} = \overrightarrow{OP}$ is a variable vector of constant length, P will describe a curve Γ on the surface of a sphere when O is held fast; hence $d\mathbf{u}/dt$, being tangent to Γ at P, will be perpendicular to the radius OP of the sphere. In brief: *If* $|\mathbf{u}|$ *is constant,* $d\mathbf{u}/dt$ *is perpendicular to* $\mathbf{u}$.

If $\mathbf{u}$ is a constant vector, that is, constant in both length and direction,

$$\Delta\mathbf{u} = 0, \quad \frac{\Delta\mathbf{u}}{\Delta t} = 0, \quad \text{and} \quad \frac{d\mathbf{u}}{dt} = 0.$$

The derivative of a constant vector is zero.

When $\mathbf{u}$ is a function of a scalar variable s, and s in turn a function of t, a change of Δt in t will produce a change Δs in s and therefore a change $\Delta\mathbf{u}$ in $\mathbf{u}$. On passing to the limit $\Delta t \to 0$ in the identity,

$$\frac{\Delta\mathbf{u}}{\Delta t} = \frac{\Delta\mathbf{u}}{\Delta s}\frac{\Delta s}{\Delta t}, \quad \text{we get} \quad \frac{d\mathbf{u}}{dt} = \frac{d\mathbf{u}}{ds}\frac{ds}{dt},$$

the familiar "chain" rule.

The higher derivatives of $\mathbf{u}(t)$ are defined as in the calculus:

$$\mathbf{u}''(t) = \frac{d\mathbf{u}'}{dt}, \qquad \mathbf{u}'''(t) = \frac{d\mathbf{u}''}{dt}, \text{ etc.}$$

42. Derivatives of Sums and Products. Let $\mathbf{u}(t)$ and $\mathbf{v}(t)$ be two differentiable vector functions of a scalar t. When t changes by an amount Δt, let $\Delta\mathbf{u}$, $\Delta\mathbf{v}$, and $\Delta(\mathbf{u} + \mathbf{v})$ denote the vectorial changes in $\mathbf{u}$, $\mathbf{v}$, and $\mathbf{u} + \mathbf{v}$. Then

$$\mathbf{u} + \mathbf{v} + \Delta(\mathbf{u} + \mathbf{v}) = \mathbf{u} + \Delta\mathbf{u} + \mathbf{v} + \Delta\mathbf{v},$$

$$\Delta(\mathbf{u} + \mathbf{v}) = \Delta\mathbf{u} + \Delta\mathbf{v},$$

$$\frac{\Delta(\mathbf{u} + \mathbf{v})}{\Delta t} = \frac{\Delta\mathbf{u}}{\Delta t} + \frac{\Delta\mathbf{v}}{\Delta t};$$

and passing to the limit $\Delta t \to 0$ gives

$$\text{(1)} \qquad \frac{d}{dt}(\mathbf{u} + \mathbf{v}) = \frac{d\mathbf{u}}{dt} + \frac{d\mathbf{v}}{dt}.$$

Consequently, *the derivative of the sum of two vectors is equal to the sum of their derivatives.* This result may be generalized to the sum of any number of vectors.

Consider next the product $f(t)\mathbf{u}(t)$ of a differentiable scalar and a vector function. When t changes by an amount Δt, let Δf, $\Delta\mathbf{u}$, and $\Delta(f\mathbf{u})$ denote the increments of f, $\mathbf{u}$, and $f\mathbf{u}$, respectively. Then, since multiplication of vectors by scalars is distributive (§ 4),

$$f\mathbf{u} + \Delta(f\mathbf{u}) = (f + \Delta f)(\mathbf{u} + \Delta\mathbf{u}) = f\mathbf{u} + f\,\Delta\mathbf{u} + \Delta f\,\mathbf{u} + \Delta f\,\Delta\mathbf{u},$$

$$\Delta(f\mathbf{u}) = f\,\Delta\mathbf{u} + \Delta f\,\mathbf{u} + \Delta f\,\Delta\mathbf{u},$$

$$\frac{\Delta(f\mathbf{u})}{\Delta t} = f\frac{\Delta\mathbf{u}}{\Delta t} + \frac{\Delta f}{\Delta t}\mathbf{u} + \Delta f\frac{\Delta\mathbf{u}}{\Delta t};$$

passing to the limit $\Delta t \to 0$, and noting that $\Delta f \to 0$, we have

$$\text{(2)} \qquad \frac{d}{dt}(f\mathbf{u}) = f\frac{d\mathbf{u}}{dt} + \frac{df}{dt}\mathbf{u}.$$

This is formally the same as the rule for differentiating a product of scalar functions.

Important special cases of (2) arise when either f or $\mathbf{u}$ is constant:

$$\frac{d}{dt}(c\mathbf{u}) = c\frac{d\mathbf{u}}{dt}, \qquad \frac{d}{dt}(f\mathbf{c}) = \frac{df}{dt}\mathbf{c}.$$

If the components of a vector $\mathbf{u}(t)$ are $u^i(t)$ when referred to a constant basis $\mathbf{e}_i$,

$$\text{(3)} \qquad \mathbf{u}(t) = \Sigma u^i(t)\,\mathbf{e}_i, \qquad \frac{d\mathbf{u}}{dt} = \Sigma \frac{du^i}{dt}\mathbf{e}_i.$$

The components of the derivative of a vector are the derivatives of its components.

If we pass now to the products $\mathbf{u} \cdot \mathbf{v}$ and $\mathbf{u} \times \mathbf{v}$, where $\mathbf{u}$ and $\mathbf{v}$ are vector functions of t, the same type of argument used in proving (2) shows that

$$\text{(4)} \qquad \frac{d}{dt}(\mathbf{u} \cdot \mathbf{v}) = \mathbf{u} \cdot \frac{d\mathbf{v}}{dt} + \frac{d\mathbf{u}}{dt} \cdot \mathbf{v},$$

$$\text{(5)} \qquad \frac{d}{dt}(\mathbf{u} \times \mathbf{v}) = \mathbf{u} \times \frac{d\mathbf{v}}{dt} + \frac{d\mathbf{u}}{dt} \times \mathbf{v}.$$

The proofs depend essentially upon the distributive laws for the dot and cross product. In (5) the order of the factors must be preserved.

If l is an axis with unit vector $\mathbf{e}$, the orthogonal component of $\mathbf{u}$ on l is $\mathbf{e} \cdot \mathbf{u}$ (15.6); hence

$$\frac{d}{dt}\operatorname{comp}_l \mathbf{u} = \mathbf{e} \cdot \frac{d\mathbf{u}}{dt} = \operatorname{comp}_l \frac{d\mathbf{u}}{dt} \cdot$$

THEOREM 1. *A necessary and sufficient condition that a proper vector* $\mathbf{u}$ *be of constant length is that*

$$\text{(6)} \qquad \mathbf{u} \cdot \frac{d\mathbf{u}}{dt} = 0.$$

Proof. Since $|\mathbf{u}|^2 = \mathbf{u} \cdot \mathbf{u}$, we have, from (4),

$$\frac{d}{dt}|\mathbf{u}|^2 = 2\mathbf{u} \cdot \frac{d\mathbf{u}}{dt} \cdot$$

If $|\mathbf{u}|$ is constant, the condition follows: conversely, the condition implies that $|\mathbf{u}|$ is constant.

THEOREM 2. *A necessary and sufficient condition that a proper vector always remain parallel to a fixed line is that*

$$\mathbf{u} \times \frac{d\mathbf{u}}{dt} = 0. \tag{7}$$

Proof. Let $\mathbf{u} = u(t)\mathbf{e}$ where $\mathbf{e}$ is a *unit* vector; then

$$\mathbf{u} \times \frac{d\mathbf{u}}{dt} = u\mathbf{e} \times \left(\frac{du}{dt}\mathbf{e} + u\frac{d\mathbf{e}}{dt}\right) = u^2\mathbf{e} \times \frac{d\mathbf{e}}{dt}.$$

If $\mathbf{e}$ is constant, $d\mathbf{e}/dt = 0$, and the condition follows. Conversely, since $u \neq 0$, the condition implies that

$$\mathbf{e} \times \frac{d\mathbf{e}}{dt} = 0. \qquad \text{Also} \quad \mathbf{e} \cdot \frac{d\mathbf{e}}{dt} = 0$$

from theorem 1. These equations are contradictory unless $d\mathbf{e}/dt = 0$; that is, $\mathbf{e}$ is constant.

43. Space Curves. Consider a space curve whose parametric equations are

$$x = x(t), \qquad y = y(t), \qquad z = z(t), \tag{1}$$

where $x(t)$, $y(t)$, $z(t)$ are analytic functions of the real variable t defined in a certain interval T of t values. To avoid having the curve degenerate into a point, we explicitly exclude the case in which all three functions are constants. We also restrict the interval T so that there is just one value of t corresponding to each point P of the curve. Then equations (1) set up a one-to-one correspondence between the points of the curve and the values of t in the interval T. The requirement that the functions $x(t)$, $y(t)$, $z(t)$ be analytic in T ensures the continuity of the functions and their derivatives of all orders and also guarantees a Taylor expansion for each function about any point of T. A curve which admits a representation (1) with functions thus restricted is called an *analytic space curve.* Moreover, if all three derivatives dx/dt, dy/dt, dz/dt do not vanish simultaneously for any value of t in the interval T, the curve is said to be *regular,* and the parameter t is said to be a *regular parameter.*

Let us make a change of parameter,

$$t = \varphi(u),$$

where $\varphi(u)$ is an analytic function of u in a certain interval U; we assume also that t just covers T as u ranges over U. Then the equations,

$$x = x[\varphi(u)], \qquad y = y[\varphi(u)], \qquad z = z[\varphi(u)]$$

constitute a new parametric representation of the curve. Writing $dt/du = \varphi'(u)$, we have

$$\frac{dx}{du} = \frac{dx}{dt}\varphi'(u), \qquad \frac{dy}{du} = \frac{dy}{dt}\varphi'(u), \qquad \frac{dz}{du} = \frac{dz}{dt}\varphi'(u);$$

hence if t is a regular parameter, u is a regular parameter when and only when $\varphi'(u) \neq 0$ in U. When this condition is fulfilled, the implicit-function theorem ensures the existence of the inverse function $u = \psi(t)$ which is also single valued and analytic in T. Thus the points of the curve not only correspond one to one to the t values in T, but also to the u values in U:

$$P \to t \to \psi(t) = u, \qquad u \to \varphi(u) = t \to P.$$

Let equations (1) define a regular analytic space curve. The position vector of the point $P(t)$ is

$$\text{(2)} \qquad \mathbf{r} = \mathbf{i}\,x(t) + \mathbf{j}\,y(t) + \mathbf{k}\,z(t) = \mathbf{r}(t);$$

and, if we write $\dot{\mathbf{r}}$, $\dot{x}$ for $d\mathbf{r}/dt$, dx/dt,

$$\text{(3)} \qquad \dot{\mathbf{r}} = \mathbf{i}\,\dot{x}(t) + \mathbf{j}\,\dot{y}(t) + \mathbf{k}\,\dot{z}(t) \neq 0.$$

If $P_0(t = t_0)$ is a fixed point on the curve, the length of the arc s from P_0 to P is defined as

$$\text{(4)} \qquad s = \int_{t_0}^{t} \sqrt{\dot{\mathbf{r}} \cdot \dot{\mathbf{r}}}\, dt,$$

an analytic function of t which is positive or negative, according as $t > t_0$ or $t < t_0$; moreover

$$\text{(5)} \qquad ds/dt = \sqrt{\dot{\mathbf{r}} \cdot \dot{\mathbf{r}}} > 0,$$

The inverse function $t = \varphi(s)$ is single valued and analytic, and $dt/ds > 0$. Putting $t = \varphi(s)$ in (2), we obtain $\mathbf{r} = \mathbf{r}(s)$, in which the arc s is a regular parameter.

The integrand in (4) is unity only when $|\dot{\mathbf{r}}| = 1$; hence we have the

THEOREM. *For the curve* $\mathbf{r} = \mathbf{r}(t)$ *the parameter* t *is the length of arc measured from a fixed point when and only when* $d\mathbf{r}/dt$ *is a unit vector.*

44. Unit Tangent Vector. Let $\mathbf{r} = \mathbf{r}(t)$ be a regular analytic space curve on which $s = \text{arc } P_0P$ is reckoned positive in the direction of increasing t. Then (Fig. 44a)

$$\frac{d\mathbf{r}}{ds} = \lim_{\Delta s \to 0} \frac{\Delta \mathbf{r}}{\Delta s} = \lim_{Q \to P} \frac{\overrightarrow{PQ}}{\text{arc } PQ}$$

is a unit vector (§ 43), and hence

$$\lim_{Q \to P} \frac{\text{chord } PQ}{\text{arc } PQ} = 1.$$

The tangent line to the curve at P is defined as the limiting position of the secant PQ as Q approaches P. Hence from its definition

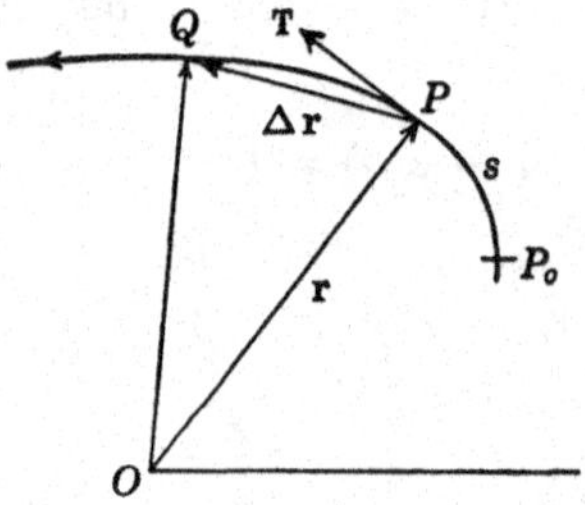

FIG. 44a

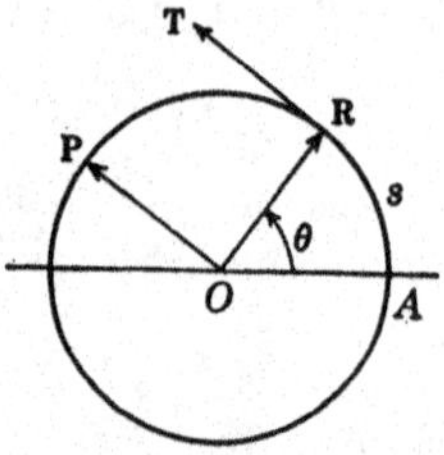

FIG. 44b

we conclude that $d\mathbf{r}/ds$ is a unit vector $\mathbf{T}$ tangent to the curve at P and pointing in the direction of increasing arcs:

$$\frac{d\mathbf{r}}{ds} = \mathbf{T}. \tag{1}$$

An important special case arises when $\mathbf{r}$ is a *unit vector revolving in a plane.* If we imagine this vector $\mathbf{R}$ always drawn from the same initial point O, its end point will describe a circle of unit radius (Fig. 44b) and $s = \theta$, where θ denotes the angle in radians between a fixed line OA and $\mathbf{R}$. If $\mathbf{P}$ is a unit vector perpendicular to $\mathbf{R}$ in the direction of increasing angles, (1) now becomes $d\mathbf{R}/d\theta = \mathbf{P}$. If $\mathbf{k}$ is a unit vector normal to the plane and pointing in the direction a right-handed screw advances when revolved in the positive sense of θ, $\mathbf{P} = \mathbf{k} \times \mathbf{R}$, and

$$\frac{d\mathbf{R}}{d\theta} = \mathbf{k} \times \mathbf{R} = \mathbf{P}. \tag{2}$$

In Fig. 44b, **k** points upward from the paper. Moreover, since $d\mathbf{P}/d\theta = \mathbf{k} \times (d\mathbf{R}/d\theta)$, we have

$$\frac{d\mathbf{P}}{d\theta} = \mathbf{k} \times \mathbf{P} = -\mathbf{R}. \tag{3}$$

We state these results in the

THEOREM. *The derivative of a unit vector revolving in a plane, with respect to the angle that it makes with a fixed direction, is another unit vector perpendicular to the first in the direction of increasing angles.*

If $P(x, y)$ is a variable point on a plane curve, $\mathbf{r} = x\mathbf{i} + y\mathbf{j}$,

$$\mathbf{T} = \frac{d\mathbf{r}}{ds} = \frac{dx}{ds}\mathbf{i} + \frac{dy}{ds}\mathbf{j},$$

and $\mathbf{i} \cdot \mathbf{T} = dx/ds$, $\mathbf{j} \cdot \mathbf{T} = dy/ds$; hence

$$\frac{dx}{ds} = \cos(\mathbf{i}, \mathbf{T}), \qquad \frac{dy}{ds} = \sin(\mathbf{i}, \mathbf{T}). \tag{4}$$

If (r, θ) are the polar coordinates of P, we write $\mathbf{r} = r\mathbf{R}$, where the polar distance r and the unit vector $\mathbf{R}$ are functions of θ. Now

$$\mathbf{T} = \frac{dr}{ds}\mathbf{R} + r\frac{d\mathbf{R}}{d\theta}\frac{d\theta}{ds} = \frac{dr}{ds}\mathbf{R} + r\frac{d\theta}{ds}\mathbf{P},$$

and $\mathbf{R} \cdot \mathbf{T} = dr/ds$, $\mathbf{P} \cdot \mathbf{T} = r\, d\theta/ds$; hence

$$\frac{dr}{ds} = \cos(\mathbf{R}, \mathbf{T}), \qquad r\frac{d\theta}{ds} = \sin(\mathbf{R}, \mathbf{T}). \tag{5}$$

Example 1. If θ is measured counterclockwise from the x axis,

$$\mathbf{R} = \mathbf{i}\cos\theta + \mathbf{j}\sin\theta, \qquad \mathbf{P} = \mathbf{i}\cos(\theta + \tfrac{1}{2}\pi) + \mathbf{j}\sin(\theta + \tfrac{1}{2}\pi).$$

From (2), the corresponding components of $d\mathbf{R}/d\theta$ and $\mathbf{P}$ must be equal; hence

$$\frac{d}{d\theta}\cos\theta = \cos(\theta + \tfrac{1}{2}\pi) = -\sin\theta, \qquad \frac{d}{d\theta}\sin\theta = \sin(\theta + \tfrac{1}{2}\pi) = \cos\theta.$$

Example 2. If r_1, r_2 are the distances of a point P on an ellipse from the foci $r_1 + r_2 = \text{const}$. On differentiating this equation with respect to s, we have from (5),

$$\frac{dr_1}{ds} + \frac{dr_2}{ds} = (\mathbf{R}_1 + \mathbf{R}_2) \cdot \mathbf{T} = 0.$$

Since $\mathbf{R}_1 + \mathbf{R}_2$ is perpendicular to $\mathbf{T}$, the normal to the ellipse at P has the direction of $\mathbf{R}_1 + \mathbf{R}_2$. The normal therefore bisects the angle between the focal radii.

45. Frenet's Formulas. Let $\mathbf{r} = \mathbf{r}(s)$ be a space curve, $s =$ arc P_0P, and $\mathbf{T} = d\mathbf{r}/ds$ the unit tangent vector at P. Since $\mathbf{T}$ is of constant length, $d\mathbf{T}/ds$, if not zero, must be perpendicular to $\mathbf{T}$. A directed line through P in the direction of $d\mathbf{T}/ds$ is called the *principal normal* of the curve at P. Let $\mathbf{N}$ denote a unit vector in the direction of the principal normal; then we may write

$$\frac{d\mathbf{T}}{ds} = \kappa\mathbf{N}, \tag{1}$$

where κ is a non-negative scalar called the *curvature* of the curve at P.

Now $\mathbf{B} = \mathbf{T} \times \mathbf{N}$ is a third unit vector perpendicular to both $\mathbf{T}$ and $\mathbf{N}$. Thus at each point of the curve we have a right-handed set of orthogonal unit vectors, $\mathbf{T}$, $\mathbf{N}$, $\mathbf{B}$, such that

$$\mathbf{T} \times \mathbf{N} = \mathbf{B}, \qquad \mathbf{N} \times \mathbf{B} = \mathbf{T}, \qquad \mathbf{B} \times \mathbf{T} = \mathbf{N}.$$

As P traverses the curve, we speak of the *moving trihedral* $\mathbf{TNB}$. A directed line through P in the direction of $\mathbf{B}$ is called the *binormal* to the curve at P.

Since $\mathbf{B}$ is a unit vector, $d\mathbf{B}/ds$, if not zero, must be perpendicular to $\mathbf{B}$. Differentiating $\mathbf{B} = \mathbf{T} \times \mathbf{N}$, we have

$$\frac{d\mathbf{B}}{ds} = \frac{d\mathbf{T}}{ds} \times \mathbf{N} + \mathbf{T} \times \frac{d\mathbf{N}}{ds} = \mathbf{T} \times \frac{d\mathbf{N}}{ds},$$

in view of (1). Hence $d\mathbf{B}/ds$ is perpendicular to $\mathbf{T}$ as well as $\mathbf{B}$ and therefore must be parallel to $\mathbf{B} \times \mathbf{T} = \mathbf{N}$. We therefore may write

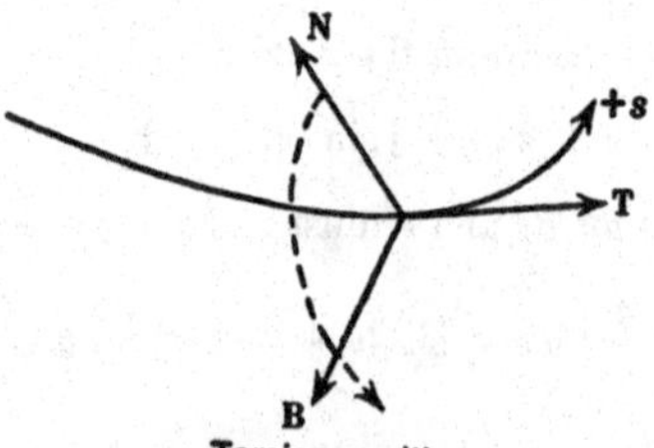

Torsion positive

FIG. 45

$$\frac{d\mathbf{B}}{ds} = -\tau\mathbf{N}, \tag{2}$$

where τ is a scalar called the *torsion* of the curve at P. The minus sign is introduced in (2) so that, when τ is positive, $d\mathbf{B}/ds$ has the direction of $-\mathbf{N}$; then, as P moves along the curve in the positive direction, $\mathbf{B}$ revolves about $\mathbf{T}$ in the same sense as a right-handed screw advancing in the direction of $\mathbf{T}$ (Fig. 45).

We may now compute $d\mathbf{N}/ds$ from $\mathbf{N} = \mathbf{B} \times \mathbf{T}$ by use of (1) and (2); thus

$$(3) \qquad \frac{d\mathbf{N}}{ds} = \frac{d\mathbf{B}}{ds} \times \mathbf{T} + \mathbf{B} \times \frac{d\mathbf{T}}{ds} = -\tau\mathbf{N} \times \mathbf{T} + \kappa\mathbf{B} \times \mathbf{N}.$$

Collecting (1), (2), and (3), we have the set of equations:

$$(4) \qquad \begin{aligned} \frac{d\mathbf{T}}{ds} &= \kappa\mathbf{N}, \\ \frac{d\mathbf{N}}{ds} &= -\kappa\mathbf{T} + \tau\mathbf{B}, \\ \frac{d\mathbf{B}}{ds} &= -\tau\mathbf{N}, \end{aligned}$$

known as *Frenet's Formulas*, which are fundamental in the theory of space curves.

If we write (3) in the form,

$$\frac{d\mathbf{N}}{ds} = (\tau\mathbf{T} + \kappa\mathbf{B}) \times \mathbf{N},$$

and introduce the *Darboux vector*,

$$(5) \qquad \boldsymbol{\delta} = \tau\mathbf{T} + \kappa\mathbf{B},$$

we have

$$\boldsymbol{\delta} \times \mathbf{T} = \kappa\mathbf{N}, \qquad \boldsymbol{\delta} \times \mathbf{B} = -\tau\mathbf{N}.$$

Hence Frenet's Formulas may be put in the symmetric form:

$$(6) \qquad \frac{d\mathbf{T}}{ds} = \boldsymbol{\delta} \times \mathbf{T}, \qquad \frac{d\mathbf{N}}{ds} = \boldsymbol{\delta} \times \mathbf{N} \qquad \frac{d\mathbf{B}}{ds} = \boldsymbol{\delta} \times \mathbf{B}.$$

Since $\boldsymbol{\delta} \cdot \mathbf{N} = 0$, we may write the Darboux vector in the form:

$$(7) \qquad \boldsymbol{\delta} = \mathbf{N} \times (\boldsymbol{\delta} \times \mathbf{N}) = \mathbf{N} \times \frac{d\mathbf{N}}{ds} \cdot$$

From (5), we see that the curvature κ and torsion τ are the components of the Darboux vector on the binormal and tangent, respectively. From (4), it is clear that both κ and τ have the dimensions of the reciprocal of length; hence

$$\rho = 1/\kappa, \qquad \sigma = 1/\tau$$

have the dimensions of length and are called the *radius of curvature* and the *radius of torsion*, respectively.

Since $\mathbf{N}$, by definition, has the same direction as $d\mathbf{T}/ds$, the curvature κ is never negative. If κ vanishes identically,

$$\frac{d\mathbf{T}}{ds} = 0, \qquad \mathbf{T} = \frac{d\mathbf{r}}{ds} = \mathbf{a}, \qquad \mathbf{r} = \mathbf{a}s + \mathbf{b},$$

where $\mathbf{a}$ and $\mathbf{b}$ are vector constants. The curve is then a straight line. Conversely, for a straight line, $\mathbf{T}$ is constant and $\kappa = 0$. *The only curves of zero curvature are straight lines.* For a straight line the preceding definition fails to determine $\mathbf{N}$. We therefore agree to give $\mathbf{N}$ any *fixed* direction normal to $\mathbf{T}$, and as before define $\mathbf{B} = \mathbf{T} \times \mathbf{N}$. Since $\mathbf{B}$ is also constant, $d\mathbf{B}/ds = 0$ and $\tau = 0$. For a straight line the Darboux vector is zero.

The torsion may be positive or negative. As P traverses the curve in the positive direction, the trihedral $\mathbf{TNB}$ will revolve about $\mathbf{T}$ as a right-handed or left-handed screw, according as τ is positive or negative. The sign of τ is independent of the choice of positive direction along the curve; for, if we reverse the positive direction, we must replace

$$s, \mathbf{T}, \frac{d\mathbf{T}}{ds}, \mathbf{N}, \frac{d\mathbf{N}}{ds}, \mathbf{B}, \frac{d\mathbf{B}}{ds} \quad \text{by} \quad -s, -\mathbf{T}, \frac{d\mathbf{T}}{ds}, \mathbf{N}, -\frac{d\mathbf{N}}{ds}, -\mathbf{B}, \frac{d\mathbf{B}}{ds},$$

and equations (4) maintain their form with unaltered κ and τ.

If τ vanishes identically, $d\mathbf{B}/ds = 0$, and $\mathbf{B}$ is a constant vector; hence, from

$$\mathbf{B}\cdot\mathbf{T} = \mathbf{B}\cdot\frac{d\mathbf{r}}{ds} = 0, \qquad \mathbf{B}\cdot(\mathbf{r} - \mathbf{r}_0) = 0,$$

that is, the curve lies in a plane normal to $\mathbf{B}$. Conversely, for a plane curve, $\mathbf{T}$ and $\mathbf{N}$ always lie in a fixed plane, while $\mathbf{B}$ is a unit vector normal to that plane; hence $d\mathbf{B}/ds = 0$ at all points where $\mathbf{N}$ and $\mathbf{B}$ are defined ($\kappa \neq 0$) and $\tau = 0$. *The only curves of zero torsion are plane.*

Example. Parallel Curves. Two curves Γ and Γ_1 are called *parallel* if a plane normal to one at any point is also normal to the other. The common normal plane cuts Γ and Γ_1 in *corresponding points* P and P_1; then $\overrightarrow{PP_1} = \mathbf{r}_1 - \mathbf{r}$ lies in the plane of $\mathbf{N}$ and $\mathbf{B}$ and may be written $\lambda\mathbf{N} + \mu\mathbf{B}$, where λ, μ are scalars. Thus we have

(i) $$\mathbf{r}_1 = \mathbf{r} + \lambda\mathbf{N} + \mu\mathbf{B};$$

on differentiating (i) with respect to s and using Frenet's Formulas, we have

$$\text{(ii)} \qquad \mathbf{T}_1 \frac{ds_1}{ds} = \mathbf{T} + \frac{d\lambda}{ds}\mathbf{N} + \lambda(-\kappa\mathbf{T} + \tau\mathbf{B}) + \frac{d\mu}{ds}\mathbf{B} - \mu\tau\mathbf{N}$$

$$= (1 - \lambda\kappa)\mathbf{T} + (\lambda' - \mu\tau)\mathbf{N} + (\mu' + \lambda\tau)\mathbf{B}.$$

Since $\mathbf{T}_1$ and $\mathbf{T}$ are normal to the same plane, we may choose the positive sense on Γ_1 so that $\mathbf{T}_1 = \mathbf{T}$; hence, from (ii), we conclude that

$$\text{(iii)} \qquad \frac{ds_1}{ds} = 1 - \lambda\kappa, \qquad \frac{d\lambda}{ds} = \mu\tau, \qquad \frac{d\mu}{ds} = -\lambda\tau.$$

Moreover

$$\lambda\frac{d\lambda}{ds} + \mu\frac{d\mu}{ds} = \frac{1}{2}\frac{d}{ds}(\lambda^2 + \mu^2) = 0,$$

and $\lambda^2 + \mu^2$ is constant. Consequently, *the distance PP_1 between corresponding points of parallel curves is always the same.*

On differentiating $\mathbf{T}_1 = \mathbf{T}$ with respect to s, we have also

$$\text{(iv)} \qquad \kappa_1\mathbf{N}_1\frac{ds_1}{ds} = \kappa\mathbf{N}; \qquad \text{hence} \quad \frac{ds_1}{ds} = \frac{\kappa}{\kappa_1},$$

and $\mathbf{N}_1 = \mathbf{N}$, $\mathbf{B}_1 = \mathbf{B}$. Comparison with (iii) gives

$$\kappa/\kappa_1 = 1 - \lambda\kappa \quad \text{or} \quad \rho_1 = \rho - \lambda.$$

Finally, on differentiating $\mathbf{B}_1 = \mathbf{B}$ with respect to s, we have

$$\text{(v)} \qquad -\tau_1\mathbf{N}_1\frac{ds_1}{ds} = -\tau\mathbf{N}; \qquad \text{hence, if } \tau \neq 0, \quad \frac{ds_1}{ds} = \frac{\tau}{\tau_1}.$$

From (iv) and (v), $\kappa/\tau = \kappa_1/\tau_1$ at corresponding points.

46. Curvature and Torsion. By use of Frenet's Formulas, the curvature and torsion are readily computed from the parametric equations of the curve. Thus, on differentiating $\mathbf{r} = \mathbf{r}(t)$ three times and denoting t derivatives by dots, we have

$$\dot{\mathbf{r}} = \frac{d\mathbf{r}}{ds}\frac{ds}{dt} = \dot{s}\mathbf{T},$$

$$\ddot{\mathbf{r}} = \ddot{s}\mathbf{T} + \dot{s}^2\frac{d\mathbf{T}}{ds}$$

$$= \ddot{s}\mathbf{T} + \dot{s}^2\kappa\mathbf{N},$$

$$\dddot{\mathbf{r}} = \dddot{s}\mathbf{T} + \ddot{s}\dot{s}\frac{d\mathbf{T}}{ds} + (2\dot{s}\ddot{s}\kappa + \dot{s}^2\dot{\kappa})\mathbf{N} + \dot{s}^3\kappa\frac{d\mathbf{N}}{ds}$$

$$= \dddot{s}\mathbf{T} + \ddot{s}\dot{s}\kappa\mathbf{N} + (2\dot{s}\ddot{s}\kappa + \dot{s}^2\dot{\kappa})\mathbf{N} + \dot{s}^3\kappa(-\kappa\mathbf{T} + \tau\mathbf{B})$$

$$= (\dddot{s} - \dot{s}^3\kappa^2)\mathbf{T} + (3\dot{s}\ddot{s}\kappa + \dot{s}^2\dot{\kappa})\mathbf{N} + \dot{s}^3\kappa\tau\mathbf{B}.$$

Hence

$$\dot{\mathbf{r}} \times \ddot{\mathbf{r}} = \dot{s}^3 \kappa \mathbf{B}, \qquad \dot{\mathbf{r}} \times \ddot{\mathbf{r}} \cdot \dddot{\mathbf{r}} = \dot{s}^6 \kappa^2 \tau,$$

and, since $|\dot{\mathbf{r}}| = |\dot{s}| \neq 0$,

(1), (2)
$$\kappa = \frac{|\dot{\mathbf{r}} \times \ddot{\mathbf{r}}|}{|\dot{\mathbf{r}}|^3}, \qquad \tau = \frac{\dot{\mathbf{r}} \times \ddot{\mathbf{r}} \cdot \dddot{\mathbf{r}}}{|\dot{\mathbf{r}} \times \ddot{\mathbf{r}}|^2}.$$

If the positive direction on the curve is that of increasing t, $\dot{s} = ds/dt > 0$; and the preceding equations show that

T has the direction of $\dot{\mathbf{r}}$,

B has the direction of $\dot{\mathbf{r}} \times \ddot{\mathbf{r}}$,

and, since $\mathbf{N} = \mathbf{B} \times \mathbf{T}$,

N has the direction of $(\dot{\mathbf{r}} \times \ddot{\mathbf{r}}) \times \dot{\mathbf{r}}$.

If the parametric equations of a plane curve are $x = x(t)$, $y = y(t)$, we have

$$\mathbf{r} = x\mathbf{i} + y\mathbf{j}, \qquad \dot{\mathbf{r}} = \dot{x}\mathbf{i} + \dot{y}\mathbf{j}, \qquad \ddot{\mathbf{r}} = \ddot{x}\mathbf{i} + \ddot{y}\mathbf{j},$$

and, from (1), its curvature is

(3)
$$\kappa = \frac{|\dot{x}\ddot{y} - \dot{y}\ddot{x}|}{(\dot{x}^2 + \dot{y}^2)^{\frac{3}{2}}}.$$

If the curve has the Cartesian equation $y = y(x)$, we can regard x as parameter: $x = t$, $y = y(t)$. Then (3) becomes

(4)
$$\kappa = \frac{|y''|}{(1 + y'^2)^{\frac{3}{2}}},$$

where the primes indicate differentiation with respect to x.

If the curve has the polar equation $r = r(\theta)$, we may write $\mathbf{r} = r\mathbf{R}$, where $\mathbf{R}$ is a unit radial vector. Then regarding θ as the parameter t, we have, from (44.2) and (44.3),

$$\dot{\mathbf{r}} = \dot{r}\mathbf{R} + r\mathbf{P}, \qquad \ddot{\mathbf{r}} = (\ddot{r} - r)\mathbf{R} + 2\dot{r}\mathbf{P};$$

hence, from (1),

(5)
$$\kappa = \frac{|r^2 + 2\dot{r}^2 - r\ddot{r}|}{(r^2 + \dot{r}^2)^{\frac{3}{2}}}.$$

Example. Find the vectors T, N, B and the curvature and torsion of the twisted cubic

$$x = 2t, \qquad y = t^2, \qquad z = t^3/3$$

at the point where $t = 1$.

Since $\mathbf{r}(t) = [2t, t^2, t^3/3]$, we have

$$\dot{\mathbf{r}} = [2, 2t, t^2], \qquad \ddot{\mathbf{r}} = [0, 2, 2t], \qquad \dddot{\mathbf{r}} = [0, 0, 2];$$

hence, when $t = 1$,

$$\dot{\mathbf{r}} = [2, 2, 1], \qquad \ddot{\mathbf{r}} = [0, 2, 2], \qquad \dddot{\mathbf{r}} = [0, 0, 2],$$

$$\dot{\mathbf{r}} \times \ddot{\mathbf{r}} = [2, -4, 4], \qquad \dot{\mathbf{r}} \times \ddot{\mathbf{r}} \cdot \dddot{\mathbf{r}} = 8.$$

Since $\mathbf{T}$, $\mathbf{B}$ are unit vectors in the directions of $\dot{\mathbf{r}}$, $\dot{\mathbf{r}} \times \ddot{\mathbf{r}}$, and $\mathbf{N} = \mathbf{B} \times \mathbf{T}$,

$$\mathbf{T} = \tfrac{1}{3}[2, 2, 1], \qquad \mathbf{N} = -\tfrac{1}{3}[2, -1, -2], \qquad \mathbf{B} = \tfrac{1}{3}[1, -2, 2].$$

Moreover, from (1) and (2), $\kappa = \frac{2}{9}$, $\tau = \frac{2}{9}$.

47. Fundamental Theorem. *Two curves for which the curvature and torsion are the same functions of the arc are congruent.*

Proof. Let the curves have the equations $\mathbf{r} = \mathbf{r}_1(s)$, $\mathbf{r} = \mathbf{r}_2(s)$. Bring the origins of arc and the trihedrals $\mathbf{T}_1\mathbf{N}_1\mathbf{B}_1$, $\mathbf{T}_2\mathbf{N}_2\mathbf{B}_2$ at these points into coincidence. At the points P_1, P_2 of the curves, corresponding to the same value of s, consider the function,

$$(1) \qquad f(s) = \mathbf{T}_1 \cdot \mathbf{T}_2 + \mathbf{N}_1 \cdot \mathbf{N}_2 + \mathbf{B}_1 \cdot \mathbf{B}_2.$$

By Frenet's Formulas,

$$\begin{aligned} \frac{df}{ds} = {} & \kappa_1 \mathbf{N}_1 \cdot \mathbf{T}_2 + \kappa_2 \mathbf{N}_2 \cdot \mathbf{T}_1 \\ & + (-\kappa_1 \mathbf{T}_1 + \tau_1 \mathbf{B}_1) \cdot \mathbf{N}_2 + (-\kappa_2 \mathbf{T}_2 + \tau_2 \mathbf{B}_2) \cdot \mathbf{N}_1 \\ & - \tau_1 \mathbf{N}_1 \cdot \mathbf{B}_2 - \tau_2 \mathbf{N}_2 \cdot \mathbf{B}_1; \end{aligned}$$

and, since $\kappa_1 = \kappa_2$, $\tau_1 = \tau_2$ for the same value of s,

$$\frac{df}{ds} = 0, \qquad f(s) = C, \quad \text{a const.}$$

But at the point $s = 0$, the trihedrals coincide, and

$$C = f(0) = 1 + 1 + 1 = 3.$$

Hence for any value of s, $f(s) = 3$, an equation that implies that each scalar product in (1) equals one; consequently,

$$\mathbf{T}_1 = \mathbf{T}_2, \qquad \mathbf{N}_1 = \mathbf{N}_2, \qquad \mathbf{B}_1 = \mathbf{B}_2.$$

From the first of these equations,

$$\frac{d\mathbf{r}_1}{ds} = \frac{d\mathbf{r}_2}{ds}, \qquad \mathbf{r}_1 = \mathbf{r}_2 + \mathbf{a};$$

and $\mathbf{a}$, the vector constant of integration, is zero, since $\mathbf{r}_1 = \mathbf{r}_2$ when $s = 0$. Therefore $\mathbf{r}_1(s) = \mathbf{r}_2(s)$, and the curves coincide.

48. Osculating Plane. The *osculating plane* of a curve at a point P_1 is defined as the limiting plane through three of its points P_1, P_2, P_3 as P_2 and P_3 approach P_1.

Let the points P_1, P_2, P_3 of the curve $\mathbf{r} = \mathbf{r}(s)$ correspond to the arc values s_1, s_2, s_3 $(s_1 < s_2 < s_3)$. Then, if $\mathbf{n}$ is the unit normal to the plane $P_1P_2P_3$, the function,

$$f(s) = [\mathbf{r}(s) - \mathbf{r}_1] \cdot \mathbf{n}$$

vanishes when $s = s_1, s_2, s_3$. Hence from Rolle's Theorem,

$$f'(s) = \mathbf{r}'(s) \cdot \mathbf{n}$$

vanishes twice, and

$$f''(s) = \mathbf{r}''(s) \cdot \mathbf{n}$$

vanishes once in the interval $s_1 < s < s_3$. Consequently, as P_2 and P_3 approach P_1, $\mathbf{n}$ approaches a limiting vector $\mathbf{n}_1$, such that

$$\mathbf{r}'(s_1) \cdot \mathbf{n}_1 = \mathbf{T}_1 \cdot \mathbf{n}_1 = 0,$$

$$\mathbf{r}''(s_1) \cdot \mathbf{n}_1 = \kappa_1 \mathbf{N}_1 \cdot \mathbf{n}_1 = 0.$$

If $\kappa_1 \neq 0$, $\mathbf{n}_1$ is perpendicular to both $\mathbf{T}_1$ and $\mathbf{N}_1$, that is, parallel to $\mathbf{B}_1$. *The osculating plane at a point of a curve is the plane of the tangent and principal normal at this point.* For a plane curve $\mathbf{T}$ and $\mathbf{N}$ lie in the plane of the curve—the osculating plane at all of its points.

For the curve $\mathbf{r} = \mathbf{r}(t)$, $\mathbf{B}$ is parallel to $\dot{\mathbf{r}} \times \ddot{\mathbf{r}}$ (§ 46); hence the equation of the osculating plane at any point $\mathbf{r}$ is

$$(1) \qquad (\mathbf{R} - \mathbf{r}) \cdot \dot{\mathbf{r}} \times \ddot{\mathbf{r}} = 0;$$

here $\mathbf{R}$ is the position vector to any point of the osculating plane.

The osculating planes of a curve form a one-parameter family. The osculating planes at the points P_1 and P_2 intersect in a straight line; and, as P_2 approaches P_1, the limiting position of this line is called the *characteristic* of the osculating plane at P_1. To find the characteristics of the osculating planes to the curve $\mathbf{r} = \mathbf{r}(s)$, we must adjoin to their equation $(\mathbf{R} - \mathbf{r}) \cdot \mathbf{B} = 0$, the equation obtained by differentiating it with respect to s, namely,

$$-\mathbf{T} \cdot \mathbf{B} - \tau(\mathbf{R} - \mathbf{r}) \cdot \mathbf{N} = 0, \quad \text{or} \quad (\mathbf{R} - \mathbf{r}) \cdot \mathbf{N} = 0,$$

if $\tau \neq 0$. The two equations,

$$(\mathbf{R} - \mathbf{r}) \cdot \mathbf{B} = 0, \qquad (\mathbf{R} - \mathbf{r}) \cdot \mathbf{N} = 0,$$

represent planes through the point $\mathbf{r}$ of the curve perpendicular to $\mathbf{B}$ and $\mathbf{N}$, respectively; together, they represent a line through the point $\mathbf{r}$ parallel to $\mathbf{N} \times \mathbf{B} = \mathbf{T}$. *The characteristics of the osculating planes of a skew curve are tangents to the curve.*

Example. For the twisted cubic

$$x = 2t, \qquad y = t^2, \qquad z = t^3/3$$

of the example in § 46, $\dot{\mathbf{r}} \times \ddot{\mathbf{r}} = 2[1, -2, 2]$ at the point $(2, 1, \frac{1}{3})$ for which $t = 1$. Hence, from (1), the equation of the osculating plane to the curve at this point is

$$(x - 2) - 2(y - 1) + 2(z - \tfrac{1}{3}) = 0.$$

49. Center of Curvature. The center of curvature of a curve at the point P_1 is defined as the center of the limiting circle through three of its points P_1, P_2, P_3, as P_2 and P_3 approach P_1.

Since the limiting plane of P_1, P_2, P_3 is the osculating plane through P_1, the limiting circle lies in this plane. If $\mathbf{c}$ and a denote the position vector of the center and the radius of the circle through the points P_1, P_2, P_3 of the curve $\mathbf{r} = \mathbf{r}(s)$, the function,

$$f(s) = [\mathbf{r}(s) - \mathbf{c}] \cdot [\mathbf{r}(s) - \mathbf{c}] - a^2$$

vanishes for $s = s_1, s_2, s_3$ $(s_1 < s_2 < s_3)$. Hence

$$f'(s) = 2(\mathbf{r} - \mathbf{c}) \cdot \mathbf{r}' = 2(\mathbf{r} - \mathbf{c}) \cdot \mathbf{T}$$

vanishes twice, and

$$f''(s) = 2\mathbf{r}' \cdot \mathbf{r}' + 2(\mathbf{r} - \mathbf{c}) \cdot \mathbf{r}'' = 2[1 + \kappa(\mathbf{r} - \mathbf{c}) \cdot \mathbf{N}]$$

vanishes once in the interval $s_1 < s < s_3$.

As P_2 and P_3 approach P_1, $\mathbf{c}$ and a approach limits $\mathbf{c}_1, a_1$, such that

$$(\mathbf{r}_1 - \mathbf{c}_1) \cdot (\mathbf{r}_1 - \mathbf{c}_1) - a_1^2 = 0,$$
$$(\mathbf{r}_1 - \mathbf{c}_1) \cdot \mathbf{T}_1 = 0,$$
$$(\mathbf{r}_1 - \mathbf{c}_1) \cdot \mathbf{N}_1 + \rho_1 = 0.$$

Since $\mathbf{r}_1 - \mathbf{c}_1$ lies in the osculating plane, its components on $\mathbf{T}_1, \mathbf{N}_1, \mathbf{B}_1$ are $0, -\rho_1, 0$, respectively, and $\mathbf{r}_1 - \mathbf{c}_1 = -\rho_1\mathbf{N}_1$. Thus the center and radius of the limiting circle are

$$\mathbf{c}_1 = \mathbf{r}_1 + \rho_1\mathbf{N}_1 \quad \text{and} \quad a_1 = \rho_1.$$

The center of curvature at any point P of the curve has the position vector,

(1) $$\mathbf{c} = \mathbf{r} + \rho\mathbf{N}.$$

This is a point on the principal normal at a distance ρ from P in its positive direction.

Example. Curves of Constant Curvature. Let $\mathbf{r} = \mathbf{r}(s)$ be a twisted curve Γ of constant curvature κ. The locus Γ_1 of its centers of curvature has the equation,

$$\mathbf{r}_1 = \mathbf{r} + \rho\mathbf{N}. \tag{2}$$

Differentiating with respect to s, we have

$$\mathbf{T}_1 \frac{ds_1}{ds} = \mathbf{T} + \rho(\tau\mathbf{B} - \kappa\mathbf{T}) = \frac{\tau}{\kappa}\mathbf{B}.$$

Choose the positive direction on Γ_1 so that $\mathbf{T}_1 = \mathbf{B}$; then

$$\frac{ds_1}{ds} = \frac{\tau}{\kappa}. \tag{3}$$

Differentiating $\mathbf{T}_1 = \mathbf{B}$ with respect to s now gives

$$\kappa_1\mathbf{N}_1 \frac{ds_1}{ds} = -\tau\mathbf{N}, \quad \text{or} \quad \kappa_1\mathbf{N}_1 = -\kappa\mathbf{N};$$

hence $\mathbf{N}_1 = -\mathbf{N}$ and $\kappa_1 = \kappa$. Thus Γ_1 is also a curve of constant curvature. Since (2) may be written

$$\mathbf{r} = \mathbf{r}_1 + \rho_1\mathbf{N}_1, \tag{4}$$

Γ is also the locus of the centers of curvature of Γ_1. Consequently from (3) we also have $ds/ds_1 = \tau_1/\kappa_1$; hence $\tau\tau_1 = \kappa\kappa_1$. The two curves thus have the following relations:

$$\mathbf{T}_1 = \mathbf{B}, \quad \mathbf{N}_1 = -\mathbf{N}, \quad \mathbf{B}_1 = \mathbf{T}; \qquad \kappa_1 = \kappa, \quad \tau\tau_1 = \kappa\kappa_1. \tag{5}$$

Twisted curves of constant curvature may be associated in pairs, each curve being the locus of the centers of curvature of the other.

50. Plane Curves. For plane curves the torsion is zero and Frenet's Formulas reduce to

$$\frac{d\mathbf{T}}{ds} = \kappa\mathbf{N} \qquad \frac{d\mathbf{N}}{ds} = -\kappa\mathbf{T}. \tag{1}$$

Since $d\mathbf{T}/ds$ is always directed towards the concave side of a plane curve, the same is true of $\mathbf{N}$. The osculating plane at any point P is the plane of the curve; and, if $\kappa \neq 0$, the center of curvature P_1 is given by

$$\mathbf{r}_1 = \mathbf{r} + \rho\mathbf{N}. \tag{2}$$

At points of inflection $d\mathbf{T}/ds = 0$, $\kappa = 0$, and $\mathbf{N}$ ceases to be defined. As we pass through a point of inflection, $\mathbf{N}$ abruptly reverses its direction, and hence $\mathbf{B} = \mathbf{T} \times \mathbf{N}$ does the same.

To remedy this discontinuous behavior of **N** and **B** at points of inflection, the following convention often is adopted in the differential geometry of plane curves. Take **B** once and for all as a *fixed* unit vector normal to the plane of the curve, and define $\mathbf{N} = \mathbf{B} \times \mathbf{T}$. As before **T**, **N**, **B** form a right-handed set of orthogonal unit vectors. The curvature κ is defined by (1) and is posi-

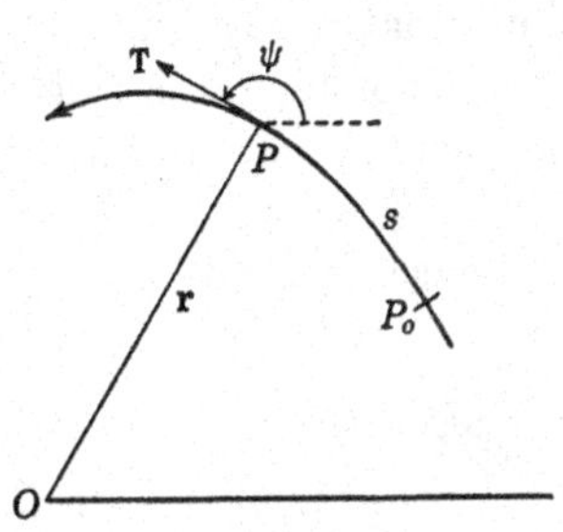

FIG. 50a

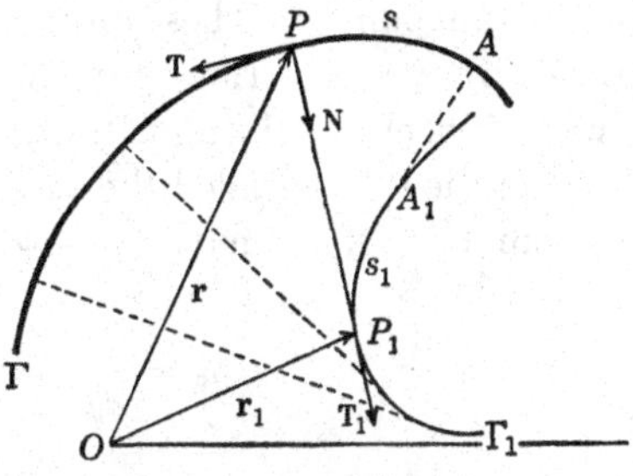

FIG. 50b

tive or negative, according as **N** has the direction of $d\mathbf{T}/ds$ or the opposite. Equations (1) still hold good; for

$$\frac{d\mathbf{N}}{ds} = \mathbf{B} \times \frac{d\mathbf{T}}{ds} = \mathbf{B} \times \kappa\mathbf{N} = -\kappa\mathbf{T}.$$

Let ψ be the angle from a fixed line in the plane to the tangent at P, taken positive in the sense determined by B (Fig. 50a). Then, from (44.2),

$$\frac{d\mathbf{T}}{ds} = \frac{d\mathbf{T}}{d\psi}\frac{d\psi}{ds} = \mathbf{B} \times \mathbf{T}\frac{d\psi}{ds} = \mathbf{N}\frac{d\psi}{ds},$$

and hence, from (1),

$$\kappa = \frac{d\psi}{ds}, \qquad \rho = \frac{ds}{d\psi}. \tag{3}$$

The locus of the centers of curvature of a plane curve is called its *evolute*. Let P and P_1 be corresponding points of a curve Γ and its evolute Γ_1 (Fig. 50b); and $s = AP$ and $s_1 = A_1P_1$ denote corresponding arcs. On differentiating the equation (2) of Γ_1 with respect to s, we have

$$\mathbf{T}_1\frac{ds_1}{ds} = \mathbf{T} + \rho\frac{d\mathbf{N}}{ds} + \frac{d\rho}{ds}\mathbf{N} = \frac{d\rho}{ds}\mathbf{N}.$$

Choose the positive direction on Γ_1 so that $\mathbf{T}_1 = \mathbf{N}$; then

$$\frac{ds_1}{ds} = \frac{d\rho}{ds}, \quad \text{and} \quad s_1 = \rho + \text{const.}$$

Hence the tangent to Γ_1 is normal to Γ; and, since $\Delta s_1 = \Delta\rho$, an arc of Γ_1 is equal to the difference in the radii of curvature of Γ to its end points. These properties show that a curve may be traced by a taut string unwound from its evolute; the string is always tangent to Γ_1 and its free portion equal to ρ. From this point of view, Γ is called the *involute* of Γ_1.

From $\mathbf{T}_1 = \mathbf{N}$, we have $\psi_1 = \psi + \pi/2$; hence, from (3),

$$\rho_1 = \frac{ds_1}{d\psi_1} = \frac{d\rho}{d\psi} = \frac{d^2 s}{d\psi^2}.$$

Example 1. The only plane curves of constant non-zero curvature are circles. For from

$$\frac{d\mathbf{N}}{ds} = -\kappa\mathbf{T}, \quad \text{or} \quad \frac{d\mathbf{r}}{ds} = -\rho\frac{d\mathbf{N}}{ds},$$

we have, on integration,

$$\mathbf{r} - \mathbf{c} = -\rho\mathbf{N}, \quad \text{or} \quad |\mathbf{r} - \mathbf{c}|^2 = \rho^2.$$

This is the equation of a circle of radius ρ and center $\mathbf{c}$.

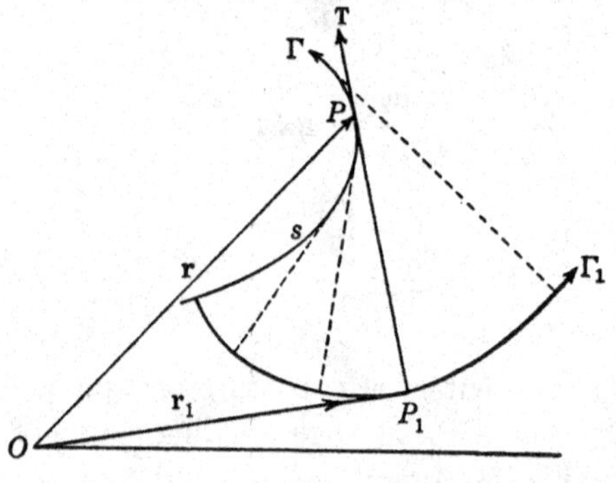

FIG. 50c

Example 2. An *involute* Γ_1 of a plane curve Γ is generated by the point P_1 of a taut string unwound from Γ (Fig. 50c). If Γ is the curve $\mathbf{r} = \mathbf{r}(s)$, the involute is given by

$$\mathbf{r}_1 = \mathbf{r} - s\mathbf{T}.$$

Hence, on differentiation with respect to s,

$$\mathbf{T}_1 \frac{ds_1}{ds} = \mathbf{T} - \mathbf{T} - s\kappa\mathbf{N} = -s\kappa\mathbf{N}.$$

If we choose the positive direction on Γ_1 so that $\mathbf{T}_1 = -\mathbf{N}$,

$$\text{(i)} \qquad \frac{ds_1}{ds} = s\kappa = s\frac{d\psi}{ds}, \quad \text{or} \quad \frac{ds_1}{d\psi} = s.$$

The *intrinsic equation* of a plane curve is the relation connecting s and ψ, say $s = s(\psi)$. For the arc s_1 along its involute we have, from (i),

$$s_1 = \int_{\psi_0}^{\psi} s(\psi)d\psi, \quad \text{provided} \quad s_1 = 0, \quad \text{when} \quad \psi = \psi_0.$$

The intrinsic equation of a circle of radius r is $s = r\psi$ when ψ is measured from the tangent at the origin of axes. For its involute we have $s_1 = r\psi^2/2$, provided $s_1 = 0$ when $\psi = 0$.

The intrinsic equation of a catenary of parameter c is $s = c \tan \psi$ when s is measured from the vertex ($\psi = 0$).† The arc s_1, measured along the involute from the vertex of the catenary, is

$$s_1 = c\int_0^{\psi} \tan \psi \, d\psi = c \log \sec \psi.$$

Example 3. *Envelopes.* Consider the plane vector function of two variables, $\mathbf{r} = \mathbf{f}(u, v)$. The one-parameter family of curves $u = c$ (constant) is given by

$$\mathbf{r}_1 = \mathbf{f}(u, v), \qquad u = \text{const.}$$

If this family has a curve envelope given by $v = \varphi(u)$, namely,

$$\text{(ii)} \qquad \mathbf{R} = \mathbf{f}[u, \varphi(u)],$$

the vectors,

$$\frac{d\mathbf{R}}{\partial u} = \frac{\partial \mathbf{f}}{\partial u} + \frac{\partial \mathbf{f}}{\partial v}\varphi'(u) \quad \text{and} \quad \frac{d\mathbf{r}_1}{dv} = \frac{\partial \mathbf{f}}{\partial v}$$

are parallel at the points of contact; that is,

$$\text{(iii)} \qquad \frac{\partial \mathbf{f}}{\partial u} \times \frac{\partial \mathbf{f}}{\partial v} = 0.$$

This condition must be fulfilled if the envelope exists. If (iii) leads to a relation $v = \varphi(u)$ (which does not make $\partial\mathbf{f}/\partial u$ or $\partial\mathbf{f}/\partial v$ zero), this relation gives the envelope (ii).

† See Brand, *Vectorial Mechanics*, equation (97.3). In Fig. 97b (p. 202), the string PA will unwrap into the position PL; as P moves along the catenary, the locus of L is the involute of the catenary. The line LQ is tangent to the involute ($\mathbf{T}_1 = -\mathbf{N}$); and, since the x-axis intercepts a constant segment $LQ = c$ on this tangent, the involute of the catenary is a *tractrix*, a curve characterized by this property.

Consider, for example, the one-parameter family of normals to the plane curve $\mathbf{r} = \mathbf{r}(s)$, namely,

$$\mathbf{r}_1 = \mathbf{r}(s) + v\,\mathbf{N}(s) \qquad (s = \text{const.})$$

If they have an envelope,

$$\frac{\partial \mathbf{r}_1}{\partial s} \times \frac{\partial \mathbf{r}_1}{\partial v} = (\mathbf{T} - v\kappa\mathbf{T}) \times \mathbf{N} = (1 - v\kappa)\mathbf{B} = 0.$$

Hence the envelope is given by $v = 1/\kappa = \rho$, or

$$\mathbf{R} = \mathbf{r}(s) + \rho\mathbf{N}(s),$$

namely the locus of the centers of curvature of the curve.

Example 4. Plane Caustic by Reflection. Light issuing from a point O is reflected from a mirror Γ (Fig. 50d). The curve enveloped by the reflected rays is called a *caustic*. If $\mathbf{R}$ and $\mathbf{e}$ are the unit vectors along the incident and reflected rays, the reflected rays are the one-parameter family of lines,

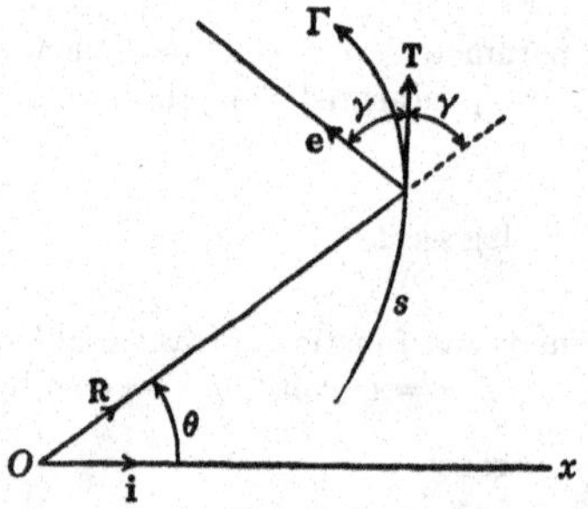

FIG. 50d

$$\mathbf{r}_1 = \mathbf{r}(s) + v\mathbf{e}(s).$$

To find their envelope, we form the equation:

$$\text{(iv)} \qquad \frac{\partial \mathbf{r}_1}{\partial s} \times \frac{\partial \mathbf{r}_1}{\partial v} = \left(\mathbf{T} + v\frac{d\mathbf{e}}{ds}\right) \times \mathbf{e} = 0.$$

Denote the angles $(\mathbf{i}, \mathbf{R}) = \theta$, $(\mathbf{R}, \mathbf{T}) = \gamma$; then $(\mathbf{T}, \mathbf{e}) = \gamma$ by the law of reflection, and

$$\psi = (\mathbf{i}, \mathbf{T}) = \theta + \gamma, \qquad \varphi = (\mathbf{i}, \mathbf{e}) = \theta + 2\gamma = 2\psi - \theta.$$

Now

$$\frac{d\mathbf{e}}{ds} = \frac{d\mathbf{e}}{d\varphi}\frac{d\varphi}{ds} = \mathbf{k} \times \mathbf{e}\left(2\frac{d\psi}{ds} - \frac{d\theta}{ds}\right) = \mathbf{k} \times \mathbf{e}\left(2\kappa - \frac{\sin\gamma}{r}\right)$$

from (44.5). Substitution in (iv) now gives

$$\mathbf{k}\sin\gamma - \mathbf{k}v\left(2\kappa - \frac{\sin\gamma}{r}\right) = 0;$$

whence

$$\text{(v)} \qquad \frac{1}{v} + \frac{1}{r} = \frac{2}{\rho\sin\gamma}.$$

This equation determines v and the caustic curve.

Letting $r \to \infty$, we obtain a beam of parallel rays; then $v = \frac{1}{2}\rho\sin\gamma$.

When Γ is a circular arc of radius ρ with center on Ox, the caustic has a cusp on Ox at a distance v_1 from Γ given by $1/v_1 + 1/r = 2/\rho$.

Example 5. The *tractrix* is a plane curve for which the segment of any tangent between the point of contact and a fixed line is constant. In Fig. 50e the fixed line is the x-axis, the constant length c, and, if $\mathbf{r} = \overrightarrow{OP}$, $\mathbf{r}_1 = \overrightarrow{OP_1}$,

$$\mathbf{r}_1 = \mathbf{r} + c\mathbf{T}.$$

Hence if we differentiate with respect to s,

$$\mathbf{i}\frac{ds_1}{ds} = \mathbf{T} + c\kappa\mathbf{N};$$

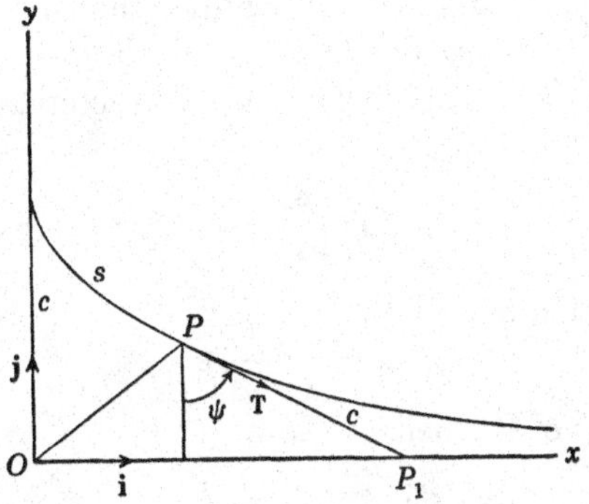

FIG. 50e

on multiplying by $\mathbf{j}\cdot$ we have

$$0 = \mathbf{j}\cdot\mathbf{T} + c\kappa\mathbf{j}\cdot\mathbf{N} = -\cos\psi + c\frac{d\psi}{ds}\sin\psi,$$

or

$$\frac{ds}{d\psi} = c\tan\psi.$$

If $s = 0$ when $\psi = 0$, this gives

$$s = c\log\sec\psi$$

for the intrinsic equation of the tractrix (cf. ex. 2).

51. Helices. *A helix is a twisted curve whose tangent makes a constant angle with a fixed direction.* If $\mathbf{e}$ is a unit vector in the fixed direction, and α is the constant angle, the defining equation of a helix is

(1) $$\mathbf{e}\cdot\mathbf{T} = \cos\alpha, \qquad (0 < \alpha < \tfrac{1}{2}\pi)\cdot$$

Differentiating (1) twice with respect to s gives

$$\mathbf{e} \cdot \mathbf{N} = 0, \qquad \mathbf{e} \cdot \frac{d\mathbf{N}}{ds} = 0;$$

hence the Darboux vector $\boldsymbol{\delta} = \mathbf{N} \times (d\mathbf{N}/ds)$ is parallel to $\mathbf{e}$. Conversely, if $\boldsymbol{\delta}$ for a curve is always parallel to a fixed vector $\mathbf{e}$,

$$\mathbf{e} \cdot \mathbf{N} = 0, \qquad \mathbf{e} \cdot \frac{d\mathbf{T}}{ds} = 0, \qquad \text{and} \quad \mathbf{e} \cdot \mathbf{T} = \cos \alpha,$$

where $\cos \alpha$ is a constant of integration; the curve is therefore a helix. *The only twisted curves whose Darboux vector has a fixed direction are helices.* Thus helices are characterized by the property,

$$\text{(2)} \qquad \boldsymbol{\delta} \times \frac{d\boldsymbol{\delta}}{ds} = 0 \qquad \text{(§ 42, theorem 2).}$$

The fixed direction of

$$\boldsymbol{\delta} = \tau \mathbf{T} + \kappa \mathbf{B}$$

is along the *axis* of the helix. Now

$$\text{(3)} \qquad \frac{d\boldsymbol{\delta}}{ds} = \frac{d\tau}{ds} \mathbf{T} + \frac{d\kappa}{ds} \mathbf{B}$$

for $\tau \, d\mathbf{T}/ds + \kappa \, d\mathbf{B}/ds = 0$ (45.4); hence

$$\boldsymbol{\delta} \times \frac{d\boldsymbol{\delta}}{ds} = \left(\kappa \frac{d\tau}{ds} - \tau \frac{d\kappa}{ds} \right) \mathbf{N} = -\tau^2 \mathbf{N} \frac{d}{ds} \left(\frac{\kappa}{\tau} \right)$$

Thus $\boldsymbol{\delta} \times (d\boldsymbol{\delta}/ds) = 0$ implies that the ratio κ/τ is constant, and conversely. *Helices are the only twisted curves for which the ratio of curvature to torsion is constant.*

If we put $\mathbf{T} = d\mathbf{r}/ds$ in (1) and integrate, we obtain

$$\mathbf{e} \cdot \mathbf{r} = s \cos \alpha + c$$

for the component of $\mathbf{r}$ in the direction of $\mathbf{e}$. Hence, if $\mathbf{r}_1$ denotes the projection of $\mathbf{r}$ on a plane perpendicular to $\mathbf{e}$,

$$\text{(4)} \qquad \mathbf{r} = \mathbf{r}_1 + (s \cos \alpha + c)\mathbf{e}.$$

Every helix therefore lies on a cylinder with generators parallel to $\mathbf{e}$. The plane curve Γ_1 traced by $\mathbf{r}_1$ is a normal section of the cylinder.

If $s = P_0P$ is the arc along the helix, and $s_1 = P_0P_1$ the arc along the normal section Γ_1 through P_0, we have, on differentiating (4) twice with respect to s,

$$\mathbf{T} = \mathbf{T}_1 \frac{ds_1}{ds} + \cos\alpha\, \mathbf{e} \tag{5}$$

$$\kappa\mathbf{N} = \kappa_1\mathbf{N}_1 \left(\frac{ds_1}{ds}\right)^2 + \mathbf{T}_1 \frac{d^2s_1}{ds^2}\,. \tag{6}$$

From (5),

$$\left(\frac{ds_1}{ds}\right)^2 = (\mathbf{T} - \mathbf{e}\cos\alpha)\cdot(\mathbf{T} - \mathbf{e}\cos\alpha)$$
$$= 1 - 2\cos^2\alpha + \cos^2\alpha = \sin^2\alpha;$$

and, if we choose the positive direction on Γ_1 so that s_1 increases with s,

$$\frac{ds_1}{ds} = \sin\alpha. \tag{7}$$

Remembering that κ and κ_1 are essentially positive, we now have, from (6), $\mathbf{N} = \mathbf{N}_1$, and

$$\kappa = \kappa_1 \sin^2\alpha. \tag{8}$$

On differentiating $\mathbf{e} = \mathbf{T}\cos\alpha + \mathbf{B}\sin\alpha$ we find $\kappa/\tau = \tan\alpha$; hence

$$\tau = \kappa_1 \sin\alpha\cos\alpha. \tag{9}$$

Finally, on integrating (7), we obtain

$$s_1 = s\sin\alpha, \tag{10}$$

the constant of integration being zero, since both s and s_1 are measured from the same point P_0.

The only twisted curve of constant curvature and torsion is the circular helix. For, since κ/τ is constant, such curves are helices; and, from (8), κ_1 is constant; that is, the normal section is a circle.

The circular helix is the only twisted curve for which the Darboux vector is constant. For from (3) we see that $\boldsymbol{\delta}$ is constant when and only when κ and τ are constant.

Example. The curve,

$$x = a \cos t, \qquad y = a \sin t, \qquad z = bt,$$

is a *circular helix;* for the curve lies on the cylinder $x^2 + y^2 = a^2$, and, since

$$\dot{\mathbf{r}} = [-a \sin t,\ a \cos t,\ b],$$

$$\cos(\mathbf{k}, \mathbf{T}) = \mathbf{k} \cdot \dot{\mathbf{r}}/|\dot{\mathbf{r}}| = b/\sqrt{a^2 + b^2} \text{ (const.)}.$$

Moreover

$$\ddot{\mathbf{r}} = [-a \cos t,\ -a \sin t,\ 0],$$

$$\dddot{\mathbf{r}} = [a \sin t,\ -a \cos t,\ 0],$$

and, from (46.1) and (46.2),

$$\kappa = \frac{|\dot{\mathbf{r}} \times \ddot{\mathbf{r}}|}{|\dot{\mathbf{r}}|^3} = \frac{a}{a^2 + b^2}, \qquad \tau = \frac{\dot{\mathbf{r}} \times \ddot{\mathbf{r}} \cdot \dddot{\mathbf{r}}}{|\dot{\mathbf{r}} \times \ddot{\mathbf{r}}|^2} = \frac{b}{a^2 + b^2}.$$

We now find the Darboux vector $\boldsymbol{\delta} = \mathbf{k}/\sqrt{a^2 + b^2}$; it is constant in magnitude and direction.

52. Kinematics of a Particle. To define the position P of a particle moving along a curve Γ, choose a point P_0 of Γ from which to measure arcs, and take a definite direction along Γ as positive. Then, if the arc $s = P_0P$ is given as a function of the time, $s = f(t)$, the motion of the particle is determined; for its position is given at every instant. The *speed* v of the particle at the instant t then is defined as

$$(1) \qquad v = \frac{ds}{dt}.$$

Thus the speed will be positive or negative, according as s is increasing or decreasing at the instant in question.

The speed measures the instantaneous rate at which the particle is moving along its path, but gives no information about its instantaneous direction of motion. We therefore introduce a vector quantity, the *velocity*, which gives the rate at which the particle is changing its position in both magnitude and direction. Let O be a definite point of a reference frame $\mathfrak{F}$, say the origin of a system of rectangular axes fixed in a rigid body. Then if $\mathbf{r} = \overrightarrow{OP}$ denotes the position vector of P, the velocity $\mathbf{v}$ of P, relative to $\mathfrak{F}$, is defined as the time derivative of its position vector:

$$(2) \qquad \mathbf{v} = \frac{d\mathbf{r}}{dt}.$$

As P moves along Γ, $\mathbf{r}$ may be regarded as a function of the arc s; hence

$$\frac{d\mathbf{r}}{dt} = \frac{d\mathbf{r}}{ds}\frac{ds}{dt} = \mathbf{T}\frac{ds}{dt},$$

where $\mathbf{T}$ is the unit tangent vector to Γ at P in the direction of increasing arcs. Thus the velocity and speed are connected by the relation,

$$\mathbf{v} = v\mathbf{T}. \tag{3}$$

The velocity of P is represented by a vector tangent to the path at P in the direction of instantaneous motion and of length numerically equal to the speed.

Finally we define the *acceleration* $\mathbf{a}$ of the particle as the time derivative of its velocity:

$$\mathbf{a} = \frac{d\mathbf{v}}{dt} = \frac{d^2\mathbf{r}}{dt^2}. \tag{4}$$

From (3), we find

$$\mathbf{a} = \frac{dv}{dt}\mathbf{T} + v\frac{d\mathbf{T}}{dt};$$

or, since $\mathbf{T}$ may be regarded as a function of s,

$$\frac{d\mathbf{T}}{dt} = \frac{d\mathbf{T}}{ds}\frac{ds}{dt} = (\kappa\mathbf{N})v = \frac{v}{\rho}\mathbf{N}$$

$$\mathbf{a} = \frac{dv}{dt}\mathbf{T} + \frac{v^2}{\rho}\mathbf{N}. \tag{5}$$

The components of the acceleration in the positive direction of the tangent, principal normal, and binormal, are therefore

$$a_t = \frac{dv}{dt}, \qquad a_n = \frac{v^2}{\rho}, \qquad a_b = 0. \tag{6}$$

The acceleration of a particle P is a vector lying in the plane to the tangent and principal normal to the path at P. The tangential component is the time derivative of the speed, the normal component the square of the speed divided by the radius of curvature at P.

The acceleration will be purely tangential when the motion is rectilinear ($\rho = \infty$); it will be purely normal when speed is constant ($dv/dt = 0$).

The velocity and acceleration vectors are regarded as localized at the moving particle.

With rectangular axes, $\mathbf{r} = x\mathbf{i} + y\mathbf{j} + z\mathbf{k}$, and

$$\text{(7)} \qquad \mathbf{v} = \frac{d\mathbf{r}}{dt} = \frac{dx}{dt}\mathbf{i} + \frac{dy}{dt}\mathbf{j} + \frac{dz}{dt}\mathbf{k},$$

$$\text{(8)} \qquad \mathbf{a} = \frac{d\mathbf{v}}{dt} = \frac{d^2x}{dt^2}\mathbf{i} + \frac{d^2y}{dt^2}\mathbf{j} + \frac{d^2z}{dt^2}\mathbf{k}.$$

The rectangular components of $\mathbf{v}$ and $\mathbf{a}$ are the first and second time derivatives of x, y, z.

Example 1. *Circular Motion.* In the case of motion in a circle of radius r,

$$s = r\theta, \qquad v = r\frac{d\theta}{dt}, \qquad \frac{dv}{dt} = r\frac{d^2\theta}{dt^2},$$

the angle θ being expressed in radians. On writing $\omega = d\theta/dt$ for the *angular speed*, we have

$$v = r\omega, \qquad a_t = r\frac{d\omega}{dt}, \qquad a_n = r\omega^2.$$

These results may also be deduced directly from the equation $\mathbf{r} = r\mathbf{R}$ of the circle if we make use of (44.2) and (44.3).

Example 2. *Uniformly Accelerated Motion.* If a particle has a *constant* acceleration $\mathbf{a}$, and $\mathbf{r} = \mathbf{r}_0$, $\mathbf{v} = \mathbf{v}_0$ when $t = 0$, we have, on integrating $d\mathbf{v}/dt = \mathbf{a}$ twice;

$$\mathbf{v} = \mathbf{a}t + \mathbf{v}_0, \qquad \mathbf{r} = \tfrac{1}{2}\mathbf{a}t^2 + \mathbf{v}_0t + \mathbf{r}_0.$$

The path is the result of superposing the displacement $\frac{1}{2}\mathbf{a}t^2$, due to the acceleration, upon $\mathbf{r}_0 + \mathbf{v}_0t$ due to rectilinear motion at constant velocity $\mathbf{v}_0$. It is easily shown to be a parabola having its axis parallel to $\mathbf{a}$. At the vertex of the parabola, $\mathbf{v}$ is perpendicular to $\mathbf{a}$; the condition $\mathbf{v} \cdot \mathbf{a} = 0$ gives $t = -\mathbf{a} \cdot \mathbf{v}_0/\mathbf{a} \cdot \mathbf{a}$ for the time of passing the vertex.

53. Relative Velocity. In § 52, we have seen how to find the velocity $\mathbf{v}$ of a particle P relative to any given reference frame $\mathfrak{F}$. If $\mathfrak{F}'$ is a second reference frame, in motion with respect to $\mathfrak{F}$, how is the velocity $\mathbf{v}'$ of P relative to $\mathfrak{F}'$ related to $\mathbf{v}$?

At any instant t let P coincide with the point Q of $\mathfrak{F}'$. At a later instant $t_1 = t + \Delta t$, let P and Q have the positions P_1, Q_1;

here Q is a fixed point of $\mathfrak{F}'$, and its motion is due to the motion of $\mathfrak{F}'$ relative to $\mathfrak{F}$. Then

$$\overrightarrow{PP_1} \text{ is the displacement of } P \text{ relative to } \mathfrak{F},$$

$$\overrightarrow{QQ_1} \text{ is the displacement of } Q \text{ relative to } \mathfrak{F},$$

$$\overrightarrow{Q_1P_1} \text{ is the displacement of } P \text{ relative to } \mathfrak{F}',$$

and

$$\overrightarrow{PP_1} = \overrightarrow{PQ_1} + \overrightarrow{Q_1P_1} = \overrightarrow{QQ_1} + \overrightarrow{Q_1P_1}.$$

If we divide this equation by Δt and pass to the limit $\Delta t \to 0$, we obtain

$$\text{(1)} \qquad \mathbf{v}_P = \mathbf{v}_Q + \mathbf{v}'_P,$$

where $\mathbf{v}_Q$, the velocity of the point Q of $\mathfrak{F}'$ relative to $\mathfrak{F}$, is called the *transfer velocity* of P.

If we regard the frame $\mathfrak{F}$ as "fixed" and velocities referred to it as "absolute," while velocities referred to $\mathfrak{F}'$ are "relative," we may state (1) as follows:

The absolute velocity of a particle is equal to the sum of its transfer and relative velocities.

In many applications all points of the frame $\mathfrak{F}'$ have the same velocity relative to $\mathfrak{F}$; then $\mathbf{v}_Q$ is the *velocity of translation* of $\mathfrak{F}'$, and we may write

$$\text{(2)} \qquad \mathbf{v}_P = \mathbf{v}_{\mathfrak{F}'} + \mathbf{v}'_P.$$

Example 1. *Wind Triangle.* An airplane p has the velocity $\mathbf{v}$ relative to the ground (the *earth* e), $\mathbf{v}'$ relative to the air; and the air (the *wind* w) has the velocity $\mathbf{V}$ relative to the ground; then $\mathbf{v} = \mathbf{V} + \mathbf{v}'$, from (2). In Fig. 53a,

$$\mathbf{v} = \overrightarrow{ep}, \qquad \mathbf{V} = \overrightarrow{ew}, \qquad \mathbf{v}' = \overrightarrow{wp};$$

thus vectors from e and w represent velocities relative to the earth and wind, respectively. The magnitudes of $\mathbf{v}$ and $\mathbf{v}'$ give the *ground speed* (ep) and *air speed* (wp) of the plane. The directions of $\mathbf{v}$ and $\mathbf{v}'$, given as angles θ and θ' measured from the north around through the east (clockwise), determine the *track* and *heading* of the plane. The plane is pointed along its heading but travels over the ground along its track. The angle $(\mathbf{v}', \mathbf{v})$ from heading to track is the *drift angle*.

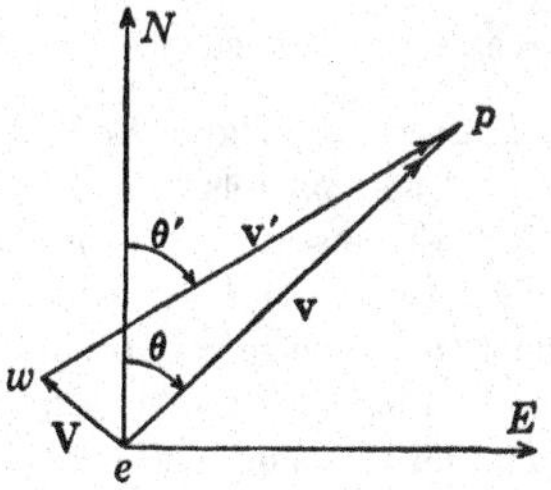

FIG. 53a

Example 2. *Interception.* A plane p is flying over the track PX with the ground speed ep (Fig. 53b). As plane p passes the point P, a plane q departs from Q to intercept plane p. If the air speed of plane q is given, over what track shall q fly in order to intercept p?

Solution. In order that plane q may intercept plane p, the velocity of q relative to p must have the direction $\overrightarrow{QP}$.

Draw the vectors $\overrightarrow{ep}$ and $\overrightarrow{ew}$, giving the velocities of plane p and the wind w relative to earth e. With w as center describe a circle having the known air speed of plane q as radius. If a *ray* drawn through point p in the direction of

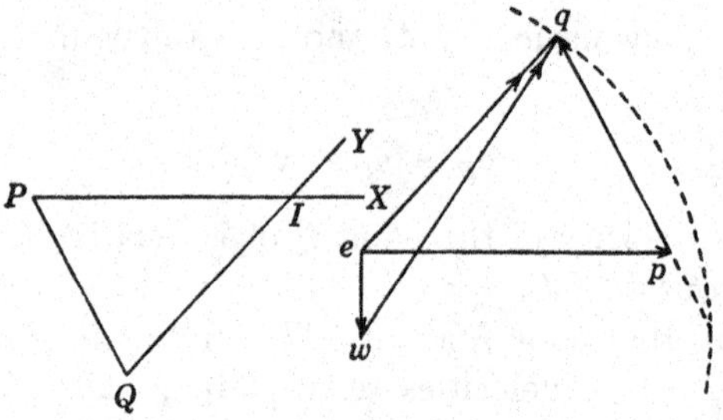

FIG. 53b

$\overrightarrow{QP}$ cuts the circle at point q, plane q will intercept plane p on flying with the ground speed eq over the track QY parallel to $\overrightarrow{eq}$; for $\overrightarrow{pq}$, the velocity of plane q relative to plane p, has the direction $\overrightarrow{QP}$. Interception occurs at I after a flying time of QI/eq (or PI/ep) hours.

Interception is impossible if the air-speed circle of plane q fails to cut the ray. If the circle cuts the ray in just one point, as in Fig. 53b, plane q can intercept p on only one track. But if the circle cuts the ray in two points, say q_1 and q_2, plane q can intercept p along two different tracks, parallel to $\overrightarrow{eq_1}$ and $\overrightarrow{eq_2}$, respectively.

Example 3. *Plane Returning to a Carrier.* An airplane p leaves a carrier s at O and patrols along the track OY while the carrier follows the course OX with constant speed of v miles per hour (Fig. 53c). If the fuel in the tank allows the plane T hours of flying time at a given air speed, at what point B must the plane turn in order to rejoin the carrier at A, T hours after its departure? Find also the time t_1, the heading, and the ground speed v_1 on the leg out; and the time t_2, the heading, and the ground speed v_2 on the leg back.

Solution. Let the vectors $\overrightarrow{es}$ and $\overrightarrow{ew}$ give the velocities of carrier and wind. With w as center describe a circle having the known air speed of p as radius—the *air-speed circle*. If this circle cuts the ray through e in the direction OY at p_1, $\overrightarrow{ep_1}$ is the velocity of plane p on the leg out (ground speed $v_1 = ep_1$).

The course of plane p relative to the carrier s is out and back along the same straight line. Now $\overrightarrow{sp_1}$ is the velocity of p relative to s on the leg out; hence the velocity of p relative to s on the leg back is a vector $\overrightarrow{sp_2}$ whose direction is opposed to $\overrightarrow{sp_1}$. Thus the point p_2 is at the intersection of the air-speed circle with the line p_1s prolonged, and $\overrightarrow{ep_2}$ is the velocity of p on the leg back (ground speed $v_2 = ep_2$).

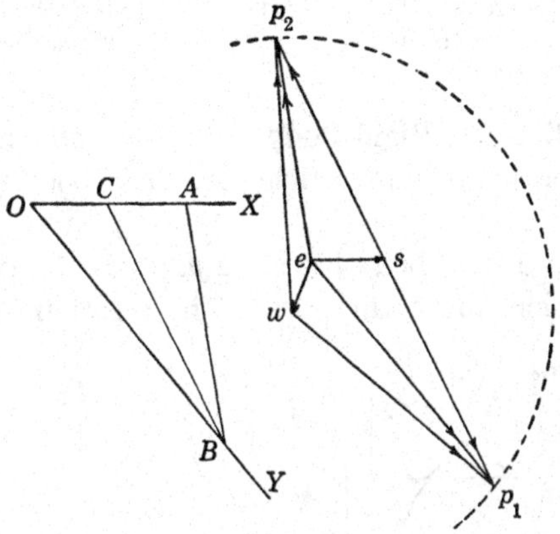

FIG. 53c

After T hours the carrier will travel the course $\overrightarrow{OA} = T\,\overrightarrow{es}$. The return track of p is along a line BA parallel to $\overrightarrow{ep_2}$. Thus p travels $t_1 = OB/ep_1$ hours on the track OB, $t_2 = BA/ep_2$ on the track BA, rejoining the carrier after $t_1 + t_2 = T$ hours.

On the two legs the plane has the speeds $u_1 = sp_1$, $u_2 = sp_2$, relative to the carrier. At any time carrier and plane lie on a line parallel to p_1sp_2. When the plane is at B, the farthest point out, the carrier is at $C(BC \parallel p_1sp_2)$. The distance $r = CB$ is called the *radius of action* ot the plane. Relative to the carrier the plane travels the course CB out, BC back; hence

$$\frac{r}{u_1} + \frac{r}{u_2} = T, \qquad r = T\frac{u_1u_2}{u_1 + u_2};$$

$$t_1 = \frac{r}{u_1} = T\frac{u_2}{u_1 + u_2}, \qquad t_2 = \frac{r}{u_2} = T\frac{u_1}{u_1 + u_2}.$$

These times agree with those previously given; for

$$t_1 = \frac{CB}{sp_1} = \frac{OB}{ep_1}, \qquad t_2 = \frac{BC}{sp_2} = \frac{BA}{ep_2}.$$

Example 4. *Alternative Airport.* Let a plane p depart from the airport O along the track OY (Fig. 53c). If a landing at an airport Y is rendered dangerous by local bad weather, the plane may be directed to land at an alternative airport A. If the fuel supply allows T hours flying time to the plane, how far may the plane fly on the track OY in order still to reach the port A by changing its course?

This problem is reduced to the one preceding if the airport A is regarded as a carrier traveling from O to A with the uniform velocity $\overrightarrow{OA}/T$. Lay off $\overrightarrow{es} = \overrightarrow{OA}/T$ and proceed precisely as in the carrier problem. The line CB, drawn parallel to p_1sp_2, fixes the farthest point B at which the plane can change its course and still reach the airport A. The time t_1 when the turning point is reached is given in ex. 3.

54. Kinematics of a Rigid Body. We shall now investigate the velocity distribution of the particles of a rigid body moving in any manner.

Consider first a rigid body having a *fixed line* or axis; its motion is then a rotation about this axis. The position of the body at

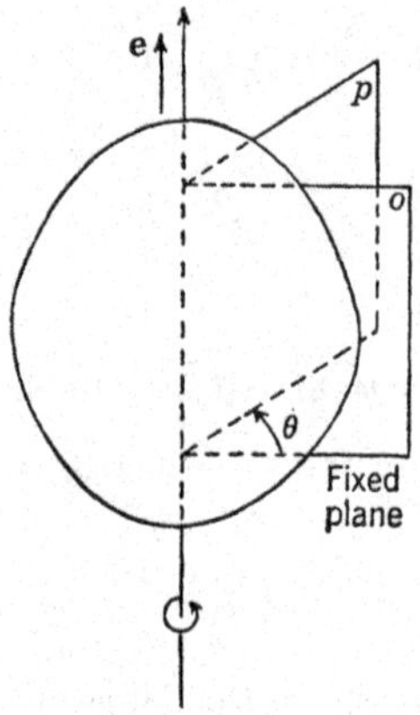

FIG. 54a

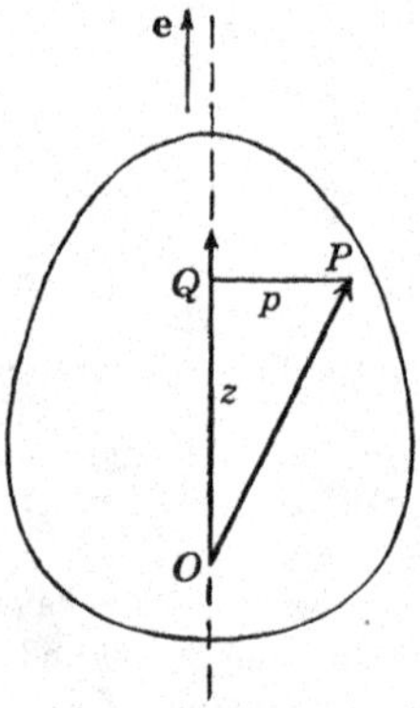

FIG. 54b

any instant may be specified by the angle θ between an axial plane o fixed in our frame of reference $\mathfrak{F}$ and an axial plane p fixed in the body (Fig. 54a). By choosing a positive direction on the axis (unit vector $\mathbf{e}$), we fix the positive sense of θ by the right-handed screw convention. Then the *angular speed* ω of the body at any instant is defined as

$$\omega = \frac{d\theta}{dt}. \tag{1}$$

Thus ω is positive or negative, according as θ is increasing or decreasing at the instant in question.

The velocity distribution in the revolving body may be simply expressed if we define the *angular velocity* as the vector,

$$\boldsymbol{\omega} = \frac{d\theta}{dt}\mathbf{e}.$$

Note that $\boldsymbol{\omega}$ always is related to the instantaneous sense of rotation by the rule of the right-hand screw.

Choose an origin O on the axis and let $\mathbf{r} = \overrightarrow{OP}$ be the position vector of any particle of the body. Then (Fig. 54b)

$$\mathbf{r} = \overrightarrow{OQ} + \overrightarrow{QP} = z\mathbf{e} + p\mathbf{R},$$

where $\mathbf{R}$ is a unit vector perpendicular to the axis and revolving with the body. Since $z\mathbf{e}$ and p are constant during the motion of P, the velocity of P is

$$\mathbf{v} = \frac{d\mathbf{r}}{dt} = p\frac{d\mathbf{R}}{d\theta}\frac{d\theta}{dt} = p\frac{d\theta}{dt}\mathbf{e}\times\mathbf{R} = \left(\frac{d\theta}{dt}\mathbf{e}\right)\times(z\mathbf{e} + p\mathbf{R})$$

that is,

$$\mathbf{v} = \boldsymbol{\omega}\times\mathbf{r}. \tag{2}$$

The velocity of any particle of a body revolving about a fixed axis is equal to the vector product of the angular velocity and the position vector of the particle referred to any origin on the axis.

Let us next consider a rigid body having one fixed point O. Let $\mathbf{i}$ be a unit vector fixed in the body, $\mathbf{j}$ a unit vector in the direction of $d\mathbf{i}/dt$ (perpendicular to $\mathbf{i}$) and $\mathbf{k} = \mathbf{i}\times\mathbf{j}$. Then we may write

$$\frac{d\mathbf{i}}{dt} = \alpha\mathbf{j}, \qquad \frac{d\mathbf{k}}{dt} = \frac{d\mathbf{i}}{dt}\times\mathbf{j} + \mathbf{i}\times\frac{d\mathbf{j}}{dt} = \mathbf{i}\times\frac{d\mathbf{j}}{dt}\cdot$$

Hence $d\mathbf{k}/dt$ (perpendicular to $\mathbf{k}$) is also perpendicular to $\mathbf{i}$ and therefore parallel to $\mathbf{k}\times\mathbf{i} = \mathbf{j}$. Thus we have

$$\frac{d\mathbf{k}}{dt} = \beta\mathbf{j}, \qquad \frac{d\mathbf{j}}{dt} = \frac{d}{dt}(\mathbf{k}\times\mathbf{i}) = \beta\mathbf{j}\times\mathbf{i} + \mathbf{k}\times\alpha\mathbf{j},$$

or, on collecting results

$$\frac{d\mathbf{i}}{dt} = \alpha\mathbf{j}, \qquad \frac{d\mathbf{j}}{dt} = (\alpha\mathbf{k} - \beta\mathbf{i})\times\mathbf{j}, \qquad \frac{d\mathbf{k}}{dt} = \beta\mathbf{j}.$$

If we now write

$$\boldsymbol{\omega} = \alpha\mathbf{k} - \beta\mathbf{i},$$

these equations assume the same form:

$$\frac{d\mathbf{i}}{dt} = \boldsymbol{\omega} \times \mathbf{i}, \qquad \frac{d\mathbf{j}}{dt} = \boldsymbol{\omega} \times \mathbf{j}, \qquad \frac{d\mathbf{k}}{dt} = \boldsymbol{\omega} \times \mathbf{k}.$$

Now the position vector $\mathbf{r}$ of any particle P in the body may be referred to the orthogonal triple $\mathbf{i}$, $\mathbf{j}$, $\mathbf{k}$ fixed in the body; thus

$$\mathbf{r} = \overrightarrow{OP} = x\mathbf{i} + y\mathbf{j} + z\mathbf{k},$$

where x, y, z remain constant during the motion. Hence the velocity of P is given by

$$\frac{d\mathbf{r}}{dt} = x\frac{d\mathbf{i}}{dt} + y\frac{d\mathbf{j}}{dt} + z\frac{d\mathbf{k}}{dt} = \boldsymbol{\omega} \times (x\mathbf{i} + y\mathbf{j} + z\mathbf{k}),$$

(3) $$\mathbf{v} = \boldsymbol{\omega} \times \overrightarrow{OP}.$$

Thus, at any instant, the velocity distribution in a rigid body with one point O fixed is the same as if it were revolving about an axis through O with angular velocity $\boldsymbol{\omega}$. The line through O in the direction of $\boldsymbol{\omega}$ is called the *instantaneous axis of rotation*, and $\boldsymbol{\omega}$ is called, as before, the *angular velocity*. Now, however, $\boldsymbol{\omega}$ may change in direction as well as in magnitude. With a fixed axis of rotation,

$$\boldsymbol{\omega} = \frac{d\theta}{dt}\mathbf{e} = \frac{d}{dt}(\theta\mathbf{e}),$$

so that $\boldsymbol{\omega}$ may be regarded as the time derivative of the vector angle $\theta\mathbf{e}$. But with a variable axis of rotation $\boldsymbol{\omega}$ no longer can be expressed as a time derivative.

Finally let us consider the general motion of a free rigid body. If, at any instant, all points of the body have the same velocity $\mathbf{v}$, the motion is said to be an *instantaneous translation*. When the velocity distribution is given by $\mathbf{v} = \boldsymbol{\omega} \times \overrightarrow{AP}$, the motion is said to be an *instantaneous rotation* about an axis through A in the direction of $\boldsymbol{\omega}$. We now shall show that, in the most general motion of a rigid body, the velocities may be regarded as compounded of an instantaneous translation and rotation.

Let A be any point of the rigid body, and denote its velocity relative to $\mathfrak{F}$ by $\mathbf{v}_A$. Consider a second reference frame $\mathfrak{F}'$ having a translation of velocity $\mathbf{v}_A$ relative to $\mathfrak{F}$. Then the motion of the body relative to $\mathfrak{F}'$ is an instantaneous rotation about an axis through A, since A has zero velocity relative to $\mathfrak{F}'$. The velocity of any particle P of the body is therefore

$$\mathbf{v}'_P = \boldsymbol{\omega} \times \overrightarrow{AP},$$

relative to $\mathfrak{F}'$, and, consequently,

$$\mathbf{v}_P = \mathbf{v}_A + \boldsymbol{\omega} \times \overrightarrow{AP}, \tag{4}$$

relative to $\mathfrak{F}$. Moreover, for any other point Q of the body,

$$\mathbf{v}_Q = \mathbf{v}_A + \boldsymbol{\omega} \times \overrightarrow{AQ}; \tag{4$'$}$$

and, on subtracting this from (4), we get

$$\mathbf{v}_P = \mathbf{v}_Q + \boldsymbol{\omega} \times \overrightarrow{QP}.$$

The content of these equations is stated in the following

THEOREM 1. *If A is any point of a free rigid body, the velocities of its points are the same as if they were compounded of an instantaneous translation $\mathbf{v}_A$ and an instantaneous rotation $\boldsymbol{\omega}$ about an axis through A; and $\boldsymbol{\omega}$ is the same for any choice of A.*

Thus the instantaneous velocity distribution of a rigid body is determined by two vectors, its angular velocity $\boldsymbol{\omega}$ and the velocity $\mathbf{v}_A$ of any point A of the body. These may be combined into the *velocity motor*,

$$\mathbf{V} = \boldsymbol{\omega} + \epsilon \mathbf{v}_A; \tag{5}$$

for, from (4),

$$\mathbf{v}_P = \mathbf{v}_A + \overrightarrow{PA} \times \boldsymbol{\omega}, \tag{6}$$

in accordance with (31.3).

If $\boldsymbol{\omega} \cdot \mathbf{v}_A \neq 0$, the motor $\mathbf{V}$ is a screw. Since $\boldsymbol{\omega} \cdot \mathbf{v}_P = \boldsymbol{\omega} \cdot \mathbf{v}_A$, the velocities of all particles of the body have the same projection on $\boldsymbol{\omega}$. The axis of $\mathbf{V}$ is called the *instantaneous axis of velocity;* its equation, from (31.8), is

$$\mathbf{r} \times \boldsymbol{\omega} = \frac{(\boldsymbol{\omega} \times \mathbf{v}_A) \times \boldsymbol{\omega}}{\boldsymbol{\omega} \cdot \boldsymbol{\omega}} \quad \text{(origin } A\text{)}, \tag{7}$$

a line parallel to $\boldsymbol{\omega}$ and passing through the point Q given by $\overrightarrow{AQ} = \boldsymbol{\omega} \times \mathbf{v}_A / \boldsymbol{\omega} \cdot \boldsymbol{\omega}$; and, from (4)′,

$$\mathbf{v}_Q = \mathbf{v}_A + \boldsymbol{\omega} \times \overrightarrow{AQ} = \frac{\boldsymbol{\omega} \cdot \mathbf{v}_A}{\boldsymbol{\omega} \cdot \boldsymbol{\omega}} \boldsymbol{\omega} = \text{proj}_{\boldsymbol{\omega}} \mathbf{v}_A. \tag{8}$$

All points of the instantaneous axis have the velocity $\mathbf{v}_Q$; for, if R is another of its points,

$$\mathbf{v}_R = \mathbf{v}_A + \boldsymbol{\omega} \times (\overrightarrow{AQ} + \overrightarrow{QR}) = \mathbf{v}_Q \quad \text{since} \quad \boldsymbol{\omega} \times \overrightarrow{QR} = 0.$$

Moreover $\mathbf{v}_Q$ is characterized by being parallel to $\boldsymbol{\omega}$; and, since all particle velocities have the same projection on $\boldsymbol{\omega}$, their least numerical value at any instant is $|\mathbf{v}_Q|$. Referred to a point Q on the instantaneous axis, the velocity motor becomes

$$\mathbf{V} = \boldsymbol{\omega} + \epsilon \mathbf{v}_Q. \tag{9}$$

This form, in which $\boldsymbol{\omega}$ and $\mathbf{v}_Q$ are parallel, symbolizes

THEOREM 2. *At any instant, the particle velocities of a rigid body may be represented by a screw motion—a rotation about an instantaneous axis combined with a velocity of translation along this axis.*

If $\boldsymbol{\omega} \neq 0$, $\boldsymbol{\omega} \cdot \mathbf{v}_A = 0$, we have $\mathbf{v}_Q = 0$. The motion is then a pure rotation of angular velocity $\boldsymbol{\omega}$ about the instantaneous axis. $\mathbf{V}$ becomes a line vector along this axis.

If $\boldsymbol{\omega} = 0$, $\mathbf{v}_A \neq 0$, we see from (4) that all points of the body have the same velocity. The instantaneous motion is then a pure translation, and $\mathbf{V} = \epsilon \mathbf{v}_A$ is a pure dual motor.

Finally let us consider the case of *plane motion* of a rigid body. If $\boldsymbol{\omega} \neq 0$, the plane of the motion is perpendicular to $\boldsymbol{\omega}$ and $\mathbf{v}_Q = 0$; the motion is then an instantaneous rotation about an axis. The point Q where the instantaneous axis cuts the (reference) plane of motion is called the *instantaneous center;* and, as previously,

$$\overrightarrow{AQ} = \frac{\boldsymbol{\omega} \times \mathbf{v}_A}{\boldsymbol{\omega} \cdot \boldsymbol{\omega}}. \tag{10}$$

If the velocities of the points A, B of the body are known and $\mathbf{v}_A$, $\mathbf{v}_B$ are not parallel, Q is at the intersection of the lines AQ, BQ drawn perpendicular to $\mathbf{v}_A$ and $\mathbf{v}_B$, respectively.

Example. Rolling Curve. Let the curve Γ roll without slipping over a fixed curve Γ_1 (Fig. 54c). The points A and A_1 were originally in contact; and, if I is the instantaneous point of contact, let $s = \text{arc } AI$, $s_1 = \text{arc } A_1I$. The conditions for pure rolling are

(i) $\quad s = s_1, \qquad \mathbf{T} = \mathbf{T}_1 \qquad \text{at } I.$

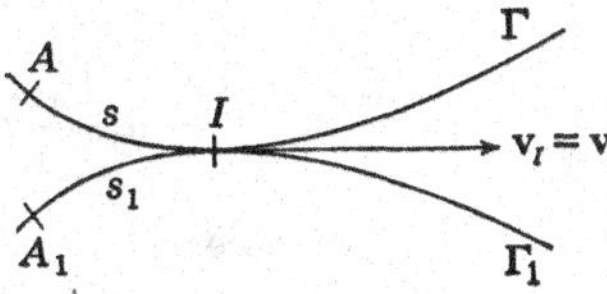

FIG. 54c

If I is regarded as a moving particle, its velocity relative to fixed and moving frames attached to Γ_1 and Γ is the same; for

$$\mathbf{v}_I = \frac{ds_1}{dt}\mathbf{T}_1 = \frac{ds}{dt}\mathbf{T} = \mathbf{v}_I'.$$

Hence, if I coincides with the fixed point Q of Γ, we have, from (53.1),

$$\mathbf{v}_I = \mathbf{v}_Q + \mathbf{v}_I', \qquad \mathbf{v}_Q = 0.$$

The instantaneous center of the rolling curve is at its point of contact with the fixed curve.

The *speed of rolling,*

(ii) $$v = \frac{ds}{dt} = \frac{ds_1}{dt}.$$

The angular speed is $\omega = d\theta/dt$, where θ is the angle between the tangents at A and A_1. We now differentiate the equation $\mathbf{T}_1 = \mathbf{T}$ with respect to t. Since $\mathbf{T}$ is a function of s and θ, we have

$$\frac{d\mathbf{T}_1}{ds_1}\frac{ds_1}{dt} = \frac{\partial\mathbf{T}}{\partial s}\frac{ds}{dt} + \frac{\partial\mathbf{T}}{\partial\theta}\frac{d\theta}{dt},$$

$$\kappa_1\mathbf{N}_1 v = \kappa\mathbf{N}v + \mathbf{N}\omega,$$

or, since $\mathbf{N}_1 = \mathbf{N}$,

(iii) $$\frac{1}{\rho_1} - \frac{1}{\rho} = \frac{\omega}{v}.$$

Here $\mathbf{N}$ is $\pi/2$ in advance of $\mathbf{T}$ and ρ and ρ_1 are positive when $\mathbf{N}$ points to the respective centers of curvature (in Fig. 54c, $\rho_1 < 0$, $\rho > 0$ and $\omega < 0$).

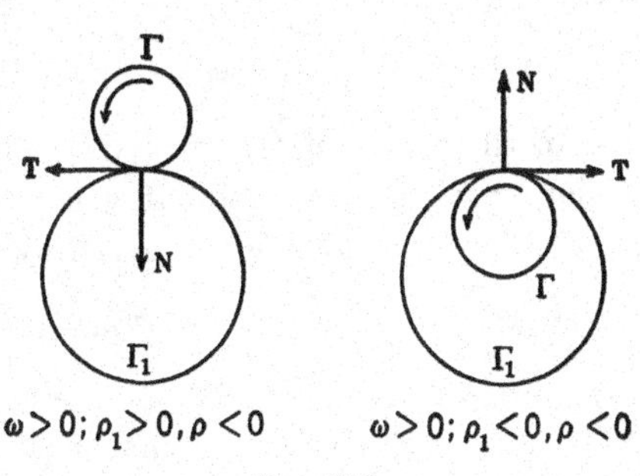

FIG. 54d

If Γ and Γ_1 are circles of radius r and r_1 (Fig. 54d),

$$\frac{\omega}{v} = \frac{1}{r_1} - \left(-\frac{1}{r}\right) = \frac{1}{r} + \frac{1}{r_1} \quad \text{for external contact,}$$

$$\frac{\omega}{v} = -\frac{1}{r_1} - \left(-\frac{1}{r}\right) = \frac{1}{r} - \frac{1}{r_1} \quad \text{for internal contact.}$$

When $r_1 \to \infty$, Γ_1 becomes a straight line and $v = \omega r$.

55. Composition of Velocities. Let a rigid body have the velocity motor,

$$\mathbf{V}' = \boldsymbol{\omega}' + \epsilon \mathbf{v}_A',$$

relative to a frame $\mathfrak{F}'$; and let the frame $\mathfrak{F}'$ have the motor,

$$\mathbf{V} = \boldsymbol{\omega} + \epsilon \mathbf{v}_B,$$

relative to a "fixed" frame $\mathfrak{F}$, where B is the point of $\mathfrak{F}'$ which coincides for the instant with the point A of the body. What is the motion of the body relative to $\mathfrak{F}$?

The velocity of any point P of the body relative to $\mathfrak{F}'$ is

$$\text{(1)} \qquad \mathbf{v}_P' = \mathbf{v}_A' + \boldsymbol{\omega}' \times \overrightarrow{AP}.$$

At the instant in question let P coincide with the point Q of $\mathfrak{F}'$. Then the velocity of Q relative to $\mathfrak{F}$ is

$$\text{(2)} \qquad \mathbf{v}_Q = \mathbf{v}_B + \boldsymbol{\omega} \times \overrightarrow{BQ}.$$

Now the velocities of P and A relative to $\mathfrak{F}$ are (53.1)

$$\mathbf{v}_P = \mathbf{v}_Q + \mathbf{v}_P', \qquad \mathbf{v}_A = \mathbf{v}_B + \mathbf{v}_A'.$$

On adding (1) and (2) and observing that $\overrightarrow{AP} = \overrightarrow{BQ}$ at the instant considered, we get

$$\mathbf{v}_P = \mathbf{v}_A + (\boldsymbol{\omega} + \boldsymbol{\omega}') \times \overrightarrow{AP}.$$

Thus the motion of the body relative to $\mathfrak{F}$ is compounded of the velocity of translation $\mathbf{v}_A$ and the angular velocity $\boldsymbol{\omega} + \boldsymbol{\omega}'$ about an axis through A; that is, the motion is given by the motor,

$$\mathbf{V} + \mathbf{V}' = \boldsymbol{\omega} + \boldsymbol{\omega}' + \epsilon \mathbf{v}_P.$$

THEOREM. *If the motion of a rigid body relative to* $\mathfrak{F}'$ *is given by the motor* $\mathbf{V}' = \boldsymbol{\omega}' + \epsilon \mathbf{v}_A'$, *and the motion of* $\mathfrak{F}'$ *relative to* $\mathfrak{F}$ *by* $\mathbf{V} = \boldsymbol{\omega} + \epsilon \mathbf{v}_B$, *the motion of the body relative to* $\mathfrak{F}$ *is given by the motor sum* $\mathbf{V} + \mathbf{V}'$ *provided* A *and* B *coincide at the instant in question.*

Two translations $\epsilon \mathbf{v}_A'$ and $\epsilon \mathbf{v}_B$ thus compound into a translation $\epsilon(\mathbf{v}_A' + \mathbf{v}_B)$.

If $\mathbf{V}' = \boldsymbol{\omega}' + \epsilon \mathbf{v}_A'$ and $\mathbf{V} = \boldsymbol{\omega} + \epsilon \mathbf{v}_B$ represent pure rotations, these motors are line vectors; hence the motion of the body relative to $\mathfrak{F}$ will be a pure rotation when and only when $\boldsymbol{\omega} + \boldsymbol{\omega}' \neq 0$ and the axes of rotation are coplanar (§ 32, theorem 1). If the axes of rotation intersect at A, the motors referred to A are

$$\mathbf{V}' = \boldsymbol{\omega}', \qquad \mathbf{V} = \boldsymbol{\omega}; \qquad \text{and} \quad \mathbf{V} + \mathbf{V}' = \boldsymbol{\omega} + \boldsymbol{\omega}'.$$

Angular velocities about intersecting axes may be compounded by vector addition.

If $\mathbf{V}$ and $\mathbf{V}'$ represent pure rotations about parallel axes and $\boldsymbol{\omega} + \boldsymbol{\omega}' = 0$, $\mathbf{V} + \mathbf{V}' = \epsilon(\mathbf{v}_B + \mathbf{v}_A')$; the motion of the body relative to $\mathfrak{F}$ is then a pure translation of velocity $\mathbf{v}_B + \mathbf{v}_A'$.

56. Rate of Change of a Vector. Referred to a fixed origin O in the frame $\mathfrak{F}$, the vector $\mathbf{u} = \overrightarrow{PQ} = \overrightarrow{OQ} - \overrightarrow{OP}$; hence

$$\frac{d\mathbf{u}}{dt} = \mathbf{v}_Q - \mathbf{v}_P, \tag{1}$$

where $\mathbf{v}_P$, $\mathbf{v}_Q$ are velocities relative to $\mathfrak{F}$.

Similarly, if $\mathfrak{F}'$ is a second frame in motion with respect to $\mathfrak{F}$, the rate of change of u relative to $\mathfrak{F}'$ is

$$\frac{d'\mathbf{u}}{dt} = \mathbf{v}_Q' - \mathbf{v}_P', \tag{2}$$

where $\mathbf{v}_P'$, $\mathbf{v}_Q'$ are velocities relative to $\mathfrak{F}'$.

Let the motion of $\mathfrak{F}'$ relative to $\mathfrak{F}$ be given by the motor $\boldsymbol{\omega} + \epsilon \mathbf{v}_A$, A being a fixed point of $\mathfrak{F}'$; and, at the instant in question, let P and Q coincide with the points R and S of $\mathfrak{F}'$. Then, from (53.1) and (54.4),

$$\mathbf{v}_P = \mathbf{v}_P' + \mathbf{v}_R = \mathbf{v}_P' + \mathbf{v}_A + \boldsymbol{\omega} \times \overrightarrow{AR},$$

$$\mathbf{v}_Q = \mathbf{v}_Q' + \mathbf{v}_S = \mathbf{v}_Q' + \mathbf{v}_A + \boldsymbol{\omega} \times \overrightarrow{AS};$$

and, on subtraction,

$$\frac{d\mathbf{u}}{dt} = \frac{d'\mathbf{u}}{dt} + \boldsymbol{\omega} \times (\overrightarrow{AS} - \overrightarrow{AR}).$$

But, since $\overrightarrow{AS} - \overrightarrow{AR} = \overrightarrow{RS} = \overrightarrow{PQ} = \mathbf{u}$,

$$\frac{d\mathbf{u}}{dt} = \frac{d'\mathbf{u}}{dt} + \boldsymbol{\omega} \times \mathbf{u}; \tag{3}$$

in particular

$$\frac{d\mathbf{u}}{dt} = \boldsymbol{\omega} \times \mathbf{u} \quad \text{if } \mathbf{u} \text{ is fixed in } \mathfrak{F}'. \tag{4}$$

When $\mathbf{u} = \boldsymbol{\omega}$, the angular velocity of the frame $\mathfrak{F}'$, we have, from (3),

$$\frac{d\boldsymbol{\omega}}{dt} = \frac{d'\boldsymbol{\omega}}{dt}. \tag{5}$$

The vector $d\boldsymbol{\omega}/dt$ is denoted by $\boldsymbol{\alpha}$ and called the *angular acceleration* vector of $\mathfrak{F}$.

Example. Kinematic Interpretation of Frenet's Formulas. At any point P of a space curve, the trihedral **TNB** may be used as a frame of reference. If P moves along the curve with unit speed, $ds/dt = 1$ and $s = t$ if $t = 0$ at the origin of arcs. Then the arc s may be interpreted as the time, and $d\mathbf{r}/ds = \mathbf{T}$ is the velocity of P.

If $\boldsymbol{\omega}$ is the angular velocity of **TNB** referred to a "fixed" frame $\mathfrak{F}$, we have, from (4),

$$\frac{d\mathbf{T}}{ds} = \boldsymbol{\omega} \times \mathbf{T}, \qquad \frac{d\mathbf{N}}{ds} = \boldsymbol{\omega} \times \mathbf{N}, \qquad \frac{d\mathbf{B}}{ds} = \boldsymbol{\omega} \times \mathbf{B}.$$

A comparison with Frenet's Formulas (44.6) shows that $\boldsymbol{\omega} = \boldsymbol{\delta}$, the Darboux vector. *The Darboux vector of a space curve is the angular-velocity vector of its moving trihedral* **TNB**.

Since the vertex P of the trihedral has the velocity $\mathbf{T}$, its motion is represented completely by the velocity motor,

$$\mathbf{V} = \boldsymbol{\delta} + \epsilon\mathbf{T} = \tau\mathbf{T} + \kappa\mathbf{B} + \epsilon\mathbf{T}.$$

From (54.7) and (54.8), we see that the axis of this motor passes through the point Q given by

$$\overrightarrow{PQ} = \frac{\boldsymbol{\delta} \times \mathbf{T}}{\boldsymbol{\delta} \cdot \boldsymbol{\delta}} = \frac{\kappa}{\kappa^2 + \tau^2}\mathbf{N} \quad \text{and} \quad \mathbf{v}_Q = \frac{\boldsymbol{\delta} \cdot \mathbf{T}}{\boldsymbol{\delta} \cdot \boldsymbol{\delta}}\boldsymbol{\delta} = \frac{\tau}{\kappa^2 + \tau^2}\boldsymbol{\delta}.$$

In general, the trihedral **TNB** has, at every instant, a screw motion: a combination of the angular velocity $\boldsymbol{\delta}$ about an axis through Q and the velocity $\mathbf{v}_Q$ along this axis. The velocity of translation vanishes when and only when the curve is plane ($\tau = 0$); then $\boldsymbol{\delta} = \kappa\mathbf{B}$, $\overrightarrow{PQ} = \rho\mathbf{N}$. For plane curves **TNB** has, at every instant, a pure rotation of angular velocity $\kappa\mathbf{B}$ about the center of curvature.

57. Theorem of Coriolis. Let $\mathbf{v}$ and $\mathbf{a}$ denote the velocity and acceleration of a particle P, relative to a frame $\mathfrak{F}$, while $\mathbf{v}'$ and $\mathbf{a}'$ denote these vectors relative to a frame $\mathfrak{F}'$ in motion with respect to $\mathfrak{F}$. Then if O and A are origins fixed in $\mathfrak{F}$ and $\mathfrak{F}'$, we have

$$\mathbf{r} = \overrightarrow{OP}, \qquad \mathbf{v} = \frac{d\mathbf{r}}{dt}, \qquad \mathbf{a} = \frac{d\mathbf{v}}{dt};$$

$$\mathbf{r}' = \overrightarrow{AP} \qquad \mathbf{v}' = \frac{d'\mathbf{r}'}{dt}, \qquad \mathbf{a}' = \frac{d'\mathbf{v}'}{dt}.$$

If $\boldsymbol{\omega}$ is the angular velocity of $\mathfrak{F}'$ with respect to $\mathfrak{F}$, we have, on differentiating,

$$\mathbf{r} = \overrightarrow{OA} + \mathbf{r}',$$

twice with respect to the time, and, on making use of (56.3),

$$\mathbf{v} = \mathbf{v}_A + \frac{d\mathbf{r}'}{dt}$$

$$= \mathbf{v}_A + \boldsymbol{\omega} \times \mathbf{r}' + \mathbf{v}',$$

$$\mathbf{a} = \mathbf{a}_A + \boldsymbol{\alpha} \times \mathbf{r}' + \boldsymbol{\omega} \times \frac{d\mathbf{r}'}{dt} + \frac{d\mathbf{v}'}{dt} \qquad \left(\boldsymbol{\alpha} = \frac{d\boldsymbol{\omega}}{dt}\right)$$

$$= \mathbf{a}_A + \boldsymbol{\alpha} \times \mathbf{r}' + \boldsymbol{\omega} \times (\boldsymbol{\omega} \times \mathbf{r}' + \mathbf{v}') + \boldsymbol{\omega} \times \mathbf{v}' + \mathbf{a}'$$

$$= \mathbf{a}_A + \boldsymbol{\alpha} \times \mathbf{r}' + \boldsymbol{\omega} \times (\boldsymbol{\omega} \times \mathbf{r}') + 2\,\boldsymbol{\omega} \times \mathbf{v}' + \mathbf{a}'.$$

The velocity and acceleration of the point Q of the frame $\mathfrak{F}'$ with which P momentarily coincides are called the *transfer velocity* and *transfer acceleration* of P. To find them, put $\mathbf{v}' = 0$, $\mathbf{a}' = 0$ in the foregoing equations; thus

$$\mathbf{v}_Q = \mathbf{v}_A + \boldsymbol{\omega} \times \mathbf{r}',$$

$$\mathbf{a}_Q = \mathbf{a}_A + \boldsymbol{\alpha} \times \mathbf{r}' + \boldsymbol{\omega} \times (\boldsymbol{\omega} \times \mathbf{r}').$$

Our equations now read

(1) $$\mathbf{v} = \mathbf{v}_Q + \mathbf{v}',$$

(2) $$\mathbf{a} = \mathbf{a}_Q + 2\boldsymbol{\omega} \times \mathbf{v}' + \mathbf{a}'.$$

Equation (1) restates the theorem on the composition of velocities already proved in § 55. Equation (2) shows that an analogous theorem for the composition of accelerations is not in general true;

we have, in fact, the additional term $2\boldsymbol{\omega} \times \mathbf{v}'$, known as the *Coriolis acceleration.* If we regard the frame $\mathfrak{F}$ as "fixed" and rates of change referred to it as "absolute," while the corresponding rates referred to $\mathfrak{F}'$ are "relative," we may state (2) as the

THEOREM OF CORIOLIS. *The absolute acceleration of a particle is equal to the sum of its transfer acceleration, Coriolis acceleration, and relative acceleration.*

The Coriolis acceleration, $2\boldsymbol{\omega} \times \mathbf{v}'$, vanishes in three cases only:
(*a*) $\boldsymbol{\omega} = 0$; the motion of $\mathfrak{F}'$ relative to $\mathfrak{F}$ is a translation.
(*b*) $\mathbf{v}' = 0$; the particle is at rest relative to $\mathfrak{F}'$.
(*c*) $\mathbf{v}'$ is parallel to $\boldsymbol{\omega}$.

A particle of mass m, acted on by forces of vector sum $\mathbf{F}$, has the equation of motion $\mathbf{F} = m\mathbf{a}$. When the motion is referred to a rotating frame $\mathfrak{F}'$, the equation of motion becomes

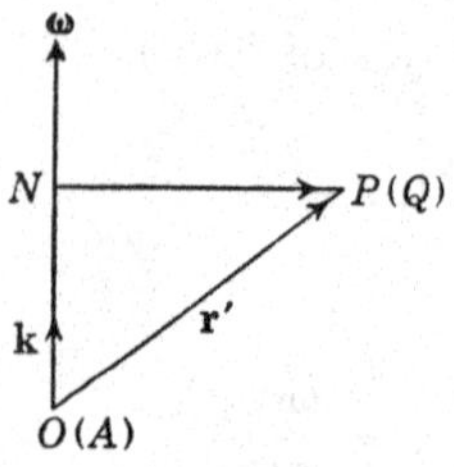

FIG. 57a

(3) $$m\mathbf{a}' = \mathbf{F} - m\mathbf{a}_Q - 2m\boldsymbol{\omega} \times \mathbf{v}'.$$

If we regard $-m\mathbf{a}_Q$ and $-2m\boldsymbol{\omega} \times \mathbf{v}'$ as fictitious forces, $m\mathbf{a}'$ equals a sum of forces just as in the case of a fixed frame. The term $-2m\boldsymbol{\omega} \times \mathbf{v}'$ is called the *Coriolis force.* When $\mathfrak{F}'$ has the constant angular velocity $\boldsymbol{\omega}$ about a fixed axis through A, $\boldsymbol{\alpha} = 0$, $\mathbf{a}_A = 0$, and $\mathbf{a}_Q = \boldsymbol{\omega} \times (\boldsymbol{\omega} \times \mathbf{r}')$. Taking O at A and the z-axis along $\boldsymbol{\omega}$ (Fig. 57a), we have

$$-m\mathbf{a}_Q = m\omega^2\,(\mathbf{r}' - \mathbf{k}\cdot\mathbf{r}'\,\mathbf{k}) = m\omega^2\,(\overrightarrow{OP} - \overrightarrow{ON}) = m\omega^2\,\overrightarrow{NP};$$

this vector perpendicular to the axis and directed *outward* is called the *centrifugal force* on the particle.

Example. Particle Falling from Rest. Refer the particle, originally at O, to the revolving frame $Oxyz$ attached to the earth: Oz points to the zenith (along a plumb line), Ox to the south, Oy to the east (Fig. 57b). The earth revolves from west to east about the axis $\overrightarrow{SN}$ and its angular velocity at north latitude λ is

$$\boldsymbol{\omega} = \frac{2\pi}{24 \times 60^2}\,(\mathbf{k}\sin\lambda - \mathbf{i}\cos\lambda)\ \text{radians/sec.}$$

When the particle is at rest relative to the earth, the force acting upon it is its local weight $m\mathbf{g}$; hence in (3) $\mathbf{F} - m\mathbf{a}_Q = m\mathbf{g}$. Therefore the equation of motion becomes

(i) $$\mathbf{a}' = \mathbf{g} - 2\boldsymbol{\omega} \times \mathbf{v}',$$

where $\mathbf{a}'$ and $\mathbf{v}'$ are the acceleration and velocity relative to the earth. We shall integrate (i) by successive approximations under the initial conditions $\mathbf{r}' = 0$, $\mathbf{v}' = 0$ when $t = 0$ and with $\mathbf{g}$ regarded as constant.

1. If we neglect the term in $\boldsymbol{\omega}$,

$$\mathbf{a}' = \mathbf{g}, \qquad \mathbf{v}' = \mathbf{g}t, \qquad \mathbf{r}' = \tfrac{1}{2}\mathbf{g}t^2.$$

The displacement $\mathbf{r}'$ is along the plumb line.

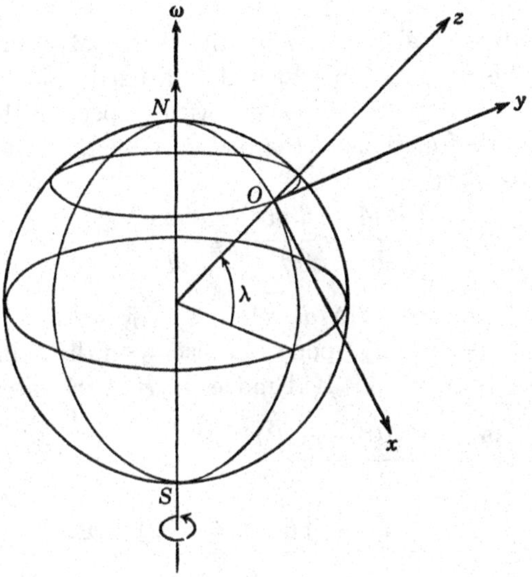

Fig. 57b

2. With $\mathbf{v}' = \mathbf{g}t$, (i) gives

$$\mathbf{a}' = \mathbf{g} - 2t\boldsymbol{\omega} \times \mathbf{g},$$

$$\mathbf{v}' = \mathbf{g}t - t^2\boldsymbol{\omega} \times \mathbf{g},$$

$$\mathbf{r}' = \tfrac{1}{2}\mathbf{g}t^2 - \tfrac{1}{3}t^3\boldsymbol{\omega} \times \mathbf{g}.$$

Since $\mathbf{g} = -g\mathbf{k}$, the second term gives an eastward deflection,

$$-\tfrac{1}{3}t^3\boldsymbol{\omega} \times \mathbf{g} = \tfrac{1}{3}t^3\omega g \cos \lambda \mathbf{j}.$$

3. With the last value of $\mathbf{v}'$,

$$\mathbf{a}' = \mathbf{g} - 2t\boldsymbol{\omega} \times \mathbf{g} + 2t^2\boldsymbol{\omega} \times (\boldsymbol{\omega} \times \mathbf{g}),$$

$$\mathbf{v}' = \mathbf{g}t - t^2\boldsymbol{\omega} \times \mathbf{g} + \tfrac{2}{3}t^3\boldsymbol{\omega} \times (\boldsymbol{\omega} \times \mathbf{g}),$$

$$\mathbf{r}' = \tfrac{1}{2}\mathbf{g}t^2 - \tfrac{1}{3}t^3\boldsymbol{\omega} \times \mathbf{g} + \tfrac{1}{6}t^4\boldsymbol{\omega} \times (\boldsymbol{\omega} \times \mathbf{g}).$$

The last term is a deflection in the meridian or xz-plane. Since $\boldsymbol{\omega} \times \mathbf{g} = -\omega g \cos \lambda\, \mathbf{j}$,

$$\boldsymbol{\omega} \times (\boldsymbol{\omega} \times \mathbf{g}) = \omega^2 g \cos \lambda\, (\mathbf{i} \sin \lambda + \mathbf{k} \cos \lambda),$$

$$\mathbf{r}' = \tfrac{1}{6} t^4 \omega^2 g \sin \lambda \cos \lambda\, \mathbf{i} + \tfrac{1}{3} t^3 \omega g \cos \lambda\, \mathbf{j} - (\tfrac{1}{2} g t^2 - \tfrac{1}{6} t^4 \omega^2 g \cos^2 \lambda)\, \mathbf{k}.$$

The first and second terms give deflections to the south and east; the third shows that the particle in t seconds falls through a distance

$$-z = \tfrac{1}{2} g t^2 - \tfrac{1}{6} t^4 \omega^2 g \cos^2 \lambda.$$

58. Derivative of a Motor. Let the motor $\mathbf{M} = \mathbf{m} + \epsilon \mathbf{m}_A$ be referred to the frame $\mathfrak{F}'$ in which A is a fixed point; and let $\mathfrak{F}'$ itself have the motion $\mathbf{V} = \boldsymbol{\omega} + \epsilon \mathbf{v}_A$ with respect to the frame $\mathfrak{F}$. If $\mathbf{m}$ and $\mathbf{m}_A$ are functions of the real variable t, the derivative of $\mathbf{M}$ relative to $\mathfrak{F}'$ is

$$\frac{d'\mathbf{M}}{dt} = \frac{d'\mathbf{m}}{dt} + \epsilon \frac{d'\mathbf{m}_A}{dt}. \tag{1}$$

In order to compute $d\mathbf{M}/dt$ relative to $\mathfrak{F}$ we must remember that $\mathbf{m}_A$ depends not only upon t but also upon the point A, which is in motion relative to $\mathfrak{F}$. If A moves to B in the interval Δt,

$$\begin{aligned}\frac{\Delta \mathbf{m}_A}{\Delta t} &= \frac{\mathbf{m}_A(t + \Delta t) - \mathbf{m}_A(t)}{\Delta t} \\ &= \frac{\mathbf{m}_B(t + \Delta t) + \overrightarrow{AB} \times \mathbf{m}(t + \Delta t) - \mathbf{m}_A(t)}{\Delta t} \qquad (31.3) \\ &= \frac{\mathbf{m}_B(t + \Delta t) - \mathbf{m}_A(t)}{\Delta t} + \frac{\overrightarrow{OB} - \overrightarrow{OA}}{\Delta t} \times \mathbf{m}(t + \Delta t),\end{aligned}$$

and hence, if $r_A = \overrightarrow{OA}$,

$$\lim_{\Delta t \to 0} \frac{\Delta \mathbf{m}_A}{\Delta t} = \frac{d\mathbf{m}_A}{dt} + \frac{d\mathbf{r}_A}{dt} \times \mathbf{m}.$$

Relative to the frame $\mathfrak{F}$, we therefore define

$$\frac{d\mathbf{M}}{dt} = \frac{d\mathbf{m}}{dt} + \epsilon \left(\frac{d\mathbf{m}_A}{dt} + \frac{d\mathbf{r}_A}{dt} \times \mathbf{m} \right). \tag{2}$$

To verify that $d\mathbf{M}/dt$ is a motor, we must show that

$$\frac{d\mathbf{m}_P}{dt} + \frac{d\mathbf{r}_P}{dt} \times \mathbf{m} = \frac{d\mathbf{m}_A}{dt} + \frac{d\mathbf{r}_A}{dt} \times \mathbf{m} + \overrightarrow{PA} \times \frac{d\mathbf{m}}{dt}.$$

This equation, in fact, follows from

$$\mathbf{m}_P = \mathbf{m}_A + (\mathbf{r}_A - \mathbf{r}_P) \times \mathbf{m}$$

on differentiation.

If t denotes time,

$$\frac{d\mathbf{r}_A}{dt} = \mathbf{v}_A;$$

and

$$\frac{d\mathbf{m}}{dt} = \boldsymbol{\omega} \times \mathbf{m} + \frac{d'\mathbf{m}}{dt}, \qquad \frac{d\mathbf{m}_A}{dt} = \boldsymbol{\omega} \times \mathbf{m}_A + \frac{d'\mathbf{m}_A}{dt},$$

from (56.3). Substituting these results in (2) gives

$$(3) \qquad \frac{d\mathbf{M}}{dt} = \boldsymbol{\omega} \times \mathbf{m} + \epsilon(\boldsymbol{\omega} \times \mathbf{m}_A + \mathbf{v}_A \times \mathbf{m}) + \frac{d'\mathbf{M}}{dt},$$

or, in view of (34.1),

$$(4) \qquad \frac{d\mathbf{M}}{dt} = \mathbf{V} \times \mathbf{M} + \frac{d'\mathbf{M}}{dt}.$$

This is the motor analogue of (56.3).

Making use of the definition (2), we may verify that the following rules of differentiation are valid:

$$(5) \qquad \frac{d}{dt}(\mathbf{M} + \mathbf{N}) = \frac{d\mathbf{M}}{dt} + \frac{d\mathbf{N}}{dt},$$

$$(6) \qquad \frac{d}{dt}(\lambda\mathbf{M}) = \lambda\frac{d\mathbf{M}}{dt} + \frac{d\lambda}{dt}\mathbf{M},$$

$$(7) \qquad \frac{d}{dt}(\mathbf{M} \cdot \mathbf{N}) = \mathbf{M} \cdot \frac{d\mathbf{N}}{dt} + \mathbf{N} \cdot \frac{d\mathbf{M}}{dt},$$

$$(8) \qquad \frac{d}{dt}(\mathbf{M} \times \mathbf{N}) = \mathbf{M} \times \frac{d\mathbf{N}}{dt} + \frac{d\mathbf{M}}{dt} \times \mathbf{N}.$$

Example 1. If the motion of a rigid body is given by the velocity motor $\mathbf{V} = \boldsymbol{\omega} + \epsilon\mathbf{v}_A$ (A a fixed point of the body), its acceleration motor is

$$\frac{d\mathbf{V}}{dt} = \boldsymbol{\alpha} + \epsilon(\mathbf{a}_A + \mathbf{v}_A \times \boldsymbol{\omega}).$$

The axis of this motor is called the *instantaneous axis of acceleration;* its equation may be written from (31.8).

Example 2. Let $\mathbf{F} = \mathbf{f} + \epsilon \mathbf{f}_A$ be a force acting on a rigid body whose velocity motor is $\mathbf{V} = \boldsymbol{\omega} + \epsilon \mathbf{v}_A$. Then

$$\text{(i)} \qquad \frac{d\mathbf{F}}{dt} = \frac{d\mathbf{f}}{dt} + \epsilon \left(\frac{d\mathbf{f}_A}{dt} + \mathbf{v}_A \times \mathbf{f} \right).$$

If $\mathbf{F}$ acts at the point P of the body, $\mathbf{f}_P = 0$ and

$$\text{(ii)} \qquad \frac{d\mathbf{F}}{dt} = \frac{d\mathbf{f}}{dt} + \epsilon \mathbf{v}_P \times \mathbf{f}, \quad \text{referred to } P.$$

Hence, referred to A,

$$\text{(iii)} \qquad \frac{d\mathbf{F}}{dt} = \frac{d\mathbf{f}}{dt} + \epsilon \left(\mathbf{v}_P \times \mathbf{f} + \overrightarrow{AP} \times \frac{d\mathbf{f}}{dt} \right);$$

let the reader show that (i) and (iii) are consistent.

Example 3. The moving trihedral TNB at the point P of a space curve $\mathbf{r} = \mathbf{r}(s)$ may be regarded as a rigid body having the velocity motor,

$$\mathbf{V} = \boldsymbol{\delta} + \epsilon \mathbf{v}_P = \boldsymbol{\delta} + \epsilon \mathbf{T},$$

as the point P traverses the curve with unit speed. The three line vectors $\mathbf{T} = \mathrm{T}$, $\mathbf{N} = \mathrm{N}$, $\mathbf{B} = \mathrm{B}$ through P are fixed in the trihedral; hence, from (4),

$$\frac{d\mathbf{T}}{ds} = \mathbf{V} \times \mathbf{T}, \qquad \frac{d\mathbf{N}}{ds} = \mathbf{V} \times \mathbf{N}, \qquad \frac{d\mathbf{B}}{ds} = \mathbf{V} \times \mathbf{B}.$$

These are Frenet's Formulas in motor form. Written out in full they become

$$\frac{d\mathbf{T}}{ds} = \kappa \mathbf{N}, \qquad \frac{d\mathbf{N}}{ds} = -\tau \mathbf{T} + \kappa \mathbf{B} + \epsilon \mathbf{B}, \qquad \frac{d\mathbf{B}}{ds} = -\tau \mathbf{N} - \epsilon \mathbf{N};$$

for example

$$\frac{d\mathbf{B}}{ds} = (\boldsymbol{\delta} + \epsilon \mathbf{T}) \times \mathbf{B} = \frac{d\mathrm{B}}{ds} - \epsilon \mathbf{N} = -\tau \mathbf{N} - \epsilon \mathbf{N}.$$

59. Summary: Vector Derivatives. The derivative $d\mathbf{u}/dt$ of a vector function $\mathbf{u}(t)$ is defined as the limit of $\Delta\mathbf{u}/\Delta t$ as Δt approaches zero. If $\mathbf{u} = \overrightarrow{OP}$ is drawn from a fixed origin and P describes the curve C as t varies, $d\mathbf{u}/dt$ is a vector tangent to C at P in the direction of increasing t. If $|\mathbf{u}|$ is constant, C is a circle and $d\mathbf{u}/dt$ is perpendicular to $\mathbf{u}$.

The derivative of a constant vector is zero. The derivatives of the sum $\mathbf{u} + \mathbf{v}$ and the products $f\mathbf{u}$, $\mathbf{u} \cdot \mathbf{v}$, $\mathbf{u} \times \mathbf{v}$ are found by the familiar rules of calculus; but for $\mathbf{u} \times \mathbf{v}$ the order of the factors must be preserved.

If s is the arc along a curve $\mathbf{r} = \mathbf{r}(s)$, $d\mathbf{r}/ds = \mathbf{T}$, a unit vector tangent to the curve the direction of increasing s. The unit *principal normal* $\mathbf{N}$ to the curve has the direction of $d\mathbf{T}/ds$; and the unit binormal $\mathbf{B} = \mathbf{T} \times \mathbf{N}$; then $[\mathbf{TNB}] = 1$. The vectors of the moving trihedral $\mathbf{TNB}$ change conformably to *Frenet's Formulas:*

$$\frac{d\mathbf{T}}{ds} = \kappa\mathbf{N} = \boldsymbol{\delta} \times \mathbf{T},$$

$$\frac{d\mathbf{N}}{ds} = -\kappa\mathbf{T} + \tau\mathbf{B} = \boldsymbol{\delta} \times \mathbf{N},$$

$$\frac{d\mathbf{B}}{ds} = -\tau\mathbf{N} = \boldsymbol{\delta} \times \mathbf{B};$$

κ is the *curvature*, τ the *torsion* of the curve; and the *Darboux vector* $\boldsymbol{\delta} = \tau\mathbf{T} + \kappa\mathbf{B}$ is the angular velocity of the moving trihedral as its vertex traverses the curve with unit speed $(ds/dt = 1)$. For plane curves $\tau = 0$.

If t denotes time, a particle P traveling along the curve $\mathbf{r} = \mathbf{r}(t)$ has the *velocity* and *acceleration*,

$$\mathbf{v} = \frac{d\mathbf{r}}{dt}, \qquad \mathbf{a} = \frac{d\mathbf{v}}{dt} = \frac{d^2\mathbf{r}}{dt^2}.$$

If $v = ds/dt$ is the *speed*, and $\rho = 1/\kappa$ the *radius of curvature* of the path,

$$\mathbf{v} = v\mathbf{T}, \qquad \mathbf{a} = \frac{dv}{dt}\mathbf{T} + \frac{v^2}{\rho}\mathbf{N}.$$

If A is any point of a free rigid body, the velocities of its points are given by the velocity motor,

$$\mathsf{V} = \boldsymbol{\omega} + \epsilon\mathbf{v}_A;$$

here $\boldsymbol{\omega}$, the *angular velocity* vector, is the same for any choice of A. Since V is a motor, for any point P of the body,

$$\mathbf{v}_P = \mathbf{v}_A + \overrightarrow{PA} \times \boldsymbol{\omega} = \mathbf{v}_A + \boldsymbol{\omega} \times \overrightarrow{AP}.$$

If, at any instant, $\boldsymbol{\omega} \neq 0$, the body has a screw motion about the axis of V, the *instantaneous axis* of velocity; this reduces to a *pure rotation* if V is a line vector $(\boldsymbol{\omega} \cdot \mathbf{v}_A = 0)$. If $\boldsymbol{\omega} = 0$ the motion is an instantaneous *translation* of velocity $\mathbf{v}_A$.

If the frame $\mathfrak{F}'$ has the angular velocity relative to $\mathfrak{F}$, the rates of change of a vector $\mathbf{u}$ relative to these frames are connected by

$$\frac{d\mathbf{u}}{dt} = \frac{d'\mathbf{u}}{dt} + \boldsymbol{\omega} \times \mathbf{u}.$$

A particle P has the velocity and acceleration $\mathbf{v}'$, $\mathbf{a}'$ relative to frame $\mathfrak{F}'$, $\mathbf{v}$, $\mathbf{a}$ relative to the frame $\mathfrak{F}$; then, if the motion of $\mathfrak{F}'$ relative to $\mathfrak{F}$ is given by the motor $\mathbf{V} = \boldsymbol{\omega} + \epsilon\mathbf{v}_Q$, where Q is the point of $\mathfrak{F}'$ coinciding at the instant with P,

$$\mathbf{v} = \mathbf{v}_Q + \mathbf{v}', \qquad \mathbf{a} = \mathbf{a}_Q + 2\boldsymbol{\omega} \times \mathbf{v}' + \mathbf{a}'.$$

The term $2\boldsymbol{\omega} \times \mathbf{v}'$ is the *acceleration of Coriolis.* When $\mathfrak{F}'$ is in translation relative to $\mathfrak{F}$, the velocity equation may be written $\mathbf{v} = \mathbf{v}_{\mathfrak{F}'} + \mathbf{v}'$.

The derivative of the motor $\mathbf{M} = \mathbf{m}(t) + \epsilon\mathbf{m}_A(t)$ is defined as

$$\frac{d\mathbf{M}}{dt} = \frac{d\mathbf{m}}{dt} + \epsilon\left(\frac{d\mathbf{m}_A}{dt} + \frac{d\mathbf{r}_A}{dt} \times \mathbf{m}\right).$$

If t denotes time, and $d'\mathbf{M}/dt$ refers to a frame $\mathfrak{F}'$ having the motion $\mathbf{V} = \boldsymbol{\omega} + \epsilon\mathbf{v}_A$, relative to a frame $\mathfrak{F}$, then

$$\frac{d\mathbf{M}}{dt} = \mathbf{V} \times \mathbf{M} + \frac{d'\mathbf{M}}{dt}.$$

PROBLEMS

1. If r and x are the distances of a point on a parabola from the focus and directrix, $r - x = 0$. Show that $(\mathbf{R} - \mathbf{i}) \cdot \mathbf{T} = 0$, and interpret the equation.

2. Prove that the tangent to a hyperbola bisects the angle between the focal radii to the point of tangency. [Cf. § 44, ex. 2.]

3. An *equiangular spiral* cuts all vectors from its pole 0 at the same angle α ($\mathbf{R} \cdot \mathbf{T} = \cos\alpha$). If (r, θ) are the polar coordinates of a variable point P on the spiral, show that

(a) $$ds/dr = \sec\alpha, \qquad s - s_0 = (r - r_0)\sec\alpha;$$

(b) $$\frac{1}{r}\frac{dr}{d\theta} = \cot\alpha, \qquad \log\frac{r}{r_0} = (\theta - \theta_0)\cot\alpha;$$

(c) $$\rho = ds/d\theta = r/\sin\alpha;$$

and that the center of curvature is the point where the perpendicular to OP at O cuts the normal at P.

4. If $\mathbf{r} = \mathbf{r}(s)$ is a plane curve Γ, show that

(*a*) $\mathbf{r}_1 = \mathbf{r} + c\mathbf{N}$ (c const.) is a *parallel curve* Γ_1 (§ 45, ex.) at a normal distance c from Γ;

(*b*) $s_1 = s - c\psi$, provided $s_1 = s = 0$, when $\psi = 0$;

(*c*) $\rho_1 = \rho - c$.

5. A curve $\mathbf{r} = \mathbf{r}(s)$ has the property that the locus $\mathbf{r}_1 = \mathbf{r} + c\mathbf{T}$ (c const.) is a straight line. Prove that the curve is *plane*, in fact, a *tractrix*. [Cf. § 50, ex. 5.]

6. An *involute* of a curve Γ, $\mathbf{r} = \mathbf{r}(s)$, is a curve Γ_1 which cuts the tangents of Γ at right angles. Prove that Γ has the one-parameter family of involutes,

$$\mathbf{r}_1 = \mathbf{r} + (c - s)\mathbf{T} \qquad (c = \text{const.});$$

and, if we take $\mathbf{T}_1 = \mathbf{N}$, $ds_1/ds = (c - s)\kappa$.

7. Show that the involute of the circle,

$$x = a \cos t, \qquad y = a \sin t,$$

obtained by unwrapping a string from the point $t = 0$ is

$$x_1 = a(\cos t + t \sin t), \qquad y_1 = a(\sin t - t \cos t);$$

and that $s_1 = \frac{1}{2}at^2$ gives the arc along the involute.

8. The cylinders $x^2 + y^2 = a^2$, $y^2 + z^2 = a^2$ intersect in two ellipses, one of which is

$$x = a \cos t, \quad y = a \sin t, \quad z = a \cos t.$$

Show that its radius of curvature is

$$\rho = a(1 + \sin^2 t)^{\frac{3}{2}}/\sqrt{2}.$$

9. The equation of a cycloid is

$$x = a(t - \sin t), \qquad y = a(1 - \cos t).$$

Prove that

(*a*) $$\mathbf{T} = \left[\sin\frac{t}{2}, \cos\frac{t}{2}\right], \qquad ds/dt = 2a \sin\frac{t}{2};$$

(*b*) $$\psi = (\mathbf{i}, \mathbf{T}) = \frac{\pi}{2} - \frac{t}{2}, \qquad \rho = -4a \sin\frac{t}{2} \qquad (44.3);$$

(*c*) The equation of its evolute is

$$x_1 = a(t + \sin t), \qquad y_1 = -a(1 - \cos t).$$

10. Find the vectors $\mathbf{T}$, $\mathbf{N}$, $\mathbf{B}$ and the curvature and torsion of the twisted cubic,

$$x = 3t, \quad y = 3t^2, \quad z = 2t^3,$$

at the points where $t = 0$ and $t = 1$. Write the equations for the normal and osculating planes to the curve at the point $t = 1$.

11. Show that curvature and torsion of the curve,

$$x = a(3t - t^3), \quad y = 3at^2, \quad z = a(3t + t^3),$$

are

$$\kappa = \tau = 1/3a(1 + t^2)^2.$$

12. Find the envelope of the family of straight lines in the xy-plane,

$$\mathbf{r} = p(\theta)\mathbf{R} + \lambda\mathbf{P},$$

where $\mathbf{R}$ and $\mathbf{P} = \mathbf{k} \times \mathbf{R}$ are the unit vectors of § 44, θ = angle ($\mathbf{i}$, $\mathbf{R}$), and p is the perpendicular distance from the origin to the line. Show that the curve,

$$\mathbf{r}_1 = p\mathbf{R} + p'\mathbf{P}$$

is the envelope, and that

$$\mathbf{T}_1 = \mathbf{T}, \qquad ds_1/d\theta = p + p''.$$

13. Find the envelope of the family of lines for which the segment included between the x-axis and y-axis is of constant length c. [In Problem 11 put $p = c \sin\theta \cos\theta$.]

Show that the envelope has the parametric equations,

$$x = c \sin^3 \theta, \qquad y = c \cos^3 \theta;$$

and that the entire length of the curve is $6c$.

14. Show that the curvature and torsion of the curve,

$$x = e^t, \quad y = e^{-t}, \quad z = \sqrt{2}\, t,$$

are

$$\kappa = -\tau = \sqrt{2}/(e^t + e^{-t})^2.$$

15. Verify that the curve,

$$x = a \sin^2 t, \quad y = a \sin t \cos t, \quad z = a \cos t,$$

lies on a sphere. Show that the curve has a double point at $(a, 0, 0)$ for the parameter values $t = \pm\pi/2$, and that the tangents to the curve at this point are perpendicular.

16. If a curve $\mathbf{r} = \mathbf{r}(s)$ lies on a sphere $(\mathbf{r} - \mathbf{c}) \cdot (\mathbf{r} - \mathbf{c}) = a^2$, show that

$$\mathbf{r} - \mathbf{c} = -\rho\mathbf{N} - \frac{1}{\tau}\frac{d\rho}{ds}\mathbf{B}.$$

Hence, prove that

$$\rho\tau + \frac{d}{ds}\left(\frac{1}{\tau}\frac{d\rho}{ds}\right) = 0$$

is a necessary and sufficient condition that a twisted curve ($\tau \neq 0$) lie on a sphere.

17. The points on two curves Γ and Γ_1 are in one-to-one correspondence. If $\mathbf{T} = \mathbf{T}_1$ at corresponding points, prove that

$$\mathbf{N} = \mathbf{N}_1, \quad \mathbf{B} = \mathbf{B}_1, \quad ds_1/ds = \kappa/\kappa_1 = \tau/\tau_1.$$

When both curves cut the rays from 0 at the same angle, show that $\mathbf{r}_1 = c\mathbf{r}$; and that $s_1 = cs$, if both arcs are measured from the same ray.

18. A curve Γ is called a *Bertrand curve* if its principal normals are principal normals of another curve Γ_1; then Γ_1 is also a Bertrand curve, and

$$\mathbf{r}_1 = \mathbf{r} + \lambda\mathbf{N}, \quad \mathbf{N}_1 = \epsilon\mathbf{N}, \quad \text{where} \quad \epsilon = 1 \text{ or } -1.$$

Show, in turn, that

(*a*) $$\lambda = c \text{ (const.)}, \qquad \mathbf{T}_1 \frac{ds_1}{ds} = (1 - c\kappa)\mathbf{T} + c\tau\mathbf{B};$$

(*b*) If φ = angle $(\mathbf{B}, \mathbf{B}_1)$, taken positive relative to $\mathbf{N}$,

$$\mathbf{B}_1 = \mathbf{B}\cos\varphi + \mathbf{T}\sin\varphi, \quad \epsilon\mathbf{T}_1 = \mathbf{T}\cos\varphi - \mathbf{B}\sin\varphi;$$

(*c*) φ = const., and

$$-\epsilon\tau_1 \frac{ds_1}{ds} = \kappa\sin\varphi - \tau\cos\varphi, \quad \kappa_1 \frac{ds_1}{ds} = \kappa\cos\varphi + \tau\sin\varphi;$$

(*d*) $$\epsilon \frac{ds_1}{ds} = \frac{1 - c\kappa}{\cos\varphi} = \frac{-c\tau}{\sin\varphi};$$

(*e*) $$c^2\tau\tau_1 = \sin^2\varphi, \qquad \kappa_1 = -\epsilon(\tau_1\cot\varphi + 1/c);$$

since $\kappa_1 = 0$, the last equation determines ϵ.

(*f*) As to the Darboux vectors, $\boldsymbol{\delta}_1\, ds_1/ds = \boldsymbol{\delta}$.

19. The *characteristics* of a one-parameter family of planes whose homogeneous coordinates (§ 26) are $(\mathbf{a}(t), \alpha_O(t))$, are the limiting lines of intersection of the planes corresponding to the parameter values t and $t + h$ as $h \to 0$. The line of intersection of the two planes has the homogeneous coordinates (26.10),

$$\{\mathbf{a}(t) \times \mathbf{a}(t + h), \quad \alpha_O(t + h)\mathbf{a}(t) - \alpha_O(t)\mathbf{a}(t + h)\};$$

and, if we divide both by h and pass to the limit $h \to 0$, show that we obtain

$$(\mathbf{a} \times \mathbf{a}', \; \alpha_O'\mathbf{a} - \alpha_O\mathbf{a}')$$

as the coordinates of the characteristics.

20. From the result of Problem 19 show that the three families of planes associated with the twisted ($\tau \neq 0$) curve $\mathbf{r} = \mathbf{r}(s)$, namely,

$(\mathbf{T}, \mathbf{r} \cdot \mathbf{T})$, normal planes,

$(\mathbf{N}, \mathbf{r} \cdot \mathbf{N})$, rectifying planes,

$(\mathbf{B}, \mathbf{r} \cdot \mathbf{B})$, osculating planes,

have as characteristics the respective families of lines:

$[\mathbf{B}, (\mathbf{r} + \rho\mathbf{N}) \times \mathbf{B}]$, parallel to $\mathbf{B}$ through the centers of curvature;

$(\boldsymbol{\delta}, \mathbf{r} \times \boldsymbol{\delta})$, parallel to $\boldsymbol{\delta}$ through points of the curve;

$(\mathbf{T}, \mathbf{r} \times \mathbf{T})$, tangent to the curve (§ 48).

21. If a particle P of mass m is subject to a *central force* $\mathbf{F} = mf(r)\mathbf{R}$, where $r = OP$ and $\mathbf{R}$ is the unit vector along $\overrightarrow{OP}$, its equation of motion is

$$m\frac{d^2\mathbf{r}}{dt^2} = \mathbf{F} \quad \text{or} \quad \frac{d\mathbf{v}}{dt} = f(r)\mathbf{R}.$$

Prove that $\mathbf{r} \times \mathbf{v} = \mathbf{h}$, a vector constant; hence, show that the motion is *plane* and that $\mathbf{r} = \overrightarrow{OP}$ sweeps out area at a constant rate (*Law of Areas*).

22. When $mf(r) = -\gamma mM/r^2$ in Problem 21 the particle P is attracted towards a mass M at O according to the law of inverse squares:

(a) $$\frac{d\mathbf{v}}{dt} = -\frac{k}{r^2}\mathbf{R} \qquad (k = \gamma M).$$

Show, in turn, that

(b) $$\mathbf{r} \times \mathbf{v} = \mathbf{h}$$

(c) $$\mathbf{h} = r^2\, \mathbf{R} \times \frac{d\mathbf{R}}{dt},$$

(d) $$\frac{d\mathbf{v}}{dt} \times \mathbf{h} = k\frac{d\mathbf{R}}{dt},$$

(e) $$\mathbf{v} \times \mathbf{h} = k(\mathbf{R} + \boldsymbol{\epsilon}),$$

where $\boldsymbol{\epsilon}$ is a constant vector. From (b) and (e),

$$\mathbf{r} \times \mathbf{v} \cdot \mathbf{h} = h^2, \qquad \mathbf{r} \cdot \mathbf{v} \times \mathbf{h} = kr(1 + \epsilon \cos\theta)$$

where θ = angle $(\boldsymbol{\epsilon}, \mathbf{r})$; thus obtain the equation of the *orbit*,

(f) $$r = \frac{h^2/k}{1 + \epsilon\cos\theta},$$

a conic section of eccentricity ϵ referred to a focus as pole. When $\epsilon < 1$, prove that the orbital *ellipse* is described in the periodic time,

$$T = \frac{\text{area of ellipse}}{h/2} = \frac{2\pi}{\sqrt{k}}a^{\frac{3}{2}},$$

and, hence,

(g) $$T^2/a^3 = 4\pi^2/\gamma M \qquad \text{(Kepler's Third Law).}$$

(See Brand's *Vectorial Mechanics*, § 177.)

23. If the plane $Ax + By + Cz = 1$, fixed in space, is referred to rectangular axes rotating with the angular velocity $\boldsymbol{\omega} = [\omega_1, \omega_2, \omega_3]$, show that

$$\left[\frac{dA}{dt}, \frac{dB}{dt}, \frac{dC}{dt}\right] = [A, B, C] \times [\omega_1, \omega_2, \omega_3].$$

24. A particle P has the cylindrical coordinates ρ, φ, z (cf. § 89, ex. 1). If P moves in the ρz-plane so that $d\rho/dt$ and dz/dt are constant while the plane itself revolves about the z-axis with the constant angular speed $d\varphi/dt = \omega$, find the acceleration of P,

(a) By direct calculation.
(b) By use of the Theorem of Coriolis.

25. A real, everywhere convex, closed plane curve with a unique tangent at each point is called an *oval*. It can be shown that an oval has just two tangents parallel to every direction in the plane. The distance between these tangents is the width of the curve in the direction of the perpendicular. Prove that the perimeter of an oval of constant width b is πb (*Barbier's Theorem*).

CHAPTER IV

LINEAR VECTOR FUNCTIONS

60. Vector Functions of a Vector. A vector $\mathbf{v}$ is said to be a function of a vector $\mathbf{r}$ if $\mathbf{v}$ is determined when $\mathbf{r}$ is given; and we write $\mathbf{v} = \mathbf{f}(\mathbf{r})$. Since $\mathbf{r}$ is determined by its components, $\mathbf{f}(\mathbf{r})$ is a function of two or three scalar variables according as $\mathbf{r}$ varies in a plane or in space. A vector function may be given by a formula, as $\mathbf{f}(\mathbf{r}) = (\mathbf{a} \times \mathbf{r}) \times \mathbf{r}$; or it may be defined geometrically. Thus if $\mathbf{r} = \overrightarrow{OP}$ varies over the points P of a given surface, and $\mathbf{v} = \overrightarrow{OQ}$ is the vector perpendicular on the tangent plane to the surface at P, $\mathbf{v} = \mathbf{f}(\mathbf{r})$.

A vector function $\mathbf{f}(\mathbf{r})$ is said to be *continuous* for $\mathbf{r} = \mathbf{r}_0$ if

$$\lim_{\mathbf{r} \to \mathbf{r}_0} \mathbf{f}(\mathbf{r}) = \mathbf{f}(\mathbf{r}_0). \tag{1}$$

This means that, when the components of $\mathbf{r}$ approach those of $\mathbf{r}_0$ in any manner, the components of $\mathbf{f}(\mathbf{r})$ approach those of $\mathbf{f}(\mathbf{r}_0)$.

A vector function is said to be *linear* when

$$\mathbf{f}(\mathbf{r} + \mathbf{s}) = \mathbf{f}(\mathbf{r}) + \mathbf{f}(\mathbf{s}), \tag{2}$$

$$\mathbf{f}(\lambda\mathbf{r}) = \lambda\mathbf{f}(\mathbf{r}), \tag{3}$$

for arbitrary $\mathbf{r}$, $\mathbf{s}$, λ. For example, linear vector functions are defined by the formulas $k\mathbf{r}$, $\mathbf{a} \times \mathbf{r}$, $\mathbf{a}\,\mathbf{b} \cdot \mathbf{r}$, in which k, $\mathbf{a}$, $\mathbf{b}$ are constant. It can be shown that, when a *continuous* vector function satisfies the relation (2), it also satisfies (3) and is therefore linear.

Since we assume that (2) holds when $\mathbf{r} = \mathbf{s} = 0$, $\mathbf{f}(0) = 2\mathbf{f}(0)$; hence, for any linear vector function,

$$\mathbf{f}(0) = 0. \tag{4}$$

A linear vector function is completely determined when $\mathbf{f}(\mathbf{a}_1)$, $\mathbf{f}(\mathbf{a}_2)$, $\mathbf{f}(\mathbf{a}_3)$ *are given for any three non-coplanar vectors* $\mathbf{a}_1$, $\mathbf{a}_2$, $\mathbf{a}_3$. For, if we express $\mathbf{r}$ in terms of $\mathbf{a}_1$, $\mathbf{a}_2$, $\mathbf{a}_3$ as a basis,

$$\mathbf{r} = x^1\mathbf{a}_1 + x^2\mathbf{a}_2 + x^3\mathbf{a}_3,$$

we have, from (2) and (3),

$$\mathbf{f}(\mathbf{r}) = x^1\mathbf{f}(\mathbf{a}_1) + x^2\mathbf{f}(\mathbf{a}_2) + x^3\mathbf{f}(\mathbf{a}_3). \tag{5}$$

Let $\mathbf{a}^1$, $\mathbf{a}^2$, $\mathbf{a}^3$ denote the set reciprocal to $\mathbf{a}_1$, $\mathbf{a}_2$, $\mathbf{a}_3$ (§ 23); then, if we write

$$\mathbf{f}(\mathbf{a}_i) = \mathbf{b}_i, \qquad x^i = \mathbf{r} \cdot \mathbf{a}^i,$$

$\mathbf{f}(\mathbf{r})$ may be written in either of the forms:

$$\mathbf{f}(\mathbf{r}) = \mathbf{r} \cdot (\mathbf{a}^1\mathbf{b}_1 + \mathbf{a}^2\mathbf{b}_2 + \mathbf{a}^3\mathbf{b}_3), \tag{6}$$

$$\mathbf{f}(\mathbf{r}) = (\mathbf{b}_1\mathbf{a}^1 + \mathbf{b}_2\mathbf{a}^2 + \mathbf{b}_3\mathbf{a}^3) \cdot \mathbf{r}. \tag{7}$$

These formulas represent the most general linear vector function.

61. Dyadics. In linear vector functions of the form,

$$\mathbf{f}(\mathbf{r}) = \mathbf{a}_1\mathbf{b}_1 \cdot \mathbf{r} + \mathbf{a}_2\mathbf{b}_2 \cdot \mathbf{r} + \cdots + \mathbf{a}_n\mathbf{b}_n \cdot \mathbf{r},$$

we now regard $f(\mathbf{r})$ as the scalar product of $\mathbf{r}$ and the operator,

$$\Phi = \mathbf{a}_1\mathbf{b}_1 + \mathbf{a}_2\mathbf{b}_2 + \cdots + \mathbf{a}_n\mathbf{b}_n. \tag{1}$$

Assuming the distributive law for such products, we now write

$$\mathbf{f}(\mathbf{r}) = \Phi \cdot \mathbf{r}. \tag{2}$$

Following Willard Gibbs, we call the operator Φ a *dyadic* and each of its terms $\mathbf{a}_i\mathbf{b}_i$ a *dyad*. The vectors $\mathbf{a}_i$ are called *antecedents*, the vectors $\mathbf{b}_i$ *consequents*.

While $\mathbf{a} \cdot \mathbf{b}$ is a scalar and $\mathbf{a} \times \mathbf{b}$ a vector, the dyad $\mathbf{ab}$ represents a new mathematical entity. Gibbs regarded $\mathbf{ab}$ as a new species of product, the "indeterminate product." We shall find indeed that this product conforms to the distributive and associative laws, but is not commutative (in general $\mathbf{ab} \neq \mathbf{ba}$).

Since $\mathbf{r}$ *follows* Φ in (2), we call $\mathbf{r}$ a *postfactor*. If we use $\mathbf{r}$ as a *prefactor*, we get, in general, a different linear vector function:

$$\mathbf{g}(\mathbf{r}) = \mathbf{r} \cdot \Phi.$$

We proceed to develop an algebra for dyadics, laying down for this purpose definitions for equality, addition, and multiplication of dyadics.

Definition of Equality. We write $\Phi = \Psi$ when

$$\Phi \cdot \mathbf{r} = \Psi \cdot \mathbf{r} \quad \text{for every vector } \mathbf{r}. \tag{3}$$

If $\mathbf{s}$ is an arbitrary vector, we have from (3)

$$\mathbf{s} \cdot (\Phi \cdot \mathbf{r}) = \mathbf{s} \cdot (\Psi \cdot \mathbf{r}), \quad \text{or} \quad (\mathbf{s} \cdot \Phi) \cdot \mathbf{r} = (\mathbf{s} \cdot \Psi) \cdot \mathbf{r}.$$

Since $\mathbf{s} \cdot \Phi$ and $\mathbf{s} \cdot \Psi$ are two vectors that yield the same scalar product with every vector $\mathbf{r}$, these vectors must be equal; thus

$$(4) \qquad \mathbf{s} \cdot \Phi = \mathbf{s} \cdot \Psi \quad \text{for every vector } \mathbf{s}.$$

Conversely, from (4) we may deduce (3). Therefore $\Phi = \Psi$ when either (3) or (4) is fulfilled.

The Zero Dyadic. We write $\Phi = 0$ when

$$(5) \qquad \Phi \cdot \mathbf{r} = 0 \quad \text{for every } \mathbf{r}.$$

The preceding argument shows that $\Phi = 0$ also when

$$(6) \qquad \mathbf{s} \cdot \Phi = 0 \quad \text{for every } \mathbf{s}.$$

Definition of Addition. The sum $\Phi + \Psi$ of two dyadics is defined by the property,

$$(7) \qquad (\Phi + \Psi) \cdot \mathbf{r} = \Phi \cdot \mathbf{r} + \Psi \cdot \mathbf{r} \quad \text{for every } \mathbf{r}.$$

If Φ is given by (1), and

$$\Psi = \mathbf{c}_1\mathbf{d}_1 + \mathbf{c}_2\mathbf{d}_2 + \cdots + \mathbf{c}_m\mathbf{d}_m,$$

$$\Phi + \Psi = \mathbf{a}_1\mathbf{b}_1 + \cdots + \mathbf{a}_n\mathbf{b}_n + \mathbf{c}_1\mathbf{d}_1 + \cdots + \mathbf{c}_m\mathbf{d}_m.$$

In the sense of this definition, Φ (or Ψ) is the sum of its dyad terms, thus justifying our notation. The order of these dyads is immaterial; but the order of the vectors in each dyad must not be altered, for, in general, $\mathbf{ab} \cdot \mathbf{r} \neq \mathbf{ba} \cdot \mathbf{r}$ and hence $\mathbf{ab} \neq \mathbf{ba}$.

The Distributive Laws,

$$(8), (9) \qquad \mathbf{a}(\mathbf{b} + \mathbf{c}) = \mathbf{ab} + \mathbf{ac}, \qquad (\mathbf{a} + \mathbf{b})\mathbf{c} = \mathbf{ac} + \mathbf{bc},$$

are valid. The proof follows at once from the definition of equality; thus from

$$\mathbf{a}(\mathbf{b} + \mathbf{c}) \cdot \mathbf{r} = (\mathbf{ab} + \mathbf{ac}) \cdot \mathbf{r},$$

we deduce (8). We may now perform expansions as in ordinary algebra, if the order of the vectors is not altered; for example:

$$(\mathbf{a} + \mathbf{b})(\mathbf{c} + \mathbf{d}) = \mathbf{ac} + \mathbf{ad} + \mathbf{bc} + \mathbf{bd}.$$

If λ is any scalar,

(10) $$(\lambda\mathbf{a})\mathbf{b} = \mathbf{a}(\lambda\mathbf{b}),$$

and we shall write simply $\lambda\mathbf{ab}$ for either member.

From (60.6) or (60.7) we conclude that any dyadic Φ can be reduced to the sum of *three* dyads. To effect this reduction on Φ as given by (1), express each antecedent $\mathbf{a}_i$ in terms of the basis $\mathbf{e}_1, \mathbf{e}_2, \mathbf{e}_3$,

$$\mathbf{a}_i = a_i^1\mathbf{e}_1 + a_i^2\mathbf{e}_2 + a_i^3\mathbf{e}_3,$$

expand by the distributive law, and collect the terms which have the same antecedent: thus

$$\Phi = \Sigma(a_i^1\mathbf{e}_1 + a_i^2\mathbf{e}_2 + a_i^3\mathbf{e}_3)\mathbf{b}_i = \mathbf{e}_1\mathbf{f}_1 + \mathbf{e}_2\mathbf{f}_2 + \mathbf{e}_3\mathbf{f}_3,$$

where $\mathbf{f}_j = a_1^j\mathbf{b}_1 + \cdots + a_n^j\mathbf{b}_n$.

We also may reduce Φ to the sum of three dyads by expressing each consequent $\mathbf{b}_i$ in terms of the basis $\mathbf{e}_1, \mathbf{e}_2, \mathbf{e}_3$, and then expanding and collecting terms which have the same consequent; then Φ assumes the form,

$$\Phi = \mathbf{g}_1\mathbf{e}_1 + \mathbf{g}_2\mathbf{e}_2 + \mathbf{g}_3\mathbf{e}_3.$$

Thus it is always possible to express any dyadic so that its antecedents or its consequents are any three non-coplanar vectors chosen at pleasure.

If $\mathbf{a}_1, \mathbf{a}_2, \mathbf{a}_3$ are non-coplanar, a dyadic Φ is completely determined by giving their transforms (§ 60). If

$$\Phi \cdot \mathbf{a}_1 = \mathbf{b}_1, \qquad \Phi \cdot \mathbf{a}_2 = \mathbf{b}_2, \qquad \Phi \cdot \mathbf{a}_3 = \mathbf{b}_3,$$

and the set $\mathbf{a}^1, \mathbf{a}^2, \mathbf{a}^3$ is reciprocal to $\mathbf{a}_1, \mathbf{a}_2, \mathbf{a}_3$, we have explicitly

(11) $$\Phi = \mathbf{b}_1\mathbf{a}^1 + \mathbf{b}_2\mathbf{a}^2 + \mathbf{b}_3\mathbf{a}^3.$$

For a physical example of a dyadic the reader may turn to § 116, where the *stress dyadic* is introduced. The name *tensor*, now used in a much more general sense, originally was applied to this dyadic.

62. Affine Point Transformation. If we draw the position vectors,

$$\mathbf{r} = \overrightarrow{OP}, \qquad \mathbf{r}' = \Phi \cdot \mathbf{r} = \overrightarrow{OQ},$$

from a common origin O, the dyadic Φ defines a certain transformation of the points of space: to each point P corresponds a definite point Q. If, when P ranges over all space, Q does likewise, this transformation is called *affine;* the dyadic Φ then is called *complete.*

Important properties of an affine point transformation follow at once from the equations,

$$\Phi \cdot (\mathbf{a} + \mathbf{b}) = \Phi \cdot \mathbf{a} + \Phi \cdot \mathbf{b}, \qquad \Phi \cdot (\lambda \mathbf{a}) = \lambda \Phi \cdot \mathbf{a},$$

which characterize a linear vector function. Since $\Phi \cdot 0 = 0$, the transformation leaves the origin invariant. *Lines and planes are transformed into lines and planes.* Thus, for variable x, the

$$\text{Line} \quad \mathbf{r} = \mathbf{a} + x\mathbf{b} \rightarrow \text{Line} \quad \mathbf{r}' = \mathbf{a}' + x\mathbf{b}';$$

and, for variable x, y, the

$$\text{Plane} \quad \mathbf{r} = \mathbf{a} + x\mathbf{b} + y\mathbf{c} \rightarrow \text{Plane} \quad \mathbf{r}' = \mathbf{a}' + x\mathbf{b}' + y\mathbf{c}'.$$

The transformed equations always represent lines and planes when Φ is complete; for we shall show in § 70 that $\mathbf{b} \neq 0$ implies $\mathbf{b}' \neq 0$, and $\mathbf{b} \times \mathbf{c} \neq 0$ implies $\mathbf{b}' \times \mathbf{c}' \neq 0$.

63. Complete and Singular Dyadics. Given an arbitrary basis, $\mathbf{e}_1, \mathbf{e}_2, \mathbf{e}_3$, we can express any dyadic in the form,

$$\Phi = \mathbf{g}_1\mathbf{e}_1 + \mathbf{g}_2\mathbf{e}_2 + \mathbf{g}_3\mathbf{e}_3. \tag{1}$$

If we express $\mathbf{r}$ in terms of the reciprocal basis,

$$\mathbf{r} = x_1\mathbf{e}^1 + x_2\mathbf{e}^2 + x_3\mathbf{e}^3,$$

then, by virtue of the equations, $\mathbf{e}_i \cdot \mathbf{e}^j = \delta_i^j$,

$$\Phi \cdot \mathbf{r} = x_1\mathbf{g}_1 + x_2\mathbf{g}_2 + x_3\mathbf{g}_3.$$

When $\mathbf{g}_1, \mathbf{g}_2, \mathbf{g}_3$ are *non-coplanar,* $\mathbf{r}' = \Phi \cdot \mathbf{r}$ assumes all possible vector values as $\mathbf{r}$ ranges over the whole of space. If we put $\mathbf{r} = \overrightarrow{OP}$, $\mathbf{r}' = \overrightarrow{OQ}$, the dyadic defines an affine transformation $\mathbf{r}' = \Phi \cdot \mathbf{r}$ of space into itself; 3-dimensional P-space goes into 3-dimensional Q-space. A dyadic having this property is said to be *complete.* A complete dyadic cannot be reduced to a sum of less than three dyads; if, for example, we could reduce Φ to the sum of two dyads, $\mathbf{ab} + \mathbf{cd}$, all vectors $\mathbf{r}$ would transform into vectors $\mathbf{r}' = \mathbf{a}\,\mathbf{b} \cdot \mathbf{r} + \mathbf{c}\,\mathbf{d} \cdot \mathbf{r}$ parallel to the plane of $\mathbf{a}$ and $\mathbf{c}$.

If, however, $\mathbf{g}_1$, $\mathbf{g}_2$, $\mathbf{g}_3$, are *coplanar, but not collinear,* we can express each $\mathbf{g}_i$ in terms of two non-parallel vectors $\mathbf{f}_1$, $\mathbf{f}_2$, and reduce Φ to the sum of two dyads:

$$\Phi = \mathbf{f}_1\mathbf{h}_1 + \mathbf{f}_2\mathbf{h}_2. \tag{2}$$

This dyadic transforms all vectors $\mathbf{r}$ into vectors $\mathbf{r}' = \Phi \cdot \mathbf{r}$ in the plane of $\mathbf{f}_1$ and $\mathbf{f}_2$; 3-dimensional P-space goes into 2-dimensional Q-space. A dyadic having this property is said to be *planar.* A planar dyadic cannot be reduced to a single dyad $\mathbf{ab}$; for then all vectors $\mathbf{r}$ would transform into vectors $\mathbf{a}\,\mathbf{b} \cdot \mathbf{r}$ parallel to $\mathbf{a}$.

If $\mathbf{g}_1$, $\mathbf{g}_2$, $\mathbf{g}_3$ are *collinear,* we can replace each $\mathbf{g}_i$ by a multiple of a single vector $\mathbf{f}$ and reduce Φ to a single dyad:

$$\Phi = \mathbf{fh}. \tag{3}$$

This dyadic transforms all vectors $\mathbf{r}$ into vectors $\mathbf{r}' = \Phi \cdot \mathbf{r}$ parallel to $\mathbf{f}$; 3-dimensional P-space goes into 1-dimensional Q-space. Such a dyadic is called *linear.*

Finally, if $\mathbf{g}_1$, $\mathbf{g}_2$, $\mathbf{g}_3$ are all zero, $\Phi = 0$.

Planar, linear, and zero dyadics collectively are called *singular.* The point transformation,

$$\overrightarrow{OQ} = \Phi \cdot \overrightarrow{OP},$$

corresponding to a singular dyadic Φ reduces 3-dimensional P-space to a 2-, 1-, or 0-dimensional Q-space.

This discussion shows that, when Φ is reduced to the form (1) in which the consequents are non-coplanar, then Φ is complete, planar, linear, or zero, according as the antecedents are non-coplanar, coplanar but not collinear, collinear, or zero.

In particular, we have the

THEOREM. *A necessary and sufficient condition that a dyadic* $\mathbf{al} + \mathbf{bm} + \mathbf{cn}$ *be complete is that the antecedents* $\mathbf{a}$, $\mathbf{b}$, $\mathbf{c}$ *and consequents* $\mathbf{l}$, $\mathbf{m}$, $\mathbf{n}$ *be two sets of non-coplanar vectors.*

As a corollary,

$$\Phi \cdot \mathbf{r} = 0 \quad \text{implies} \quad \mathbf{r} = 0 \quad \text{when } \Phi \text{ is complete.} \tag{4}$$

For, if $\Phi = \mathbf{al} + \mathbf{bm} + \mathbf{cn}$,

$$\mathbf{a}\,\mathbf{l} \cdot \mathbf{r} + \mathbf{b}\,\mathbf{m} \cdot \mathbf{r} + \mathbf{c}\,\mathbf{n} \cdot \mathbf{r} = 0, \qquad \mathbf{l} \cdot \mathbf{r} = \mathbf{m} \cdot \mathbf{r} = \mathbf{n} \cdot \mathbf{r} = 0,$$

and hence $\mathbf{r} = 0$.

64. Conjugate Dyadics. The dyadics,

$$\Phi = \mathbf{a}_1\mathbf{b}_1 + \mathbf{a}_2\mathbf{b}_2 + \cdots + \mathbf{a}_n\mathbf{b}_n,$$

$$\Phi_c = \mathbf{b}_1\mathbf{a}_1 + \mathbf{b}_2\mathbf{a}_2 + \cdots + \mathbf{b}_n\mathbf{a}_n,$$

are said to be *conjugates* of each other. In general Φ and Φ_c are different dyadics; but evidently

(1) $$\Phi \cdot \mathbf{r} = \mathbf{r} \cdot \Phi_c$$

define the same linear vector function. If two dyadics are equal, their conjugates are equal; for $\Phi = \Psi$ implies that

$$\Phi \cdot \mathbf{r} = \Psi \cdot \mathbf{r}, \quad \text{or} \quad \mathbf{r} \cdot \Phi_c = \mathbf{r} \cdot \Psi_c,$$

for every $\mathbf{r}$, and hence $\Phi_c = \Psi_c$, by definition.

A dyadic is called:

$$\textit{Symmetric} \quad \text{if} \quad \Phi_c = \Phi,$$

$$\textit{Antisymmetric} \quad \text{if} \quad \Phi_c = -\Phi.$$

The importance of these special types of dyadics is due to the

THEOREM. *Every dyadic can be expressed in just one way as the sum of a symmetric and an antisymmetric dyadic.*

Proof. For any dyadic Φ, we have the identity,

(1) $$\Phi = \frac{\Phi + \Phi_c}{2} + \frac{\Phi - \Phi_c}{2} = \Psi + \Omega;$$

since

$$\Psi_c = \frac{\Phi_c + \Phi}{2} = \Psi, \qquad \Omega_c = \frac{\Phi_c - \Phi}{2} = -\Omega,$$

Ψ and Ω are, respectively, symmetric and antisymmetric. Moreover Φ can be so expressed in only one way. For, if

$$\Psi + \Omega = \Psi' + \Omega'$$

gave two such decompositions, we have, on taking conjugates,

$$\Psi - \Omega = \Psi' - \Omega',$$

and hence $\Psi = \Psi'$, $\Omega = \Omega'$.

65. Product of Dyadics. If the transformations corresponding to two linear vector functions,

$$\mathbf{v} = \Phi \cdot \mathbf{r}, \qquad \mathbf{w} = \Psi \cdot \mathbf{v},$$

are applied in succession, their resultant,

$$\mathbf{w} = \Psi \cdot (\Phi \cdot \mathbf{r}),$$

is a third linear vector function; for $\mathbf{w}_1 + \mathbf{w}_2$ corresponds to $\mathbf{r}_1 + \mathbf{r}_2$, and $\lambda\mathbf{w}$ to $\lambda\mathbf{r}$. This function is written

$$\mathbf{w} = (\Psi \cdot \Phi) \cdot \mathbf{r},$$

and $\Psi \cdot \Phi$ is called the *product* of the dyadics Ψ and Φ, taken in this order. The defining equation for the product $\Psi \cdot \Phi$ is therefore

(1) $$(\Psi \cdot \Phi) \cdot \mathbf{r} = \Psi \cdot (\Phi \cdot \mathbf{r}) \quad \text{for every } \mathbf{r}.$$

From (1) we find that the distributive and associative laws hold for the products of dyadics:

(2) $$\Phi \cdot (\Psi + \Omega) = \Phi \cdot \Psi + \Phi \cdot \Omega,$$

(3) $$(\Phi + \Psi) \cdot \Omega = \Phi \cdot \Omega + \Psi \cdot \Omega;$$

(4) $$\Phi \cdot (\Psi \cdot \Omega) = (\Phi \cdot \Psi) \cdot \Omega.$$

Proofs. For every vector $\mathbf{r}$,

$$\begin{aligned}
[\Phi \cdot (\Psi + \Omega)] \cdot \mathbf{r} &= \Phi \cdot [(\Psi + \Omega) \cdot \mathbf{r}] \\
&= \Phi \cdot [\Psi \cdot \mathbf{r} + \Omega \cdot \mathbf{r}] \\
&= \Phi \cdot (\Psi \cdot \mathbf{r}) + \Phi \cdot (\Omega \cdot \mathbf{r}) \\
&= (\Phi \cdot \Psi) \cdot \mathbf{r} + (\Phi \cdot \Omega) \cdot \mathbf{r}; \\
[(\Phi + \Psi) \cdot \Omega] \cdot \mathbf{r} &= (\Phi + \Psi) \cdot (\Omega \cdot \mathbf{r}) \\
&= \Phi \cdot (\Omega \cdot \mathbf{r}) + \Psi \cdot (\Omega \cdot \mathbf{r}) \\
&= (\Phi \cdot \Omega) \cdot \mathbf{r} + (\Psi \cdot \Omega) \cdot \mathbf{r}; \\
[\Phi \cdot (\Psi \cdot \Omega)] \cdot \mathbf{r} &= \Phi \cdot [(\Psi \cdot \Omega) \cdot \mathbf{r}] \\
&= \Phi \cdot [\Psi \cdot (\Omega \cdot \mathbf{r})] \\
&= (\Phi \cdot \Psi) \cdot (\Omega \cdot \mathbf{r}) \\
&= [(\Phi \cdot \Psi) \cdot \Omega] \cdot \mathbf{r}.
\end{aligned}$$

Equations (2), (3), (4) follow from these results by the definition of equality. In the proofs of (2) and (3), definition (1) justifies the first and last steps; in the proof of (3), (1) is applied in *every* step.

In order to compute a product $\Phi \cdot \Psi$ explicitly, we first find the product of two dyads. By definition,

$$[(\mathbf{ab}) \cdot (\mathbf{cd})] \cdot \mathbf{r} = \mathbf{ab} \cdot (\mathbf{cd} \cdot \mathbf{r}) = \mathbf{ab} \cdot \mathbf{c}\,\mathbf{d} \cdot \mathbf{r} = (\mathbf{b} \cdot \mathbf{c})\mathbf{ad} \cdot \mathbf{r},$$

for every $\mathbf{r}$, and hence

$$(5) \qquad (\mathbf{ab}) \cdot (\mathbf{cd}) = (\mathbf{b} \cdot \mathbf{c})\, \mathbf{ad}.$$

The product of $\mathbf{ab}$ and $\mathbf{cd}$, in this order, is the scalar $\mathbf{b} \cdot \mathbf{c}$ times the dyad $\mathbf{ad}$. Similarly,

$$(6) \qquad (\mathbf{cd}) \cdot (\mathbf{ab}) = (\mathbf{d} \cdot \mathbf{a})\, \mathbf{cb},$$

which in general differs from (5).

Making use of the distributive law, we now may form the product of any two dyadics,

$$\Phi = \sum_{i=1}^{n} \mathbf{a}_i\mathbf{b}_i, \qquad \Psi = \sum_{j=1}^{m} \mathbf{c}_j\mathbf{d}_j,$$

by expanding into nm dyads.

$$(7) \qquad \Phi \cdot \Psi = \sum_{i=1}^{n} \sum_{j=1}^{m} (\mathbf{b}_i \cdot \mathbf{c}_j)\mathbf{a}_i\mathbf{d}_j.$$

The conjugate of $\Phi \cdot \Psi$ is

$$(8) \qquad (\Phi \cdot \Psi)_c = \sum_{j=1}^{m} \sum_{i=1}^{n} (\mathbf{c}_j \cdot \mathbf{b}_i)\mathbf{d}_j\mathbf{a}_i = \Psi_c \cdot \Phi_c.$$

The conjugate of the product of two dyadics is the product of their conjugates taken in reverse order.

Making use of (8), we now find that

$$\mathbf{r} \cdot (\Phi \cdot \Psi) = (\Phi \cdot \Psi)_c \cdot \mathbf{r} = (\Psi_c \cdot \Phi_c) \cdot \mathbf{r} = \Psi_c \cdot (\Phi_c \cdot \mathbf{r});$$

hence

$$(9) \qquad \mathbf{r} \cdot (\Phi \cdot \Psi) = (\mathbf{r} \cdot \Phi) \cdot \Psi \quad \text{for every } \mathbf{r},$$

an associative law analogous to (1) but with $\mathbf{r}$ as prefactor.

If $\mathbf{e}_1, \mathbf{e}_2, \mathbf{e}_3$ form a basis and $\mathbf{e}^1, \mathbf{e}^2, \mathbf{e}^3$ the reciprocal basis, we can express any two dyadics in the form,

$$\Phi = \mathbf{f}_1\mathbf{e}_1 + \mathbf{f}_2\mathbf{e}_2 + \mathbf{f}_3\mathbf{e}_3, \qquad \Psi = \mathbf{e}^1\mathbf{g}_1 + \mathbf{e}^2\mathbf{g}_2 + \mathbf{e}^3\mathbf{g}_3.$$

In the product $\Phi \cdot \Psi$, six dyads vanish, and we find

$$\Phi \cdot \Psi = \mathbf{f}_1\mathbf{g}_1 + \mathbf{f}_2\mathbf{g}_2 + \mathbf{f}_3\mathbf{g}_3. \tag{10}$$

If Φ and Ψ are complete, $\mathbf{f}_1, \mathbf{f}_2, \mathbf{f}_3$ and $\mathbf{g}_1, \mathbf{g}_2, \mathbf{g}_3$ are non-coplanar sets; then $\Phi \cdot \Psi$ is also complete. But if Φ or Ψ is singular, one of these sets must be coplanar, and $\Phi \cdot \Psi$ is likewise singular. Therefore *the product of two dyadics is complete when and only when both dyadic factors are complete.*

If Φ is complete, it may be "canceled" from equations such as $\Phi \cdot \Psi = 0$, $\Psi \cdot \Phi = 0$, to give $\Psi = 0$. Thus, if $\Phi \cdot \Psi = 0$, $\mathbf{f}_1, \mathbf{f}_2, \mathbf{f}_3$ are non-coplanar in (10), and hence $\mathbf{g}_1 = \mathbf{g}_2 = \mathbf{g}_3 = 0$. We also may "cancel" Φ in $\Phi \cdot \Psi = \Phi \cdot \Omega$; for this is equivalent to $\Phi \cdot (\Psi - \Omega) = 0$, and hence $\Psi - \Omega = 0$.

66. Idemfactor and Reciprocal. The unit dyadic or *idemfactor* $\mathbf{I}$ is defined by the equation,

$$\mathbf{I} \cdot \mathbf{r} = \mathbf{r} \quad \text{for every } \mathbf{r}. \tag{1}$$

The idemfactor is unique (§ 60). For any basis $\mathbf{e}_1, \mathbf{e}_2, \mathbf{e}_3$, we have $\mathbf{I} \cdot \mathbf{e}_j = \mathbf{e}_j$, $\mathbf{I} \cdot \mathbf{e}^j = \mathbf{e}^j$; hence, from (60.7),

$$\mathbf{I} = \mathbf{e}_1\mathbf{e}^1 + \mathbf{e}_2\mathbf{e}^2 + \mathbf{e}_3\mathbf{e}^3 = \mathbf{e}^1\mathbf{e}_1 + \mathbf{e}^2\mathbf{e}_2 + \mathbf{e}^3\mathbf{e}_3; \tag{2}$$

in particular the self-reciprocal basis $\mathbf{i}, \mathbf{j}, \mathbf{k}$ gives

$$\mathbf{I} = \mathbf{ii} + \mathbf{jj} + \mathbf{kk}. \tag{3}$$

Evidently $\mathbf{I}$ is symmetric and complete.

THEOREM. *In order that* $\mathbf{a}, \mathbf{b}, \mathbf{c}$ *and* $\mathbf{u}, \mathbf{v}, \mathbf{w}$ *form reciprocal sets, it is necessary and sufficient that*

$$\mathbf{au} + \mathbf{bv} + \mathbf{cw} = \mathbf{I}. \tag{4}$$

When the sets are reciprocal, we have just proved (4). Conversely, if (4) holds, $\mathbf{a}, \mathbf{b}, \mathbf{c}$ are non-coplanar, since $\mathbf{I}$ is complete; let $\mathbf{a}', \mathbf{b}', \mathbf{c}'$ denote the reciprocal set. Using these vectors as prefactors on (4) gives $\mathbf{u} = \mathbf{a}'$, $\mathbf{v} = \mathbf{b}'$, $\mathbf{w} = \mathbf{c}'$.

Multiplying a dyadic by $\mathbf{I}$ leaves it unaltered; for, from

$$(\Phi \cdot \mathbf{I}) \cdot \mathbf{r} = \Phi \cdot (\mathbf{I} \cdot \mathbf{r}) = \Phi \cdot \mathbf{r}, \qquad \mathbf{r} \cdot (\mathbf{I} \cdot \Phi) = (\mathbf{r} \cdot \mathbf{I}) \cdot \Phi = \mathbf{r} \cdot \Phi,$$

we conclude that

$$\Phi \cdot \mathbf{I} = \mathbf{I} \cdot \Phi = \Phi. \tag{5}$$

If $\Phi \cdot \Psi = \mathbf{I}$, Φ and Ψ are both complete, since $\mathbf{I}$ is complete (§ 65). From

$$\Psi \cdot (\Phi \cdot \Psi) = \Psi \cdot \mathbf{I} \quad \text{or} \quad (\Psi \cdot \Phi) \cdot \Psi = \mathbf{I} \cdot \Psi,$$

we have also $\Psi \cdot \Phi = \mathbf{I}$. Two dyadics Φ, Ψ are said to be *reciprocals* of each other when

$$\Phi \cdot \Psi = \Psi \cdot \Phi = \mathbf{I}, \tag{6}$$

and we write $\Psi = \Phi^{-1}$, $\Phi = \Psi^{-1}$.

A complete dyadic Φ has a unique reciprocal Φ^{-1}. If

$$\Phi = \mathbf{e}_1\mathbf{f}_1 + \mathbf{e}_2\mathbf{f}_2 + \mathbf{e}_3\mathbf{f}_3, \qquad \Phi^{-1} = \mathbf{f}^1\mathbf{e}^1 + \mathbf{f}^2\mathbf{e}^2 + \mathbf{f}^3\mathbf{e}^3, \tag{7}$$

for their products in either order give $\mathbf{I}$:

$$\Phi \cdot \Phi^{-1} = \mathbf{e}_1\mathbf{e}^1 + \mathbf{e}_2\mathbf{e}^2 + \mathbf{e}_3\mathbf{e}^3, \qquad \Phi^{-1} \cdot \Phi = \mathbf{f}^1\mathbf{f}_1 + \mathbf{f}^2\mathbf{f}_2 + \mathbf{f}^3\mathbf{f}_3.$$

Since dyadic multiplication is associative

$$(\Phi \cdot \Psi) \cdot (\Psi^{-1} \cdot \Phi^{-1}) = \Phi \cdot (\Psi \cdot \Psi^{-1}) \cdot \Phi^{-1} = \Phi \cdot \Phi^{-1} = \mathbf{I};$$

$$(\Phi \cdot \Psi)^{-1} = \Psi^{-1} \cdot \Phi^{-1}. \tag{8}$$

Making use of (8), we have

$$(\Phi \cdot \Psi \cdot \Omega)^{-1} = \Omega^{-1} \cdot (\Phi \cdot \Psi)^{-1} = \Omega^{-1} \cdot \Psi^{-1} \cdot \Phi^{-1};$$

and, in general, *the reciprocal of the product of n dyadics is the product of their reciprocals taken in reverse order.*

For positive integral n we define

$$\Phi^n = \Phi \cdot \Phi \cdot \Phi \cdots \text{to } n \text{ factors}; \tag{9}$$

$$\Phi^{-n} = (\Phi^{-1})^n = (\Phi^n)^{-1}; \tag{10}$$

and also

$$\Phi^0 = \mathbf{I}. \tag{11}$$

With these definitions,

$$\Phi^m \cdot \Phi^n = \Phi^{m+n}, \qquad (\Phi^m)^n = \Phi^{mn}, \tag{12}$$

for all integral exponents. But owing to the non-commutativity of the factors in a dyadic product, $(\Phi \cdot \Psi)^n$ is not in general equal to $\Phi^n \cdot \Psi^n$.

By means of reciprocal dyadics, we readily may solve certain vector and dyadic equations. Thus, if Φ is complete, the equations,

$$\Phi \cdot \mathbf{r} = \mathbf{v}, \qquad \mathbf{r} \cdot \Phi = \mathbf{v}, \qquad \Phi \cdot \Psi = \Omega, \qquad \Psi \cdot \Phi = \Omega,$$

have the respective unique solutions:

$$\mathbf{r} = \Phi^{-1} \cdot \mathbf{v}, \qquad \mathbf{r} = \mathbf{v} \cdot \Phi^{-1}, \qquad \Psi = \Phi^{-1} \cdot \Omega, \qquad \Psi = \Omega \cdot \Phi^{-1}.$$

67. The Dyadic $\Phi \times \mathbf{v}$. If $\Phi = \Sigma \mathbf{a}_i \mathbf{b}_i$, we define the dyadics:

(1) $$\Phi \times \mathbf{v} = \Sigma \mathbf{a}_i \mathbf{b}_i \times \mathbf{v}, \qquad \mathbf{v} \times \Phi = \Sigma \mathbf{v} \times \mathbf{a}_i \mathbf{b}_i.$$

From these definitions we have the relations:

(2) $$(\Phi \times \mathbf{v}) \cdot \mathbf{r} = \Phi \cdot (\mathbf{v} \times \mathbf{r}), \qquad \mathbf{r} \cdot (\mathbf{v} \times \Phi) = (\mathbf{r} \times \mathbf{v}) \cdot \Phi;$$

(3) $$\mathbf{r} \cdot (\Phi \times \mathbf{v}) = (\mathbf{r} \cdot \Phi) \times \mathbf{v}, \qquad (\mathbf{v} \times \Phi) \cdot \mathbf{r} = \mathbf{v} \times (\Phi \cdot \mathbf{r}).$$

When $\Phi = \mathrm{I}$, these give

(4) $$(\mathrm{I} \times \mathbf{v}) \cdot \mathbf{r} = \mathbf{v} \times \mathbf{r}, \qquad \mathbf{r} \cdot (\mathbf{v} \times \mathrm{I}) = \mathbf{r} \times \mathbf{v};$$

(5) $$\mathbf{r} \cdot (\mathrm{I} \times \mathbf{v}) = \mathbf{r} \times \mathbf{v}, \qquad (\mathbf{v} \times \mathrm{I}) \cdot \mathbf{r} = \mathbf{v} \times \mathbf{r}.$$

Since these equations hold for any $\mathbf{r}$,

(6) $$\mathrm{I} \times \mathbf{v} = \mathbf{v} \times \mathrm{I},$$

(7) $$(\mathrm{I} \times \mathbf{v})_c = -\mathrm{I} \times \mathbf{v}.$$

Thus $\mathrm{I} \times \mathbf{v}$ is *antisymmetric* and *planar;* it transforms all vectors r into vectors perpendicular to $\mathbf{v}$.

Since

$$\mathbf{a}_i \mathbf{b}_i \cdot (\mathrm{I} \times \mathbf{v}) = \mathbf{a}_i \mathbf{b}_i \times \mathbf{v}, \qquad (\mathrm{I} \times \mathbf{v}) \cdot \mathbf{a}_i \mathbf{b}_i = \mathbf{v} \times \mathbf{a}_i \mathbf{b}_i,$$

the definitions (1) give

(8) $$\Phi \cdot (\mathrm{I} \times \mathbf{v}) = \Phi \times \mathbf{v}, \qquad (\mathrm{I} \times \mathbf{v}) \cdot \Phi = \mathbf{v} \times \Phi.$$

Consequently, the operations $\mathbf{v} \times$ and $\times \mathbf{v}$ on vectors or dyadics may be replaced by dyadic products $(\mathrm{I} \times \mathbf{v}) \cdot$ and $\cdot (\mathrm{I} \times \mathbf{v})$.

For any vectors $\mathbf{u}, \mathbf{v}, \mathbf{r}$ we have

$$[\mathrm{I} \times (\mathbf{u} \times \mathbf{v})] \cdot \mathbf{r} = (\mathbf{u} \times \mathbf{v}) \times \mathbf{r} = (\mathbf{vu} - \mathbf{uv}) \cdot \mathbf{r},$$

and hence

(9) $$\mathrm{I} \times (\mathbf{u} \times \mathbf{v}) = \mathbf{vu} - \mathbf{uv}.$$

From this identity we arrive at the general form of any antisymmetric dyadic.

THEOREM. *If the dyadic* $\Omega = \Sigma \mathbf{a}_i \mathbf{b}_i$ *is antisymmetric, and the vector* $\boldsymbol{\omega} = \Sigma \mathbf{a}_i \times \mathbf{b}_i$,

$$\Omega = -\tfrac{1}{2}\mathbf{I} \times \boldsymbol{\omega}. \tag{10}$$

Proof. The conjugate of Ω is $-\Omega = \Sigma \mathbf{b}_i \mathbf{a}_i$; hence

$$-2\Omega = \Sigma(\mathbf{b}_i\mathbf{a}_i - \mathbf{a}_i\mathbf{b}_i) = \Sigma \mathbf{I} \times (\mathbf{a}_i \times \mathbf{b}_i) = \mathbf{I} \times \boldsymbol{\omega}.$$

Corollary. Every antisymmetric dyadic is planar.

68. First Scalar and Vector Invariant. We may express a dyadic $\Phi = \Sigma \mathbf{a}_i \mathbf{b}_i$ in various forms by substituting vector sums for $\mathbf{a}_i$, $\mathbf{b}_i$, and expanding and collecting terms by applying the distributive law. For all these forms there are certain functions of the vectors, in terms of which Φ is expressed, which remain the same. These functions, which may be scalar, vector, or dyadic, are called *invariants* of the dyadic.

Each step of the process in changing $\Phi = \Sigma \mathbf{a}_i \mathbf{b}_i$ from one form to another may be paralleled by the same step in transforming the scalar and the vector,

$$\varphi_1 = \Sigma \mathbf{a}_i \cdot \mathbf{b}_i, \tag{1}$$

$$\boldsymbol{\phi} = \Sigma \mathbf{a}_i \times \mathbf{b}_i, \tag{2}$$

obtained by placing a dot or cross between the vectors of each dyad of Φ. Each application of the distributive law in the transformation of Φ is also valid in the corresponding transformations of φ_1 and $\boldsymbol{\phi}$; and, just as Φ is not altered by these changes, the same is true of the scalar φ_1 and vector $\boldsymbol{\phi}$. These quantities are therefore *invariants* of Φ with respect to the transformations in question. They are called, respectively, the *scalar* (or *first scalar invariant*) and *vector* of the dyadic.

For example, if $\Phi = \mathbf{ij} + \mathbf{jk} + \mathbf{kk}$,

$$\varphi_1 = \mathbf{i} \cdot \mathbf{j} + \mathbf{j} \cdot \mathbf{k} + \mathbf{k} \cdot \mathbf{k} = 1, \qquad \boldsymbol{\phi} = \mathbf{i} \times \mathbf{j} + \mathbf{j} \times \mathbf{k} + \mathbf{k} \times \mathbf{k} = \mathbf{k} + \mathbf{i}.$$

For the idemfactor $\mathbf{I} = \mathbf{ii} + \mathbf{jj} + \mathbf{kk}$, the scalar is 3 and the vector 0; and we obtain these same values if $\mathbf{I}$ is expressed in terms of an arbitrary basis: $\mathbf{I} = \mathbf{e}_1\mathbf{e}^1 + \mathbf{e}_2\mathbf{e}^2 + \mathbf{e}_3\mathbf{e}^3$. Again, for the dyadic,

$$\Omega = \mathbf{I} \times \mathbf{v} = \mathbf{i}\,\mathbf{i} \times \mathbf{v} + \mathbf{j}\,\mathbf{j} \times \mathbf{v} + \mathbf{k}\,\mathbf{k} \times \mathbf{v},$$

we have $\omega_1 = 0$, and

$$\boldsymbol{\omega} = \mathbf{i} \times (\mathbf{i} \times \mathbf{v}) + \mathbf{j} \times (\mathbf{j} \times \mathbf{v}) + \mathbf{k} \times (\mathbf{k} \times \mathbf{v}) = \mathbf{I} \cdot \mathbf{v} - 3\mathbf{v} = -2\mathbf{v};$$

thus $\Omega = -\frac{1}{2}\mathbf{I} \times \boldsymbol{\omega}$, as in (67.10).

The scalar or vector of the sum of two dyadics is the sum of their scalars or vectors: thus, if

$$(3) \qquad \Phi = \Psi + \Omega; \qquad \varphi_1 = \psi_1 + \omega_1, \quad \boldsymbol{\phi} = \boldsymbol{\psi} + \boldsymbol{\omega}.$$

It is principally to this property that these invariants owe their importance.

In (64.1), we have expressed any dyadic Φ as the sum of a symmetric and antisymmetric dyadic:

$$(4) \qquad \Phi = \tfrac{1}{2}(\Phi + \Phi_c) + \tfrac{1}{2}(\Phi - \Phi_c).$$

The antisymmetric part $\Omega = \frac{1}{2}(\Phi - \Phi_c)$ has the vector,

$$\boldsymbol{\omega} = \tfrac{1}{2}\boldsymbol{\phi} - \tfrac{1}{2}(-\boldsymbol{\phi}) = \boldsymbol{\phi}.$$

From the theorem of § 67 we now have

$$(5) \qquad \Omega = -\tfrac{1}{2}\mathbf{I} \times \boldsymbol{\phi}.$$

Hence, from (4),

$$2\Phi = \Phi + \Phi_c - \mathbf{I} \times \boldsymbol{\phi},$$

$$(6) \qquad \Phi_c = \Phi + \mathbf{I} \times \boldsymbol{\phi}.$$

From (6), we have the

THEOREM. *A necessary and sufficient condition that a dyadic be symmetric is that its vector invariant vanish.*

69. Further Invariants. We may obtain further invariants of the dyadic $\Phi = \Sigma \mathbf{a}_i \mathbf{b}_i$ by processes that are distributive with respect to addition. The most important of these are the *dyadic*,

$$(1) \qquad \Phi_2 = \tfrac{1}{2}\sum_{i,j} \mathbf{a}_i \times \mathbf{a}_j \; \mathbf{b}_i \times \mathbf{b}_j,$$

called by Gibbs the *second* of Φ; its scalar invariant,

$$(2) \qquad \varphi_2 = \tfrac{1}{2}\sum_{i,j} (\mathbf{a}_i \times \mathbf{a}_j) \cdot (\mathbf{b}_i \times \mathbf{b}_j);$$

and the scalar,

$$(3) \qquad \varphi_3 = \tfrac{1}{6} \sum_{i,j,k} (\mathbf{a}_i \times \mathbf{a}_j \cdot \mathbf{a}_k)(\mathbf{b}_i \times \mathbf{b}_j \cdot \mathbf{b}_k).$$

In (1) the summation is taken over all permutations i, j. When $i = j$, the dyad vanishes; when $j \neq i$, the permutations i, j and j, i give the same dyad, so that each dyad occurs twice in the final sum; this doubling is avoided by the factor $\frac{1}{2}$.

In (3) the summation is taken over all permutations i, j, k. When two subscripts are the same, the term vanishes; when i, j, k all differ, the $3! = 6$ permutations of these subscripts give the same term, so that each term occurs six times in the final sum; this multiplication of terms is avoided by the factor $\frac{1}{6}$.

Let Φ be reduced to the three-term form:

$$\Phi = \mathbf{al} + \mathbf{bm} + \mathbf{cn}. \tag{4}$$

The invariants considered thus far are now

$$\varphi_1 = \mathbf{a} \cdot \mathbf{l} + \mathbf{b} \cdot \mathbf{m} + \mathbf{c} \cdot \mathbf{n}, \tag{5}$$

$$\boldsymbol{\phi} = \mathbf{a} \times \mathbf{l} + \mathbf{b} \times \mathbf{m} + \mathbf{c} \times \mathbf{n}, \tag{6}$$

$$\Phi_2 = \mathbf{b} \times \mathbf{c}\, \mathbf{m} \times \mathbf{n} + \mathbf{c} \times \mathbf{a}\, \mathbf{n} \times \mathbf{l} + \mathbf{a} \times \mathbf{b}\, \mathbf{l} \times \mathbf{m}, \tag{7}$$

$$\varphi_2 = (\mathbf{b} \times \mathbf{c}) \cdot (\mathbf{m} \times \mathbf{n}) + (\mathbf{c} \times \mathbf{a}) \cdot (\mathbf{n} \times \mathbf{l}) + (\mathbf{a} \times \mathbf{b}) \cdot (\mathbf{l} \times \mathbf{m}), \tag{8}$$

$$\varphi_3 = (\mathbf{a} \times \mathbf{b} \cdot \mathbf{c})(\mathbf{l} \times \mathbf{m} \cdot \mathbf{n}). \tag{9}$$

The numbers $\varphi_1, \varphi_2, \varphi_3$ often are called the *first, second,* and *third scalars* of Φ.

If Φ is *singular*, the antecedents $\mathbf{a}, \mathbf{b}, \mathbf{c}$ or the consequents $\mathbf{l}, \mathbf{m}, \mathbf{n}$ in (4) will be coplanar, and $\varphi_3 = 0$. Conversely, if $\varphi_3 = 0$, $\mathbf{a}, \mathbf{b}, \mathbf{c}$ or $\mathbf{l}, \mathbf{m}, \mathbf{n}$ are coplanar sets and Φ is singular. Therefore *a dyadic is singular when and only when its third scalar is zero.*

If $\varphi_3 = 0$, Φ must be planar, linear, or zero. When Φ is linear it can be reduced to the form $\mathbf{al}$ and $\Phi_2 = 0$. Conversely, if $\Phi_2 = 0$, Φ will be linear or zero; for, if we choose a non-coplanar set $\mathbf{a}, \mathbf{b}, \mathbf{c}$ as antecedents in (4), $\mathbf{b} \times \mathbf{c}, \mathbf{c} \times \mathbf{a}, \mathbf{a} \times \mathbf{b}$ are also non-coplanar, and $\Phi_2 = 0$ implies that

$$\mathbf{m} \times \mathbf{n} = \mathbf{n} \times \mathbf{l} = \mathbf{l} \times \mathbf{m} = 0;$$

then $\mathbf{l}, \mathbf{m}, \mathbf{n}$ are parallel or zero.

Therefore we may state the

THEOREM. *Necessary and sufficient conditions that a dyadic Φ be*

Complete		$\varphi_3 \neq 0,$	
Planar	*are that*	$\varphi_3 = 0,$	$\Phi_2 \neq 0,$
Linear		$\Phi_2 = 0,$	$\Phi \neq 0.$

We next compute the invariants of the dyadic $\Psi = \Phi_2$. From (7),

$$\Psi_2 = (\mathbf{c} \times \mathbf{a}) \times (\mathbf{a} \times \mathbf{b})\ (\mathbf{n} \times \mathbf{l}) \times (\mathbf{l} \times \mathbf{m}) + \text{cyclical terms}$$

$$= [\mathbf{abc}][\mathbf{lmn}](\mathbf{al} + \mathbf{bm} + \mathbf{cn});$$

hence, from (9),

$$\Psi_2 = \varphi_3 \Phi. \tag{10}$$

We now may compute the three scalars of Ψ:

$$\psi_1 = \varphi_2, \qquad \psi_2 = \varphi_3 \varphi_1, \qquad \psi_3 = {\varphi_3}^2; \tag{11}$$

the first follows from (8), the second from (10), and the third from

$$(\mathbf{b} \times \mathbf{c}) \cdot (\mathbf{c} \times \mathbf{a}) \times (\mathbf{a} \times \mathbf{b}) = [\mathbf{abc}]^2, \qquad (\mathbf{m} \times \mathbf{n}) \cdot (\mathbf{n} \times \mathbf{l}) \times (\mathbf{l} \times \mathbf{m}) = [\mathbf{lmn}]^2.$$

Finally, the vector invariant of Ψ is

$$\boldsymbol{\psi} = (\mathbf{b} \times \mathbf{c}) \times (\mathbf{m} \times \mathbf{n}) + (\mathbf{c} \times \mathbf{a}) \times (\mathbf{n} \times \mathbf{l}) + (\mathbf{a} \times \mathbf{b}) \times (\mathbf{l} \times \mathbf{m}).$$

If we express $\boldsymbol{\psi}$ in terms of **a**, **b**, **c**, the term in **a** is

$$\mathbf{a}\ \mathbf{c} \cdot \mathbf{n} \times \mathbf{l} - \mathbf{a}\ \mathbf{b} \cdot \mathbf{l} \times \mathbf{m} = \mathbf{al} \cdot (\mathbf{b} \times \mathbf{m} + \mathbf{c} \times \mathbf{n}) = \mathbf{al} \cdot \boldsymbol{\phi}$$

from (6); hence

$$\boldsymbol{\psi} = (\mathbf{al} + \mathbf{bm} + \mathbf{cn}) \cdot \boldsymbol{\phi} = \Phi \cdot \boldsymbol{\phi}. \tag{12}$$

It can be shown that all scalar invariants of Φ may be expressed in terms of the six scalars,

$$\varphi_1, \varphi_2, \varphi_3; \quad \boldsymbol{\phi} \cdot \boldsymbol{\phi}, \boldsymbol{\phi} \cdot \boldsymbol{\psi}, \boldsymbol{\psi} \cdot \boldsymbol{\psi}. \tag{13}$$

This property is expressed by saying that these six scalars form a *complete* system of invariants. When Φ is symmetric, $\boldsymbol{\phi} = \boldsymbol{\psi} = 0$, and the last three scalars vanish.

The third scalar of the product of two dyadics is equal to the product of their third scalars. For any two dyadics Φ, Ψ can be put in the form,

$$\Phi = \mathbf{f}_1\mathbf{e}_1 + \mathbf{f}_2\mathbf{e}_2 + \mathbf{f}_3\mathbf{e}_3, \qquad \Psi = \mathbf{e}^1\mathbf{g}_1 + \mathbf{e}^2\mathbf{g}_2 + \mathbf{e}^3\mathbf{g}_3,$$

where $\mathbf{e}_1, \mathbf{e}_2, \mathbf{e}_3$ and $\mathbf{e}^1, \mathbf{e}^2, \mathbf{e}^3$ are reciprocal sets (§ 65); hence

$$\varphi_3\psi_3 = [\mathbf{f}_1\mathbf{f}_2\mathbf{f}_3][\mathbf{e}_1\mathbf{e}_2\mathbf{e}_3][\mathbf{e}^1\mathbf{e}^2\mathbf{e}^3][\mathbf{g}_1\mathbf{g}_2\mathbf{g}_3] = [\mathbf{f}_1\mathbf{f}_2\mathbf{f}_3][\mathbf{g}_1\mathbf{g}_2\mathbf{g}_3],$$

which is the third scalar of

$$\Phi \cdot \Psi = \mathbf{f}_1\mathbf{g}_1 + \mathbf{f}_2\mathbf{g}_2 + \mathbf{f}_3\mathbf{g}_3.$$

Example. For the idemfactor $\mathbf{I} = \mathbf{ii} + \mathbf{jj} + \mathbf{kk}$, the second $\mathbf{I}_2 = \mathbf{I}$, the vector is zero, and the three scalars are 3, 3, and 1. From

$$\text{(i)} \qquad \mathbf{I} = \mathbf{e}_1\mathbf{e}^1 + \mathbf{e}_2\mathbf{e}^2 + \mathbf{e}_3\mathbf{e}^3,$$

we have, for $\mathbf{I}_2$,

$$\text{(ii)} \qquad \mathbf{I} = \mathbf{e}_2 \times \mathbf{e}_3\, \mathbf{e}^2 \times \mathbf{e}^3 + \mathbf{e}_3 \times \mathbf{e}_1\, \mathbf{e}^3 \times \mathbf{e}^1 + \mathbf{e}_1 \times \mathbf{e}_2\, \mathbf{e}^1 \times \mathbf{e}^2,$$

and the vector,

$$\mathbf{e}_1 \times \mathbf{e}^1 + \mathbf{e}_2 \times \mathbf{e}^2 + \mathbf{e}_3 \times \mathbf{e}^3 = 0.$$

The third scalar gives

$$[\mathbf{e}_1\mathbf{e}_2\mathbf{e}_3][\mathbf{e}^1\mathbf{e}^2\mathbf{e}^3] = 1.$$

From (ii) we have, for example,

$$\mathbf{I} \cdot \mathbf{e}^3 = \mathbf{e}_1 \times \mathbf{e}_2[\mathbf{e}^1\mathbf{e}^2\mathbf{e}^3], \qquad \mathbf{e}^3 = \mathbf{e}_1 \times \mathbf{e}_2/[\mathbf{e}_1\mathbf{e}_2\mathbf{e}_3].$$

Thus (i) and (ii) give a handy compendium of the properties of reciprocal vector sets.

70. Second and Adjoint Dyadic. The *second* of the dyadic $\Phi = \Sigma \mathbf{a}_i\mathbf{b}_i$, namely,

$$\text{(1)} \qquad \Phi_2 = \tfrac{1}{2}\sum_{i,j} \mathbf{a}_i \times \mathbf{a}_j\, \mathbf{b}_i \times \mathbf{b}_j,$$

has the property,

$$\text{(2)} \qquad (\Phi \cdot \mathbf{u}) \times (\Phi \cdot \mathbf{v}) = \Phi_2 \cdot (\mathbf{u} \times \mathbf{v}).$$

Proof. We have

$$\begin{aligned} (\Phi \cdot \mathbf{u}) \times (\Phi \cdot \mathbf{v}) &= (\sum_i \mathbf{a}_i\mathbf{b}_i \cdot \mathbf{u}) \times (\sum_j \mathbf{a}_j\mathbf{b}_j \cdot \mathbf{v}) \\ &= \sum_{ij} \mathbf{a}_i \times \mathbf{a}_j\, (\mathbf{b}_i \cdot \mathbf{u})(\mathbf{b}_j \cdot \mathbf{v}) \\ &= \sum_{ij} \mathbf{a}_j \times \mathbf{a}_i\, (\mathbf{b}_j \cdot \mathbf{u})(\mathbf{b}_i \cdot \mathbf{v}), \end{aligned}$$

on interchanging i and j. The left member also equals half the sum of the two last expressions:

$$\tfrac{1}{2}\sum_{ij} (\mathbf{a}_i \times \mathbf{a}_j)\{(\mathbf{b}_i \cdot \mathbf{u})(\mathbf{b}_j \cdot \mathbf{v}) - (\mathbf{b}_i \cdot \mathbf{v})(\mathbf{b}_j \cdot \mathbf{u})\},$$

or, with regard to (20.1),

$$\tfrac{1}{2}\sum_{ij} (\mathbf{a}_i \times \mathbf{a}_j)(\mathbf{b}_i \times \mathbf{b}_j) \cdot (\mathbf{u} \times \mathbf{v}) = \Phi_2 \cdot (\mathbf{u} \times \mathbf{v}).$$

The conjugate of Φ_2 is called the *adjoint* of Φ and written Φ_a. The adjoint satisfies the important relation,

(3) $$\Phi \cdot \Phi_a = \Phi_a \cdot \Phi = \varphi_3 I.$$

Proof. Write $\Phi = \mathbf{al} + \mathbf{bm} + \mathbf{cn}$; then

(4) $$\Phi_a = \mathbf{m} \times \mathbf{n}\, \mathbf{b} \times \mathbf{c} + \mathbf{n} \times \mathbf{l}\, \mathbf{c} \times \mathbf{a} + \mathbf{l} \times \mathbf{m}\, \mathbf{a} \times \mathbf{b}.$$

Choose for the antecedents $\mathbf{a}$, $\mathbf{b}$, $\mathbf{c}$ of Φ a non-coplanar set (§ 61). Then, if $\mathbf{a}'$, $\mathbf{b}'$, $\mathbf{c}'$ denote the reciprocal set, we have, by direct multiplication,

$$\begin{aligned}\Phi \cdot \Phi_a &= (\mathbf{l} \cdot \mathbf{m} \times \mathbf{n})(\mathbf{a}\, \mathbf{b} \times \mathbf{c} + \mathbf{b}\, \mathbf{c} \times \mathbf{a} + \mathbf{c}\, \mathbf{a} \times \mathbf{b}) \\ &= (\mathbf{l} \cdot \mathbf{m} \times \mathbf{n})(\mathbf{a} \cdot \mathbf{b} \times \mathbf{c})(\mathbf{aa}' + \mathbf{bb}' + \mathbf{cc}') \\ &= \varphi_3 \mathbf{I}.\end{aligned}$$

If we choose the consequents $\mathbf{l}$, $\mathbf{m}$, $\mathbf{n}$ of Φ as a non-coplanar set (then $\mathbf{a}$, $\mathbf{b}$, $\mathbf{c}$ may or may not be coplanar), let $\mathbf{l}'$, $\mathbf{m}'$, $\mathbf{n}'$ denote the reciprocal set. Then

$$\begin{aligned}\Phi_a \cdot \Phi &= (\mathbf{a} \cdot \mathbf{b} \times \mathbf{c})(\mathbf{m} \times \mathbf{n}\, \mathbf{l} + \mathbf{n} \times \mathbf{l}\, \mathbf{m} + \mathbf{l} \times \mathbf{m}\, \mathbf{n}) \\ &= (\mathbf{a} \cdot \mathbf{b} \times \mathbf{c})(\mathbf{l} \cdot \mathbf{m} \times \mathbf{n})(\mathbf{l}'\mathbf{l} + \mathbf{m}'\mathbf{m} + \mathbf{n}'\mathbf{n}) \\ &= \varphi_3 \mathbf{I}.\end{aligned}$$

If Φ is complete, $\varphi_3 \neq 0$; then (3) shows that

(5) $$\Phi^{-1} = \Phi_a / \varphi_3.$$

From (3) we also may show that, if $\mathbf{u}$, $\mathbf{v}$, $\mathbf{w}$ are any three vectors,

(6) $$[\Phi \cdot \mathbf{u},\ \Phi \cdot \mathbf{v},\ \Phi \cdot \mathbf{w}] = \varphi_3 [\mathbf{uvw}].$$

Proof. Since Φ_a is the conjugate of Φ_2, we have, from (2),

$$(\Phi \cdot \mathbf{u}) \times (\Phi \cdot \mathbf{v}) = (\mathbf{u} \times \mathbf{v}) \cdot \Phi_a,$$

$$(\Phi \cdot \mathbf{u}) \times (\Phi \cdot \mathbf{v}) \cdot (\Phi \cdot \mathbf{w}) = (\mathbf{u} \times \mathbf{v}) \cdot \Phi_a \cdot \Phi \cdot \mathbf{w}.$$

On replacing $\Phi_a \cdot \Phi$ by $\varphi_3 \mathbf{I}$, we obtain (6).

We now can deduce important properties of the affine transformation (§ 62):

(7) $$\mathbf{r}' = \Phi \cdot \mathbf{r} \qquad (\varphi_3 \neq 0).$$

The vector area $\mathbf{u} \times \mathbf{v}$ (§ 17) is transformed into

(8) $$\mathbf{u}' \times \mathbf{v}' = \Phi_2 \cdot (\mathbf{u} \times \mathbf{v}).$$

From (4), the third scalars of Φ_a and Φ_2 equal $[\mathbf{abc}]^2[\mathbf{lmn}]^2 = \varphi_3^2 \neq 0$. Since Φ and Φ_2 are complete, $\mathbf{r} \neq 0$ and $\mathbf{u} \times \mathbf{v} \neq 0$ imply $\mathbf{r}' \neq 0$, $\mathbf{u}' \times \mathbf{v}' \neq 0$ (thus filling a gap in § 62). An affine transformation *invariably* changes lines into lines, planes into planes.

Moreover (7) transforms any parallelepiped $[\mathbf{uvw}]$ into another $[\mathbf{u}'\mathbf{v}'\mathbf{w}']$ whose volume is $\varphi_3[\mathbf{uvw}]$ according to (6). As any volume can be regarded as the limit of a sum of parallelepiped elements, *the affine transformation alters all volumes in the constant ratio of* $\varphi_3/1$.

71. Invariant Directions. We next seek those vectors $\mathbf{r}$ which are transformed by Φ into scalar multiples of $\mathbf{r}$, say

$$\Phi \cdot \mathbf{r} = \lambda \mathbf{r}. \tag{1}$$

If we write this equation,

$$\Psi \cdot \mathbf{r} = 0 \quad \text{where} \quad \Psi = \Phi - \lambda \mathrm{I},$$

it is clear that the *multiplier* λ must make the dyadic Ψ singular; its third scalar ψ_3 is then zero (§ 69). Conversely, if λ is a root of the equation $\psi_3 = 0$, λ is a multiplier of Φ; for, when Ψ is planar or linear, $\Psi \cdot \mathbf{r} = 0$ for all vectors $\mathbf{r}$ normal to the plane of the consequents of Ψ.

Let us write

$$\Phi = \mathbf{al} + \mathbf{bm} + \mathbf{cn}, \qquad \mathrm{I} = \mathbf{aa}' + \mathbf{bb}' + \mathbf{cc}',$$

where $\mathbf{a}, \mathbf{b}, \mathbf{c}$ are non-coplanar and $\mathbf{a}', \mathbf{b}', \mathbf{c}'$ the reciprocal set; then

$$\Psi = \mathbf{a}(\mathbf{l} - \lambda \mathbf{a}') + \mathbf{b}(\mathbf{m} - \lambda \mathbf{b}') + \mathbf{c}(\mathbf{n} - \lambda \mathbf{c}'),$$

$$\psi_3 = [\mathbf{abc}][(\mathbf{l} - \lambda \mathbf{a}')(\mathbf{m} - \lambda \mathbf{b}')(\mathbf{n} - \lambda \mathbf{c}')].$$

The second box product in ψ_3 gives, on expansion,

$$[\mathbf{lmn}] - \lambda\{[\mathbf{a}'\mathbf{mn}] + [\mathbf{b}'\mathbf{nl}] + [\mathbf{c}'\mathbf{lm}]\}$$
$$+ \lambda^2\{[\mathbf{b}'\mathbf{c}'\mathbf{l}] + [\mathbf{c}'\mathbf{a}'\mathbf{m}] + [\mathbf{a}'\mathbf{b}'\mathbf{n}]\} - \lambda^3[\mathbf{a}'\mathbf{b}'\mathbf{c}'].$$

Substituting for the primed vectors gives

$$\mathbf{a}' = \frac{\mathbf{b} \times \mathbf{c}}{[\mathbf{abc}]}, \cdots; \qquad \mathbf{b}' \times \mathbf{c}' = \frac{\mathbf{a}}{[\mathbf{abc}]}, \cdots;$$

and then multiplying the entire expansion by [abc] yields

$$\psi_3 = [\mathbf{abc}][\mathbf{lmn}] - \lambda\{(\mathbf{b}\times\mathbf{c})\cdot(\mathbf{m}\times\mathbf{n}) + (\mathbf{c}\times\mathbf{a})\cdot(\mathbf{n}\times\mathbf{l}) + (\mathbf{a}\times\mathbf{b})\cdot(\mathbf{l}\times\mathbf{m})\} + \lambda^2(\mathbf{a}\cdot\mathbf{l} + \mathbf{b}\cdot\mathbf{m} + \mathbf{c}\cdot\mathbf{n}) - \lambda^3,$$

or, on making use of (69.5), (69.8), (69.9),

$$\psi_3 = \varphi_3 - \lambda\varphi_2 + \lambda^2\varphi_1 - \lambda^3. \tag{2}$$

Denote the right member of (2) by $f(\lambda)$; then the cubic equation,

$$\psi_3 = f(\lambda) = 0, \tag{3}$$

is called the *characteristic equation* of Φ. Since $\Phi - \lambda I$ is singular when $f(\lambda) = 0$, the coefficients of $f(\lambda)$ depend only upon the nature of the dyadic Φ and not upon the particular form in which it is expressed. We thus have an independent proof of the invariance of $\varphi_1, \varphi_2, \varphi_3$.

The three roots $\lambda_1, \lambda_2, \lambda_3$ of (3) are called the *multipliers* or *characteristic numbers* of Φ. From the relations between the roots and coefficients of an algebraic equation, we have

$$\varphi_1 = \lambda_1 + \lambda_2 + \lambda_3, \quad \varphi_2 = \lambda_2\lambda_3 + \lambda_3\lambda_1 + \lambda_1\lambda_2, \quad \varphi_3 = \lambda_1\lambda_2\lambda_3. \tag{4}$$

The cubic (3) may have three real roots (not necessarily distinct) or one real root and two conjugate complex roots. To find the invariant direction corresponding to a root λ_1, we consider in turn the three cases in which $\Phi - \lambda_1\mathbf{I}$ is singular.

1. If $\Phi - \lambda_1\mathbf{I}$ is *planar*, let $\Phi - \lambda_1\mathbf{I} = \mathbf{ch} + \mathbf{dk}$; then the postfactor $\mathbf{r}_1 = \mathbf{h}\times\mathbf{k}$ reduces the right member to zero and gives the invariant direction for λ_1. If $\Phi - \lambda_1\mathbf{I}$ has more than two dyads, the cross product of any two non-parallel consequents will be normal to their plane and give a vector parallel to $\mathbf{r}_1$.

If λ_1 is real, $\mathbf{r}_1$ is a real vector. But if λ_1 is complex, say $\lambda_1 = \alpha + i\beta$, then $\mathbf{r}_1 = \mathbf{a} + i\mathbf{b}$ is also complex. Then, on equating the real and imaginary parts of

$$\Phi\cdot(\mathbf{a} + i\mathbf{b}) = (\alpha + i\beta)(\mathbf{a} + i\mathbf{b}), \tag{5}$$

we have

$$\Phi\cdot\mathbf{a} = \alpha\mathbf{a} - \beta\mathbf{b}, \qquad \Phi\cdot\mathbf{b} = \beta\mathbf{a} + \alpha\mathbf{b}. \tag{6}$$

In this case we know a second multiplier,

$$\lambda_2 = \alpha - i\beta, \quad \text{with} \quad \mathbf{r}_2 = \mathbf{a} - i\mathbf{b},$$

as invariant direction. For, from (6),

$$\Phi \cdot (\mathbf{a} - i\mathbf{b}) = (\alpha - i\beta)(\mathbf{a} - i\mathbf{b}).$$

From (5), we conclude that $\mathbf{a}$ and $\mathbf{b}$ are not parallel; for, if $\mathbf{b} = k\mathbf{a}$, $\Phi \cdot \mathbf{a} = (\alpha + i\beta)\mathbf{a}$, which is impossible unless $\beta = 0$.

2. If $\Phi - \lambda_1 \mathrm{I}$ is *linear*, it may be reduced to a single dyad $\mathbf{ch}$. Then, if $\mathbf{r}_1$ is any vector in the plane perpendicular to $\mathbf{h}$, $\Phi \cdot \mathbf{r}_1 = \lambda_1 \mathbf{r}_1$. In this case there is a whole plane of invariant directions corresponding to λ_1.

3. If $\Phi - \lambda_1 \mathrm{I}$ is *zero*, $\Phi \cdot \mathbf{r}_1 = \lambda_1 \mathbf{r}_1$ for any vector $\mathbf{r}_1$. All directions are invariant with the multiplier λ_1.

If we write $\mathbf{r}' = \Phi \cdot \mathbf{r}$, we have, from (70.2),

$$\text{(7)} \qquad \mathbf{u}' \times \mathbf{v}' = (\Phi \cdot \mathbf{u}) \times (\Phi \cdot \mathbf{v}) = \Phi_2 \cdot (\mathbf{u} \times \mathbf{v}).$$

Now Φ transforms all vectors in the plane of $\mathbf{u}$, $\mathbf{v}$ into vectors of the plane $\mathbf{u}'$, $\mathbf{v}'$. These planes will be the same if $\mathbf{u}' \times \mathbf{v}' = \lambda\, \mathbf{u} \times \mathbf{v}$, that is, if $\mathbf{u} \times \mathbf{v}$ is an invariant direction of Φ_2. Hence *the invariant planes of Φ are normal to the invariant directions of Φ_2.*

Example 1. $\Phi = \mathbf{ii} + \mathbf{j}(\mathbf{i} + 2\mathbf{j}) + \mathbf{k}(\mathbf{j} + 2\mathbf{k})$. Then

$$\Phi_2 = \mathbf{i}(4\mathbf{i} - 2\mathbf{j} + \mathbf{k}) + \mathbf{j}(2\mathbf{j} - \mathbf{k}) + 2\mathbf{kk},$$

$$\varphi_1 = 1 + 2 + 2 = 5, \qquad \varphi_2 = 4 + 2 + 2 = 8, \qquad \varphi_3 = 4.$$

The characteristic equation,

$$f(\lambda) = 4 - 8\lambda + 5\lambda^2 - \lambda^3 = (1 - \lambda)(2 - \lambda)^2 = 0,$$

has the roots, $\lambda_1 = 1$, $\quad \lambda_2 = \lambda_3 = 2$.

For the root, $\lambda_1 = 1$,

$$\Phi - \lambda_1 \mathrm{I} = \mathbf{ji} + \mathbf{jj} + \mathbf{kj} + \mathbf{kk} = \mathbf{j}(\mathbf{i} + \mathbf{j}) + \mathbf{k}(\mathbf{j} + \mathbf{k}),$$

and the corresponding invariant direction is

$$\mathbf{r}_1 = (\mathbf{i} + \mathbf{j}) \times (\mathbf{j} + \mathbf{k}) = \mathbf{i} - \mathbf{j} + \mathbf{k}.$$

For the double root, $\lambda_2 = \lambda_3 = 2$,

$$\Phi - \lambda_2 \mathrm{I} = -\mathbf{ii} + \mathbf{ji} + \mathbf{kj} = (-\mathbf{i} + \mathbf{j})\mathbf{i} + \mathbf{kj},$$

and $\mathbf{r}_2 = \mathbf{i} \times \mathbf{j} = \mathbf{k}$.

Example 2. $\Phi = \mathbf{ij} + \mathbf{jk} + \mathbf{ki}$. Then

$$\Phi_2 = \Phi; \qquad \varphi_1 = 0, \qquad \varphi_2 = 0, \qquad \varphi_3 =$$

The characteristic equation, $1 - \lambda^3 = 0$, has the roots,

$$\lambda_1 = 1, \qquad \lambda_2 = \frac{-1 + i\sqrt{3}}{2} = \omega, \qquad \lambda_3 = \frac{-1 - i\sqrt{3}}{2} = \omega^2.$$

For $\lambda_1 = 1$,

$$\Phi - \lambda_1 \mathbf{I} = \mathbf{i}(\mathbf{j} - \mathbf{i}) + \mathbf{j}(\mathbf{k} - \mathbf{j}) + \mathbf{k}(\mathbf{i} - \mathbf{k})$$

is planar· its consequents are all normal to

$$\mathbf{r}_1 = (\mathbf{j} - \mathbf{i}) \times (\mathbf{k} - \mathbf{j}) = \mathbf{i} + \mathbf{j} + \mathbf{k},$$

which is the corresponding invariant direction.

For $\lambda_2 = \omega$, the consequents of

$$\Phi - \lambda_2 \mathbf{I} = \mathbf{i}(\mathbf{j} - \omega\mathbf{i}) + \mathbf{j}(\mathbf{k} - \omega\mathbf{j}) + \mathbf{k}(\mathbf{i} - \omega\mathbf{k})$$

are all normal to

$$\mathbf{r}_2 = (\mathbf{j} - \omega\mathbf{i}) \times (\mathbf{k} - \omega\mathbf{j}) = \mathbf{i} + \omega\mathbf{j} + \omega^2\mathbf{k}.$$

For $\lambda_3 = \omega^2$, the consequents of

$$\Phi - \lambda_3 \mathbf{I} = \mathbf{i}(\mathbf{j} - \omega^2\mathbf{i}) + \mathbf{j}(\mathbf{k} - \omega^2\mathbf{j}) + \mathbf{k}(\mathbf{i} - \omega^2\mathbf{k})$$

are all normal to

$$\mathbf{r}_3 = (\mathbf{j} - \omega^2\mathbf{i}) \times (\mathbf{k} - \omega^2\mathbf{j}) = \mathbf{i} + \omega^2\mathbf{j} + \omega\mathbf{k}.$$

72. Symmetric Dyadics. *The characteristic numbers of a symmetric dyadic are all real.*

Proof. If $\lambda_1 = \alpha + i\beta$ is a complex multiplier and $\mathbf{r}_1 = \mathbf{a} + i\mathbf{b}$ the corresponding invariant direction of the symmetric dyadic Φ, we have (§ 71)

$$\Phi \cdot (\mathbf{a} + i\mathbf{b}) = (\alpha + i\beta)(\mathbf{a} + i\mathbf{b});$$

$$\Phi \cdot \mathbf{a} = \alpha\mathbf{a} - \beta\mathbf{b}, \qquad \Phi \cdot \mathbf{b} = \beta\mathbf{a} + \alpha\mathbf{b}.$$

Since Φ is symmetric, $\mathbf{b} \cdot \Phi \cdot \mathbf{a} = \mathbf{a} \cdot \Phi \cdot \mathbf{b}$; hence

$$\alpha\, \mathbf{b} \cdot \mathbf{a} - \beta\, \mathbf{b} \cdot \mathbf{b} = \beta\, \mathbf{a} \cdot \mathbf{a} + \alpha\, \mathbf{a} \cdot \mathbf{b}, \quad \text{or} \quad \beta(\mathbf{a} \cdot \mathbf{a} + \mathbf{b} \cdot \mathbf{b}) = 0.$$

But $\mathbf{a}$ and $\mathbf{b}$ are real vectors; and $\mathbf{a} \cdot \mathbf{a} + \mathbf{b} \cdot \mathbf{b}$ is positive; hence $\beta = 0$, and λ_1 is real.

The invariant directions corresponding to two distinct characteristic numbers of a symmetric dyadic are perpendicular.

Proof. From the equations,

$$\Phi \cdot \mathbf{r}_1 = \lambda_1 \mathbf{r}_1, \qquad \Phi \cdot \mathbf{r}_2 = \lambda_2 \mathbf{r}_2 \qquad (\lambda_1 \neq \lambda_2),$$

and

$$\mathbf{r}_2 \cdot \Phi \cdot \mathbf{r}_1 = \mathbf{r}_1 \cdot \Phi \cdot \mathbf{r}_2,$$

we deduce

$$(\lambda_1 - \lambda_2)\mathbf{r}_1 \cdot \mathbf{r}_2 = 0, \qquad \mathbf{r}_1 \cdot \mathbf{r}_2 = 0.$$

Consider now the following cases.

1. If $f(\lambda) = 0$ has three distinct roots, the corresponding invariant directions are mutually perpendicular and may be denoted by **i**, **j**, **k**. Then

$$\Phi \cdot \mathbf{i} = \lambda_1 \mathbf{i}, \qquad \Phi \cdot \mathbf{j} = \lambda_2 \mathbf{j}, \qquad \Phi \cdot \mathbf{k} = \lambda_3 \mathbf{k},$$

and, from (61.11),

$$\Phi = \lambda_1 \mathbf{ii} + \lambda_2 \mathbf{jj} + \lambda_3 \mathbf{kk}. \tag{1}$$

2. If two roots of $f(\lambda) = 0$ are equal, let $\lambda_1 \neq \lambda_2 = \lambda_3$. Then, if **i**, **j** are the perpendicular invariant directions corresponding to λ_1 and λ_2, we may write

$$\Phi \cdot \mathbf{i} = \lambda_1 \mathbf{i}, \qquad \Phi \cdot \mathbf{j} = \lambda_2 \mathbf{j}, \qquad \Phi \cdot \mathbf{k} = \mathbf{v},$$

v being the (unknown) transform of **k**; then, from (61.11),

$$\Phi = \lambda_1 \mathbf{ii} + \lambda_2 \mathbf{jj} + \mathbf{vk}.$$

Now the symmetry of Φ requires that $\mathbf{vk} = \mathbf{kv}$, and hence **v** must be a multiple of **k**, say $\gamma\mathbf{k}$. Thus

$$\Phi = \lambda_1 \mathbf{ii} + \lambda_2 \mathbf{jj} + \gamma \mathbf{kk}, \qquad \varphi_1 = \lambda_1 + \lambda_2 + \gamma;$$

but, since $\varphi_1 = \lambda_1 + 2\lambda_2$ by hypothesis, $\gamma = \lambda_2$, and

$$\Phi = \lambda_1 \mathbf{ii} + \lambda_2 \mathbf{jj} + \lambda_2 \mathbf{kk}. \tag{2}$$

Corresponding to the double root $\lambda_2 = \lambda_3$, *we have a whole plane of invariant directions.*

3. If all three roots of $f(\lambda) = 0$ are equal, $\lambda_1 = \lambda_2 = \lambda_3$, let **i** denote an invariant direction corresponding to λ_1, and write

$$\Phi \cdot \mathbf{i} = \lambda_1 \mathbf{i}, \qquad \Phi \cdot \mathbf{j} = \mathbf{u}, \qquad \Phi \cdot \mathbf{k} = \mathbf{v};$$

then, from (61.11),

$$\Phi = \lambda_1 \mathbf{ii} + \mathbf{uj} + \mathbf{vk}.$$

Now the symmetry of Φ requires that $\mathbf{uj} + \mathbf{vk} = \mathbf{ju} + \mathbf{kv}$, and this in turn shows that **u** and **v** have the form,

$$\mathbf{u} = \alpha \mathbf{j} + \gamma \mathbf{k}, \qquad \mathbf{v} = \gamma \mathbf{j} + \beta \mathbf{k}.$$

Thus

$$\Phi = \lambda_1 \mathbf{ii} + (\alpha \mathbf{j} + \gamma \mathbf{k})\mathbf{j} + (\gamma \mathbf{j} + \beta \mathbf{k})\mathbf{k},$$

$$\varphi_1 = \lambda_1 + \alpha + \beta, \qquad \varphi_3 = \lambda_1(\alpha\beta - \gamma^2);$$

but, since $\varphi_1 = 3\lambda_1$, $\varphi_3 = \lambda_1^3$ by hypothesis,

$$\alpha + \beta = 2\lambda_1, \qquad \alpha\beta - \gamma^2 = \lambda_1^2.$$

Elimination of λ_1 gives the relation,

$$(\alpha + \beta)^2 - 4(\alpha\beta - \gamma^2) = (\alpha - \beta)^2 + 4\gamma^2 = 0;$$

and, since neither $(\alpha - \beta)^2$ nor $4\gamma^2$ can be negative, their sum can vanish only if $\alpha - \beta = 0$, $\gamma = 0$. Hence $\alpha = \beta = \lambda_1$, and Φ reduces to

$$\Phi = \lambda_1\mathbf{ii} + \lambda_1\mathbf{jj} + \lambda_1\mathbf{kk} = \lambda_1\mathbf{I}. \tag{3}$$

In the case of a triple root all directions are invariant.

Evidently all cases are included in (1); by making two or three of the roots equal, we obtain (2) and (3).

THEOREM. *Every symmetric dyadic may be reduced to the form,*

$$\Phi = \alpha\mathbf{ii} + \beta\mathbf{jj} + \gamma\mathbf{kk}, \tag{4}$$

in which α, β, γ *are the real multipliers and* **i, j, k** *corresponding invariant directions.*

The reciprocal of Φ is

$$\Phi^{-1} = \frac{1}{\alpha}\mathbf{ii} + \frac{1}{\beta}\mathbf{jj} + \frac{1}{\gamma}\mathbf{kk}; \tag{5}$$

for $\Phi \cdot \Phi^{-1} = \mathbf{I}$. Moreover,

$$\Phi_2 = \beta\gamma\mathbf{ii} + \gamma\alpha\mathbf{jj} + \alpha\beta\mathbf{kk} = \Phi_a; \tag{6}$$

and, by direct multiplication,

$$\Phi^n = \alpha^n\mathbf{ii} + \beta^n\mathbf{jj} + \gamma^n\mathbf{kk}. \tag{7}$$

If Φ is symmetric and $\Phi^n = 0$, then $\Phi = 0$. Moreover, if $\mathbf{r} \cdot \Phi \cdot \mathbf{r} = 0$ for every $\mathbf{r}$, $\Phi = 0$; for, if we choose $\mathbf{r} = \mathbf{i}, \mathbf{j}, \mathbf{k}$ in turn, $\alpha = \beta = \gamma = 0$.

Example 1. For the symmetric dyadic,

$$\Phi = \mathbf{i}(\mathbf{j} + \mathbf{k}) + \mathbf{j}(\mathbf{k} + \mathbf{i}) + \mathbf{k}(\mathbf{i} + \mathbf{j}),$$

$\varphi_1 = 0$, $\varphi_2 = -3$, $\varphi_3 = 2$, and the characteristic equation,

$$2 + 3\lambda - \lambda^3 = (2 - \lambda)(1 + \lambda)^2 = 0,$$

has the roots, $\lambda_1 = 2$, $\lambda_2 = \lambda_3 = -1$.

For $\lambda_1 = 2$,

$$\Phi - \lambda_1 \mathbf{I} = \mathbf{i}(-2\mathbf{i} + \mathbf{j} + \mathbf{k}) + \mathbf{j}(-2\mathbf{j} + \mathbf{k} + \mathbf{i}) + \mathbf{k}(-2\mathbf{k} + \mathbf{i} + \mathbf{j}),$$

The consequents are all normal to

$$(-2\mathbf{i} + \mathbf{j} + \mathbf{k}) \times (-2\mathbf{j} + \mathbf{k} + \mathbf{i}) = 3(\mathbf{i} + \mathbf{j} + \mathbf{k});$$

hence $\mathbf{r}_1 = \mathbf{i} + \mathbf{j} + \mathbf{k}$.

For $\lambda_2 = \lambda_3 = -1$,

$$\Phi - \lambda_2 \mathbf{I} = (\mathbf{i} + \mathbf{j} + \mathbf{k})(\mathbf{i} + \mathbf{j} + \mathbf{k})$$

is linear. Hence any vector in the plane perpendicular to $\mathbf{i} + \mathbf{j} + \mathbf{k}$ is an invariant direction.

If we choose the *unit* vectors,

$$\mathbf{i}' = (\mathbf{i} + \mathbf{j} + \mathbf{k})/\sqrt{3}, \qquad \mathbf{j}' = (\mathbf{i} - \mathbf{j})/\sqrt{2},$$

as invariant directions for the roots 2 and -1, respectively, then the unit vector,

$$\mathbf{k}' = \mathbf{i}' \times \mathbf{j}' = (\mathbf{i} + \mathbf{j} - 2\mathbf{k})/\sqrt{6},$$

gives a second invariant direction for -1. Therefore

$$\Phi = 2\mathbf{i}'\mathbf{i}' - \mathbf{j}'\mathbf{j}' - \mathbf{k}'\mathbf{k}',$$

as we may readily verify.

Example 2. *Inertia Dyadic.* Let O be any point of a rigid body and s an axis through O in the direction of the unit vector $\mathbf{e}$. Then, if dm is an element of mass at P, at a distance p from the axis s, the moment of inertia of the body about s is defined as the integral $\int p^2\, dm$ over the body. If $\mathbf{r} = \overrightarrow{OP}$,

$$p^2 = r^2 - (\mathbf{e} \cdot \mathbf{r})^2 = \mathbf{e} \cdot (r^2\mathbf{I} - \mathbf{r}\mathbf{r}) \cdot \mathbf{e};$$

hence, if we introduce the dyadic,

$$\text{(8)} \qquad \mathbf{K} = \mathbf{I} \int r^2\, dm - \int \mathbf{r}\mathbf{r}\, dm,$$

known as the *inertia dyadic* of the body for the point O,

$$\text{(9)} \qquad \mathbf{e} \cdot \mathbf{K} \cdot \mathbf{e} = \int p^2\, dm.$$

For example,

$$\mathbf{i} \cdot \mathbf{K} \cdot \mathbf{i} = \int (r^2 - x^2)\, dm = \int (y^2 + z^2)\, dm$$

is the moment of inertia about the x-axis. Thus $\mathbf{K}$ effects a synthesis of the moments of inertia of a body about all axes through O.

This dyadic also has the property that, for any pair of perpendicular unit vectors $\mathbf{e}_1$, $\mathbf{e}_2$,

$$-\mathbf{e}_1 \cdot \mathbf{K} \cdot \mathbf{e}_2 = \int (\mathbf{e}_1 \cdot \mathbf{r})(\mathbf{r} \cdot \mathbf{e}_2)\, dm$$

is the *product of inertia* for the corresponding axes; thus

$$-\mathbf{i} \cdot \mathbf{K} \cdot \mathbf{j} = \int (\mathbf{i} \cdot \mathbf{r})(\mathbf{r} \cdot \mathbf{j})\, dm = \int xy\, dm.$$

The inertia dyadic **K** is evidently symmetric. Hence we always can find three mutually perpendicular axes x, y, z through O such that

$$\mathbf{K} = A\mathbf{ii} + B\mathbf{jj} + C\mathbf{kk}. \tag{10}$$

These axes are called the *principal axes of inertia* at O, and A, B, C are the moments of inertia of the body about these principal axes. The principal axes are characterized by the property that the product of inertia for any pair is zero.

The ellipsoid,

$$\mathbf{r} \cdot \mathbf{K} \cdot \mathbf{r} = 1, \quad \text{or} \quad Ax^2 + By^2 + Cz^2 = 1, \tag{11}$$

is called the *ellipsoid of inertia* at O; its principal axes are the principal axes of inertia at O. It has the property that the moment of inertia about any axis s through O and cutting the ellipsoid at P is $1/(OP)^2$. For, if $\overrightarrow{OP} = \mathbf{r} = r\mathbf{e}$,

$$\mathbf{r} \cdot \mathbf{K} \cdot \mathbf{r} = r^2\mathbf{e} \cdot \mathbf{K} \cdot \mathbf{e} = 1, \qquad \mathbf{e} \cdot \mathbf{K} \cdot \mathbf{e} = 1/r^2.$$

73. The Hamilton–Cayley Equation. The identity (70.3) applied to the dyadic $\Psi = \Phi - \lambda \mathbf{I}$ gives

$$\Psi \cdot \Psi_a = f(\lambda)\mathbf{I}. \tag{1}$$

The form of Ψ given in § 71 shows that we may write

$$\Psi_a = \mathbf{A} + \mathbf{B}\lambda + \mathbf{C}\lambda^2, \tag{2}$$

where **A**, **B**, **C** are dyadics independent of λ; hence, from (1),

$$(\Phi - \lambda\mathbf{I}) \cdot (\mathbf{A} + \mathbf{B}\lambda + \mathbf{C}\lambda^2) = (\varphi_3 - \varphi_2\lambda + \varphi_1\lambda^2 - \lambda^3)\mathbf{I}. \tag{3}$$

Since (3) is an identity in λ, the dyadic coefficients of like powers of λ in the two members must be equal, hence

$$\begin{aligned} \Phi \cdot \mathbf{A} &= \varphi_3\mathbf{I}, \\ \Phi \cdot \mathbf{B} - \mathbf{A} &= -\varphi_2\mathbf{I}, \\ \Phi \cdot \mathbf{C} - \mathbf{B} &= \varphi_1\mathbf{I}, \\ -\mathbf{C} &= -\mathbf{I}. \end{aligned} \tag{4}$$

If we multiply these equations in order by **I**, Φ, Φ^2, Φ^3 and add, the first members cancel, and we get

$$\varphi_3\mathbf{I} - \varphi_2\Phi + \varphi_1\Phi^2 - \Phi^3 = 0. \tag{5}$$

Every dyadic Φ satisfies this cubic equation, the *Hamilton–Cayley Equation;* it evidently is formed by replacing λ by Φ in the characteristic equation,

$$f(\lambda) = \varphi_3 - \varphi_2\lambda + \varphi_1\lambda^2 - \lambda^3 = 0, \tag{6}$$

and inserting **I** in the constant term. If the λ_1, λ_2, λ_3 are the characteristic numbers of Φ,

$$f(\lambda) = (\lambda_1 - \lambda)(\lambda_2 - \lambda)(\lambda_3 - \lambda);$$

hence (5) also may be written

$$(\Phi - \lambda_1\mathbf{I}) \cdot (\Phi - \lambda_2\mathbf{I}) \cdot (\Phi - \lambda_3\mathbf{I}) = 0, \tag{7}$$

in which the dyadic factors are commutative.

From equations (4), we find

$$\mathbf{C} = \mathbf{I}, \qquad \mathbf{B} = \Phi - \varphi_1\mathbf{I}, \qquad \mathbf{A} = \Phi^2 - \varphi_1\Phi + \varphi_2\mathbf{I};$$

hence, from (2),

$$\Psi_a = \Phi^2 - (\varphi_1 - \lambda)\Phi + (\varphi_2 - \varphi_1\lambda + \lambda^2)\mathbf{I}. \tag{8}$$

When $\lambda = \lambda_1$, a characteristic number,

$$\varphi_1 - \lambda_1 = \lambda_2 + \lambda_3, \qquad \varphi_2 - \varphi_1\lambda_1 + \lambda_1^2 = \lambda_2\lambda_3,$$

and (8) becomes

$$(\Phi - \lambda_1\mathbf{I})_a = (\Phi - \lambda_2\mathbf{I}) \cdot (\Phi - \lambda_3\mathbf{I}). \tag{9}$$

Although every dyadic Φ satisfies the cubic (5), Φ will satisfy an equation of lower degree when $\Psi_a = (\Phi - \lambda\mathbf{I})_a$ vanishes for a characteristic number λ_i. The equation of lowest degree satisfied by Φ is called its *minimum equation.**

If $\Psi = \Phi - \lambda\mathbf{I} = 0$ for a characteristic number λ_1 (then also $\Psi_a = 0$), the minimum equation is linear, namely,

$$\Phi - \lambda_1\mathbf{I} = 0. \tag{10}$$

When $\Phi = \lambda_1\mathbf{I}$, $\Psi = (\lambda_1 - \lambda)\mathbf{I}$, and $\psi_3 = f(\lambda) = (\lambda_1 - \lambda)^3$; the Hamilton–Cayley Equation is therefore

$$(\Phi - \lambda_1\mathbf{I})^3 = 0. \tag{11}$$

* For its formation and properties see Macduffee, C. C., *An Introduction to Abstract Algebra,* New York, 1940, pp. 224–6. This treatment for $n \times n$ matrices also applies to dyadics, regarded as 3×3 matrices.

If Ψ does not vanish for any characteristic number, but $\Psi_a = 0$ when $\lambda = \lambda_1$, we see from (9) that the minimum equation is the quadratic,

$$(12) \qquad (\Phi - \lambda_2 \mathbf{I}) \cdot (\Phi - \lambda_3 \mathbf{I}) = 0.$$

When $\Psi_a = 0$, $\Psi_2 = 0$, and, consequently, Ψ is linear (theorem, § 69); hence we may write $\Phi = \lambda_1 \mathbf{I} + \mathbf{uv}$. If we take $\mathbf{u} = \mathbf{i}$, $\mathbf{v} = \alpha\mathbf{i} + \beta\mathbf{j} + \gamma\mathbf{k}$, $\mathbf{I} = \mathbf{ii} + \mathbf{jj} + \mathbf{kk}$, we have

$$\Psi = (\lambda_1 - \lambda)\mathbf{I} + \alpha\mathbf{ii} + \beta\mathbf{ij} + \gamma\mathbf{ik},$$

and the determinant of Ψ's matrix is

$$\psi_3 = f(\lambda) = (\lambda_1 - \lambda)^2(\lambda_1 + \alpha - \lambda).$$

Thus the characteristic numbers of Φ are λ_1, $\lambda_2 = \lambda_1$, $\lambda_3 = \lambda_1 + \alpha$, and its Hamilton–Cayley Equation is

$$(13) \qquad (\Phi - \lambda_1 \mathbf{I})^2 \cdot (\Phi - \lambda_3 \mathbf{I}) = 0.$$

From $\Phi = \lambda_1 \mathbf{I} + \mathbf{uv}$, we see that all vectors perpendicular to $\mathbf{v}$ have the multiplier λ_1, whereas vectors parallel to $\mathbf{u}$ have the multiplier $\lambda_1 + \mathbf{u} \cdot \mathbf{v} = \lambda_1 + \alpha$. When $\mathbf{u} \cdot \mathbf{v} = \alpha = 0$, (13) reduces to (11).

We note that, in every case, the minimum equation and the Hamilton–Cayley Equation have the same linear factors and differ only in their degree of multiplicity.

74. Normal Form of the General Dyadic. Every complete dyadic transforms at least one set of mutually orthogonal directions (its *principal directions*) into another set of the same kind.

To find the principal directions of the complete dyadic Φ, consider the dyadic $\Phi_c \cdot \Phi$. The latter is complete (§ 65) and symmetric; for, from (65.8),

$$(\Phi_c \cdot \Phi)_c = \Phi_c \cdot \Phi_{cc} = \Phi_c \cdot \Phi.$$

We therefore may write (§ 72)

$$(1) \qquad \Phi_c \cdot \Phi = \lambda_1 \mathbf{ii} + \lambda_2 \mathbf{jj} + \lambda_3 \mathbf{kk}, \qquad (\lambda_i \neq 0).$$

Consequently,

$$\mathbf{i} \cdot \Phi_c \cdot \Phi \cdot \mathbf{j} = \mathbf{j} \cdot \Phi_c \cdot \Phi \cdot \mathbf{k} = \mathbf{k} \cdot \Phi_c \cdot \Phi \cdot \mathbf{i} = 0,$$

or

$$(\Phi \cdot \mathbf{i}) \cdot (\Phi \cdot \mathbf{j}) = (\Phi \cdot \mathbf{j}) \cdot (\Phi \cdot \mathbf{k}) = (\Phi \cdot \mathbf{k}) \cdot (\Phi \cdot \mathbf{i}) = 0.$$

The vectors $\Phi \cdot \mathbf{i}$, $\Phi \cdot \mathbf{j}$, $\Phi \cdot \mathbf{k}$ are thus mutually orthogonal, and hence $\mathbf{i}, \mathbf{j}, \mathbf{k}$ give a set of principal directions of Φ. If we write

$$(2) \qquad \Phi \cdot \mathbf{i} = \alpha \mathbf{i}', \qquad \Phi \cdot \mathbf{j} = \beta \mathbf{j}', \qquad \Phi \cdot \mathbf{k} = \gamma \mathbf{k}',$$

where $\mathbf{i}', \mathbf{j}', \mathbf{k}'$ is a second dextral set of unit vectors, we have, from (63.11),

$$(3) \qquad \Phi = \alpha \mathbf{i}'\mathbf{i} + \beta \mathbf{j}'\mathbf{j} + \gamma \mathbf{k}'\mathbf{k}.$$

Moreover, we always can arrange so that α, β, γ have the same sign. If, for example, α and β have one sign, γ the opposite, we can replace $\mathbf{i}', \mathbf{j}'$ by $-\mathbf{i}', -\mathbf{j}'$, and the set $-\mathbf{i}', -\mathbf{j}', \mathbf{k}'$ still will be dextral.

From (3) and

$$\Phi_c = \alpha \mathbf{i}\mathbf{i}' + \beta \mathbf{j}\mathbf{j}' + \gamma \mathbf{k}\mathbf{k}',$$

we have, by direct multiplication,

$$(4) \qquad \Phi_c \cdot \Phi = \alpha^2 \mathbf{i}\mathbf{i} + \beta^2 \mathbf{j}\mathbf{j} + \gamma^2 \mathbf{k}\mathbf{k},$$

$$(5) \qquad \Phi \cdot \Phi_c = \alpha^2 \mathbf{i}'\mathbf{i}' + \beta^2 \mathbf{j}'\mathbf{j}' + \gamma^2 \mathbf{k}'\mathbf{k}'.$$

These symmetric dyadics, which in general are different, have the same multipliers, evidently all positive.

We have therefore proved the

THEOREM. *If Φ is complete, any three invariant directions of $\Phi_c \cdot \Phi$ that are mutually orthogonal are principal directions of Φ, and conversely. The principal directions of Φ transform into invariant directions of $\Phi \cdot \Phi_c$. Any complete dyadic Φ can be reduced to the normal form* (3) *in which the scalars α, β, γ are square roots of the multipliers of $\Phi_c \cdot \Phi$ (or $\Phi \cdot \Phi_c$) having the same sign.*

Example. Homogeneous Strain. In distinction to the ideal rigid body, the particles of a deformable body are capable of displacements relative to one another. The totality of such relative displacements is said to constitute its state of *strain.*

Suppose that a particle at P moves to P' under the strain; then $\mathbf{r}' = \overrightarrow{OP'}$ is a continuous function of $\mathbf{r} = \overrightarrow{OP}$. The simplest type of strain occurs when $\mathbf{r}'$ is a constant linear vector function of $\mathbf{r}$,

$$(6) \qquad \mathbf{r}' = \Phi \cdot \mathbf{r} \quad (\Phi \text{ complete});$$

the strain is then said to be *homogeneous.* We have seen in § 62 and § 70 that a homogeneous strain transforms lines into lines and planes into planes; and evidently parallelism is preserved. Moreover all volumes are altered in the constant ratio of $\varphi_3/1$.

Since Φ is complete, $\mathbf{r} = \Phi^{-1} \cdot \mathbf{r}'$. The particles originally on a sphere about O are displaced so as to lie upon an ellipsoid; for $\mathbf{r} \cdot \mathbf{r} = a^2$ transforms into

$$\mathbf{r}' \cdot \Phi_c^{-1} \cdot \Phi^{-1} \cdot \mathbf{r}' = a^2.$$

When Φ is reduced to the form (3),

$$\Phi^{-1} = \frac{1}{\alpha}\,\mathbf{i}\mathbf{i}' + \frac{1}{\beta}\,\mathbf{j}\mathbf{j}' + \frac{1}{\gamma}\,\mathbf{k}\mathbf{k}',$$

$$\Phi_c^{-1} \cdot \Phi^{-1} = \frac{1}{\alpha^2}\,\mathbf{i}'\mathbf{i}' + \frac{1}{\beta^2}\,\mathbf{j}'\mathbf{j}' + \frac{1}{\gamma^2}\,\mathbf{k}'\mathbf{k}',$$

and the foregoing strain ellipsoid (with $a = 1$) has the equation,

$$\frac{x'^2}{\alpha^2} + \frac{y'^2}{\beta^2} + \frac{z'^2}{\gamma^2} = 1.$$

The principal directions $\mathbf{i}, \mathbf{j}, \mathbf{k}$ of Φ are called the *principal axes of strain;* they transform into $\alpha\mathbf{i}', \beta\mathbf{j}'\ \gamma\mathbf{k}'$, the principal semiaxes of the strain ellipsoid.

75. Rotations and Reflections. *In order that a dyadic Φ transform all vectors so that their lengths are unchanged, it is necessary and sufficient that its inverse be equal to its conjugate:*

(1) $$\Phi^{-1} = \Phi_c.$$

Proof. If, for any vector $\mathbf{r}$,

(2) $$(\Phi \cdot \mathbf{r}) \cdot (\Phi \cdot \mathbf{r}) = \mathbf{r} \cdot \mathbf{r},$$

then

$$\mathbf{r} \cdot \Phi_c \cdot \Phi \cdot \mathbf{r} = \mathbf{r} \cdot \mathbf{I} \cdot \mathbf{r}, \qquad \mathbf{r} \cdot (\Phi_c \cdot \Phi - \mathbf{I}) \cdot \mathbf{r} = 0,$$

and, since $\Phi_c \cdot \Phi - \mathbf{I}$ is symmetric, it must be zero:

$$\Phi_c \cdot \Phi = \mathbf{I}, \quad \text{or} \quad \Phi_c = \Phi^{-1}.$$

Conversely, if (1) is fulfilled,

$$(\Phi \cdot \mathbf{r}) \cdot (\Phi \cdot \mathbf{r}) = \mathbf{r} \cdot \Phi_c \cdot \Phi \cdot \mathbf{r} = \mathbf{r} \cdot \mathbf{I} \cdot \mathbf{r} = \mathbf{r} \cdot \mathbf{r}.$$

A dyadic that preserves the lengths of vectors also preserves the angles between them: for, by virtue of (1),

$$(\Phi \cdot \mathbf{r}) \cdot (\Phi \cdot \mathbf{s}) = \mathbf{r} \cdot \Phi_c \cdot \Phi \cdot \mathbf{s} = \mathbf{r} \cdot \mathbf{I} \cdot \mathbf{s} = \mathbf{r} \cdot \mathbf{s};$$

and, since lengths are unaltered,

$$\cos(\Phi \cdot \mathbf{r}, \Phi \cdot \mathbf{s}) = \cos(\mathbf{r}, \mathbf{s}).$$

Condition (1), although sufficient to ensure preservation of angles, is by no means necessary. Thus the dyadic $\lambda\mathbf{I}$ preserves angles but multiplies all lengths by λ.

Since Φ preserves lengths and angles, any orthogonal set of unit vectors is transformed into another such set. Thus there are two possible cases: the dextral set $\mathbf{i}, \mathbf{j}, \mathbf{k}$ is transformed into another dextral set $\mathbf{i}', \mathbf{j}', \mathbf{k}'$, or into a sinistral set $\mathbf{i}', \mathbf{j}', -\mathbf{k}'$. Hence we have two types of length-preserving dyadics:

$$\Phi = \mathbf{i}'\mathbf{i} + \mathbf{j}'\mathbf{j} \pm \mathbf{k}'\mathbf{k}. \tag{3}$$

Their third scalar is $\varphi_3 = \pm 1$. Moreover,

$$\Phi_2 = \pm\mathbf{i}'\mathbf{i} \pm \mathbf{j}'\mathbf{j} + \mathbf{k}'\mathbf{k} = \pm\Phi.$$

This shows that the vector invariant of Φ_2 is $\pm\boldsymbol{\phi}$; but, from (69.12), this vector is $\Phi \cdot \boldsymbol{\phi}$. Equating these values, we obtain

$$\Phi \cdot \boldsymbol{\phi} = \pm\boldsymbol{\phi}; \tag{4}$$

the vector invariant of Φ *gives an invariant direction of multiplier*, $\varphi_3 = \pm 1$. If we choose $\mathbf{k}$ in this direction, (3) becomes

$$\Phi = \mathbf{i}'\mathbf{i} + \mathbf{j}'\mathbf{j} + \varphi_3\mathbf{k}\mathbf{k}, \qquad \varphi_3 = \pm 1. \tag{5}$$

When $\varphi_3 = 1$, $\Phi \cdot \mathbf{r}$ transforms $\mathbf{i}, \mathbf{j}, \mathbf{k}$ into $\mathbf{i}', \mathbf{j}', \mathbf{k}$. But a rotation Θ about $\mathbf{k}$ as axis, through an angle λ such that $\mathbf{i}, \mathbf{j}$ revolve into $\mathbf{i}', \mathbf{j}'$, transforms $\mathbf{i}, \mathbf{j}, \mathbf{k}$ in the same way; and, since Θ is linear vector function, $\Phi = \Theta$ (§ 60). Thus

$$\Theta = \mathbf{i}'\mathbf{i} + \mathbf{j}'\mathbf{j} + \mathbf{k}\mathbf{k} \tag{6}$$

is a rotation about the axis $\mathbf{k}$ through an angle λ determined by its scalar and vector invariants:

$$\theta_1 = 1 + 2\cos\lambda, \qquad \boldsymbol{\theta} = -2\sin\lambda\,\mathbf{k}. \tag{7}$$

When $\varphi_3 = -1$,

$$\Phi = \mathbf{i}'\mathbf{i} + \mathbf{j}'\mathbf{j} - \mathbf{k}\mathbf{k} = (\mathbf{i}'\mathbf{i} + \mathbf{j}'\mathbf{j} + \mathbf{k}\mathbf{k}) \cdot (\mathbf{i}\mathbf{i} + \mathbf{j}\mathbf{j} - \mathbf{k}\mathbf{k}).$$

The dyadic in the first factor is the rotation Θ. The dyadic $\mathbf{i}\mathbf{i} + \mathbf{j}\mathbf{j} - \mathbf{k}\mathbf{k}$ in the second factor transforms $\mathbf{i}, \mathbf{j}, \mathbf{k}$ into $\mathbf{i}, \mathbf{j}, -\mathbf{k}$. But a reflection Σ in the plane of $\mathbf{i}, \mathbf{j}$ transforms $\mathbf{i}, \mathbf{j}, \mathbf{k}$ in the same way; and, since Σ is a linear vector function,

$$\Sigma = \mathbf{i}\mathbf{i} + \mathbf{j}\mathbf{j} - \mathbf{k}\mathbf{k} = \mathbf{I} - 2\mathbf{k}\mathbf{k}. \tag{8}$$

Thus, when $\varphi_3 = -1$, $\Phi = \Theta \cdot \Sigma$.

THEOREM. *A dyadic Φ that preserves lengths is a rotation Θ when $\varphi_3 = 1$, and a rotation Θ followed by a reflection Σ in the plane perpendicular to its axis when $\varphi_3 = -1$. In the latter case $\Theta \cdot \Sigma$ will reduce to Σ, a pure reflection, when $\Theta = \mathbf{I}$.*

76. Basic Dyads. If we express both antecedents and consequents of a dyadic Φ in terms of a given basis $\mathbf{e}_1, \mathbf{e}_2, \mathbf{e}_3$, we obtain upon expansion $3 \times 3 = 9$ types of basic dyads $\mathbf{e}_i\mathbf{e}_j$ $(i, j = 1, 2, 3)$. On collecting terms we may write Φ:

(1)
$$\begin{aligned}\Phi = \varphi^{11}\mathbf{e}_1\mathbf{e}_1 + \varphi^{12}\mathbf{e}_1\mathbf{e}_2 + \varphi^{13}\mathbf{e}_1\mathbf{e}_3 +\\ \varphi^{21}\mathbf{e}_2\mathbf{e}_1 + \varphi^{22}\mathbf{e}_2\mathbf{e}_2 + \varphi^{23}\mathbf{e}_2\mathbf{e}_3 +\\ \varphi^{31}\mathbf{e}_3\mathbf{e}_1 + \varphi^{32}\mathbf{e}_3\mathbf{e}_2 + \varphi^{33}\mathbf{e}_3\mathbf{e}_3.\end{aligned}$$

The nine coefficients φ^{ij} are called the *contravariant components* of Φ relative to the basis $\mathbf{e}_i$ (cf. § 23). If we drop the nine basic dyads $\mathbf{e}_i\mathbf{e}_j$ in (1), we can represent Φ by the 3×3 *matrix*,

(1)′
$$\Phi = \begin{pmatrix} \varphi^{11} & \varphi^{12} & \varphi^{13} \\ \varphi^{21} & \varphi^{22} & \varphi^{23} \\ \varphi^{31} & \varphi^{32} & \varphi^{33} \end{pmatrix},$$

a skeleton of numbers arranged in a definite order, which stands for the full expression (1). This is analogous to the use of a number triple (u^1, u^2, u^3), or 1×3 matrix, to represent the vector $\mathbf{u} = u^1\mathbf{e}_1 + u^2\mathbf{e}_2 + u^3\mathbf{e}_3$.

If we express the vectors of Φ in terms of the reciprocal basis $\mathbf{e}^i$, we write

(2)
$$\Phi = \Sigma\Sigma\varphi_{ij}\mathbf{e}^i\mathbf{e}^j.$$

The nine numbers φ_{ij} are called the *covariant components* of Φ relative to the basis $\mathbf{e}_i$. Just as before, we may represent Φ by a matrix of the components φ_{ij}. This corresponds to the use of the number triple (u_1, u_2, u_3) to represent the vector $\mathbf{u} = u_1\mathbf{e}^1 + u_2\mathbf{e}^2 + u_3\mathbf{e}^3$.

But with the same basis $\mathbf{e}_i$, we can represent Φ in two other ways. First, we may express the antecedents of Φ in terms of $\mathbf{e}_i$, the consequents in terms of $\mathbf{e}^j$; the basic dyads are then $\mathbf{e}_i\mathbf{e}^j$, and

(3)
$$\Phi = \Sigma\Sigma\varphi^{i}_{\cdot j}\,\mathbf{e}_i\mathbf{e}^j.$$

Or we may express the antecedents in terms of $\mathbf{e}^i$ and consequents in terms of $\mathbf{e}_j$, so that the basic dyads are $\mathbf{e}^i\mathbf{e}_j$; then

$$(4) \qquad \Phi = \Sigma\Sigma\varphi_i^{\cdot j}\mathbf{e}^i\mathbf{e}_j.$$

The components $\varphi^i_{\cdot j}$, $\varphi_i^{\cdot j}$ are called *mixed.* If we represent Φ by matrices of mixed components, we must indicate the *order* of the subscripts as shown in the preceding notation; for, in general, $\varphi^i_{\cdot j} \neq \varphi_j^{\cdot i}$. However, if we use the full notations (3) or (4), the components can be written φ^i_j; for the *order* of the indices then is shown by the base vectors.

77. Nonion Form. When the self-reciprocal orthogonal set **i, j, k** is used as a basis, all four representations of Φ given in § 76 become the same. Since upper and lower indices no longer are needed, we write the orthogonal components of Φ arbitrarily as φ_{ij}. When no basis is indicated, the components of the matrix (φ_{ij}) shall be regarded as orthogonal, and Φ itself is said to be in its *nonion form:*

$$(1) \qquad \Phi = \varphi_{11}\mathbf{ii} + \varphi_{12}\mathbf{ij} + \cdots + \varphi_{33}\mathbf{kk} = \begin{pmatrix} \varphi_{11} & \varphi_{12} & \varphi_{13} \\ \varphi_{21} & \varphi_{22} & \varphi_{23} \\ \varphi_{31} & \varphi_{32} & \varphi_{33} \end{pmatrix}.$$

The conjugate of Φ corresponds to the *transpose* of this matrix:

$$(2) \qquad \Phi_c = \begin{pmatrix} \varphi_{11} & \varphi_{21} & \varphi_{31} \\ \varphi_{12} & \varphi_{22} & \varphi_{32} \\ \varphi_{13} & \varphi_{23} & \varphi_{33} \end{pmatrix}.$$

Hence Φ is symmetric when $\varphi_{ij} = \varphi_{ji}$, antisymmetric when $\varphi_{ij} = -\varphi_{ij}$ $(\varphi_{ii} = 0)$.

The first scalar and vector invariant of Φ are readily computed:

$$(3) \qquad \varphi_1 = \varphi_{11} + \varphi_{22} + \varphi_{33},$$

$$(4) \qquad \boldsymbol{\phi} = (\varphi_{23} - \varphi_{32})\mathbf{i} + (\varphi_{31} - \varphi_{13})\mathbf{j} + (\varphi_{12} - \varphi_{21})\mathbf{k}.$$

In order to compute φ_2 and φ_3, write Φ in three-term form:

$$(5) \qquad \begin{aligned} \Phi = {} & \mathbf{i}(\varphi_{11}\mathbf{i} + \varphi_{12}\mathbf{j} + \varphi_{13}\mathbf{k}) + \\ & \mathbf{j}(\varphi_{21}\mathbf{i} + \varphi_{22}\mathbf{j} + \varphi_{23}\mathbf{k}) + \\ & \mathbf{k}(\varphi_{31}\mathbf{i} + \varphi_{32}\mathbf{j} + \varphi_{33}\mathbf{k}). \end{aligned}$$

Then, since $[\mathbf{ijk}] = 1$, φ_3 is the box product of three consequents, namely the determinant of matrix (φ_{ij}):

$$\varphi_3 = \begin{vmatrix} \varphi_{11} & \varphi_{12} & \varphi_{13} \\ \varphi_{21} & \varphi_{22} & \varphi_{23} \\ \varphi_{31} & \varphi_{32} & \varphi_{33} \end{vmatrix}. \tag{6}$$

Thus Φ is singular when the determinant of its matrix vanishes.

As to Φ_2, the terms with antecedent $\mathbf{i} = \mathbf{j} \times \mathbf{k}$ have as consequents,

$$(\varphi_{21}\mathbf{i} + \varphi_{22}\mathbf{j} + \varphi_{23}\mathbf{k}) \times (\varphi_{31}\mathbf{i} + \varphi_{32}\mathbf{j} + \varphi_{33}\mathbf{k}) = \Phi^{11}\mathbf{i} + \Phi^{12}\mathbf{j} + \Phi^{13}\mathbf{k},$$

where Φ^{ij} denotes the cofactor † of φ_{ij} in the determinant (6); hence

$$\Phi_2 = \begin{pmatrix} \Phi^{11} & \Phi^{12} & \Phi^{13} \\ \Phi^{21} & \Phi^{22} & \Phi^{23} \\ \Phi^{31} & \Phi^{32} & \Phi^{33} \end{pmatrix}, \tag{7}$$

$$\varphi_2 = \Phi^{11} + \Phi^{22} + \Phi^{33}, \tag{8}$$

and Φ_a is the transpose of (7). Moreover, since $\Phi^{-1} = \Phi_a/\varphi_3$ (70.5),

$$\Phi^{-1} = \begin{pmatrix} \varphi^{11} & \varphi^{21} & \varphi^{31} \\ \varphi^{12} & \varphi^{22} & \varphi^{32} \\ \varphi^{13} & \varphi^{23} & \varphi^{33} \end{pmatrix}, \tag{9}$$

where $\varphi^{ij} = \Phi^{ij}/\varphi_3$ is the *reduced cofactor* of φ_{ij}. Making use of the well-known relations in determinant theory,

$$\varphi_{i1}\varphi^{j1} + \varphi_{i2}\varphi^{j2} + \varphi_{i3}\varphi^{j3} = \varphi_{1i}\varphi^{1j} + \varphi_{2i}\varphi^{2j} + \varphi_{3i}\varphi^{3j} = \delta_i^j,$$

we may verify that $\Phi \cdot \Phi^{-1}$ or $\Phi^{-1} \cdot \Phi$ give the idemfactor:

$$\mathbf{I} = \begin{pmatrix} 1 & 0 & 0 \\ 0 & 1 & 0 \\ 0 & 0 & 1 \end{pmatrix}. \tag{10}$$

The characteristic equation of Φ is obtained by equating the third scalar of $\Phi - \lambda\mathbf{I}$ to zero; hence, with Φ in nonion form, it

† The cofactor of φ_{ij} in the determinant $|\varphi_{ij}|$ is defined as the coefficient of φ_{ij} in the expansion of the determinant; it equals the minor obtained by striking out the ith row and jth column with the sign $(-1)^{i+j}$ affixed.

becomes

$$
(11) \qquad f(\lambda) = \begin{vmatrix} \varphi_{11} - \lambda & \varphi_{12} & \varphi_{13} \\ \varphi_{21} & \varphi_{22} - \lambda & \varphi_{23} \\ \varphi_{31} & \varphi_{32} & \varphi_{33} - \lambda \end{vmatrix} = 0.
$$

Example. If the dyadic Φ in (1) is symmetric, and $\mathbf{r} = x\mathbf{i} + y\mathbf{j} + \mathbf{k}$, then

$$
\text{(i)} \qquad \mathbf{r} \cdot \Phi \cdot \mathbf{r} = \varphi_{11}x^2 + 2\varphi_{12}xy + \varphi_{22}y^2 + 2\varphi_{13}x + 2\varphi_{23}y + \varphi_{33} = 0
$$

represents a conic section—an ellipse, parabola, or hyperbola, according as $\varphi_{12}^2 - \varphi_{11}\varphi_{22} < 1$, $= 1$, or >1. Let this conic cut the sides BC, CA, AB of a triangle ABC in the points R_1, R_1'; R_2, R_2'; R_3, R_3', respectively, and let the corresponding division ratios be ρ_1, ρ_1'; ρ_2, ρ_2'; ρ_3, ρ_3'. If we put $\mathbf{r} = (\mathbf{b} + \rho\mathbf{c})/(1 + \rho)$ in (i) we find that ρ_1, ρ_1' are the roots of the quadratic equation

$$
\mathbf{b} \cdot \Phi \cdot \mathbf{b} + 2\rho\, \mathbf{b} \cdot \Phi \cdot \mathbf{c} + \rho^2 \mathbf{c} \cdot \Phi \cdot \mathbf{c} = 0;
$$

hence

$$
\rho_1\rho_1' = \frac{\mathbf{b} \cdot \Phi \cdot \mathbf{b}}{\mathbf{c} \cdot \Phi \cdot \mathbf{c}}; \qquad \text{and} \quad \rho_2\rho_2' = \frac{\mathbf{c} \cdot \Phi \cdot \mathbf{c}}{\mathbf{a} \cdot \Phi \cdot \mathbf{a}}, \quad \rho_3\rho_3' = \frac{\mathbf{a} \cdot \Phi \cdot \mathbf{a}}{\mathbf{b} \cdot \Phi \cdot \mathbf{b}}
$$

follow in the same way. On multiplying these equations, we have

$$
\text{(ii)} \qquad \rho_1\rho_1'\rho_2\rho_2'\rho_3\rho_3' = 1.
$$

Now let $R_2'R_3$, $R_3'R_1$, $R_1'R_2$ meet BC, CA, AB in the points S_1, S_2, S_3 which divide the respective sides in the ratios σ_1, σ_2, σ_3. Then by the Theorem of Menelaus (§ 7, ex. 2)

$$
\rho_2'\rho_3\sigma_1 = -1, \qquad \rho_3'\rho_1\sigma_2 = -1, \qquad \rho_1'\rho_2\sigma_3 = -1.
$$

On multiplying these equations together and making use of (ii), we have $\sigma_1\sigma_2\sigma_3 = -1$; hence S_1, S_2, S_3 are collinear. We thus have proved *Pascal's Theorem:* ‡ *The opposite sides of a hexagon* ($R_1R_1'R_2R_2'R_3R_3'$) *inscribed in a conic meet in three collinear points* (S_1, S_2, S_3).

78. Matric Algebra. The sum of two dyadics $\mathbf{A} + \mathbf{B}$ in nonion form obviously is obtained by adding corresponding elements of their matrices. As to the product $\mathbf{C} = \mathbf{A} \cdot \mathbf{B}$, let us consider the formation of a certain dyad of $\mathbf{C}$, say $c_{12}\mathbf{ij}$. This evidently results from the product of terms in $\mathbf{A}$ with antecedent $\mathbf{i}$ and terms in $\mathbf{B}$ with consequent $\mathbf{j}$: thus

$$
c_{12} = a_{11}b_{12} + a_{12}b_{22} + a_{13}b_{32} = \sum_r a_{1r}b_{r2}.
$$

The general result is therefore

$$
c_{ij} = \sum_r a_{ir}b_{rj};
$$

‡ This proof is due to Wedderburn, Am. Math. Monthly, vol. 52, 1945, p. 383.

the element in the ith row and jth column of $\mathbf{A} \cdot \mathbf{B}$ *is the sum of the products of the elements in the ith row of A by the corresponding elements in the jth column of* **B**—the "row-column rule."

The foregoing rules for the sum and product of dyadics in nonion form are precisely the classic definitions for the sum and product of *square matrices.* These definitions may be extended to *rectangular matrices* with m rows and n columns ($m \times n$ matrices).

1. The sum $A + B$ of two $m \times n$ matrices is the $m \times n$ matrix obtained by adding their corresponding elements.

2. The product of an $m \times p$ matrix A and a $p \times n$ matrix B is the $m \times n$ matrix $C = AB$, whose element in the ith row and jth column is

$$c_{ij} = \sum_{r=1}^{p} a_{ir} b_{rj}. \tag{1}$$

Note that only similar matrices can be added; whereas in a product the second matrix must have the same number of rows as the first has columns.

This extension enables us to interpret scalar products of dyadics and vectors in terms of matric algebra. We regard a vector $\mathbf{u}$ with *rectangular* components u_i either as 1×3 matrix (row vector) or a 3×1 matrix (column vector). Since a dyadic $\mathbf{A}$ in nonion form is a 3×3 matrix, $\mathbf{u}$ must be a row vector in $\mathbf{u} \cdot \mathbf{A}$, a column vector in $\mathbf{A} \cdot \mathbf{u}$. With this proviso, the rules of matric algebra give values of vector components in full agreement with vector algebra:

$$\mathbf{v} = \mathbf{u} \cdot \mathbf{A}, \qquad v_i = u_1 a_{1i} + u_2 a_{2i} + u_3 a_{3i}; \tag{2}$$

$$\mathbf{w} = \mathbf{A} \cdot \mathbf{u}, \qquad w_i = a_{i1} u_1 + a_{i2} u_2 + a_{i3} u_3. \tag{3}$$

The matric product of a row vector $\mathbf{u}$ into a column vector $\mathbf{v}$ is a 1×1 matrix consisting of a single element, the scalar product:

$$\mathbf{u} \cdot \mathbf{v} = u_1 v_1 + u_2 v_2 + u_3 v_3. \tag{4}$$

But the matric product $\mathbf{vu}$ of a column vector into a row vector is a 3×3 matrix, namely the dyad,

$$\mathbf{vu} = \begin{pmatrix} v_1 u_1 & v_1 u_2 & v_1 u_3 \\ v_2 u_1 & v_2 u_2 & v_2 u_3 \\ v_3 u_1 & v_3 u_2 & v_3 u_3 \end{pmatrix}, \tag{5}$$

in nonion form. *Matric multiplication, although associative and distributive with respect to addition, is in general not commutative.* The proofs follow readily from (1).

79. Differentiation of Dyadics. If a dyadic Φ is a function of a scalar variable t, we define

$$\frac{d\Phi}{dt} = \lim_{\Delta t \to 0} \frac{\Phi(t + \Delta t) - \Phi(t)}{\Delta t}.$$

For a single dyad $\mathbf{ab}$, let $\mathbf{a}$ and $\mathbf{b}$ become $\mathbf{a} + \Delta\mathbf{a}$, $\mathbf{b} + \Delta\mathbf{b}$ when t becomes $t + \Delta t$; then, since

$$\frac{(\mathbf{a} + \Delta\mathbf{a})(\mathbf{b} + \Delta\mathbf{b}) - \mathbf{ab}}{\Delta t} = \mathbf{a}\frac{\Delta\mathbf{b}}{\Delta t} + \frac{\Delta\mathbf{a}}{\Delta t}\mathbf{b} + \frac{\Delta\mathbf{a}}{\Delta t}\Delta\mathbf{b},$$

we have, on passing to the limit $\Delta t \to 0$,

$$(1) \qquad \frac{d}{dt}(\mathbf{ab}) = \mathbf{a}\frac{d\mathbf{b}}{dt} + \frac{d\mathbf{a}}{dt}\mathbf{b}.$$

This formula suggests the usual product rule *with the order of the factors preserved.*

The derivative of the dyadic $\Phi = \Sigma \mathbf{a}_i\mathbf{b}_i$ is evidently the sum of the derivatives of its dyads. If Φ is given in the nonion form (77.1),

$$(2) \qquad \frac{d\Phi}{dt} = \begin{pmatrix} \varphi'_{11} & \varphi'_{12} & \varphi'_{13} \\ \varphi'_{21} & \varphi'_{22} & \varphi'_{23} \\ \varphi'_{31} & \varphi'_{32} & \varphi'_{33} \end{pmatrix},$$

where the primes denote derivatives with respect to t. Note that the derivative of φ_3, the determinant of the matrix (φ_{ij}), is not the determinant of the matrix (φ'_{ij}).

The derivatives of products such as $\Phi \cdot \mathbf{r}$, $\Phi \times \mathbf{r}$, $\mathbf{s} \cdot \Phi \cdot \mathbf{r}$, which conform to the distributive law, are computed just as in the calculus when the order of the factors is preserved. For example,

$$(3) \qquad \frac{d}{dt}(\Phi \cdot \mathbf{r}) = \frac{d\Phi}{dt} \cdot \mathbf{r} + \Phi \cdot \frac{d\mathbf{r}}{dt};$$

$$(4) \qquad \frac{d}{dt}(\mathbf{s} \cdot \Phi \cdot \mathbf{r}) = \frac{d\mathbf{s}}{dt} \cdot \Phi \cdot \mathbf{r} + \mathbf{s} \cdot \frac{d\Phi}{dt} \cdot \mathbf{r} + \mathbf{s} \cdot \Phi \cdot \frac{d\mathbf{r}}{dt}.$$

80. Triadics. A *triadic* is defined as a sum of *triads*, $\Sigma \mathbf{a}_i\mathbf{b}_i\mathbf{c}_i$. A triad **abc** consists of three vectors written in a definite order. We may regard a triadic as an operator which converts vectors **r** into dyadics; thus

$$\Phi \cdot \mathbf{r} = \Sigma \mathbf{a}_i\mathbf{b}_i\mathbf{c}_i \cdot \mathbf{r}, \qquad \mathbf{r} \cdot \Phi = \Sigma \mathbf{r} \cdot \mathbf{a}_i\mathbf{b}_i\mathbf{c}_i.$$

If Φ and Ψ are two triadics, we write $\Phi = \Psi$, when

$$(1) \qquad \Phi \cdot \mathbf{r} = \Psi \cdot \mathbf{r} \quad \text{for every vector } \mathbf{r}.$$

Then, for any vector **s**, $\mathbf{s} \cdot \Phi \cdot \mathbf{r} = \mathbf{s} \cdot \Psi \cdot \mathbf{r}$, and, from the definition of equality for dyadics (61.3),

$$(2) \qquad \mathbf{s} \cdot \Phi = \mathbf{s} \cdot \Psi \quad \text{for every vector } \mathbf{s}.$$

Conversely, from (2) we may deduce (1). Thus from either (1) or (2) we may conclude that $\Phi = \Psi$.

Using a given basis $\mathbf{e}_i$ and its reciprocal $\mathbf{e}^i$ to form basic dyads, we have seen in § 76 that the $3^2 = 9$ components of a dyadic are of $2^2 = 4$ types. For a triadic Φ, the $3^3 = 27$ components are of $2^3 = 8$ types; for, for each vector in the basic, triads may be chosen from the set $\mathbf{e}_i$ or $\mathbf{e}^i$ giving 2^3 types; and, for a given type, each index on the base vectors may be chosen in three ways, giving 3^3 components.

Similarly, we define a *tetradic* as the sum of *tetrads* $\Sigma \mathbf{a}_i\mathbf{b}_i\mathbf{c}_i\mathbf{d}_i$. Two tetradics Φ, Ψ are equal when either (1) or (2) holds good. A tetradic has $3^4 = 81$ components of $2^4 = 16$ types.

We shall speak of scalars, vectors, dyadics, triadics, $\cdots$ collectively as *tensors* of *valence* 0, 1, 2, 3, $\cdots$. The equality of two tensors of valence n, say $\Phi = \Psi$, depends upon the equality of two tensors of valence $n - 1$, as required by equations (1) or (2).

81. Summary: Dyadic Algebra. A *linear vector function* $\mathbf{f}(\mathbf{r})$ is characterized by the properties,

$$\mathbf{f}(\mathbf{a} + \mathbf{b}) = \mathbf{f}(\mathbf{a}) + \mathbf{f}(\mathbf{b}), \qquad \mathbf{f}(\lambda\mathbf{a}) = \lambda\mathbf{f}(\mathbf{a}).$$

A *dyadic* $\Phi = \Sigma \mathbf{a}_i\mathbf{b}_i$ is the sum of *dyads* $\mathbf{a}_i\mathbf{b}_i$; the vectors $\mathbf{a}_i$ are *antecedents*, $\mathbf{b}_i$ are *consequents*. The *conjugate* of Φ is $\Phi_c = \Sigma \mathbf{b}_i\mathbf{a}_i$. Any linear vector function may be expressed as $\Phi \cdot \mathbf{r}$ (or $\mathbf{r} \cdot \Phi_c$); and, if $[\mathbf{a}_1\mathbf{a}_2\mathbf{a}_3] \neq 0$ and $f(\mathbf{a}_i) = \mathbf{b}_i$ $(i = 1, 2, 3)$,

$$f(\mathbf{r}) = \Phi \cdot \mathbf{r}, \quad \text{where} \quad \Phi = \mathbf{b}_1\mathbf{a}^1 + \mathbf{b}_2\mathbf{a}^2 + \mathbf{b}_3\mathbf{a}^3.$$

Basic definitions ($\mathbf{r}$ an arbitrary vector):

$\Phi = \Psi$: $\quad \Phi \cdot \mathbf{r} = \Psi \cdot \mathbf{r}$;

$\Phi = 0$: $\quad \Phi \cdot \mathbf{r} = 0$;

$\Phi + \Psi$: $\quad (\Phi + \Psi) \cdot \mathbf{r} = \Phi \cdot \mathbf{r} + \Psi \cdot \mathbf{r}$;

$\Phi \cdot \Psi$: $\quad (\Phi \cdot \Psi) \cdot \mathbf{r} = \Phi \cdot (\Psi \cdot \mathbf{r})$;

$\mathbf{I}$ (idemfactor): $\quad \mathbf{I} \cdot \mathbf{r} = \mathbf{r}$;

Φ^{-1} (reciprocal): $\quad \Phi \cdot \Phi^{-1} = \Phi^{-1} \cdot \Phi = \mathbf{I}$;

$\Phi \times \mathbf{v}$: $\quad \Sigma \mathbf{a}_i \mathbf{b}_i \times \mathbf{v}$ $\quad$ if $\quad \Phi = \Sigma \mathbf{a}_i \mathbf{b}_i$;

$\mathbf{v} \times \Phi$: $\quad \Sigma \mathbf{v} \times \mathbf{a}_i \mathbf{b}_i$.

Every dyadic may be reduced to the sum of three dyads,

$$\Phi = \mathbf{al} + \mathbf{bm} + \mathbf{cn},$$

in which either antecedents or consequents may be an arbitrary non-coplanar set. In this form, the principal *invariants* of Φ are:

Dyadic: $\quad \Psi = \mathbf{b} \times \mathbf{c}\, \mathbf{m} \times \mathbf{n} + \mathbf{c} \times \mathbf{a}\, \mathbf{n} \times \mathbf{l} + \mathbf{a} \times \mathbf{b}\, \mathbf{l} \times \mathbf{m} = \Phi_2$;

Vector: $\quad \boldsymbol{\phi} = \mathbf{a} \times \mathbf{l} + \mathbf{b} \times \mathbf{m} + \mathbf{c} \times \mathbf{n}$,

$\quad \boldsymbol{\psi} = (\mathbf{b} \times \mathbf{c}) \times (\mathbf{m} \times \mathbf{n}) + \text{cycl} = \Phi \cdot \boldsymbol{\phi}$;

Scalar: $\quad \varphi_1 = \mathbf{a} \cdot \mathbf{l} + \mathbf{b} \cdot \mathbf{m} + \mathbf{c} \cdot \mathbf{n}$,

$\quad \varphi_2 = \psi_1 = (\mathbf{b} \times \mathbf{c}) \cdot (\mathbf{m} \times \mathbf{n}) + \text{cycl}$,

$\quad \varphi_3 = [\mathbf{abc}][\mathbf{lmn}]$,

$\quad \varphi_4 = \boldsymbol{\phi} \cdot \boldsymbol{\phi}, \qquad \varphi_5 = \boldsymbol{\phi} \cdot \boldsymbol{\psi}, \qquad \varphi_6 = \boldsymbol{\psi} \cdot \boldsymbol{\psi}$.

Φ is *complete* if $\varphi_3 \neq 0$; only complete dyadics have reciprocals. Φ is *planar* if $\varphi_3 = 0$, $\Phi_2 \neq 0$; a *planar* dyadic may be reduced to *two* dyads. Φ is *linear* if $\Phi_2 = 0$, $\Phi \neq 0$; a linear dyadic may be reduced to *one* dyad. Planar and linear dyadics are called *singular*.

Fundamental identities:

$$(\Phi \cdot \Psi)_c = \Psi_c \cdot \Phi_c, \qquad (\Phi \cdot \Psi)^{-1} = \Psi^{-1} \cdot \Phi^{-1};$$

$$\mathbf{I} \times \mathbf{v} = \mathbf{v} \times \mathbf{I}, \qquad (\mathbf{I} \times \mathbf{v})_c = -\mathbf{I} \times \mathbf{v}, \qquad \mathbf{I} \times (\mathbf{u} \times \mathbf{v}) = \mathbf{vu} - \mathbf{uv};$$

$$\Phi \cdot \Phi_a = \Phi_a \cdot \Phi = \varphi_3 \mathbf{I} \quad (\text{the } \textit{adjoint } \Phi_a = \Phi_{2c});$$

$$\varphi_3 \mathbf{I} - \varphi_2 \Phi + \varphi_1 \Phi^2 - \Phi^3 = 0 \quad \text{(Hamilton–Cayley Equation)}.$$

Φ is *symmetric* if $\Phi_c = \Phi$, *antisymmetric* if $\Phi_c = -\Phi$. Every dyadic Φ can be expressed uniquely as the sum of a symmetric and antisymmetric dyadic; the latter is $-\frac{1}{2}\mathbf{I} \times \boldsymbol{\phi}$.

The vector $\mathbf{r}_1$ is an *invariant direction* of Φ with multiplier λ_1 if $\Phi \cdot \mathbf{r}_1 = \lambda_1 \mathbf{r}_1$. The multipliers of Φ satisfy the cubic,

$$\varphi_3 - \varphi_2\lambda + \varphi_1\lambda^2 - \lambda^3 = 0 \quad \text{(characteristic equation)}.$$

The invariant direction $\mathbf{r}_1$ makes $(\Phi - \lambda_1\mathbf{I}) \cdot \mathbf{r}_1 = 0$.

If Φ is symmetric its multipliers λ_i are all real; and, if $\lambda_1 \neq \lambda_2$, $\mathbf{r}_1 \perp \mathbf{r}_2$. A symmetric Φ always may be reduced to the form,

$$\Phi = \lambda_1\mathbf{ii} + \lambda_2\mathbf{jj} + \lambda_3\mathbf{kk};$$

the multipliers λ_i need not be distinct.

Any dyadic Φ can be reduced to the normal form,

$$\Phi = \alpha\mathbf{i}'\mathbf{i} + \beta\mathbf{j}'\mathbf{j} + \gamma\mathbf{k}'\mathbf{k},$$

in which α, β, γ all have the same sign; $\mathbf{i}, \mathbf{j}, \mathbf{k}$ and $\mathbf{i}', \mathbf{j}', \mathbf{k}'$ are two dextral sets of orthogonal unit vectors.

When Φ is complete, the point transformation, $\mathbf{r}' = \Phi \cdot \mathbf{r}$, changes lines into lines and planes into planes, preserves parallelism, and alters all volumes in the ratio $\varphi_3/1$. It preserves lengths when and only when $\Phi^{-1} = \Phi_c$; it is then a rotation if $\varphi_3 = 1$, a rotation followed by a reflection if $\varphi_3 = -1$.

PROBLEMS

1. Prove that

$$\mathbf{a}\,\mathbf{b} \times \mathbf{c} + \mathbf{b}\,\mathbf{c} \times \mathbf{a} + \mathbf{c}\,\mathbf{a} \times \mathbf{b} = [\mathbf{abc}]\,\mathbf{I}.$$

2. For any dyadic Φ show that

$$\mathbf{u} \cdot \Phi \cdot \mathbf{v} - \mathbf{v} \cdot \Phi \cdot \mathbf{u} = \boldsymbol{\phi} \cdot \mathbf{u} \times \mathbf{v}.$$

3. If Φ has the characteristic numbers $\lambda_1, \lambda_2, \lambda_3$ and the corresponding invariant directions $\mathbf{r}_1, \mathbf{r}_2, \mathbf{r}_3$, prove that Φ^n (n an integer) has the characteristic numbers $\lambda_1^n, \lambda_2^n, \lambda_3^n$ corresponding to the same invariant directions.

4. If Φ has the characteristic numbers $\lambda_1, \lambda_2, \lambda_3$ for the directions $\mathbf{r}_1, \mathbf{r}_2, \mathbf{r}_3$, prove that Φ_2 has the characteristic numbers $\lambda_2\lambda_3, \lambda_3\lambda_1, \lambda_1\lambda_2$ corresponding to the directions $\mathbf{r}_2 \times \mathbf{r}_3, \mathbf{r}_3 \times \mathbf{r}_1, \mathbf{r}_1 \times \mathbf{r}_2$. [Cf. (70.2).]

5. If $\Psi = \Phi_c$, prove that $\Psi_2 = \Phi_{2c}$, $\boldsymbol{\psi} = -\boldsymbol{\phi}$ and $\psi_1 = \varphi_1, \psi_2 = \varphi_2, \psi_3 = \varphi_3$.

6. Prove that the scalar invariants of $\Psi = \Phi^2$ are

$$\psi_1 = \varphi_1^2 - 2\varphi_2, \quad \psi_2 = \varphi_2^2 - 2\varphi_1\varphi_3, \quad \psi_3 = \varphi_3^2.$$

[Use Prob. 3 and (71.4).]

7. Given the invariants Φ_2, $\boldsymbol{\phi}$, φ_1, φ_2, φ_3 of Φ, find the corresponding invariants for

$$(a)\ k\Phi, \quad (b)\ \Phi \times \mathbf{u}, \quad (c)\ \Phi^{-1}.$$

8. Compute Φ_2, $\boldsymbol{\phi}$, and the six scalar invariants of (69.13), namely,

$$\varphi_1,\ \varphi_2,\ \varphi_3; \qquad \boldsymbol{\phi} \cdot \boldsymbol{\phi},\ \boldsymbol{\phi} \cdot \Phi \cdot \boldsymbol{\phi},\ \boldsymbol{\phi} \cdot \Phi^2 \cdot \boldsymbol{\phi},$$

for the dyadic

$$\Phi = \begin{pmatrix} 1 & 3 & -2 \\ 2 & 0 & 4 \\ -1 & 2 & 3 \end{pmatrix}.$$

9. If $\Phi = \mathbf{I} \times \mathbf{e}$ and $\mathbf{e}$ is a unit vector, prove that

$$\Phi^2 = -(\mathbf{I} - \mathbf{ee}),\ \Phi^3 = -\Phi,\ \Phi^4 = \mathbf{I} - \mathbf{ee},\ \Phi^5 = \Phi.$$

10. Show that the symmetric part of Φ, namely, $\Psi = \frac{1}{2}(\Phi + \Phi_c)$, has the scalar invariants

$$\psi_1 = \varphi_1, \quad \psi_2 = \varphi_2 - \tfrac{1}{4}\boldsymbol{\phi} \cdot \boldsymbol{\phi}, \quad \psi_3 = \varphi_3 - \tfrac{1}{4}\boldsymbol{\phi} \cdot \Phi \cdot \boldsymbol{\phi}.$$

11. Prove that $(\Phi \cdot \Psi)_2 = \Phi_2 \cdot \Psi_2$.

12. Prove that the first three scalar invariants of $\Phi \cdot \Psi$ and $\Psi \cdot \Phi$ are the same.

13. If $\Phi = \sum_{i=1}^{m} \mathbf{a}_i\mathbf{b}_i$, $\Psi = \sum_{j=1}^{n} \mathbf{c}_j\mathbf{d}_j$,

we define the *double-dot* product $\Phi{:}\Psi$ as the scalar

$$\Phi{:}\Psi = \sum_{i=1}^{m} \sum_{j=1}^{n} (\mathbf{a}_i \cdot \mathbf{c}_j)(\mathbf{b}_i \cdot \mathbf{d}_j).$$

Prove that

(*a*) $\qquad \Phi{:}\Psi = \Psi{:}\Phi, \qquad \Phi{:}(\Psi + \Omega) = \Phi{:}\Psi + \Phi{:}\Omega;$

(*b*) $\qquad (\mathbf{uv}){:}\Phi = \mathbf{u} \cdot \Phi \cdot \mathbf{v}, \qquad \Phi{:}\mathbf{I} = \varphi_1;$

(*c*) If Φ is given in the nonion form (77.1),

$$\Phi{:}\Phi = \varphi_{11}^2 + \varphi_{12}^2 + \varphi_{13}^2 + \varphi_{21}^2 + \varphi_{22}^2 + \varphi_{23}^2 + \varphi_{31}^2 + \varphi_{32}^2 + \varphi_{33}^2.$$

Hence $\Phi = 0$ when and only when $\Phi{:}\Phi = 0$.

14. Any central quadric surface with its center at the origin has an equation of the form,

(1) $$\mathbf{r} \cdot \Phi \cdot \mathbf{r} = 1,$$

where Φ is a symmetric dyadic; for, if Φ is reduced to the standard form (72.4), we have

$$\alpha x^2 + \beta y^2 + \gamma z^2 = 1.$$

This represents an ellipsoid, an hyperboloid of one sheet, or an hyperboloid of two sheets, according as α, β. γ include no, one, or two *negative* constants.

If $\mathbf{p} \neq 0$, the equation (1) associates with every point $(1, \mathbf{p})$ the plane $(\Phi \cdot \mathbf{p}, 1)$, its *polar plane*. Prove that:

(*a*) If the point $(1, \mathbf{p})$ lies on the quadric surface, its polar plane $(\Phi \cdot \mathbf{p}, 1)$ is tangent to the quadric at $(1, \mathbf{p})$. [Find the points where the line $\mathbf{r} = \mathbf{p} + \lambda\mathbf{e}$ cuts the quadric when $\mathbf{e} \cdot \Phi \cdot \mathbf{p} = 0$.]

(*b*) If the polar plane of $(1, \mathbf{p})$ passes through $(1, \mathbf{q})$, the polar plane of $(1, \mathbf{q})$ passes through $(1, \mathbf{p})$.

15. The *diametral plane* of any point $(1, \mathbf{p})$ on the quadric surface (1) is the locus of the mid-points of all chords parallel to $\mathbf{p}$. Show that the diametral plane of $(1, \mathbf{p})$ is $(\Phi \cdot \mathbf{p}, 0)$.

16. If three points $(1, \mathbf{u})$, $(1, \mathbf{v})$, $(1, \mathbf{w})$ on the ellipsoid $\mathbf{r} \cdot \Phi \cdot \mathbf{r} = 1$, satisfy the equations,

$$\mathbf{u} \cdot \Phi \cdot \mathbf{v} = \mathbf{v} \cdot \Phi \cdot \mathbf{w} = \mathbf{w} \cdot \Phi \cdot \mathbf{u} = 0;$$

the position vectors $\mathbf{u}$, $\mathbf{v}$, $\mathbf{w}$ are said to form a *conjugate* set. Show that:

(*a*) The vectors $\mathbf{u}$, $\mathbf{v}$, $\mathbf{w}$ and $\Phi \cdot \mathbf{u}$, $\Phi \cdot \mathbf{v}$, $\Phi \cdot \mathbf{w}$ form reciprocal sets.

(*b*) $\Phi^{-1} = \mathbf{uu} + \mathbf{vv} + \mathbf{ww}$.

(*c*) For any conjugate set $\mathbf{u}$, $\mathbf{v}$, $\mathbf{w}$, of the ellipsoid $\mathbf{r} \cdot \Phi \cdot \mathbf{r} = 1$, the sum $\mathbf{u} \cdot \mathbf{u} + \mathbf{v} \cdot \mathbf{v} + \mathbf{w} \cdot \mathbf{w}$ and the product $\mathbf{u} \times \mathbf{v} \cdot \mathbf{w}$ are constant.

17. Verify by direct computation that the dyadic Φ in Problem 8 satisfies its Hamilton–Cayley Equation.

18. A rigid body with one point O fixed has the inertia dyadic $\mathbf{K}$ relative to O (§ 72, ex. 2). If the angular velocity of the body at any instant is $\boldsymbol{\omega}$, show that its moment of momentum $\mathbf{H}$ (defined as $\int \mathbf{r} \times \mathbf{v}\, dm$) and kinetic energy T (defined as $\frac{1}{2} \int \mathbf{v} \cdot \mathbf{v}\, dm$) are given by

$$\mathbf{H} = \mathbf{K} \cdot \boldsymbol{\omega}, \qquad T = \tfrac{1}{2}\boldsymbol{\omega} \cdot \mathbf{K} \cdot \boldsymbol{\omega}.$$

19. If the forces acting on the body of Problem 18 have the moment sum $\mathbf{M}$ about O, the equation of motion is $d\mathbf{H}/dt = \mathbf{M}$. Show, from (56.3), that

$$\frac{d\mathbf{H}}{dt} = \mathbf{K} \cdot \frac{d\boldsymbol{\omega}}{dt} + \boldsymbol{\omega} \times \mathbf{K} \cdot \boldsymbol{\omega}.$$

Let $\mathbf{K} = A\mathbf{ii} + B\mathbf{jj} + C\mathbf{kk}$ (72.10) when referred to the principal axes of inertia (fixed in the body). Then if $\boldsymbol{\omega} = [\omega_1, \omega_2, \omega_3]$, $\mathbf{M} = [M_1, M_2, M_3]$ referred to these axes, deduce *Euler's Equations of Motion:*

$$A\dot{\omega}_1 - (B - C)\omega_2\omega_3 = M_1,$$

$$B\dot{\omega}_2 - (C - A)\omega_3\omega_1 = M_2,$$

$$C\dot{\omega}_3 - (A - B)\omega_1\omega_2 = M_3.$$

20. In Problem 19 show that $dT/dt = \mathbf{M} \cdot \boldsymbol{\omega}$ (the *energy equation*).

21. In Problem 18, suppose that the only forces acting on the body are its weight $\mathbf{W}$ and the reaction $\mathbf{R}$ at the support O; then *if the center of mass is at O*, $\mathbf{W}$ passes through O and $\mathbf{M} = 0$. Prove in turn that

(*a*) $\mathbf{H}$ is a constant vector.

(*b*) T is a constant scalar.

(*c*) If $\boldsymbol{\omega} = \overrightarrow{OP}$, the locus of P in space is the *invariable plane* $\boldsymbol{\omega} \cdot \mathbf{H} = 2T$; and the locus of P in the body is the *energy ellipsoid* $\boldsymbol{\omega} \cdot \mathbf{K} \cdot \boldsymbol{\omega} = 2T$.

(*d*) The energy ellipsoid is always tangent to the invariable plane at P.

(*e*) The body moves so that its energy ellipsoid rolls without slipping on the invariable plane (*Poinsot's Theorem*). [See Brand's *Vectorial Mechanics*, § 219.]

22. The vectors of a dyadic Φ are fixed in a rigid body having the instantaneous angular velocity $\boldsymbol{\omega}$, relative to a "fixed" frame. Show that, relative to this frame,

$$d\Phi/dt = \boldsymbol{\omega} \times \Phi - \Phi \times \boldsymbol{\omega}.$$

23. A rigid body revolving about its fixed point O has the angular velocity $\boldsymbol{\omega} = [\omega_1, \omega_2, \omega_3]$ referred to fixed rectangular axes through O. If I_1, I_2, I_3 are the moments of inertia about the fixed axes x, y, z and I_{23}, I_{31}, I_{12} are the products of inertia for yz, zx, xy, show that

$$dI_1/dt = 2(I_{12}\omega_3 - I_{13}\omega_2), \cdots,$$

$$dI_{23}/dt = (I_3 - I_2)\omega_1 + I_{13}\omega_3 - I_{12}\omega_2, \cdots,$$

[If $\mathbf{K}$ is the inertia dyadic of the body relative to O, $I_1 = \mathbf{i} \cdot \mathbf{K} \cdot \mathbf{i}$, $I_{23} = -\mathbf{j} \cdot \mathbf{K} \cdot \mathbf{k}$, etc. See § 72, ex. 2.]

24. The axis of a homogeneous solid of revolution has the direction of the unit vector $\mathbf{e}$. If its principal moments of inertia at the mass center G are A, A, C, show that the inertia dyadic at G is

$$\mathbf{K}_G = A\mathbf{I} + (C - A)\mathbf{ee}.$$

Hence find the moments and products of inertia with respect to fixed rectangular axes x, y, z through G if $\mathbf{e} = [l, m, n]$.

25. If G is the mass center of a rigid body of mass m and $\mathbf{r}^* = \overrightarrow{OG}$, show that the inertia dyadic at O is

$$\mathbf{K}_O = m(\mathbf{r}^* \cdot \mathbf{r}^* \mathbf{I} - \mathbf{r}^*\mathbf{r}^*) + \mathbf{K}_G.$$

Hence compare moments and products of inertia for parallel axes at O and G.

26. If λ is arbitrary parameter $< a^2$ but $\neq b^2$ or c^2, the dyadic,

$$\Phi_\lambda = \frac{\mathbf{ii}}{a^2 - \lambda} + \frac{\mathbf{jj}}{b^2 - \lambda} + \frac{\mathbf{kk}}{c^2 - \lambda} \quad (a > b > c),$$

defines a one-parameter family of *confocal* quadric surfaces $\mathbf{r} \cdot \Phi_\lambda \cdot \mathbf{r} = 1$. These are ellipsoids if $\lambda < c^2$, hyperboloids of one sheet if $c^2 < \lambda < b^2$, hyperbolas of two sheets if $b^2 < \lambda < a^2$.

Prove that

(*a*) The central quadrics $\mathbf{r} \cdot \Psi \cdot \mathbf{r} = 1$ and $\mathbf{r} \cdot \Theta \cdot \mathbf{r} = 1$ are confocal when and only when $\Psi^{-1} - \Theta^{-1} = k\mathbf{I}$.

(*b*) Two confocal central quadrics of different species intersect at right angles. [$\Theta - \Psi = k\Theta \cdot \Psi$.]

CHAPTER V

DIFFERENTIAL INVARIANTS

82. Gradient of a Scalar. The points P of a certain region may be specified by giving their position vectors $\mathbf{r} = \overrightarrow{OP}$; and we shall on occasion refer to P as the "point $\mathbf{r}$." A scalar, vector, or dyadic which is uniquely defined at every point P of a certain region is called a *point function* in this region and will be denoted by $f(\mathbf{r})$. For example, the temperature and velocity of a fluid at the points of a three-dimensional region are scalar and vector point functions respectively.

A scalar point function $f(\mathbf{r})$ is said to be *continuous* at a point P_1 if to each positive number ϵ, arbitrarily small, there corresponds a positive number δ such that

$$|f(\mathbf{r}) - f(\mathbf{r}_1)| < \epsilon \quad \text{when} \quad |\mathbf{r} - \mathbf{r}_1| < \delta;$$

then, as $\mathbf{r}$ approaches $\mathbf{r}_1$ in any manner, $\lim f(\mathbf{r}) = f(\mathbf{r}_1)$.

From a point P_1 draw a ray in the direction of the unit vector,

$$\mathbf{e} = \mathbf{i} \cos \alpha + \mathbf{j} \cos \beta + \mathbf{k} \cos \gamma.$$

Along this ray $\mathbf{r} = \mathbf{r}_1 + s\mathbf{e}$, where s denotes the distance P_1P, and $f(r)$ is a function of s. We now define

$$\frac{df}{ds} = \lim_{s \to 0} \frac{f(\mathbf{r}_1 + s\mathbf{e}) - f(\mathbf{r}_1)}{s} \tag{1}$$

as the *directional derivative* of $f(\mathbf{r})$ at P_1 in the direction $\mathbf{e}$. If this limit exists on all rays issuing from P_1, $f(\mathbf{r})$ is said to be *differentiable* at P_1.

The rectangular coordinates of any point P on the ray $\mathbf{r} = \mathbf{r}_1 + s\mathbf{e}$ $(s > 0)$ are

$$x = x_1 + s \cos \alpha, \qquad y = y_1 + s \cos \beta, \qquad z = z_1 + s \cos \gamma.$$

If $f(\mathbf{r})$ is given as a function $f(x, y, z)$,

$$\frac{df}{ds} = \frac{\partial f}{\partial x}\frac{dx}{ds} + \frac{\partial f}{\partial y}\frac{dy}{ds} + \frac{\partial f}{\partial z}\frac{dz}{ds} \tag{2}$$

$$= \frac{\partial f}{\partial x}\cos\alpha + \frac{\partial f}{\partial y}\cos\beta + \frac{\partial f}{\partial z}\cos\gamma;$$

or, since

$$\mathbf{e}\cdot\mathbf{i} = \cos\alpha, \qquad \mathbf{e}\cdot\mathbf{j} = \cos\beta, \qquad \mathbf{e}\cdot\mathbf{k} = \cos\gamma,$$

$$\frac{df}{ds} = \mathbf{e}\cdot\left(\mathbf{i}\frac{\partial f}{\partial x} + \mathbf{j}\frac{\partial f}{\partial y} + \mathbf{k}\frac{\partial f}{\partial z}\right). \tag{3}$$

The vector in parenthesis is called the *gradient* of $f(\mathbf{r})$ and is written grad f or

$$\nabla f = \mathbf{i}\frac{\partial f}{\partial x} + \mathbf{j}\frac{\partial f}{\partial y} + \mathbf{k}\frac{\partial f}{\partial z}. \tag{4}$$

∇f is a vector point function; thus, when (3) is written as

$$\frac{df}{ds} = \mathbf{e}\cdot\nabla f, \tag{5}$$

the direction enters only through the factor $\mathbf{e}$. *The directional derivative of a scalar function at a point P is the component of its gradient at P in the given direction.* The gradient ∇f at P effects a synthesis of all the directional derivatives of f at P. In effect, the *vector* ∇f replaces the infinity of *scalars* df/ds.

When P varies along a *curve* tangent to $\mathbf{e}$ at P_1, $f(\mathbf{r})$ is a function of the arc $s = P_1P$ along the curve, and df/ds is still given by (2); for, at P_1, $d\mathbf{r}/ds = \mathbf{e}$ (44.1), and hence

$$dx/ds = \cos\alpha, \qquad dy/ds = \cos\beta, \qquad dz/ds = \cos\gamma.$$

Since df/ds, as defined by (1), does not depend upon any specific choice of coordinates, (5) shows that the gradient ∇f has the same property. In fact, we determine ∇f by giving the directional derivatives df/ds_i in three non-coplanar directions $\mathbf{e}_i$. For, since $\mathbf{e}_i\cdot\nabla f = df/ds_i$ are the covariant components of ∇f, we have, from (24.2),

$$\nabla f = \mathbf{e}^1\frac{df}{ds_1} + \mathbf{e}^2\frac{df}{ds_2} + \mathbf{e}^3\frac{df}{ds_3}. \tag{6}$$

We proceed to specify the length and direction of ∇f independently of the coordinate system. The points for which f has a constant value lie on a *level surface* of f. In any direction $\mathbf{e}$ tangent to the level surface at P, $df/ds = \mathbf{e} \cdot \nabla f = 0$; hence ∇f is normal to the level surface at P. If $\mathbf{n}$ is a unit vector normal to the level surface and directed towards increasing values of f, $\mathbf{n} \cdot \nabla f > 0$. Hence ∇f has the direction of $\mathbf{n}$, and its magnitude is the value of df/ds in this direction. Writing this *normal derivative* df/dn, we have

$$\nabla f = \frac{df}{dn}\mathbf{n}. \tag{7}$$

For example, if $f = r$, the distance OP from the origin, the level surfaces are spheres about O as center, and $\mathbf{n} = \mathbf{R}$, the unit radial vector; the normal derivative $dr/dn = 1$, and

$$\nabla r = \mathbf{R}. \tag{8}$$

When f is a function $f(x, y, z)$ of rectangular coordinates, ∇f is given by (4). In particular,

$$\nabla x = \mathbf{i}, \qquad \nabla y = \mathbf{j}, \qquad \nabla z = \mathbf{k}. \tag{9}$$

If f is a function $f(u, v, w)$ of variables which themselves are functions of x, y, z, we have

$$\frac{df}{ds} = \frac{\partial f}{\partial u}\frac{du}{ds} + \frac{\partial f}{\partial v}\frac{dv}{ds} + \frac{\partial f}{\partial w}\frac{dw}{ds}$$

or, in view of (5),

$$\mathbf{e} \cdot \nabla f = \mathbf{e} \cdot \left(\nabla u \frac{\partial f}{\partial u} + \nabla v \frac{\partial f}{\partial v} + \nabla w \frac{\partial f}{\partial w}\right) \cdot$$

As this equation holds for every $\mathbf{e}$,

$$\nabla f = \nabla u \frac{\partial f}{\partial u} + \nabla v \frac{\partial f}{\partial v} + \nabla w \frac{\partial f}{\partial w} \cdot \tag{10}$$

When $f = f(u, v)$ or $f = f(u)$,

$$\nabla f = \nabla u \frac{\partial f}{\partial u} + \nabla v \frac{\partial f}{\partial v}, \qquad \nabla f = \nabla u \frac{df}{du},$$

respectively; for example,

$$\nabla(uv) = v\,\nabla u + u\,\nabla v, \qquad \nabla u^n = nu^{n-1}\,\nabla u, \qquad \nabla \log u = \frac{1}{u}\,\nabla u.$$

When f is constant, $\nabla f = 0$; conversely, $\nabla f = 0$ implies $\partial f/\partial x = \partial f/\partial y = \partial f/\partial z = 0$, and f is constant.

If $f(r)$ is *differentiable* in a region R, ∇f is defined at all points of R. If moreover ∇f is continuous in R, we say that $f(\mathbf{r})$ is *continuously differentiable* in R.

Example. Gradients in a Plane. The gradient of a point function $f(x, y)$ in the xy-plane is

$$\nabla f = \frac{\partial f}{\partial x}\mathbf{i} + \frac{\partial f}{\partial y}\mathbf{j}.$$

At any point P, ∇f is normal to the level curve $f(x, y) = c$ through P.

For a function $f(r, \theta)$ of plane polar coordinates,

$$\nabla f = \frac{\partial f}{\partial r}\mathbf{R} + \frac{\partial f}{\partial \theta}\frac{\mathbf{P}}{r},$$

in the notation of § 44; for $\nabla r = \mathbf{R}$ and $\nabla\theta = \mathbf{P}/r$ (from (44.5) $d\theta/ds = 1/r$ in the direction perpendicular to $\mathbf{R}$).

For a function $f(r_1, r_2)$ of bipolar coordinates,

$$\nabla f = \frac{\partial f}{\partial r_1}\mathbf{R}_1 + \frac{\partial f}{\partial r_2}\mathbf{R}_2,$$

where $\mathbf{R}_1$, $\mathbf{R}_2$ are unit radial vectors from O_1, O_2 directed to the point in question.

The families of curves $u = c$, $v = c$ cut at right angles when $\nabla u \cdot \nabla v = 0$. For example, the ellipses $r_1 + r_2 = c$ and hyperbolas $r_1 - r_2 = c$ cut orthogonally since $(\mathbf{R}_1 + \mathbf{R}_2) \cdot (\mathbf{R}_1 - \mathbf{R}_2) = 0$.

83. Gradient of a Vector. Let $\mathbf{f}(\mathbf{r})$ denote a vector point function in a certain region. It is said to be *continuous* at a point P_1 if to each $\epsilon > 0$ there corresponds a $\delta > 0$ such that

$$|\mathbf{f}(\mathbf{r}) - \mathbf{f}(\mathbf{r}_1)| < \epsilon \quad \text{when} \quad |\mathbf{r} - \mathbf{r}_1| < \delta.$$

If $\mathbf{f}(\mathbf{r})$ is given by its rectangular components,

$$\mathbf{f}(\mathbf{r}) = f_1\mathbf{i} + f_2\mathbf{j} + f_3\mathbf{k},$$

$\mathbf{f}(\mathbf{r})$ is continuous at P_1 when the three scalar point functions f_1, f_2, f_3 are continuous there.

Let P_1P be a ray drawn from P_1 in the direction of the unit vector,

$$\mathbf{e} = \mathbf{i} \cos \alpha + \mathbf{j} \cos \beta + \mathbf{k} \cos \gamma.$$

Along the ray $\mathbf{r} = \mathbf{r}_1 + s\mathbf{e}$ $(s > 0)$ and $\mathbf{f}(\mathbf{r})$ is a function of s. We now define

$$\frac{d\mathbf{f}}{ds} = \lim_{s \to 0} \frac{\mathbf{f}(\mathbf{r}_1 + s\mathbf{e}) - \mathbf{f}(\mathbf{r}_1)}{s} \tag{1}$$

as the *directional derivative* of $\mathbf{f}(\mathbf{r})$ at P_1 in the direction $\mathbf{e}$. If this limit exists on all rays issuing from P_1, $\mathbf{f}(\mathbf{r})$ is said to be *differentiable* at P_1. Evidently $\mathbf{f}(\mathbf{r})$ is differentiable if its rectangular components f_i are differentiable.

If $\mathbf{f}(\mathbf{r})$ is given as a function $\mathbf{f}(x, y, z)$ of rectangular coordinates, we have. just as in § 82,

$$\frac{d\mathbf{f}}{ds} = \frac{\partial \mathbf{f}}{\partial x}\frac{dx}{ds} + \frac{\partial \mathbf{f}}{\partial y}\frac{dy}{ds} + \frac{\partial \mathbf{f}}{\partial z}\frac{dz}{ds} \tag{2}$$

$$= \frac{\partial \mathbf{f}}{\partial x} \cos \alpha + \frac{\partial \mathbf{f}}{\partial y} \cos \beta + \frac{\partial \mathbf{f}}{\partial z} \cos \gamma.$$

On replacing the cosines by $\mathbf{e} \cdot \mathbf{i}$, $\mathbf{e} \cdot \mathbf{j}$, $\mathbf{e} \cdot \mathbf{k}$,

$$\frac{d\mathbf{f}}{ds} = \mathbf{e} \cdot \left(\mathbf{i}\frac{\partial \mathbf{f}}{\partial x} + \mathbf{j}\frac{\partial \mathbf{f}}{\partial y} + \mathbf{k}\frac{\partial \mathbf{f}}{\partial z}\right), \tag{3}$$

a formula entirely analogous to (82.3). The dyadic in parenthesis is called the *gradient* of $\mathbf{f}(\mathbf{r})$ and is written grad $\mathbf{f}$ or

$$\nabla \mathbf{f} = \mathbf{i}\frac{\partial \mathbf{f}}{\partial x} + \mathbf{j}\frac{\partial \mathbf{f}}{\partial y} + \mathbf{k}\frac{\partial \mathbf{f}}{\partial z}. \tag{4}$$

$\nabla \mathbf{f}$ is a dyadic point function; thus, when (3) is written

$$\frac{d\mathbf{f}}{ds} = \mathbf{e} \cdot \nabla \mathbf{f}, \tag{5}$$

the direction enters only through the prefactor $\mathbf{e}$. The gradient $\nabla \mathbf{f}$ at P effects a synthesis of all the directional derivatives of $\mathbf{f}$ at P. In effect, the *dyadic* $\nabla \mathbf{f}$ replaces the infinity of *vectors* $d\mathbf{f}/ds$.

Since $d\mathbf{f}/ds$, as defined by (1), does not depend upon any specific choice of coordinates, (5) shows that $\nabla \mathbf{f}$ has the same property.

In fact $\nabla \mathbf{f}$ is determined by giving the directional derivatives $d\mathbf{f}/ds_i$ in three non-coplanar directions $\mathbf{e}_i$:

$$\nabla \mathbf{f} = \mathbf{e}^1 \frac{d\mathbf{f}}{ds_1} + \mathbf{e}^2 \frac{d\mathbf{f}}{ds_2} + \mathbf{e}^3 \frac{d\mathbf{f}}{ds_3}. \tag{6}$$

For, if we put $d\mathbf{f}/ds_i = \mathbf{e}_i \cdot \nabla \mathbf{f}$, the right member becomes

$$(\mathbf{e}^1\mathbf{e}_1 + \mathbf{e}^2\mathbf{e}_2 + \mathbf{e}^3\mathbf{e}_3) \cdot \nabla \mathbf{f} = \mathbf{I} \cdot \nabla \mathbf{f} = \nabla \mathbf{f}.$$

When $\mathbf{f}$ is a function of rectangular coordinates $\mathbf{f}(x, y, z)$, $\nabla \mathbf{f}$ is given by (4). More generally, for $\mathbf{f}(u, v, w)$ we have

$$\frac{d\mathbf{f}}{ds} = \frac{\partial \mathbf{f}}{\partial u}\frac{du}{ds} + \frac{\partial \mathbf{f}}{\partial v}\frac{dv}{ds} + \frac{\partial \mathbf{f}}{\partial w}\frac{dw}{ds},$$

$$\mathbf{e} \cdot \nabla \mathbf{f} = \mathbf{e} \cdot \left(\nabla u \frac{\partial \mathbf{f}}{\partial u} + \nabla v \frac{\partial \mathbf{f}}{\partial v} + \nabla w \frac{\partial \mathbf{f}}{\partial w} \right),$$

for any $\mathbf{e}$; and, from the definition of dyadic equality,

$$\nabla \mathbf{f} = \nabla u \frac{\partial \mathbf{f}}{\partial u} + \nabla v \frac{\partial \mathbf{f}}{\partial v} + \nabla w \frac{\partial \mathbf{f}}{\partial w}. \tag{7}$$

84. Divergence and Rotation. The first scalar invariant of the gradient $\nabla \mathbf{f}$ of a vector $\mathbf{f}$ is called the *divergence* of $\mathbf{f}$ and is written $\nabla \cdot \mathbf{f}$ or div $\mathbf{f}$.

The vector invariant of $\nabla \mathbf{f}$ is called the *rotation* or *curl* of $\mathbf{f}$ and is written $\nabla \times \mathbf{f}$, rot $\mathbf{f}$, or curl $\mathbf{f}$.

In terms of rectangular coordinates, we therefore have the defining equations:

$$\nabla \mathbf{f} = \operatorname{grad} \mathbf{f} = \mathbf{i} \frac{\partial \mathbf{f}}{\partial x} + \mathbf{j} \frac{\partial \mathbf{f}}{\partial y} + \mathbf{k} \frac{\partial \mathbf{f}}{\partial z}, \tag{1}$$

$$\nabla \cdot \mathbf{f} = \operatorname{div} \mathbf{f} = \mathbf{i} \cdot \frac{\partial \mathbf{f}}{\partial x} + \mathbf{j} \cdot \frac{\partial \mathbf{f}}{\partial y} + \mathbf{k} \cdot \frac{\partial \mathbf{f}}{\partial z}, \tag{2}$$

$$\nabla \times \mathbf{f} = \operatorname{rot} \mathbf{f} = \mathbf{i} \times \frac{\partial \mathbf{f}}{\partial x} + \mathbf{j} \times \frac{\partial \mathbf{f}}{\partial y} + \mathbf{k} \times \frac{\partial \mathbf{f}}{\partial z}. \tag{3}$$

If $\mathbf{f}$ is resolved into rectangular components,

$$\mathbf{f} = \mathbf{i}f_1 + \mathbf{j}f_2 + \mathbf{k}f_3,$$

we obtain from (1) its gradient $\nabla \mathbf{f}$ in nonion form with the matrix:

(4) $$\nabla \mathbf{f} = \begin{pmatrix} \partial f_1/\partial x & \partial f_2/\partial x & \partial f_3/\partial x \\ \partial f_1/\partial y & \partial f_2/\partial y & \partial f_3/\partial y \\ \partial f_1/\partial z & \partial f_2/\partial z & \partial f_3/\partial z \end{pmatrix}.$$

The first scalar and vector invariants of ∇f now become

(5) $$\nabla \cdot \mathbf{f} = \frac{\partial f_1}{\partial x} + \frac{\partial f_2}{\partial y} + \frac{\partial f_3}{\partial z},$$

(6) $$\nabla \times \mathbf{f} = \mathbf{i}\left(\frac{\partial f_3}{\partial y} - \frac{\partial f_2}{\partial z}\right) + \mathbf{j}\left(\frac{\partial f_1}{\partial z} - \frac{\partial f_3}{\partial x}\right) + \mathbf{k}\left(\frac{\partial f_2}{\partial x} - \frac{\partial f_1}{\partial y}\right).$$

The last expression is easily remembered when written in determinant form:

(7) $$\nabla \times \mathbf{f} = \begin{vmatrix} \mathbf{i} & \mathbf{j} & \mathbf{k} \\ \dfrac{\partial}{\partial x} & \dfrac{\partial}{\partial y} & \dfrac{\partial}{\partial z} \\ f_1 & f_2 & f_3 \end{vmatrix}.$$

These expressions for divergence and rotation may be written down at once, if we regard $\nabla \cdot \mathbf{f}$ and $\nabla \times \mathbf{f}$ as products of the vector operator *del* (or *nabla*),

(8) $$\nabla = \mathbf{i}\frac{\partial}{\partial x} + \mathbf{j}\frac{\partial}{\partial y} + \mathbf{k}\frac{\partial}{\partial z},$$

with the vector $\mathbf{f} = \mathbf{i}f_1 + \mathbf{j}f_2 + \mathbf{k}f_3$.

For the position vector $\mathbf{r} = x\mathbf{i} + y\mathbf{j} + z\mathbf{k}$, we have

$$\nabla \mathbf{r} = \mathbf{i}\frac{\partial \mathbf{r}}{\partial x} + \mathbf{j}\frac{\partial \mathbf{r}}{\partial y} + \mathbf{k}\frac{\partial \mathbf{r}}{\partial z} = \mathbf{ii} + \mathbf{jj} + \mathbf{kk},$$

the idemfactor; hence

(9) $$\nabla \mathbf{r} = \mathbf{I}, \qquad \operatorname{div} \mathbf{r} = 3, \qquad \operatorname{rot} \mathbf{r} = 0.$$

When $\mathbf{f} = \nabla \varphi$, the gradient of a scalar,

$$f_1 = \frac{\partial \varphi}{\partial x}, \qquad f_2 = \frac{\partial \varphi}{\partial y}, \qquad f_3 = \frac{\partial \varphi}{\partial z}.$$

Using these components, (5) and (6) gives

$$(10)\qquad \nabla \cdot \nabla\varphi = \frac{\partial^2\varphi}{\partial x^2} + \frac{\partial^2\varphi}{\partial y^2} + \frac{\partial^2\varphi}{\partial z^2},$$

$$(11)\qquad \nabla \times \nabla\varphi = 0.$$

The differential operator $\nabla \cdot \nabla$ or div grad is called the Laplacian and often is written

$$(12)\qquad \nabla^2 = \frac{\partial^2}{\partial x^2} + \frac{\partial^2}{\partial y^2} + \frac{\partial^2}{\partial z^2}.$$

When $\mathbf{f} = \nabla \times \mathbf{g}$, the rotation of a vector $\mathbf{g}$,

$$f_1 = \frac{\partial g_3}{\partial y} - \frac{\partial g_2}{\partial z}, \qquad f_2 = \frac{\partial g_1}{\partial z} - \frac{\partial g_3}{\partial x}, \qquad f_3 = \frac{\partial g_2}{\partial x} - \frac{\partial g_1}{\partial y}.$$

In this case (5) and (6) give

$$(13)\qquad \nabla \cdot (\nabla \times \mathbf{g}) = 0,$$

$$(14)\qquad \nabla \times (\nabla \times \mathbf{g}) = \nabla(\nabla \cdot \mathbf{g}) - \nabla^2\mathbf{g}.$$

Proof of (14): From (6),

$$\nabla \times (\nabla \times \mathbf{g}) = \mathbf{i}\left\{\frac{\partial^2 g_2}{\partial y\,\partial x} - \frac{\partial^2 g_1}{\partial y^2} - \frac{\partial^2 g_1}{\partial z^2} + \frac{\partial^2 g_3}{\partial z\,\partial x}\right\} + \cdots$$

$$= \mathbf{i}\left\{\frac{\partial}{\partial x}\left(\frac{\partial g_1}{\partial x} + \frac{\partial g_2}{\partial y} + \frac{\partial g_3}{\partial z}\right) - \nabla^2 g_1\right\} + \cdots$$

$$= \mathbf{i}\left\{\frac{\partial}{\partial x}(\nabla \cdot \mathbf{g}) - \nabla^2 g_1\right\} + \cdots$$

$$= \nabla(\nabla \cdot \mathbf{g}) - \nabla^2\mathbf{g}.$$

Note that, if we regard ∇ as an actual vector and expand $\nabla \times (\nabla \times \mathbf{g})$, but keep the ∇'s to the left of $\mathbf{g}$, we obtain $\nabla(\nabla \cdot \mathbf{g}) - (\nabla \cdot \nabla)\mathbf{g}$, the correct result.

The proofs of (11), (13), and (14) depend upon changing the order of differentiations in mixed second derivatives; this is always valid when the derivatives in question are continuous.

The foregoing differential relations of the second order also may be written

$$\text{(10)} \qquad \operatorname{div}\operatorname{grad}\varphi = \nabla^2\varphi,$$

$$\text{(11)} \qquad \operatorname{rot}\operatorname{grad}\varphi = 0,$$

$$\text{(13)} \qquad \operatorname{div}\operatorname{rot}\mathbf{g} = 0,$$

$$\text{(14)} \qquad \operatorname{rot}\operatorname{rot}\mathbf{g} = \operatorname{grad}\operatorname{div}\mathbf{g} - \nabla^2\mathbf{g}.$$

Example. The velocity distribution in a rigid body is given by (54.4):

$$\mathbf{v}_P = \mathbf{v}_O + \boldsymbol{\omega}\times\overrightarrow{OP} = \mathbf{v}_O + \boldsymbol{\omega}\times\mathbf{r}.$$

Hence, from (3),

$$\operatorname{rot}\mathbf{v}_P = \operatorname{rot}(\boldsymbol{\omega}\times\mathbf{r}) = \mathbf{i}\times(\boldsymbol{\omega}\times\mathbf{i}) + \cdots = 3\boldsymbol{\omega} - \boldsymbol{\omega} = 2\boldsymbol{\omega};$$

the rotation of the velocity of the particles of a rigid body at any instant is equal to twice its instantaneous angular velocity.

85. Differentiation Formulas. Starting from the defining equations (84.1), (84.2), (84.3), we now can deduce some useful identities.

If λ is a scalar point function:

$$\text{(1)} \qquad \nabla(\lambda\mathbf{f}) = (\nabla\lambda)\mathbf{f} + \lambda\nabla\mathbf{f},$$

$$\text{(2)} \qquad \operatorname{div}(\lambda\mathbf{f}) = (\nabla\lambda)\cdot\mathbf{f} + \lambda\operatorname{div}\mathbf{f},$$

$$\text{(3)} \qquad \operatorname{rot}(\lambda\mathbf{f}) = (\nabla\lambda)\times\mathbf{f} + \lambda\operatorname{rot}\mathbf{f}.$$

For the vector point function $\mathbf{f}\times\mathbf{g}$:

$$\text{(4)} \qquad \nabla(\mathbf{f}\times\mathbf{g}) = (\nabla\mathbf{f})\times\mathbf{g} - (\nabla\mathbf{g})\times\mathbf{f},$$

$$\text{(5)} \qquad \operatorname{div}(\mathbf{f}\times\mathbf{g}) = \mathbf{g}\cdot\operatorname{rot}\mathbf{f} - \mathbf{f}\cdot\operatorname{rot}\mathbf{g},$$

$$\text{(6)} \qquad \operatorname{rot}(\mathbf{f}\times\mathbf{g}) = \mathbf{g}\cdot\nabla\mathbf{f} - \mathbf{f}\cdot\nabla\mathbf{g} + \mathbf{f}\operatorname{div}\mathbf{g} - \mathbf{g}\operatorname{div}\mathbf{f}.$$

We give the proof of (6):

$$\begin{aligned}\operatorname{rot}(\mathbf{f}\times\mathbf{g}) &= \mathbf{i}\times\left(\frac{\partial\mathbf{f}}{\partial x}\times\mathbf{g} + \mathbf{f}\times\frac{\partial\mathbf{g}}{\partial x}\right) + \cdots \\ &= \mathbf{g}\cdot\mathbf{i}\frac{\partial\mathbf{f}}{\partial x} - \left(\mathbf{i}\cdot\frac{\partial\mathbf{f}}{\partial x}\right)\mathbf{g} + \left(\mathbf{i}\cdot\frac{\partial\mathbf{g}}{\partial x}\right)\mathbf{f} - \mathbf{f}\cdot\mathbf{i}\frac{\partial\mathbf{g}}{\partial x} + \cdots \\ &= \mathbf{g}\cdot\nabla\mathbf{f} - (\operatorname{div}\mathbf{f})\mathbf{g} + (\operatorname{div}\mathbf{g})\mathbf{f} - \mathbf{f}\cdot\nabla\mathbf{g}.\end{aligned}$$

For the scalar point function $\mathbf{f} \cdot \mathbf{g}$:

$$(7) \qquad \nabla(\mathbf{f} \cdot \mathbf{g}) = (\nabla\mathbf{f}) \cdot \mathbf{g} + (\nabla\mathbf{g}) \cdot \mathbf{f}$$

From (68.6), $\Phi_c = \Phi + \mathbf{I} \times \boldsymbol{\phi}$; hence

$$(8) \qquad (\nabla\mathbf{f})_c = \nabla\mathbf{f} + \mathbf{I} \times \operatorname{rot} \mathbf{f},$$

$$(9) \qquad (\nabla\mathbf{f}) \cdot \mathbf{a} = \mathbf{a} \cdot \nabla\mathbf{f} + \mathbf{a} \times \operatorname{rot} \mathbf{f}.$$

Making use of (9), we also can write (7) as

$$(10) \qquad \nabla(\mathbf{f} \cdot \mathbf{g}) = \mathbf{g} \cdot \nabla\mathbf{f} + \mathbf{f} \cdot \nabla\mathbf{g} + \mathbf{g} \times \operatorname{rot} \mathbf{f} + \mathbf{f} \times \operatorname{rot} \mathbf{g}.$$

If Φ is a constant dyadic,

$$(11) \qquad \nabla(\mathbf{r} \cdot \Phi) = (\nabla\mathbf{r}) \cdot \Phi = \mathbf{I} \cdot \Phi = \Phi;$$

$$(12) \qquad \operatorname{div} (\mathbf{r} \cdot \Phi) = \varphi_1,$$

$$(13) \qquad \operatorname{rot} (\mathbf{r} \cdot \Phi) = \boldsymbol{\phi}.$$

Finally, if Φ is a *constant symmetric* dyadic,

$$(14) \qquad \nabla(\mathbf{r} \cdot \Phi \cdot \mathbf{r}) = 2\Phi \cdot \mathbf{r}.$$

From this result we can prove the

THEOREM. *A symmetric dyadic Φ transforms any vector $\mathbf{r} = \overrightarrow{OP}$ into a vector $\mathbf{r}' = \Phi \cdot \mathbf{r}$ normal to the real central quadric surface $\mathbf{r} \cdot \Phi \cdot \mathbf{r} = \pm 1$ * at the point P; and $r' = 1/p$ where p is the distance from O to the tangent plane to the quadric at P.*

Proof. From (72.4), we see that

$$\mathbf{r} \cdot \Phi \cdot \mathbf{r} = \alpha x^2 + \beta y^2 + \gamma z^2 = \pm 1$$

represents a central quadric (an ellipsoid if α, β, γ have the same sign). From (14), $\mathbf{r}'$ has the direction of $\mathbf{n}$, a unit normal to the quadric at P; hence

$$r'\mathbf{n} = \Phi \cdot \mathbf{r}, \qquad r'\mathbf{n} \cdot \mathbf{r} = \mathbf{r} \cdot \Phi \cdot \mathbf{r} = \pm 1,$$

and $r' = 1/|\mathbf{n} \cdot \mathbf{r}| = 1/p$.

86. Gradient of a Tensor. If $\mathbf{f}(\mathbf{r})$ is a tensor point function of valence ν (cf. § 80), the derivative of $\mathbf{f}(\mathbf{r})$ at P_1 in the direction of the unit vector $\mathbf{e}$ is defined by

$$(1) \qquad \frac{d\mathbf{f}}{ds} = \lim_{s \to 0} \frac{\mathbf{f}(\mathbf{r}_1 + s\mathbf{e}) - \mathbf{f}(\mathbf{r}_1)}{s}.$$

* The sign is chosen so that the quadric is real.

If this limit exists for all rays drawn from P_1, $\mathbf{f}(\mathbf{r})$ is said to be *differentiable* at P_1. When referred to a constant basis, $\mathbf{f}(\mathbf{r})$ is differentiable when all its scalar components are differentiable.

If $\mathbf{f}(\mathbf{r})$ is given as a function $\mathbf{f}(x, y, z)$ of rectangular coordinates,

$$\frac{d\mathbf{f}}{ds} = \frac{\partial \mathbf{f}}{\partial x}\frac{dx}{ds} + \frac{\partial \mathbf{f}}{\partial y}\frac{dy}{ds} + \frac{\partial \mathbf{f}}{\partial z}\frac{dz}{ds}.$$

Since $dx/ds = \mathbf{e} \cdot \mathbf{i}$, etc., we have, just as in § 83,

$$\text{(2)} \qquad \frac{d\mathbf{f}}{ds} = \mathbf{e} \cdot \nabla \mathbf{f},$$

$$\text{(3)} \qquad \nabla \mathbf{f} = \mathbf{i}\frac{\partial \mathbf{f}}{\partial x} + \mathbf{j}\frac{\partial \mathbf{f}}{\partial y} + \mathbf{k}\frac{\partial \mathbf{f}}{\partial z} = \mathbf{i}_r \frac{\partial \mathbf{f}}{\partial x_r}.\dagger$$

Here $\nabla \mathbf{f}$, the *gradient* of $\mathbf{f}$ (grad $\mathbf{f}$) is tensor point function of valence $\nu + 1$; $\nabla \mathbf{f}$ effects a synthesis of all the directional derivatives (valence ν) of the tensor $\mathbf{f}$. Thus, if $\mathbf{f}$ is dyadic, the triadic $\nabla \mathbf{f}$ replaces the infinity of dyadics $d\mathbf{f}/ds$.

$\nabla \mathbf{f}$ *is independent of the choice of coordinates.* At any point, $\nabla \mathbf{f}$ is completely determined when $d\mathbf{f}/ds_i$ is given for three non-coplanar directions $\mathbf{e}_i$:

$$\text{(4)} \qquad \nabla \mathbf{f} = \mathbf{e}^1 \frac{d\mathbf{f}}{ds_1} + \mathbf{e}^2 \frac{d\mathbf{f}}{ds_2} + \mathbf{e}^3 \frac{d\mathbf{f}}{ds_3};$$

for, if we put $d\mathbf{f}/ds_i = \mathbf{e}_i \cdot \nabla \mathbf{f}$, the right member becomes $\mathbf{I} \cdot \nabla \mathbf{f} = \nabla \mathbf{f}$.

When $\mathbf{f}(\mathbf{r})$ is a vector, we obtain from the dyadic $\nabla \mathbf{f}$ the invariants div $\mathbf{f}$ and rot $\mathbf{f}$ by putting dots and crosses, respectively, between the vectors of each term. Similarly, if $\mathbf{f}(\mathbf{r})$ is any tensor of valence $\nu > 0$, we obtain from $\nabla \mathbf{f}$ the further invariants,

$$\text{(5)} \qquad \nabla \cdot \mathbf{f} = \mathbf{i} \cdot \frac{\partial \mathbf{f}}{\partial x} + \mathbf{j} \cdot \frac{\partial \mathbf{f}}{\partial y} + \mathbf{k} \cdot \frac{\partial \mathbf{f}}{\partial z} = \mathbf{i}_r \cdot \frac{\partial \mathbf{f}}{\partial x_r},$$

$$\text{(6)} \qquad \nabla \times \mathbf{f} = \mathbf{i} \times \frac{\partial \mathbf{f}}{\partial x} + \mathbf{j} \times \frac{\partial \mathbf{f}}{\partial y} + \mathbf{k} \times \frac{\partial \mathbf{f}}{\partial z} = \mathbf{i}_r \times \frac{\partial \mathbf{f}}{\partial x_r},$$

of valence $\nu - 1$ and ν, respectively. In forming these invariants from $\nabla \mathbf{f}$, the dots and crosses are inserted between the first and second vectors to the left—*we dot and cross in the first position.*

† Here $\mathbf{i}, \mathbf{j}, \mathbf{k}$; x, y, z are written $\mathbf{i}_1, \mathbf{i}_2, \mathbf{i}_3$; x_1, x_2, x_3, and we employ the *summation convention*: a repeated index (as r) denotes summation over the index range 1, 2, 3. See § 145.

For example consider the dyadic

$$\mathbf{f}(\mathbf{r}) = f_{11}\mathbf{ii} + f_{12}\mathbf{ij} + \cdots = f_{st}\mathbf{i}_s\mathbf{i}_t$$

where s and t are summation indices. Then if D_r denotes $\partial/\partial x_r$,

$$\nabla\mathbf{f} = \mathbf{i}_r\mathrm{D}_r\mathbf{f} = \mathrm{D}_r f_{st}\mathbf{i}_r\mathbf{i}_s\mathbf{i}_t, \quad \text{a triadic;}$$

$$\nabla \cdot \mathbf{f} = \mathbf{i}_r \cdot \mathrm{D}_r\mathbf{f} = \mathrm{D}_r f_{st}\, \mathbf{i}_r \cdot \mathbf{i}_s\, \mathbf{i}_t$$

$$= (\mathrm{D}_1 f_{1t} + \mathrm{D}_2 f_{2t} + \mathrm{D}_3 f_{3t})\mathbf{i}_t, \quad \text{a vector;}$$

here $\mathbf{i}_r \cdot \mathbf{i}_s = \delta_{rs}$, the Kronecker delta (23.2);

$$\nabla \times \mathbf{f} = \mathbf{i}_r \times \mathrm{D}_r\mathbf{f} = \mathrm{D}_r f_{st}\, \mathbf{i}_r \times \mathbf{i}_s\, \mathbf{i}_t$$

$$= (\mathrm{D}_2 f_{3t} - \mathrm{D}_3 f_{2t})\mathbf{i}_1\mathbf{i}_t + (\mathrm{D}_3 f_{1t} - \mathrm{D}_1 f_{3t})\mathbf{i}_2\mathbf{i}_t$$

$$+ (\mathrm{D}_1 f_{2t} - \mathrm{D}_2 f_{1t})\mathbf{i}_3\mathbf{i}_t, \quad \text{a dyadic.}$$

Again, if $\lambda(\mathbf{r})$ is a variable scalar and Φ a *constant* dyadic,

(7) $$\nabla(\lambda\Phi) = (\nabla\lambda)\Phi, \quad \text{a triadic;}$$

(8) $$\nabla \cdot (\lambda\Phi) = (\nabla\lambda) \cdot \Phi, \quad \text{a vector;}$$

(9) $$\nabla \times (\lambda\Phi) = (\nabla\lambda) \times \Phi, \quad \text{a dyadic.}$$

For any tensor $\mathbf{f}(\mathbf{r})$ with continuous second derivatives we have identities which are generalizations of (10), (11), (13), (14) of § 84:

(10) $$\nabla \cdot \nabla\mathbf{f} = \nabla^2\mathbf{f},$$

(11) $$\nabla \times \nabla\mathbf{f} = 0;$$

and, if $\mathbf{f}(\mathbf{r})$ is not a scalar,

(12) $$\nabla \cdot (\nabla \times \mathbf{f}) = 0,$$

(13) $$\nabla \times (\nabla \times \mathbf{f}) = \nabla(\nabla \cdot \mathbf{f}) - \nabla^2\mathbf{f}.$$

The proofs are straightforward applications of (3), (5) and (6):

$$\nabla \cdot \nabla\mathbf{f} = \mathbf{i}_r \cdot \mathrm{D}_r(\mathbf{i}_s\mathrm{D}_s\mathbf{f}) = \mathbf{i}_r \cdot \mathbf{i}_s\, \mathrm{D}_r\mathrm{D}_s\mathbf{f} = \mathrm{D}_r\mathrm{D}_r\mathbf{f} = \nabla^2\mathbf{f};$$

$$\nabla \times \nabla\mathbf{f} = \mathbf{i}_r \times \mathrm{D}_r(\mathbf{i}_s\mathrm{D}_s\mathbf{f}) = \mathbf{i}_r \times \mathbf{i}_s\, \mathrm{D}_r\mathrm{D}_s\mathbf{f} = 0;$$

$$\nabla \cdot (\nabla \times \mathbf{f}) = \mathbf{i}_r \cdot \mathrm{D}_r(\mathbf{i}_s \times \mathrm{D}_s\mathbf{f}) = \mathbf{i}_r \times \mathbf{i}_s \cdot \mathrm{D}_r\mathrm{D}_s\mathbf{f} = 0;$$

$$\nabla \times (\nabla \times \mathbf{f}) = \mathbf{i}_r \times \mathrm{D}_r(\mathbf{i}_s \times \mathrm{D}_s\mathbf{f}) = \mathbf{i}_r \times (\mathbf{i}_s \times \mathrm{D}_r\mathrm{D}_s\mathbf{f})$$

$$= \mathbf{i}_s\, \mathbf{i}_r \cdot \mathrm{D}_r\mathrm{D}_s\mathbf{f} - \mathbf{i}_r \cdot \mathbf{i}_s\, \mathrm{D}_r\mathrm{D}_s\mathbf{f}$$

$$= \mathbf{i}_s\mathrm{D}_s(\mathbf{i}_r \cdot \mathrm{D}_r\mathbf{f}) - \mathrm{D}_r\mathrm{D}_r\mathbf{f}$$

$$= \nabla(\nabla \cdot \mathbf{f}) - \nabla^2\mathbf{f}.$$

Since $\mathbf{i}_r \times \mathbf{i}_s = -\mathbf{i}_s \times \mathbf{i}_r$, $\mathrm{D}_r\mathrm{D}_s\mathbf{f} = \mathrm{D}_s\mathrm{D}_r\mathbf{f}$, all non-zero terms in $\nabla \times \nabla\mathbf{f}$ and $\nabla \cdot (\nabla \times \mathbf{f})$ cancel in pairs.

87. Functional Dependence.

THEOREM 1. *A necessary and sufficient condition that two continuously differentiable functions $u(x, y)$, $v(x, y)$ satisfy identically a functional relation $f(u, v) = 0$ is that their Jacobian vanish:*

$$(1) \qquad \frac{\partial(u, v)}{\partial(x, y)} = \begin{vmatrix} \partial u/\partial x & \partial u/\partial y \\ \partial v/\partial x & \partial v/\partial y \end{vmatrix} = 0 \quad \text{or} \quad \nabla u \times \nabla v = 0.$$

Proof. The condition is necessary; for, from $f(u, v) = 0$, we have

$$\frac{\partial f}{\partial u} \nabla u + \frac{\partial f}{\partial v} \nabla v = 0, \qquad \frac{\partial f}{\partial v} \nabla u \times \nabla v = 0.$$

If u is constant, $\nabla u = 0$, and $\nabla u \times \nabla v = 0$. If u is not constant, $f(u, v)$ must contain v, $\partial f/\partial v$ is not identically zero, and again $\nabla u \times \nabla v = 0$.

Conversely, suppose that $\nabla u \times \nabla v = 0$. This relation is satisfied if either u or v is constant; thus, if $u = c$, we may take $f(u, v) = u - c$. If u and v are not constant, ∇v is parallel to ∇u; hence ∇v is normal to the curves $u = c$. Along these curves, $dv/ds = 0$, and v is constant. In other words, a level curve $u = a$ is also a level curve $v = b$; when u is given, v is determined, and v is a function of u.

THEOREM 2. *A necessary and sufficient condition that two continuously differentiable functions $u(x, y, z)$, $v(x, y, z)$ satisfy identically a functional relation $f(u, v) = 0$ is that*

$$(2) \qquad \nabla u \times \nabla v = \begin{vmatrix} \mathbf{i} & \mathbf{j} & \mathbf{k} \\ \partial u/\partial x & \partial u/\partial y & \partial u/\partial z \\ \partial v/\partial x & \partial v/\partial y & \partial v/\partial z \end{vmatrix} = 0.$$

The proof is essentially the same as in theorem 1. Instead of level *curves* of u and v, we now have level *surfaces*.

THEOREM 3. *A necessary and sufficient condition that three continuously differentiable functions $u(x, y, z)$, $v(x, y, z)$, $w(x, y, z)$ satisfy identically a functional relation $f(u, v, w) = 0$, is that their Jacobian vanish:*

$$(3) \qquad \frac{\partial(u, v, w)}{\partial(x, y, z)} = \begin{vmatrix} \partial u/\partial x & \partial u/\partial y & \partial u/\partial z \\ \partial v/\partial x & \partial v/\partial y & \partial v/\partial z \\ \partial w/\partial x & \partial w/\partial y & \partial w/\partial z \end{vmatrix} = \nabla u \times \nabla v \cdot \nabla w = 0.$$

Proof. The condition is necessary; for, from $f(u, v, w) = 0$, we have

$$\frac{\partial f}{\partial u}\nabla u + \frac{\partial f}{\partial v}\nabla v + \frac{\partial f}{\partial w}\nabla w = 0, \qquad \frac{\partial f}{\partial w}\nabla u \times \nabla v \cdot \nabla w = 0.$$

If u and v alone satisfy a relation $f(u, v) = 0$, $\nabla u \times \nabla v = 0$, and also $\nabla u \times \nabla v \cdot \nabla w = 0$. If this is not the case, $f(u, v, w)$ must contain w, $\partial f/\partial w$ is not identically zero, and again $\nabla u \times \nabla v \cdot \nabla w = 0$.

Conversely, suppose that $\nabla u \times \nabla v \cdot \nabla w = 0$. If $\nabla u \times \nabla v = 0$, $f(u, v) = 0$ from theorem 2. If $\nabla u \times \nabla v \neq 0$, consider the curve of intersection of the level surfaces $u = a$, $v = b$. ∇u and ∇v are normal to this curve; and, since ∇u, ∇v, ∇w are coplanar, ∇w is also normal to the curve. Therefore $dw/ds = 0$ and $w = c$ along the curve $u = a$, $v = b$; in other words, when u and v are given, w is determined: w is a function of u and v.

88. Curvilinear Coordinates. In a given region let

$$(1) \qquad u = u(x, y, z), \qquad v = v(x, y, z), \qquad w = w(x, y, z)$$

be three continuously differentiable functions whose Jacobian $\nabla u \times \nabla v \cdot \nabla w \neq 0$ at all points. The functions, therefore, are not connected by a relation $f(u, v, w) = 0$. Since the Jacobian is continuous, its sign cannot change in the region; and, to be explicit, we shall suppose that

$$(2) \qquad \nabla u \times \nabla v \cdot \nabla w > 0.$$

This involves no loss in generality; for, if the Jacobian were negative, an interchange of v and w (for example) would make it positive.

Under the foregoing hypotheses a well-known theorem ‡ states that in the neighborhood of any point (x_0, y_0, z_0) the equations (1) have a unique inverse

$$(3) \qquad x = x(u, v, w), \qquad y = y(u, v, w), \qquad z = z(u, v, w);$$

and that these functions are also continuously differentiable. At least in a suitably restricted region of uvw space, each set of values u, v, w yields, through equations (3), a unique set x, y, z. In this region the correspondence $(x, y, z) \sim (u, v, w)$ effected by (1) and

‡ Cf. Sokolnikoff, *Advanced Calculus*, New York, 1939, p. 434.

(3) is one-to-one. Instead of specifying a point P_0 by the coordinates (x_0, y_0, z_0), we may use instead the three numbers,

$$u_0 = u(x_0, y_0, z_0), \qquad v_0 = v(x_0, y_0, z_0), \qquad w_0 = w(x_0, y_0, z_0).$$

When (u_0, v_0, w_0) are given, P_0 is located at the point of intersection of the three *coordinate surfaces*,

$$u(x, y, z) = u_0, \qquad v(x, y, z) = v_0, \qquad w(x, y, z) = w_0.$$

These, in general, will intersect in three curves, the *coordinate curves*, along which only one of the quantities u, v, w can vary. For this reason u, v, w are called *curvilinear coordinates*, in distinction to the rectangular coordinates x, y, z, for which the coordinate curves are straight lines.

If the position vector $\mathbf{r}$ is regarded as a function of x, y, z, $\nabla \mathbf{r} = \mathbf{I}$ (84.9). But, if $\mathbf{r}$ is regarded as a function of u, v, w,

$$\nabla \mathbf{r} = \nabla u \frac{\partial \mathbf{r}}{\partial u} + \nabla v \frac{\partial \mathbf{r}}{\partial v} + \nabla w \frac{\partial \mathbf{r}}{\partial w},$$

from (83.7). We therefore have the fundamental relation,

$$\nabla u\, \mathbf{r}_u + \nabla v\, \mathbf{r}_v + \nabla w\, \mathbf{r}_w = \mathbf{I}, \tag{4}$$

on writing $\mathbf{r}_u = \partial \mathbf{r}/\partial u$, etc. From the theorem of § 66, we now conclude that:

The vector triples $\nabla u, \nabla v, \nabla w$ and $\mathbf{r}_u, \mathbf{r}_v, \mathbf{r}_w$ form reciprocal sets.

The vectors $\mathbf{r}_u$ are tangent to the *u curves*, the coordinate curves along which v and w are constant. Thus at any point $P(u, v, w)$, $\mathbf{r}_u, \mathbf{r}_v, \mathbf{r}_w$ are tangent to the three coordinate curves meeting there; and, from (23.6),

$$[\mathbf{r}_u \mathbf{r}_v \mathbf{r}_w][\nabla u\ \nabla v\ \nabla w] = 1. \tag{5}$$

Here $[\mathbf{r}_u \mathbf{r}_v \mathbf{r}_w]$ is the Jacobian $\partial(x, y, z)/\partial(u, v, w)$ of the inverse transformation (3); and, from (2),

$$J = [\mathbf{r}_u \mathbf{r}_v \mathbf{r}_w] > 0. \tag{6}$$

Moreover, from the properties of reciprocal sets,

$$\nabla u = \frac{\mathbf{r}_v \times \mathbf{r}_w}{J}, \qquad \nabla v = \frac{\mathbf{r}_w \times \mathbf{r}_u}{J}, \qquad \nabla w = \frac{\mathbf{r}_u \times \mathbf{r}_v}{J}, \tag{7}$$

$$\mathbf{r}_u = J\, \nabla v \times \nabla w, \quad \mathbf{r}_v = J\, \nabla w \times \nabla u, \quad \mathbf{r}_w = J\, \nabla u \times \nabla v. \tag{8}$$

The volume of the parallelepiped whose edges are $\mathbf{r}_u\,du$, $\mathbf{r}_v\,dv$, $\mathbf{r}_w\,dw$ is called the element of volume dV; thus

(9) $$dV = [\mathbf{r}_u\mathbf{r}_v\mathbf{r}_w]\,du\,dv\,dw = J\,du\,dv\,dw.$$

To find the gradient of a tensor $\mathbf{f}(u, v, w)$, we first compute

(10) $$\frac{d\mathbf{f}}{ds} = \frac{\partial\mathbf{f}}{\partial u}\frac{du}{ds} + \frac{\partial\mathbf{f}}{\partial v}\frac{dv}{ds} + \frac{\partial\mathbf{f}}{\partial w}\frac{dw}{ds}$$
$$= \mathbf{e}\cdot(\nabla u\,\mathbf{f}_u + \nabla v\,\mathbf{f}_v + \nabla w\,\mathbf{f}_w).$$

But, from (86.2), $d\mathbf{f}/ds = \mathbf{e}\cdot\nabla\mathbf{f}$ for every $\mathbf{e}$; hence

(11) $$\nabla\mathbf{f} = \nabla u\,\mathbf{f}_u + \nabla v\,\mathbf{f}_v + \nabla w\,\mathbf{f}_w.$$

Thus, in curvilinear coordinates the operator *del* becomes

(12) $$\nabla = \nabla u\frac{\partial}{\partial u} + \nabla v\frac{\partial}{\partial v} + \nabla w\frac{\partial}{\partial w}\cdot$$

From (11), we obtain $\nabla\cdot\mathbf{f}$ and $\nabla\times\mathbf{f}$ by dotting and crossing:

(13) $$\nabla\cdot\mathbf{f} = \nabla u\cdot\mathbf{f}_u + \nabla v\cdot\mathbf{f}_v + \nabla w\cdot\mathbf{f}_w,$$

(14) $$\nabla\times\mathbf{f} = \nabla u\times\mathbf{f}_u + \nabla v\times\mathbf{f}_v + \nabla w\times\mathbf{f}_w.$$

For purposes of computation, it is usually more convenient to eliminate ∇u, ∇v, ∇w from these formulas by use of equations (7). We thus obtain

(15) $$\nabla\mathbf{f} = \frac{1}{J}\{\mathbf{r}_v\times\mathbf{r}_w\,\mathbf{f}_u + \mathbf{r}_w\times\mathbf{r}_u\,\mathbf{f}_v + \mathbf{r}_u\times\mathbf{r}_v\,\mathbf{f}_w\}\,,$$

and corresponding equations for $\nabla\cdot\mathbf{f}$ and $\nabla\times\mathbf{f}$. In view of the identity,

$$(\mathbf{r}_v\times\mathbf{r}_w)_u + (\mathbf{r}_w\times\mathbf{r}_u)_v + (\mathbf{r}_u\times\mathbf{r}_v)_w = 0,$$

these also may be written

(16) $$\nabla\mathbf{f} = \frac{1}{J}\{(\mathbf{r}_v\times\mathbf{r}_w\,\mathbf{f})_u + (\mathbf{r}_w\times\mathbf{r}_u\,\mathbf{f})_v + (\mathbf{r}_u\times\mathbf{r}_v\,\mathbf{f})_w\},$$

(17) $$\nabla\cdot\mathbf{f} = \frac{1}{J}\{(\mathbf{r}_v\times\mathbf{r}_w\cdot\mathbf{f})_u + (\mathbf{r}_w\times\mathbf{r}_u\cdot\mathbf{f})_v + (\mathbf{r}_u\times\mathbf{r}_v\cdot\mathbf{f})_w\},$$

(18) $$\nabla\times\mathbf{f} = \frac{1}{J}\{[(\mathbf{r}_v\times\mathbf{r}_w)\times\mathbf{f}]_u + [(\mathbf{r}_w\times\mathbf{r}_u)\times\mathbf{f}]_v + [(\mathbf{r}_u\times\mathbf{r}_v)\times\mathbf{f}]_w\}.$$

When the triple products in (18) are expanded, the brace becomes

$$(\mathbf{r}_w \mathbf{r}_v \cdot \mathbf{f})_u - (\mathbf{r}_v \mathbf{r}_w \cdot \mathbf{f})_u + (\mathbf{r}_u \mathbf{r}_w \cdot \mathbf{f})_v - (\mathbf{r}_w \mathbf{r}_u \cdot \mathbf{f})_v$$

$$+ (\mathbf{r}_v \mathbf{r}_u \cdot \mathbf{f})_w - (\mathbf{r}_u \mathbf{r}_v \cdot \mathbf{f})_w$$

$$= \mathbf{r}_w(\mathbf{r}_v \cdot \mathbf{f})_u - \mathbf{r}_v(\mathbf{r}_w \cdot \mathbf{f})_u + \mathbf{r}_u(\mathbf{r}_w \cdot \mathbf{f})_v - \mathbf{r}_w(\mathbf{r}_u \cdot \mathbf{f})_v$$

$$+ \mathbf{r}_v(\mathbf{r}_u \cdot \mathbf{f})_w - \mathbf{r}_u(\mathbf{r}_v \cdot \mathbf{f})_w.$$

With this value for the brace, (18) may be written compactly in determinant form:

$$(19) \qquad \nabla \times \mathbf{f} = \frac{1}{J} \begin{vmatrix} \mathbf{r}_u & \mathbf{r}_v & \mathbf{r}_w \\ \dfrac{\partial}{\partial u} & \dfrac{\partial}{\partial v} & \dfrac{\partial}{\partial w} \\ \mathbf{r}_u \cdot \mathbf{f} & \mathbf{r}_v \cdot \mathbf{f} & \mathbf{r}_w \cdot \mathbf{f} \end{vmatrix}.$$

This equation reduces to (84.7) when $\mathbf{r}_u, \mathbf{r}_v, \mathbf{r}_w$ are replaced by $\mathbf{r}_x = \mathbf{i}, \mathbf{r}_y = \mathbf{j}, \mathbf{r}_z = \mathbf{k}$.

89. Orthogonal Coordinates. The curvilinear coordinates u, v, w are said to be *orthogonal* if the coordinate curves (along which one coordinate only can vary) cut at right angles. Since $\mathbf{r}_u, \mathbf{r}_v, \mathbf{r}_w$ are tangent to the coordinate curves, the coordinates are orthogonal when and only when

$$(1) \qquad \mathbf{r}_u \cdot \mathbf{r}_v = \mathbf{r}_v \cdot \mathbf{r}_w = \mathbf{r}_w \cdot \mathbf{r}_u = 0.$$

We choose the notation so that

$$(2) \qquad \mathbf{r}_u = U\mathbf{a}, \qquad \mathbf{r}_v = V\mathbf{b}, \qquad \mathbf{r}_w = W\mathbf{c},$$

where $\mathbf{a}, \mathbf{b}, \mathbf{c}$ are a dextral set of orthogonal unit vectors and U, V, W are all positive; then $[\mathbf{abc}] = 1$, and

$$(3) \qquad J = [\mathbf{r}_u \mathbf{r}_v \mathbf{r}_w] = UVW.$$

The set of vectors reciprocal to $\mathbf{r}_u, \mathbf{r}_v, \mathbf{r}_w$ are the gradients of the coordinates:

$$(4) \qquad \nabla u = \frac{\mathbf{a}}{U}, \qquad \nabla v = \frac{\mathbf{b}}{V}, \qquad \nabla w = \frac{\mathbf{c}}{W}.$$

From the general formulas (11), (17), (19) of the preceding article we obtain ∇f, $\nabla \cdot \mathbf{f}$ and $\nabla \times \mathbf{f}$ in orthogonal curvilinear coordinates:

(5) $$\nabla \mathbf{f} = \frac{1}{U}\,\mathbf{a}\mathbf{f}_u + \frac{1}{V}\,\mathbf{b}\mathbf{f}_v + \frac{1}{W}\,\mathbf{c}\mathbf{f}_w,$$

(6) $$\nabla \cdot \mathbf{f} = \frac{1}{UVW}\left\{\frac{\partial}{\partial u}(VW\,\mathbf{a}\cdot\mathbf{f}) + \frac{\partial}{\partial v}(WU\,\mathbf{b}\cdot\mathbf{f}) + \frac{\partial}{\partial w}(UV\,\mathbf{c}\cdot\mathbf{f})\right\},$$

(7) $$\nabla \times \mathbf{f} = \frac{1}{UVW}\begin{vmatrix} U\mathbf{a} & V\mathbf{b} & W\mathbf{c} \\ \dfrac{\partial}{\partial u} & \dfrac{\partial}{\partial v} & \dfrac{\partial}{\partial w} \\ U\,\mathbf{a}\cdot\mathbf{f} & V\,\mathbf{b}\cdot\mathbf{f} & W\,\mathbf{c}\cdot\mathbf{f} \end{vmatrix}.$$

If we put $\mathbf{f} = \nabla g$ in (6), we obtain the Laplacian $\nabla^2 g = \nabla \cdot \nabla g$ in orthogonal coordinates,

(8) $$\nabla^2 g = \frac{1}{UVW}\left\{\frac{\partial}{\partial u}\left(\frac{VW}{U}\,g_u\right) + \frac{\partial}{\partial v}\left(\frac{WU}{V}\,g_v\right) + \frac{\partial}{\partial w}\left(\frac{UV}{W}\,g_w\right)\right\}.$$

When the curvilinear coordinates are given by the equations,

$$x = x(u, v, w), \qquad y = y(u, v, w), \qquad z = z(u, v, w),$$

we may compute the (positive) functions U, V, W from equations of the type,

(9) $$U = |\mathbf{r}_u| = |x_u\mathbf{i} + y_u\mathbf{j} + z_u\mathbf{k}| = \sqrt{x_u^2 + y_u^2 + z_u^2}.$$

The element of volume (88.9) is now

(10) $$dV = UVW\,du\,dv\,dw.$$

Example 1. *Cylindrical Coordinates.* The point $P(x, y, z)$ projects into the point $Q(x, y, 0)$ in the xy-plane. If ρ, φ are polar coordinates of Q in the xy-plane, $u = \rho$, $v = \varphi$, $w = z$ are called the *cylindrical coordinates* of P (Fig. 89a). They are related to rectangular coordinates by the equations:

(10) $$x = \rho\cos\varphi, \qquad y = \rho\sin\varphi, \qquad z = z.$$

From $\mathbf{r} = \mathbf{i}x + \mathbf{j}y + \mathbf{k}z$, we have

$$\mathbf{r}_\rho = [\cos\varphi, \sin\varphi, 0],$$

$$\mathbf{r}_\varphi = [-\rho\sin\varphi, \rho\cos\varphi, 0],$$

$$\mathbf{r}_z = [0, 0, 1].$$

Since these vectors are mutually perpendicular and $[\mathbf{r}_\rho\mathbf{r}_\varphi\mathbf{r}_z] > 0$, cylindrical coordinates form an orthogonal system which is dextral in the order ρ, φ, z. Moreover,

(11) $$U = |\mathbf{r}_\rho| = 1, \qquad V = |\mathbf{r}_\varphi| = \rho, \qquad W = |\mathbf{r}_z| = 1;$$

(12) $$J = UVW = \rho.$$

The level surfaces $\rho = a$, $\varphi = b$, $z = c$ are cylinders about the z-axis, planes through the z-axis, and planes perpendicular to the z-axis. The coordinate

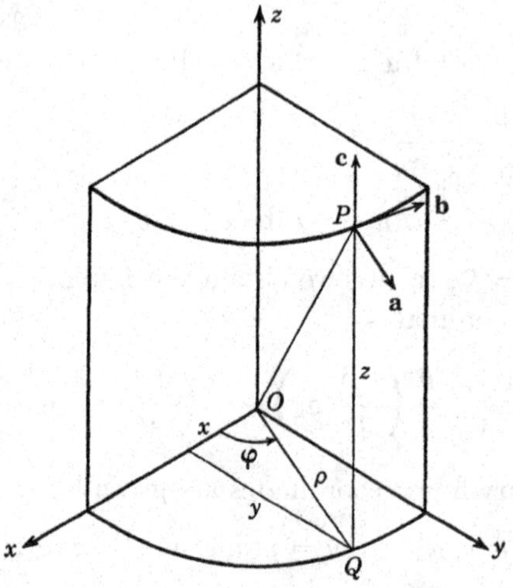

FIG. 89a

curves for ρ are rays perpendicular to the z-axis; for φ, horizontal circles centered on the z-axis; for z, lines parallel to the z-axis. The element of volume $dV = \rho\, d\rho\, d\varphi\, dz$.

From (8), the Laplacian is

(13) $$\nabla^2 g = \frac{1}{\rho}\frac{\partial}{\partial \rho}(\rho g_\rho) + \frac{1}{\rho^2} g_{\varphi\varphi} + g_{zz}.$$

If $g = \log \rho$, $\rho g_\rho = 1$; hence $\log \rho$ satisfies Laplace's Equation $\nabla^2 g = 0$. Such a function is called *harmonic*.

Example 2. *Spherical Coordinates.* The spherical coordinates of a point $P(x, y, z)$ are its distance $r = OP$ from the origin, the angle θ between OP and the z-axis, and the dihedral angle φ between the xz-plane and the plane $z\,OP$ (Fig. 89b). They are related to rectangular coordinates by the equations:

(14) $$x = r \sin\theta \cos\varphi, \qquad y = r \sin\theta \sin\varphi, \qquad z = r\cos\theta.$$

From $\mathbf{r} = \mathbf{i}x + \mathbf{j}y + \mathbf{k}z$, we have

$$\mathbf{r}_r = [\sin\theta\cos\varphi, \sin\theta\sin\varphi, \cos\theta],$$

$$\mathbf{r}_\theta = [r\cos\theta\cos\varphi, r\cos\theta\sin\varphi, -r\sin\theta],$$

$$\mathbf{r}_\varphi = [-r\sin\theta\sin\varphi, r\sin\theta\cos\varphi, 0].$$

Since these vectors are mutually perpendicular and $[\mathbf{r}_r\mathbf{r}_\theta\mathbf{r}_\varphi] > 0$, spherical coordinates form an orthogonal system which is dextral in the order r, θ, φ. Moreover,

$$(15) \qquad U = |\mathbf{r}_r| = 1, \qquad V = |\mathbf{r}_\theta| = r, \qquad W = |\mathbf{r}_\varphi| = r\sin\theta;$$

$$(16) \qquad J = UVW = r^2 \sin\theta.$$

The level surfaces $r = a$, $\theta = b$, $\varphi = c$ are spheres about O, cones about the z-axis with vertex at O, and planes through the z-axis. The coordinate curves

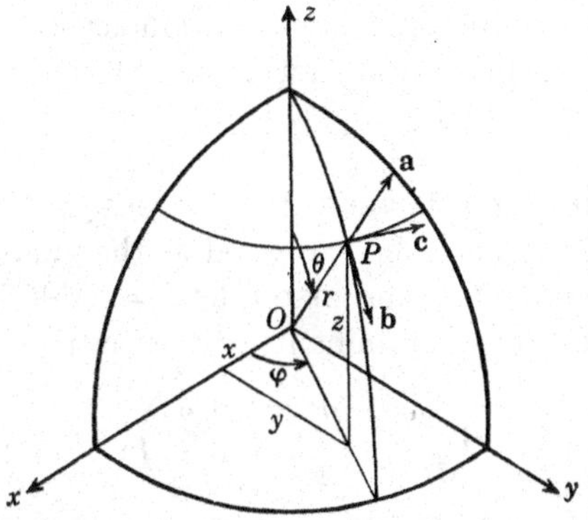

FIG. 89b

for r are rays from the origin; for θ, vertical circles centered at the origin; for φ, horizontal circles centered on the z-axis. The element of volume $dV = r^2 \sin\theta\, dr\, d\theta\, d\varphi$.

From (8), we now have

$$(17) \qquad \nabla^2 g = \frac{1}{r^2 \sin\theta}\left\{\sin\theta \frac{\partial}{\partial r}(r^2 g_r) + \frac{\partial}{\partial\theta}(\sin\theta\, g_\theta) + \frac{1}{\sin\theta} g_{\varphi\varphi}\right\}.$$

If $g = 1/r$, $r^2 g_r = -1$; hence $1/r$ satisfies Laplace's Equation $\nabla^2 g = 0$.

90. Total Differential. In passing from the point P to P', the position vector $\mathbf{r} = \overrightarrow{OP}$ changes by the increment $\Delta\mathbf{r} = \overrightarrow{PP'}$. Then, if $\mathbf{f}(\mathbf{r})$ is any differentiable tensor point function, the total differential of $\mathbf{f}(\mathbf{r})$ is defined by the equation:

$$(1) \qquad d\mathbf{f} = \Delta\mathbf{r} \cdot \nabla\mathbf{f}.$$

In particular, when $\mathbf{f} = \mathbf{r}$, $\nabla\mathbf{f} = \mathbf{I}$, and

$$(2) \qquad d\mathbf{r} = \Delta\mathbf{r}.$$

The differential of the position vector is the same as the increment The defining equation (1) therefore may be written

$$(3) \qquad d\mathbf{f} = d\mathbf{r} \cdot \nabla\mathbf{f}.$$

If $\mathbf{f}$ is a function $\mathbf{f}(u, v, w)$ of curvilinear coordinates,

$$\nabla\mathbf{f} = \nabla u\, \mathbf{f}_u + \nabla v\, \mathbf{f}_v + \nabla w\, \mathbf{f}_w,$$

and, from (3), since $d\mathbf{r} \cdot \nabla u = du$, etc.,

$$(4) \qquad d\mathbf{f} = \mathbf{f}_u\, du + \mathbf{f}_v\, dv + \mathbf{f}_w\, dw.$$

91. Irrotational Vectors. A vector function $\mathbf{f}(\mathbf{r})$ is said to be *irrotational* in a region R when $\operatorname{rot} \mathbf{f} = 0$ in R.

If $\varphi(\mathbf{r})$ is a scalar function with continuous second derivatives, its gradient $\nabla\varphi$ is irrotational; for, from (84.11),

$$(1) \qquad \operatorname{rot} \operatorname{grad} \varphi = 0.$$

Conversely, if $\operatorname{rot} \mathbf{f} = 0$, and $\mathbf{f}$ is continuously differentiable in R, we shall show that $\mathbf{f}$ may be expressed as the gradient of a scalar $\varphi(\mathbf{r})$. Using rectangular coordinates, let $\mathbf{f} = f_1\mathbf{i} + f_2\mathbf{j} + f_3\mathbf{k}$; then, if $\operatorname{rot} \mathbf{f} = 0$, the three determinants of the matrix,

$$(2) \qquad \begin{pmatrix} \partial/\partial x & \partial/\partial y & \partial/\partial z \\ f_1 & f_2 & f_3 \end{pmatrix},$$

all vanish (84.7). Under this condition we shall determine a scalar function $\varphi(\mathbf{r})$, so that $\nabla\varphi = \mathbf{f}$, that is,

$$(3) \quad \frac{\partial\varphi}{\partial x} = f_1(x, y, z), \qquad \frac{\partial\varphi}{\partial y} = f_2(x, y, z), \qquad \frac{\partial\varphi}{\partial z} = f_3(x, y, z).$$

Let (x_0, y_0, z_0) be an arbitrary point of R. On integrating $\partial\varphi/\partial x = f_1$ with respect to x and regarding y and z as constant parameters, we have

$$(4) \qquad \varphi = \int_{x_0}^{x} f_1(x, y, z)\, dx + \alpha(y, z),$$

where $\alpha(y, z)$ is a function of y and z as yet undetermined. Hence

$$\frac{\partial\varphi}{\partial y} = \int_{x_0}^{x} \frac{\partial f_1}{\partial y}\, dx + \frac{\partial\alpha}{\partial y} = \int_{x_0}^{x} \frac{\partial f_2}{\partial x}\, dx + \frac{\partial\alpha}{\partial y}, \quad \text{or}$$

$$f_2(x, y, z) = f_2(x, y, z) - f_2(x_0, y, z) + \frac{\partial\alpha}{\partial y};$$

$$\frac{\partial\varphi}{\partial z} = \int_{x_0}^{x} \frac{\partial f_1}{\partial z}\, dx + \frac{\partial\alpha}{\partial z} = \int_{x_0}^{x} \frac{\partial f_3}{\partial x}\, dx + \frac{\partial\alpha}{\partial z}, \quad \text{or}$$

$$f_3(x, y, z) = f_3(x, y, z) - f_3(x_0, y, z) + \frac{\partial\alpha}{\partial z}.$$

Instead of three equations (3) for $\varphi(x, y, z)$, we now have *two* equations,

$$\text{(5)} \qquad \frac{\partial\alpha}{\partial y} = f_2(x_0, y, z), \qquad \frac{\partial\alpha}{\partial z} = f_3(x_0, y, z)$$

for $\alpha(y, z)$, with the condition $\partial f_2/\partial z = \partial f_3/\partial y$. The problem has been reduced from three to two dimensions.

On integrating $\partial\alpha/\partial y = f_2$ with respect to y, we have

$$\text{(6)} \qquad \alpha = \int_{y_0}^{y} f_2(x_0, y, z)\, dy + \beta(z),$$

where $\beta(z)$ is a function of z to be determined. Hence

$$\frac{\partial\alpha}{\partial z} = \int_{y_0}^{y} \frac{\partial f_2}{\partial z}\, dy + \frac{d\beta}{dz} = \int_{y_0}^{y} \frac{\partial f_3}{\partial y}\, dy + \frac{d\beta}{dz}, \quad \text{or}$$

$$f_3(x_0, y, z) = f_3(x_0, y, z) - f_3(x_0, y_0, z) + \frac{d\beta}{dz};$$

$$\text{(7)} \qquad \frac{d\beta}{dz} = f_3(x_0, y_0, z).$$

We now have *one* equation (7) to determine $\beta(z)$; and

$$\text{(8)} \qquad \beta = \int_{z_0}^{z} f_3(x_0, y_0, z)\, dz.$$

On collecting the results (4), (6), and (8), we have finally

$$\text{(9)} \quad \varphi = \int_{x_0}^{x} f_1(x, y, z)\, dx + \int_{y_0}^{y} f_2(x_0, y, z)\, dy + \int_{z_0}^{z} f_3(x_0, y_0, z)\, dz.$$

Direct substitution shows that $\nabla\varphi = \mathbf{f}$; and, if ψ is a second function for which $\nabla\psi = \mathbf{f}$, $\nabla(\psi - \varphi) = 0$, and $\psi - \varphi$ is a constant. Thus (9) gives the solution of equations (3), determined uniquely except for an additive constant.

There are evidently five other forms for φ which may be obtained from (9) by permuting 1, 2, 3 and making the corresponding permutation on x, y, z; for example,

$$\text{(10)} \quad \varphi = \int_{y_0}^{y} f_2(x, y, z)\, dy + \int_{z_0}^{z} f_3(x, y_0, z)\, dz + \int_{x_0}^{x} f_1(x, y_0, z_0)\, dx.$$

Moreover, x_0, y_0, z_0 may be given any values that do not make the integrands infinite.

In mathematical physics, it is customary to express an irrotational vector $\mathbf{f}$ as the *negative* of a gradient. Thus, if $\psi = -\varphi$, we have

$$\text{(11)} \qquad \mathbf{f} = -\nabla\psi;$$

ψ is then called the *scalar potential* of $\mathbf{f}$.

Example 1. When

$$\mathbf{f} = 2xz\mathbf{i} + 2yz^2\mathbf{j} + (x^2 + 2y^2z - 1)\mathbf{k},$$

we find that $\operatorname{rot}\mathbf{f} = 0$; hence $\mathbf{f} = \nabla\varphi$. With $x_0 = y_0 = z_0 = 0$, we have, from (9),

$$\varphi = \int_0^x 2xz\,dx + \int_0^y 2yz^2\,dy - \int_0^z dz = x^2z + y^2z^2 - z.$$

Example 2. *Exact Equation.* The differential equation,

$$\text{(i)} \qquad \mathbf{f}\cdot d\mathbf{r} = f_1\,dx + f_2\,dy + f_3\,dz = 0,$$

is said to be exact when $\mathbf{f}\cdot d\mathbf{r} = d\varphi$, the differential of a scalar. If (i) is exact, we have, from (90.3), $d\mathbf{r}\cdot\mathbf{f} = d\mathbf{r}\cdot\nabla\varphi$; and, since $d\mathbf{r}$ is arbitrary, $\mathbf{f} = \nabla\varphi$. When $\mathbf{f}$ is continuously differentiable, $\mathbf{f} = \nabla\varphi$ implies $\operatorname{rot}\mathbf{f} = 0$, and conversely. Therefore we may state the

THEOREM. *If* $\mathbf{f}$ *is a continuously differentiable vector, in order that* $\mathbf{f}\cdot d\mathbf{r} = 0$ *be exact it is necessary and sufficient that* $\operatorname{rot}\mathbf{f} = 0$.

Thus, in view of ex. 1, the equation,

$$2xz\,dx + 2yz^2\,dy + (x^2 + 2y^2z - 1)\,dz = 0,$$

is exact and may be put in the form $d\varphi = 0$. Its general solution is $\varphi = c$, that is,

$$x^2z + y^2z^2 - z = c.$$

If (i) is not exact, a scalar λ which makes $\lambda\mathbf{f}\cdot d\mathbf{r} = 0$ exact is called an *integrating factor* of (i). The preceding theorem shows that λ must satisfy the equation,

$$\text{(ii)} \qquad \operatorname{rot}\lambda\mathbf{f} = 0 \quad\text{or}\quad \nabla\lambda\times\mathbf{f} + \lambda\operatorname{rot}\mathbf{f} = 0.$$

On multiplying (ii) by $\mathbf{f}\cdot$, we have

$$\text{(iii)} \qquad \mathbf{f}\cdot\operatorname{rot}\mathbf{f} = 0.$$

Hence, when $\mathbf{f}\cdot d\mathbf{r} = 0$ admits an integrating factor, $\mathbf{f}\cdot\operatorname{rot}\mathbf{f} = 0$. Conversely, when $\mathbf{f}$ is continuously differentiable, the condition (iii) ensures the existence of an integrating factor. We shall prove this in § 105. For this reason $\mathbf{f}\cdot\operatorname{rot}\mathbf{f} = 0$ is called the *integrability condition* for $\mathbf{f}\cdot d\mathbf{r} = 0$.

When $\mathbf{f}\cdot d\mathbf{r} = 0$ is integrable, $\lambda\mathbf{f} = \nabla\varphi$. Then the vector field $\mathbf{f}(\mathbf{r})$, being parallel to $\nabla\varphi$, is everywhere normal to the level surfaces $\varphi = c$. The condition (iii) therefore implies the existence of a one-parameter family of surfaces everywhere normal to $\mathbf{f}$. Thus the geometrical content of the condition $\mathbf{f}\cdot\operatorname{rot}\mathbf{f} = 0$ is that the vector field $\mathbf{f}(\mathbf{r})$ is *surface-normal*.

92. Solenoidal Vectors. A vector function $\mathbf{f}(\mathbf{r})$ is said to be *solenoidal* in a region R when $\operatorname{div} \mathbf{f} = 0$ in R.

If $\mathbf{g}(\mathbf{r})$ is a vector function whose components have continuous second derivatives, $\operatorname{rot} \mathbf{g}$ is solenoidal, for, from (84.13),

$$\operatorname{div} \operatorname{rot} \mathbf{g} = 0. \tag{1}$$

Conversely, if $\operatorname{div} \mathbf{f} = 0$, we shall show that $\mathbf{f}$ may be expressed as the rotation of a vector $\mathbf{g}$.

Using rectangular coordinates, let $\mathbf{f} = f_1\mathbf{i} + f_2\mathbf{j} + f_3\mathbf{k}$, and (84.5)

$$\operatorname{div} \mathbf{f} = \frac{\partial f_1}{\partial x} + \frac{\partial f_2}{\partial y} + \frac{\partial f_3}{\partial z} = 0. \tag{2}$$

We now shall determine a vector $\mathbf{g} = g_1\mathbf{i} + g_2\mathbf{j} + g_3\mathbf{k}$, so that

$$\mathbf{f} = \operatorname{rot} \mathbf{g} = \begin{vmatrix} \mathbf{i} & \mathbf{j} & \mathbf{k} \\ \partial/\partial x & \partial/\partial y & \partial/\partial z \\ g_1 & g_2 & g_3 \end{vmatrix}. \tag{3}$$

We first find a particular solution of (3) for which $g_3 = 0$; then (3) is equivalent to the scalar equations:

$$f_1 = -\frac{\partial g_2}{\partial z}, \qquad f_2 = \frac{\partial g_1}{\partial z}, \qquad f_3 = \frac{\partial g_2}{\partial x} - \frac{\partial g_1}{\partial y}. \tag{4}$$

The first two equations of (4) are satisfied when

$$g_2 = -\int_{z_0}^{z} f_1(x, y, z)\, dz, \qquad g_1 = \int_{z_0}^{z} f_2(x, y, z)\, dz + \alpha(x, y); \tag{5}$$

in these integrations x and y are regarded as constant parameters, and $\alpha(x, y)$ is a function as yet undetermined. In order that these functions satisfy the third equation of (4),

$$-\int_{z_0}^{z} \left(\frac{\partial f_1}{\partial x} + \frac{\partial f_2}{\partial y}\right) dz - \frac{\partial \alpha}{\partial y} = f_3,$$

or, in view of (2),

$$\int_{z_0}^{z} \frac{\partial f_3}{\partial z}\, dz - \frac{\partial \alpha}{\partial y} = f_3(x, y, z).$$

When we perform the integration, this reduces to

$$-f_3(x, y, z_0) - \frac{\partial \alpha}{\partial y} = 0,$$

an equation which is satisfied by taking

$$\alpha(x, y) = -\int_{y_0}^{y} f_3(x, y, z_0)\, dy.$$

Hence the vector g, whose components are

$$(6)\quad g_1 = \int_{z_0}^{z} f_2(x, y, z)\, dz - \int_{y_0}^{y} f_3(x, y, z_0)\, dy,$$

$$g_2 = -\int_{z_0}^{z} f_1(x, y, z)\, dz, \qquad g_3 = 0,$$

is a particular solution of our problem.

If $\mathbf{G}$ is any other solution, rot $\mathbf{G}$ = rot $\mathbf{g}$ = $\mathbf{f}$, and hence

$$(7)\qquad \text{rot}\,(\mathbf{G} - \mathbf{g}) = 0.$$

But any irrotational vector may be expressed as the gradient of a scalar φ (§ 91); hence the general solution of (7) is $\mathbf{G} - \mathbf{g} = \nabla\varphi$, where $\varphi(\mathbf{r})$ is an arbitrary twice-differentiable scalar. When div $\mathbf{f} = 0$, the general solution of rot $\mathbf{g} = \mathbf{f}$ is therefore $\mathbf{g} + \nabla\varphi$; its rectangular components are obtained by adding $\partial\varphi/\partial x$, $\partial\varphi/\partial y$, $\partial\varphi/\partial z$ to the components of $\mathbf{g}$ given in (6).

In mathematical physics, the solenoidal vector $\mathbf{f}$ = rot $\mathbf{g}$ is said to be derived from the *vector potential* $\mathbf{g}$.

Example 1. When

$$\mathbf{f} = x(z - y)\mathbf{i} + y(x - z)\mathbf{j} + z(y - x)\mathbf{k},$$

we find that div $\mathbf{f} = 0$. Therefore $\mathbf{f}$ = rot $\mathbf{g}$; if we take $z_0 = y_0 = 0$, the particular solution (6) is

$$g_1 = \int_0^z y(x - z)\, dz = xyz - \tfrac{1}{2}yz^2,$$

$$g_2 = -\int_0^z x(z - y)\, dz = -\tfrac{1}{2}xz^2 + xyz, \qquad g_3 = 0.$$

Example 2. If u and v are continuously differentiable scalars, the vector $\nabla u \times \nabla v$ is solenoidal, for, from (85.3),

$$\nabla u \times \nabla v = \text{rot}\,(u\nabla v).$$

Conversely, we shall show in § 104 that a continuously differentiable solenoidal vector always can be expressed in the form $\nabla u \times \nabla v$.

93. Surfaces. A surface is represented in parametric form by the equations:

$$(1) \qquad x = x(u, v), \qquad y = y(u, v), \qquad z = z(u, v).$$

We assume that the three functions of the *surface coordinates* u, v are continuous and have continuous first partial derivatives, a requirement briefly expressed by saying that the functions are *continuously differentiable.* In order that equations (1) represent a proper surface, we must exclude the two cases:

(i) the functions $x(u, v)$, $y(u, v)$, $z(u, v)$ are constants: equations (1) then represent a *point;*

(ii) these functions are expressible as functions of a single variable $t = t(u, v)$; equations (1) then represent a *curve.*

In case (i) all the elements of the Jacobian matrix,

$$(2) \qquad \begin{pmatrix} \dfrac{\partial x}{\partial u} & \dfrac{\partial y}{\partial u} & \dfrac{\partial z}{\partial u} \\ \dfrac{\partial x}{\partial v} & \dfrac{\partial y}{\partial v} & \dfrac{\partial z}{\partial v} \end{pmatrix},$$

vanish. In case (ii), the rows of the matrix are dx/dt, dy/dt, dz/dt multiplied by $\partial t/\partial u$ and $\partial t/\partial v$, respectively, and all of its two-rowed determinants vanish identically. We exclude these cases by requiring that the matrix (2) be, in general, of *rank two;* then, at least one of its two-rowed determinants,

$$(3) \quad A = \begin{vmatrix} y_u & z_u \\ y_v & z_v \end{vmatrix}, \qquad B = \begin{vmatrix} z_u & x_u \\ z_v & x_v \end{vmatrix}, \qquad C = \begin{vmatrix} x_u & y_u \\ x_v & y_v \end{vmatrix},$$

is not identically zero.

Even when the matrix in general is of rank two, the three determinants A, B, C, all may vanish for certain points u, v. Such points are called *singular*, in contrast to the *regular* points, where at least one determinant is not zero.

If we introduce the position vector to the surface,

$$(4) \qquad \mathbf{r} = x\mathbf{i} + y\mathbf{j} + z\mathbf{k} = \mathbf{r}(u, v),$$

$$(5) \qquad \mathbf{r}_u \times \mathbf{r}_v = A\mathbf{i} + B\mathbf{j} + C\mathbf{k},$$

and the condition for a regular point may be written

$$(6) \qquad \mathbf{r}_u \times \mathbf{r}_v \neq 0.$$

If we introduce new parameters $\bar{u}$, $\bar{v}$ by means of the equations,

$$(7) \qquad u = u(\bar{u}, \bar{v}), \qquad v = v(\bar{u}, \bar{v}),$$

we shall require that the Jacobian of this transformation,

$$(8) \qquad J = \frac{\partial(u, v)}{\partial(\bar{u}, \bar{v})} \neq 0.$$

Then equations (7) may be solved for $\bar{u}$ and $\bar{v}$, yielding

$$(9) \qquad \bar{u} = \bar{u}(u, v), \qquad \bar{v} = \bar{v}(u, v),$$

and the correspondence between u, v and $\bar{u}, \bar{v}$ will be one to one. Since

$$(10) \quad \mathbf{r}_{\bar{u}} \times \mathbf{r}_{\bar{v}} = \left(\mathbf{r}_u \frac{\partial u}{\partial \bar{u}} + \mathbf{r}_v \frac{\partial v}{\partial \bar{u}}\right) \times \left(\mathbf{r}_u \frac{\partial u}{\partial \bar{v}} + \mathbf{r}_v \frac{\partial v}{\partial \bar{v}}\right) = J\, \mathbf{r}_u \times \mathbf{r}_v,$$

the requirement $J \neq 0$ makes $\mathbf{r}_{\bar{u}} \times \mathbf{r}_{\bar{v}} \neq 0$ a consequence of (6). Thus a point which is regular with respect to the parameters u, v is also regular with respect to $\bar{u}, \bar{v}$.

94. First Fundamental Form. A curve on the surface $\mathbf{r}(u, v)$ may be obtained by setting u and v equal to functions of a single variable t:

$$(1) \qquad u = u(t), \qquad v = v(t).$$

A tangent vector along the curve (1) is given by

$$\dot{\mathbf{r}} = \frac{d\mathbf{r}}{dt} = \mathbf{r}_u \dot{u} + \mathbf{r}_v \dot{v}$$

and

$$(2) \qquad \dot{\mathbf{r}} \cdot \dot{\mathbf{r}} = \mathbf{r}_u \cdot \mathbf{r}_u\, \dot{u}^2 + 2\mathbf{r}_u \cdot \mathbf{r}_v\, \dot{u}\dot{v} + \mathbf{r}_v \cdot \mathbf{r}_v\, \dot{v}^2.$$

The arc s along the curve is defined as in (43.4):

$$s = \int_{t_0}^{t} \sqrt{\dot{\mathbf{r}} \cdot \dot{\mathbf{r}}}\, dt; \quad \text{and} \quad \frac{ds}{dt} = \sqrt{\dot{\mathbf{r}} \cdot \dot{\mathbf{r}}}.$$

Since $du = \dot{u}\, dt$, $dv = \dot{v}\, dt$ by definition, on multiplying (2) by dt^2, we have

$$(3) \qquad ds^2 = \mathbf{r}_u \cdot \mathbf{r}_u\, du^2 + 2\mathbf{r}_u \cdot \mathbf{r}_v\, du\, dv + \mathbf{r}_v \cdot \mathbf{r}_v\, dv^2.$$

This *first fundamental quadratic form* usually is written

$$(4) \qquad ds^2 = E\, du^2 + 2F\, du\, dv + G\, dv^2,$$

where

$$(5) \qquad E = \mathbf{r}_u \cdot \mathbf{r}_u, \qquad F = \mathbf{r}_u \cdot \mathbf{r}_v, \qquad G = \mathbf{r}_v \cdot \mathbf{r}_v.$$

Moreover, from (20.1),

$$(\mathbf{r}_u \times \mathbf{r}_v) \cdot (\mathbf{r}_u \times \mathbf{r}_v) = \begin{vmatrix} \mathbf{r}_u \cdot \mathbf{r}_u & \mathbf{r}_u \cdot \mathbf{r}_v \\ \mathbf{r}_v \cdot \mathbf{r}_u & \mathbf{r}_v \cdot \mathbf{r}_v \end{vmatrix} = EG - F^2;$$

hence

$$(6) \qquad |\mathbf{r}_u \times \mathbf{r}_v|^2 = EG - F^2$$

is positive at every regular point.

The curves $v = \text{const}$ (u-curves) and $u = \text{const}$ (v-curves) are called the *parametric curves* on the surface. For these curves $dv = 0$ and $du = 0$, respectively, and the corresponding elements of arc are

$$(7) \qquad ds_1 = \sqrt{E}\, du, \qquad ds_2 = \sqrt{G}\, dv.$$

Since the vectors $\mathbf{r}_u$, $\mathbf{r}_v$ are tangent to the u-curves and v-curves, respectively, the parametric curves will cut at right angles when and only when

$$(8) \qquad F = \mathbf{r}_u \cdot \mathbf{r}_v = 0.$$

The vector $\mathbf{r}_u \times \mathbf{r}_v$ is normal to the surface. The parallelogram formed by the vectors $\mathbf{r}_u\, du$, $\mathbf{r}_v\, dv$, tangent to the parametric curves and of length ds_1, ds_2, has the vector area (§ 16),

$$(9) \qquad d\mathbf{S} = \mathbf{r}_u \times \mathbf{r}_v\, du\, dv.$$

We shall call this the *vector element of area;* the *scalar element of area* is

$$(10) \qquad dS = |\mathbf{r}_u \times \mathbf{r}_v|\, du\, dv = \sqrt{EG - F^2}\, du\, dv.$$

The unit normal $\mathbf{n}$ to the surface will be chosen as

$$(11) \qquad \mathbf{n} = \frac{\mathbf{r}_u \times \mathbf{r}_v}{\sqrt{EG - F^2}} = \frac{\mathbf{r}_u \times \mathbf{r}_v}{H};$$

then $d\mathbf{S} = \mathbf{n}\, dS$. At every regular point the vectors $\mathbf{r}_u$, $\mathbf{r}_v$, $\mathbf{n}$ form a dextral set; for

$$(12) \qquad [\mathbf{r}_u \mathbf{r}_v \mathbf{n}] = H = \sqrt{EG - F^2} > 0.$$

95. Surface Gradients. Let $\mathbf{f}(u, v)$ be a differentiable function, scalar, vector, or dyadic, which is defined at the points of the surface $\mathbf{r} = \mathbf{r}(u, v)$. We shall compute the derivative of $\mathbf{f}(u, v)$ with respect to the arc s along a surface curve $u = u(t)$, $v = v(t)$. Along this curve,

$$\frac{d\mathbf{f}}{ds} = \mathbf{f}_u \frac{du}{ds} + \mathbf{f}_v \frac{dv}{ds}, \tag{1}$$

where $du/ds = \dot{u}/\dot{s}$, $dv/ds = \dot{v}/\dot{s}$, and, from (94.4),

$$\dot{s} = \frac{ds}{dt} = \sqrt{E\dot{u}^2 + 2F\dot{u}\dot{v} + G\dot{v}^2}. \tag{2}$$

If we apply (1) to the position vector $\mathbf{r}(u, v)$, we obtain the unit tangent vector $\mathbf{e}$ to the curve:

$$\mathbf{e} = \mathbf{r}_u \frac{du}{ds} + \mathbf{r}_v \frac{dv}{ds}. \tag{3}$$

Let $\mathbf{a}, \mathbf{b}, \mathbf{c}$ denote the set reciprocal to $\mathbf{r}_u, \mathbf{r}_v, \mathbf{n}$; then, since $[\mathbf{r}_u\mathbf{r}_v\mathbf{n}] = H$,

$$\mathbf{a} = \frac{\mathbf{r}_v \times \mathbf{n}}{H}, \qquad \mathbf{b} = \frac{\mathbf{n} \times \mathbf{r}_u}{H}, \qquad \mathbf{c} = \frac{\mathbf{r}_u \times \mathbf{r}_v}{H} = \mathbf{n}. \tag{4}$$

Now from (3) $\mathbf{a} \cdot \mathbf{e} = du/ds$, $\mathbf{b} \cdot \mathbf{e} = dv/ds$; hence (1) may be written

$$\frac{d\mathbf{f}}{ds} = \mathbf{e} \cdot (\mathbf{a}\mathbf{f}_u + \mathbf{b}\mathbf{f}_v).$$

We now define $\mathbf{a}\mathbf{f}_u + \mathbf{b}\mathbf{f}_v$ as the *surface gradient* of $\mathbf{f}$ and denote it by $\nabla_s\mathbf{f}$ † or Grad $\mathbf{f}$:

$$\nabla_s\mathbf{f} = \operatorname{Grad}\mathbf{f} = \mathbf{a}\,\mathbf{f}_u + \mathbf{b}\,\mathbf{f}_v. \tag{5}$$

Grad f has the characteristic properties:

$$\mathbf{e} \cdot \operatorname{Grad}\mathbf{f} = \frac{d\mathbf{f}}{ds}, \qquad \mathbf{n} \cdot \operatorname{Grad}\mathbf{f} = 0; \tag{6}$$

$$\mathbf{r}_u \cdot \operatorname{Grad}\mathbf{f} = \frac{\partial\mathbf{f}}{\partial u}, \qquad \mathbf{r}_v \cdot \operatorname{Grad}\mathbf{f} = \frac{\partial\mathbf{f}}{\partial v}. \tag{7}$$

† In *surface* geometry we shall write $\nabla\mathbf{f}$ for the surface gradient.

If $f(\mathbf{r})$ is a tensor of valence ν, Grad $\mathbf{f}$ is of valence $\nu + 1$. At any point (u, v) of the surface, Grad $\mathbf{f}$ is in effect a synthesis of all the values of $d\mathbf{f}/ds$ for surface curves through this point. Since Grad f depends only on the point (u, v), df/ds is the same for all surface curves having the unit tangent vector $\mathbf{e}$ at this point.

At any point (u, v) of the surface Grad $\mathbf{f}$ is completely determined when $d\mathbf{f}/ds_i$ is given for two directions $\mathbf{e}_i$ in the tangent plane at (u, v). If the set reciprocal to $\mathbf{e}_1, \mathbf{e}_2, \mathbf{n}$ is $\mathbf{e}^1, \mathbf{e}^2, \mathbf{n}$,*

$$\text{Grad}\,\mathbf{f} = \mathbf{e}^1 \frac{d\mathbf{f}}{ds_1} + \mathbf{e}^2 \frac{d\mathbf{f}}{ds_2}; \tag{8}$$

for, by virtue of equations (6), the right member of (8) may be written

$$(\mathbf{e}^1\mathbf{e}_1 + \mathbf{e}^2\mathbf{e}_2 + \mathbf{nn}) \cdot \text{Grad}\,\mathbf{f} = \mathbf{I} \cdot \text{Grad}\,\mathbf{f}$$

where $\mathbf{I}$ is the idemfactor (§ 66). In particular if $\mathbf{e}_1$ and $\mathbf{e}_2 = \mathbf{n} \times \mathbf{e}_1$ are perpendicular unit vectors, $\mathbf{e}^1 = \mathbf{e}_1$, $\mathbf{e}^2 = \mathbf{e}_2$. In view of (8), Grad $\mathbf{f}$ *is independent of the coordinates x, y, z and of the surface parameters u, v.*

96. Surface Divergence and Rotation. If $\mathbf{f}(\mathbf{r})$ is a tensor point function defined over the surface $\mathbf{r} = \mathbf{r}(u, v)$, its surface gradient,

$$\nabla_s\mathbf{f} = \mathbf{a}\,\mathbf{f}_u + \mathbf{b}\,\mathbf{f}_v, \tag{1}$$

has the invariants,

$$\nabla_s \cdot \mathbf{f} = \mathbf{a} \cdot \mathbf{f}_u + \mathbf{b} \cdot \mathbf{f}_v, \tag{2}$$

$$\nabla_s \times \mathbf{f} = \mathbf{a} \times \mathbf{f}_u + \mathbf{b} \times \mathbf{f}_v. \tag{3}$$

We recall that the set $\mathbf{a}, \mathbf{b}, \mathbf{n}$ is reciprocal to $\mathbf{r}_u, \mathbf{r}_v, \mathbf{n}$.

When $\mathbf{f}(\mathbf{r})$ is a vector, $\nabla_s\mathbf{f}$ is a planar dyadic; then the scalar and vector invariants of $\nabla_s\mathbf{f}$ are called the *surface divergence* and *surface rotation* and are written

$$\nabla_s \cdot \mathbf{f} = \text{Div}\,\mathbf{f}, \qquad \nabla_s \times \mathbf{f} = \text{Rot}\,\mathbf{f}. \tag{4}$$

The *second* of $\nabla_s\mathbf{f}$ is the linear dyadic,

$$(\nabla_s\mathbf{f})_2 = \mathbf{a} \times \mathbf{b}\,\mathbf{f}_u \times \mathbf{f}_v = \frac{1}{H}\mathbf{n}\,\mathbf{f}_u \times \mathbf{f}_v; \tag{5}$$

the second scalar of $\nabla_s\mathbf{f}$ is therefore $\mathbf{n} \cdot \mathbf{f}_u \times \mathbf{f}_v/H$.

* Since $\mathbf{e}_1 \times \mathbf{e}_2 = \lambda\mathbf{n}$, $\mathbf{e}_3 = \mathbf{n}$, we have $\mathbf{e}^3 = \lambda\mathbf{n}/\lambda = \mathbf{n}$.

For the position vector $\mathbf{r}$ to the surface, we have

$$\nabla_s \mathbf{r} = \mathbf{a}\mathbf{r}_u + \mathbf{b}\mathbf{r}_v = \mathbf{I} - \mathbf{n}\mathbf{n}, \tag{6}$$

$$\text{Div}\,\mathbf{r} = 2, \qquad \text{Rot}\,\mathbf{r} = 0. \tag{7}$$

For the unit normal $\mathbf{n}$,

$$\text{Rot}\,\mathbf{n} = 0. \tag{8}$$

To prove this, put $\mathbf{f} = \mathbf{n}$, $\mathbf{a} = \mathbf{r}_v \times \mathbf{n}/H$, $\mathbf{b} = \mathbf{n} \times \mathbf{r}_u/H$ in (3); then

$$\begin{aligned} H\,\text{Rot}\,\mathbf{n} &= (\mathbf{r}_v \times \mathbf{n}) \times \mathbf{n}_u + (\mathbf{n} \times \mathbf{r}_u) \times \mathbf{n}_v. \\ &= (\mathbf{r}_v \cdot \mathbf{n}_u - \mathbf{r}_u \cdot \mathbf{n}_v)\mathbf{n} - \mathbf{n} \cdot \mathbf{n}_u\,\mathbf{r}_u + \mathbf{n} \cdot \mathbf{n}_v\,\mathbf{r}_u; \end{aligned}$$

Now, from $\mathbf{n} \cdot \mathbf{n} = 1$, we have, on differentiation with respect to u and v,

$$\mathbf{n} \cdot \mathbf{n}_u = \mathbf{n} \cdot \mathbf{n}_v = 0; \tag{9}$$

and, from $\mathbf{r}_u \cdot \mathbf{n} = \mathbf{r}_v \cdot \mathbf{n} = 0$,

$$\mathbf{r}_u \cdot \mathbf{n}_v = \mathbf{r}_v \cdot \mathbf{n}_u = -\mathbf{n} \cdot \mathbf{r}_{uv}; \tag{10}$$

hence $H\,\text{Rot}\,\mathbf{n} = 0$.

From the defining equations (1), (2), (3), we may derive various expansion formulas. Thus, if λ is a scalar, $\mathbf{f}$ a tensor,

$$\nabla_s(\lambda\mathbf{f}) = (\nabla_s\lambda)\mathbf{f} + \lambda\nabla_s\mathbf{f}, \tag{11}$$

$$\nabla_s \cdot (\lambda\mathbf{f}) = (\nabla_s\lambda) \cdot \mathbf{f} + \lambda\nabla_s \cdot \mathbf{f}, \tag{12}$$

$$\nabla_s \times (\lambda\mathbf{f}) = (\nabla_s\lambda) \times \mathbf{f} + \lambda\nabla_s \times \mathbf{f}. \tag{13}$$

If $\mathbf{g}$ and $\mathbf{f}$ are vectors,

$$\text{Div}\,(\mathbf{g} \times \mathbf{f}) = (\text{Rot}\,\mathbf{g}) \cdot \mathbf{f} - \mathbf{g} \cdot \text{Rot}\,\mathbf{f}; \tag{14}$$

and, in particular,

$$\text{Div}\,(\mathbf{n} \times \mathbf{f}) = -\mathbf{n} \cdot \text{Rot}\,\mathbf{f}. \tag{15}$$

Proof of (14):

$$\begin{aligned} \text{Div}\,(\mathbf{g} \times \mathbf{f}) &= \mathbf{a} \cdot (\mathbf{g}_u \times \mathbf{f} + \mathbf{g} \times \mathbf{f}_u) + \mathbf{b} \cdot (\mathbf{g}_v \times \mathbf{f} + \mathbf{g} \times \mathbf{f}_v) \\ &= (\mathbf{a} \times \mathbf{g}_u + \mathbf{b} \times \mathbf{g}_v) \cdot \mathbf{f} - \mathbf{g} \cdot (\mathbf{a} \times \mathbf{f}_u + \mathbf{b} \times \mathbf{f}_v) \\ &= (\text{Rot}\,\mathbf{g}) \cdot \mathbf{f} - \mathbf{g} \cdot \text{Rot}\,\mathbf{f}. \end{aligned}$$

97. Spatial and Surface Invariants. If $\mathbf{f}(\mathbf{r})$ is a tensor function of valence ν defined over a 3-dimensional region including the surface $\mathbf{r} = \mathbf{r}(u, v)$, its spatial gradient at a point (u, v) of the surface may be computed from (86.4). If $\mathbf{e}_1$, $\mathbf{e}_2$ are vectors in the tangent plane at (u, v) and $\mathbf{e}_3 = \mathbf{n}$, then, at all points of the surface,

$$(1) \qquad \nabla\mathbf{f} = \mathbf{e}^1 \frac{d\mathbf{f}}{ds_1} + \mathbf{e}^2 \frac{d\mathbf{f}}{ds_2} + \mathbf{n}\frac{d\mathbf{f}}{dn} = \nabla_s\mathbf{f} + \mathbf{n}\frac{d\mathbf{f}}{dn};$$

here the set $\mathbf{e}^1$, $\mathbf{e}^2$, $\mathbf{n}$ is reciprocal to $\mathbf{e}_1$, $\mathbf{e}_2$, $\mathbf{n}$ and $d\mathbf{f}/dn$ denotes the derivative of $\mathbf{f}$ in the direction of $\mathbf{n}$. Moreover, if $\nu > 0$,

$$(2) \qquad \nabla \cdot \mathbf{f} = \mathbf{e}^1 \cdot \frac{d\mathbf{f}}{ds_1} + \mathbf{e}^2 \cdot \frac{d\mathbf{f}}{ds_2} + \mathbf{n} \cdot \frac{d\mathbf{f}}{dn} = \nabla_s \cdot \mathbf{f} + \mathbf{n} \cdot \frac{d\mathbf{f}}{dn},$$

$$(3) \qquad \nabla \times \mathbf{f} = \mathbf{e}^1 \times \frac{d\mathbf{f}}{ds_1} + \mathbf{e}^2 \times \frac{d\mathbf{f}}{ds_2} + \mathbf{n} \times \frac{d\mathbf{f}}{dn} = \nabla_s \times \mathbf{f} + \mathbf{n} \times \frac{d\mathbf{f}}{dn}.$$

From (1), we see that the tensors of valence $\nu + 1$,

$$(4) \qquad \mathbf{n} \times \nabla\mathbf{f} = \mathbf{n} \times \nabla_s\mathbf{f},$$

are the same over the surface; therefore both may be written $\mathbf{n} \times \nabla\mathbf{f}$. Thus $\mathbf{n} \times \nabla\mathbf{f}$ may be computed solely from the values of $\mathbf{f}$ on the surface.

The same is true of the invariants obtained from $\mathbf{n} \times \nabla\mathbf{f}$ by dotting and crossing in the first position. Since

$$(5) \qquad \mathbf{n} \times \nabla\mathbf{f} = \mathbf{n} \times \mathbf{a}\, \mathbf{f}_u + \mathbf{n} \times \mathbf{b}\, \mathbf{f}_v,$$

from (96.1), this process yields

$$(\mathbf{n} \times \mathbf{a}) \cdot \mathbf{f}_u + (\mathbf{n} \times \mathbf{b}) \cdot \mathbf{f}_v = \mathbf{n} \cdot (\mathbf{a} \times \mathbf{f}_u + \mathbf{b} \times \mathbf{f}_v) = \mathbf{n} \cdot \nabla_s \times \mathbf{f},$$

$$(\mathbf{n} \times \mathbf{a}) \times \mathbf{f}_u + (\mathbf{n} \times \mathbf{b}) \times \mathbf{f}_v = \mathbf{a}\,\mathbf{n} \cdot \mathbf{f}_u + \mathbf{b}\,\mathbf{n} \cdot \mathbf{f}_v - \mathbf{n}(\mathbf{a} \cdot \mathbf{f}_u + \mathbf{b} \cdot \mathbf{f}_v)$$
$$= \mathbf{n} \overset{2}{\cdot} \nabla_s\mathbf{f} - \mathbf{n}\, \nabla_s \cdot \mathbf{f}.$$

Here $\mathbf{n} \overset{2}{\cdot} \nabla_s\mathbf{f}$ means that $\mathbf{n}$ is dotted into the second vector from the left in each term of $\nabla_s\mathbf{f}$. *These invariants remain the same when computed from the corresponding spatial quantities.* Thus we verify at once, from (4),

$$(6) \qquad \mathbf{n} \cdot \nabla \times \mathbf{f} = \mathbf{n} \cdot \nabla_s \times \mathbf{f};$$

and, from (1) and (2),

$$(7) \qquad \mathbf{n} \overset{2}{\cdot} \nabla\mathbf{f} - \mathbf{n}\, \nabla \cdot \mathbf{f} = \mathbf{n} \overset{2}{\cdot} \nabla_s\mathbf{f} - \mathbf{n}\, \nabla_s \cdot \mathbf{f}.$$

When $\mathbf{f}$ is a vector, these invariants become

$$(6)' \qquad \mathbf{n} \cdot \operatorname{rot} \mathbf{f} = \mathbf{n} \cdot \operatorname{Rot} \mathbf{f},$$

$$(7)' \qquad (\operatorname{grad} \mathbf{f}) \cdot \mathbf{n} - \mathbf{n} \operatorname{div} \mathbf{f} = (\operatorname{Grad} \mathbf{f}) \cdot \mathbf{n} - \mathbf{n} \operatorname{Div} \mathbf{f}.$$

We now express $\mathbf{n} \times \nabla\mathbf{f}$ and its invariants (6) and (7) in terms of the surface coordinates u, v. Since $\mathbf{a}, \mathbf{b}, \mathbf{n}$ and $\mathbf{r}_u, \mathbf{r}_v, \mathbf{n}$ are reciprocal sets, we have, from (5),

$$(8) \qquad \mathbf{n} \times \nabla\mathbf{f} = \frac{1}{H}(\mathbf{r}_v\mathbf{f}_u - \mathbf{r}_u\mathbf{f}_v) = \frac{1}{H}\{(\mathbf{r}_v\mathbf{f})_u - (\mathbf{r}_u\mathbf{f})_v\},$$

since $\mathbf{r}_{uv} = \mathbf{r}_{vu}$ if these derivatives are continuous. Dotting and crossing now yields

$$(9) \qquad \mathbf{n} \cdot \nabla \times \mathbf{f} = \frac{1}{H}\{\mathbf{r}_v \cdot \mathbf{f})_u - (\mathbf{r}_u \cdot \mathbf{f})_v\}$$

$$(10) \qquad \mathbf{n} \overset{2}{\cdot} \nabla\mathbf{f} - \mathbf{n}\nabla \cdot \mathbf{f} = \frac{1}{H}\{(\mathbf{r}_v \times \mathbf{f})_u - (\mathbf{r}_u \times \mathbf{f})_v\}.$$

Since $\mathbf{r}_u, \mathbf{r}_v$ are tangent to the surface, (9) shows that: *If a vector* $\mathbf{f}$ *is everywhere normal to the surface,* $\mathbf{n} \cdot \operatorname{rot} \mathbf{f} = 0$.

With $\mathbf{f} = \nabla_s\mathbf{g}$ in (9) we obtain the important identity,

$$(11) \qquad \mathbf{n} \cdot \nabla_s \times (\nabla_s\mathbf{g}) = 0;$$

for then

$$\mathbf{r}_v \cdot \mathbf{f} = \mathbf{g}_v, \qquad \mathbf{r}_u \cdot \mathbf{f} = \mathbf{g}_u \qquad (95.7).$$

When $\mathbf{f} = \mathbf{pq}$, a dyad, we have, from (9),

$$\mathbf{n} \cdot \nabla \times (\mathbf{pq}) = \frac{1}{H}\{(\mathbf{r}_v \cdot \mathbf{pq})_u - (\mathbf{r}_u \cdot \mathbf{pq})_v\}$$

$$= \frac{1}{H}\{(\mathbf{r}_v \cdot \mathbf{p})_u - (\mathbf{r}_u \cdot \mathbf{p})_v\}\mathbf{q} + \frac{1}{H}\mathbf{p} \cdot (\mathbf{r}_v\mathbf{q}_u - \mathbf{r}_u\mathbf{q}_v),$$

that is,

$$(12) \qquad \mathbf{n} \cdot \nabla \times (\mathbf{pq}) = (\mathbf{n} \cdot \operatorname{rot} \mathbf{p})\mathbf{q} + \mathbf{p} \cdot \mathbf{n} \times \nabla\mathbf{q}.$$

For future use we compute the invariant (10) when $\mathbf{f} = \mathbf{n}\mathbf{g}$ and $\mathbf{g}$ is an arbitrary tensor:

$$\mathbf{n} \overset{2}{\cdot} \nabla\mathbf{f} - \mathbf{n}\,\nabla \cdot \mathbf{f} = \frac{1}{H}\{(\mathbf{r}_v \times \mathbf{n}\,\mathbf{g})_u - (\mathbf{r}_u \times \mathbf{n}\,\mathbf{g})_v\}$$

$$= \mathbf{a}\mathbf{g}_u + \mathbf{b}\mathbf{g}_v + \frac{1}{H}\{(\mathbf{r}_v \times \mathbf{n})_u - (\mathbf{r}_u \times \mathbf{n})_v\}\,\mathbf{g}.$$

Now

$$\text{Grad}\,\mathbf{g} = \mathbf{a}\mathbf{g}_u + \mathbf{b}\mathbf{g}_v \qquad (95.5),$$

$$\mathbf{n} \overset{2}{\cdot} \text{Grad}\,\mathbf{n} = (\mathbf{a}\mathbf{n}_u + \mathbf{b}\mathbf{n}_v) \cdot \mathbf{n} = 0 \qquad (96.9);$$

hence, on putting $\mathbf{f} = \mathbf{n}$ in (10), we have

$$-\mathbf{n}\,\text{Div}\,\mathbf{n} = \frac{1}{H}\{(\mathbf{r}_v \times \mathbf{n})_u - (\mathbf{r}_u \times \mathbf{n})_v\}.$$

Thus, with $\mathbf{f} = \mathbf{n}\mathbf{g}$,

$$\mathbf{n} \overset{2}{\cdot} \nabla\mathbf{f} - \mathbf{n}\,\nabla \cdot \mathbf{f} = \text{Grad}\,\mathbf{g} - (\text{Div}\,\mathbf{n})\,\mathbf{n}\mathbf{g}. \tag{13}$$

98. Summary: Differential Invariants. Let $\mathbf{f}(\mathbf{r})$ be a tensor point function of valence ν; its derivative at the point P and in the direction of the unit vector $\mathbf{e}$ is defined as

$$\frac{d\mathbf{f}}{ds} = \lim_{s \to 0} \frac{\mathbf{f}(\mathbf{r} + s\mathbf{e}) - \mathbf{f}(\mathbf{r})}{s} \qquad (s > 0).$$

If $d\mathbf{f}/ds_i$ denote the directional derivatives corresponding to three non-coplanar unit vectors $\mathbf{e}_i$, the *gradient* of $\mathbf{f}$, namely,

$$\text{grad}\,\mathbf{f} = \nabla\mathbf{f} = \mathbf{e}^1 \frac{d\mathbf{f}}{ds_1} + \mathbf{e}^2 \frac{d\mathbf{f}}{ds_2} + \mathbf{e}^3 \frac{d\mathbf{f}}{ds_3},$$

has the property,

$$\frac{d\mathbf{f}}{ds} = \mathbf{e} \cdot \nabla\mathbf{f}.$$

$\nabla\mathbf{f}$, a tensor of valence $\nu + 1$, thus gives a synthesis of all the directional derivatives of $\mathbf{f}$ at point $\mathbf{r}$. If the vectors $\mathbf{e}_i$ are $\mathbf{i}, \mathbf{j}, \mathbf{k}$,

$$\nabla\mathbf{f} = \mathbf{i}\mathbf{f}_x + \mathbf{j}\mathbf{f}_y + \mathbf{k}\mathbf{f}_z \qquad (\mathbf{f}_x = \partial\mathbf{f}/\partial x, \text{ etc.}).$$

From $\nabla\mathbf{f}$ (valence $\nu + 1$) we derive the invariants $\nabla \cdot \mathbf{f}$ (valence $\nu - 1$) and $\nabla \times \mathbf{f}$ (valence ν) by dotting and crossing in the first position. When $\mathbf{f}$ is a vector ($\nu = 1$),

$$\nabla \cdot \mathbf{f} = \text{div}\,\mathbf{f}, \quad \text{the } \textit{divergence} \text{ of } \mathbf{f},$$

$$\nabla \times \mathbf{f} = \text{rot}\,\mathbf{f}, \quad \text{the } \textit{rotation} \text{ of } \mathbf{f}.$$

When $\mathbf{r}$ is a function $\mathbf{r}(u, v, w)$ of curvilinear coordinates

$$\nabla\mathbf{f} = \nabla u\,\mathbf{f}_u + \nabla v\,\mathbf{f}_v + \nabla w\,\mathbf{f}_w,$$

$$\nabla\mathbf{r} = \nabla u\,\mathbf{r}_u + \nabla v\,\mathbf{r}_v + \nabla w\,\mathbf{r}_w = \mathbf{I}.$$

The sets ∇u, ∇v, ∇w and $\mathbf{r}_u$, $\mathbf{r}_v$, $\mathbf{r}_w$ are reciprocal. With $J = [\mathbf{r}_u\mathbf{r}_v\mathbf{r}_w]$, we have

$$J\,\nabla\mathbf{f} = (\mathbf{r}_v \times \mathbf{r}_w\,\mathbf{f})_u + \text{cycl},$$

$$J\,\nabla\cdot\mathbf{f} = (\mathbf{r}_v \times \mathbf{r}_w \cdot \mathbf{f})_u + \text{cycl},$$

$$J\,\nabla\times\mathbf{f} = ((\mathbf{r}_v \times \mathbf{r}_w) \times \mathbf{f})_u + \text{cycl} = \begin{vmatrix} \mathbf{r}_u & \mathbf{r}_v & \mathbf{r}_w \\ \partial/\partial u & \partial/\partial v & \partial/\partial w \\ \mathbf{r}_u\cdot\mathbf{f} & \mathbf{r}_v\cdot\mathbf{f} & \mathbf{r}_w\cdot\mathbf{f} \end{vmatrix}.$$

When $\mathbf{r}_u$, $\mathbf{r}_v$, $\mathbf{r}_w$ are mutually orthogonal, the coordinates u, v, w are called *orthogonal*. Then if $\mathbf{a}$, $\mathbf{b}$, $\mathbf{c}$ denote a dextral set of orthogonal unit vectors, $\mathbf{r}_u = U\mathbf{a}$, $\mathbf{r}_v = V\mathbf{b}$, $\mathbf{r}_w = W\mathbf{c}$, and $J = UVW$ in the preceding formulas.

For any tensor $\mathbf{f}(\mathbf{r})$,

$$\nabla\cdot\nabla\mathbf{f} = \nabla^2\mathbf{f}, \qquad \nabla\times\nabla\mathbf{f} = 0,$$

$$\nabla\cdot(\nabla\times\mathbf{f}) = 0, \qquad \nabla\times(\nabla\times\mathbf{f}) = \nabla(\nabla\cdot\mathbf{f}) - \nabla^2\mathbf{f}.$$

The operator $\nabla\cdot\nabla$ is called the *Laplacian.*

A vector $\mathbf{f}$ is called *irrotational* if $\operatorname{rot}\mathbf{f} = 0$, *solenoidal* if $\operatorname{div}\mathbf{f} = 0$. An irrotational vector $\mathbf{f}$ can be expressed as the gradient of a scalar ($\mathbf{f} = \nabla\varphi$); a solenoidal vector $\mathbf{f}$ can be expressed as the rotation of a vector ($\mathbf{f} = \operatorname{rot}\mathbf{g}$).

For the surface $\mathbf{r} = \mathbf{r}(u, v)$, the fundamental quadratic form is

$$ds^2 = E\,du^2 + 2F\,du\,dv + G\,dv^2,$$

where

$$E = \mathbf{r}_u\cdot\mathbf{r}_u, \qquad F = \mathbf{r}_u\cdot\mathbf{r}_v, \qquad G = \mathbf{r}_v\cdot\mathbf{r}_v.$$

At a regular point,

$$H = |\mathbf{r}_u \times \mathbf{r}_v| = \sqrt{EG - F^2} > 0,$$

and the unit surface normal is defined as $\mathbf{n} = \mathbf{r}_u \times \mathbf{r}_v/H$.

The derivative of a tensor $\mathbf{f}(\mathbf{r})$ along any surface curve tangent to the unit vector $\mathbf{e}$ at the point (u, v) is

$$\frac{d\mathbf{f}}{ds} = \mathbf{e}\cdot(\mathbf{a}\,\mathbf{f}_u + \mathbf{b}\,\mathbf{f}_v);$$

the set $\mathbf{a}$, $\mathbf{b}$, $\mathbf{n}$ is reciprocal to $\mathbf{r}_u$, $\mathbf{r}_v$, $\mathbf{n}$.

The *surface gradient,*

$$\nabla_s\mathbf{f} = \operatorname{Grad}\mathbf{f} = \mathbf{a}\,\mathbf{f}_u + \mathbf{b}\,\mathbf{f}_v,$$

has the properties,

$$\mathbf{e} \cdot \operatorname{Grad} \mathbf{f} = \frac{d\mathbf{f}}{ds}, \qquad \mathbf{n} \cdot \operatorname{Grad} \mathbf{f} = 0;$$

$$\mathbf{r}_u \cdot \operatorname{Grad} \mathbf{f} = \frac{d\mathbf{f}}{\partial u}, \qquad \mathbf{r}_v \cdot \operatorname{Grad} \mathbf{f} = \frac{d\mathbf{f}}{\partial v}.$$

From $\nabla_s \mathbf{f}$ we derive the invariants,

$$\nabla_s \cdot \mathbf{f} = \mathbf{a} \cdot \mathbf{f}_u + \mathbf{b} \cdot \mathbf{f}_v, \qquad \nabla_s \times \mathbf{f} = \mathbf{a} \times \mathbf{f}_u + \mathbf{b} \times \mathbf{f}_v,$$

by dotting and crossing in the first position. When $\mathbf{f}$ is a vector,

$$\nabla_s \cdot \mathbf{f} = \operatorname{Div} \mathbf{f}, \quad \text{the } \textit{surface divergence of } \mathbf{f},$$

$$\nabla_s \times \mathbf{f} = \operatorname{Rot} \mathbf{f}, \quad \text{the } \textit{surface rotation} \text{ of } \mathbf{f}.$$

The tensor $\mathbf{n} \times \nabla\mathbf{f}$ and its dot and cross invariants,

$$\mathbf{n} \cdot \nabla \times \mathbf{f}, \qquad \mathbf{n} \overset{2}{\cdot} \nabla\mathbf{f} - \mathbf{n}\,\nabla \cdot \mathbf{f},$$

are not altered when ∇ is replaced by ∇_s. They may be computed solely from the values $\mathbf{f}(\mathbf{r})$ assumes on the surface. Thus, if $H = [\mathbf{r}_u \mathbf{r}_v \mathbf{n}]$,

$$\mathbf{n} \times \nabla\mathbf{f} = \frac{1}{H}\{(\mathbf{r}_v\mathbf{f})_u - (\mathbf{r}_u\mathbf{f})_v\};$$

and the invariants follow by dotting and crossing between $\mathbf{r}_u$, $\mathbf{r}_v$ and $\mathbf{f}$.

PROBLEMS

1. If $\mathbf{R} = \mathbf{r}/r$ is the unit radial vector, prove that

$$\operatorname{div} \mathbf{R} = 2/r, \qquad \operatorname{rot} \mathbf{R} = 0.$$

2. For any scalar function $f(r)$ of r alone prove that

$$\nabla^2 f(r) = f_{rr} + 2f_r/r.$$

If $\nabla^2 f(r) = 0$, show that $f = A/r + B$.

3. If $\mathbf{a}$ is a constant vector, prove that

$$\operatorname{grad} (\mathbf{a} \cdot \mathbf{r}) = \mathbf{a}, \quad \operatorname{div} (\mathbf{a} \times \mathbf{r}) = 0, \quad \operatorname{rot} (\mathbf{a} \times \mathbf{r}) = 2\mathbf{a}.$$

4. If $\mathbf{a}$ is a constant vector, prove that

$$\operatorname{grad} (\mathbf{a} \cdot \mathbf{f}) = \nabla\mathbf{f} \cdot \mathbf{a} = \mathbf{a} \cdot \nabla\mathbf{f} + \mathbf{a} \times \operatorname{rot} \mathbf{f},$$

$$\operatorname{div} (\mathbf{a} \times \mathbf{f}) = -\mathbf{a} \cdot \operatorname{rot} \mathbf{f},$$

$$\operatorname{rot} (\mathbf{a} \times \mathbf{f}) = \mathbf{a} \operatorname{div} \mathbf{f} - \mathbf{a} \cdot \nabla\mathbf{f}.$$

Specialize these results when $\mathbf{f} = \mathbf{r}$.

5. For any vector point function **f** prove that

$$\tfrac{1}{2}\nabla(\mathbf{f}\cdot\mathbf{f}) = \nabla\mathbf{f}\cdot\mathbf{f} = \mathbf{f}\cdot\nabla\mathbf{f} + \mathbf{f}\times\text{rot }\mathbf{f}.$$

6. If φ and ψ are scalar point functions, prove that

$$\text{div }(\nabla\varphi\times\nabla\psi) = 0$$

7. If **e** is a unit vector, prove that

$$\text{div }(\mathbf{e}\cdot\mathbf{r})\mathbf{e} = 1, \qquad \text{rot }(\mathbf{e}\cdot\mathbf{r})\mathbf{e} = 0;$$

$$\text{div }(\mathbf{e}\times\mathbf{r})\times\mathbf{e} = 2, \qquad \text{rot }(\mathbf{e}\times\mathbf{r})\times\mathbf{e} = 0.$$

8. If **a** is a constant vector, prove that $\nabla(\mathbf{r}\times\mathbf{a}) = \mathbf{I}\times\mathbf{a}$.

9. Show that Laplace's Equation $\nabla^2 f = 0$ in cylindrical and spherical coordinates is

$$\frac{\partial^2 f}{\partial\rho^2} + \frac{1}{\rho}\frac{\partial f}{\partial\rho} + \frac{1}{\rho^2}\frac{\partial^2 f}{\partial\varphi^2} + \frac{\partial^2 f}{\partial z^2} = 0,$$

and

$$r\frac{\partial^2(rf)}{\partial r^2} + \frac{1}{\sin\theta}\frac{\partial}{\partial\theta}\left(\sin\theta\frac{\partial f}{\partial\theta}\right) + \frac{1}{\sin^2\theta}\frac{\partial^2 f}{\partial\varphi^2} = 0.$$

10. Prove that

$$\nabla\nabla\frac{1}{r} = \frac{1}{r^3}(3\mathbf{R}\mathbf{R} - \mathbf{I})$$

where **R** is a unit radial vector.

11. If **f** is a vector point function, prove that

$$\nabla\cdot(\nabla\mathbf{f})_c = \text{grad div }\mathbf{f}; \qquad \nabla\times(\nabla\mathbf{f})_c = (\nabla\text{ rot }\mathbf{f})_c.$$

12. If $\nabla\mathbf{f}$ is antisymmetric, prove that rot **f** is constant and that the dyadic $\nabla\mathbf{f}$ itself is constant.

13. If λ is a scalar point function, prove that

$$\nabla(\lambda\mathbf{I}) = \nabla\lambda\mathbf{I}, \quad \nabla\cdot(\lambda\mathbf{I}) = \nabla\lambda, \quad \nabla\times(\lambda\mathbf{I}) = \nabla\lambda\times\mathbf{I}.$$

14. If **f** is a vector point function, prove that

$$\nabla\cdot(\mathbf{I}\times\mathbf{f}) = \text{rot }\mathbf{f}, \qquad \nabla\times(\mathbf{I}\times\mathbf{f}) = (\nabla\mathbf{f})_c - \mathbf{I}\text{ div }\mathbf{f}.$$

15. For the dyad **fg** prove that

$$\nabla\cdot(\mathbf{fg}) = (\text{div }\mathbf{f})\mathbf{g} + \mathbf{f}\cdot\nabla\mathbf{g}, \qquad \nabla\times(\mathbf{fg}) = (\text{rot }\mathbf{f})\mathbf{g} - \mathbf{f}\times\nabla\mathbf{g}.$$

In particular, if **r** is the position vector,

$$\nabla\cdot(\mathbf{rr}) = 4\mathbf{r}, \qquad \nabla\times(\mathbf{rr}) = -\mathbf{I}\times\mathbf{r}.$$

16. If Φ is a constant dyadic, prove that

$$\text{div }\Phi\cdot\mathbf{r} = \varphi_1; \qquad \text{rot }\Phi\cdot\mathbf{r} = -\boldsymbol{\phi}\ (\S\ 69).$$

17. If the scalar function $f(\rho, \varphi)$ in plane polar coordinates is harmonic, prove that $f(\rho^{-1}, \varphi)$ is also harmonic.

18. If the scalar function $f(r, \theta, \varphi)$ in spherical coordinates is harmonic, prove that $r^{-1}f(r^{-1}, \theta, \varphi)$ is harmonic.

19. If f, g, h are scalar point functions, prove that

$$\nabla^2(fg) = g\nabla^2 f + 2\nabla f \cdot \nabla g + f\nabla^2 g;$$

$$\nabla^2(fgh) = gh\,\nabla^2 f + hf\,\nabla^2 g + fg\,\nabla^2 h + 2f\nabla g \cdot \nabla h + 2g\nabla h \cdot \nabla f + 2h\nabla f \cdot \nabla g.$$

20. If u, v, w are orthogona' coordinates and $f(u)$, $g(v)$, $h(w)$ are scalar functions of a single variable, show that

$$\nabla^2(fgh) = fgh\left\{\frac{\nabla^2 f}{f} + \frac{\nabla^2 g}{g} + \frac{\nabla^2 h}{h}\right\}.$$

21. If $\mathbf{a}$ is a constant vector and $\mathbf{f} = \mathbf{a}r^n$, prove that

$$\nabla \mathbf{f} = nr^{n-2}\mathbf{ra}, \qquad \nabla^2 \mathbf{f} = n(n+1)r^{n-2}\mathbf{a}.$$

22. If $\mathbf{f} = r^n\mathbf{r}$, prove that

$$\nabla \mathbf{f} = r^n\mathbf{I} + nr^{n-2}\mathbf{rr}, \qquad \nabla^2 \mathbf{f} = n(n+3)r^{n-2}\mathbf{r}.$$

23. If $\mathbf{f} = r^n\mathbf{r}$, find a scalar function φ such that $\mathbf{f} = \nabla\varphi$.

24. Prove that rot $\mathbf{f} = 0$, and find φ so that $\mathbf{f} = \nabla\varphi$:

(a) $\mathbf{f} = yz\mathbf{i} + zx\mathbf{j} + xy\mathbf{k}$; (b) $\mathbf{f} = 2xy\mathbf{i} + (x^2 + \log z)\mathbf{j} + y/z\,\mathbf{k}$;

(c) $\mathbf{f} = \Phi \cdot \mathbf{r}$, Φ a constant symmetric dyadic.

25. If $\mathbf{a}$ is a constant vector and $\mathbf{f} = r^n\,\mathbf{a} \times \mathbf{r}$, prove that

$$\operatorname{div} \mathbf{f} = 0, \qquad \operatorname{rot} \mathbf{f} = (n+2)r^n\mathbf{a} - nr^{n-2}(\mathbf{a} \cdot \mathbf{r})\mathbf{r}.$$

26. Find a vector $\mathbf{g}$ such that rot $\mathbf{g} = \mathbf{a} \times \mathbf{r}$ (cf. § 92).

27. Prove that

$$\nabla \frac{\mathbf{a} \cdot \mathbf{r}}{r^3} = -\operatorname{rot} \frac{\mathbf{a} \times \mathbf{r}}{r^3}.$$

28. The position vectors from O to the fixed points P_1, P_2 are $\mathbf{r}_1$, $\mathbf{r}_2$.

(a) If $\mathbf{f} = (\mathbf{r} - \mathbf{r}_1) \times (\mathbf{r} - \mathbf{r}_2)$, show that

$$\operatorname{div} \mathbf{f} = 0, \qquad \operatorname{rot} \mathbf{f} = 2(\mathbf{r}_2 - \mathbf{r}_1).$$

(b) Prove that

$$\nabla(\mathbf{r} - \mathbf{r}_1) \cdot (\mathbf{r} - \mathbf{r}_2) = 2\mathbf{r} - \mathbf{r}_1 - \mathbf{r}_2.$$

29. If $\mathbf{f}(u, v)$ is a vector function defined over the surface $\mathbf{r} = \mathbf{r}(u, v)$, prove that

$$H \operatorname{Div} \mathbf{f} = \mathbf{n} \cdot (\mathbf{r}_u \times \mathbf{f}_v - \mathbf{r}_v \times \mathbf{f}_u).$$

30. If the vector $\mathbf{f}(u, v)$ is always normal to the surface $\mathbf{r} = \mathbf{r}(u, v)$, prove that Rot $\mathbf{f}$ is tangent to the surface or zero.

31. If u, v, w are curvilinear coordinates, prove the operational identities:

$$[\mathbf{r}_u\mathbf{r}_v\mathbf{r}_w]\nabla = \mathbf{r}_v \times \mathbf{r}_w \frac{\partial}{\partial u} + \mathbf{r}_w \times \mathbf{r}_u \frac{\partial}{\partial v} + \mathbf{r}_u \times \mathbf{r}_v \frac{\partial}{\partial w};$$

$$(\mathbf{r}_u \times \mathbf{r}_v) \times \nabla = \mathbf{r}_v \frac{\partial}{\partial u} - \mathbf{r}_u \frac{\partial}{\partial v}; \qquad \mathbf{r}_u \cdot \nabla = \frac{\partial}{\partial u}.$$

CHAPTER VI

INTEGRAL TRANSFORMATIONS

99. Green's Theorem in the Plane. Let R be a region of the xy-plane bounded by a simple closed curve C which consists of a finite number of smooth arcs. Then, if $P(x, y)$, $Q(x, y)$ are continuous functions with continuous first partial derivatives,

$$\iint_R \left(\frac{\partial Q}{\partial x} - \frac{\partial P}{\partial y}\right) dx\, dy = \oint_C (P\, dx + Q\, dy), \tag{1}$$

where the circuit integral on the right is taken in the *positive sense;* then a person making the circuit C will always have the region R to his left.

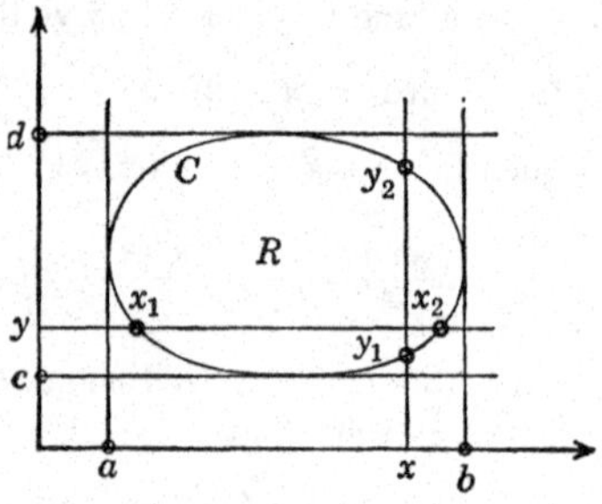

FIG. 99a

Consider first a region R in which boundary C is cut by every line parallel to the x- or y-axis in two points at most (Fig. 99a). Then, if a horizontal line cuts C in the points (x_1, y), (x_2, y),

$$\begin{aligned}\iint \frac{\partial Q}{\partial x}\, dx\, dy &= \int_c^d \{Q(x_2, y) - Q(x_1, y)\}\, dy \\ &= \int_c^d Q(x_2, y)\, dy + \int_d^c Q(x_1, y)\, dy \\ &= \oint Q(x, y)\, dy.\end{aligned}$$

And, if a vertical line cuts C in the points (x, y_1), (x, y_2),

$$\iint \frac{\partial P}{\partial y}\, dy\, dx = \int_a^b \{P(x, y_2) - P(x, y_1)\}\, dx$$

$$= -\int_a^b P(x, y_1)\, dx - \int_b^a P(x, y_2)\, dx$$

$$= -\oint P(x, y)\, dx.$$

On subtracting this equation from the preceding, we get (1).

We now can extend this formula to more general regions that may be divided into a finite number of subregions which have the property that a line parallel to the x- or y-axis cuts their boundary

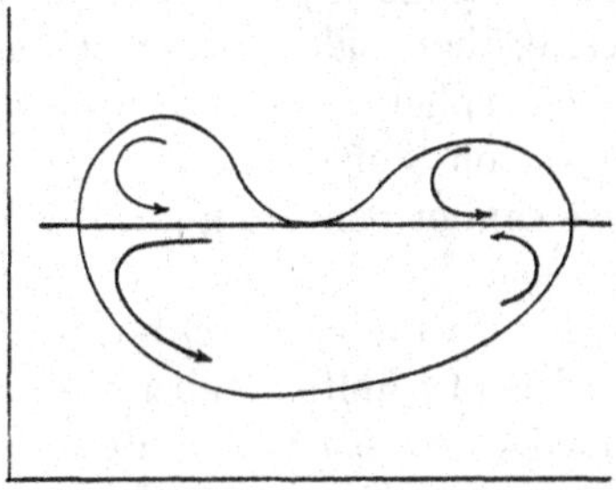

FIG. 99b

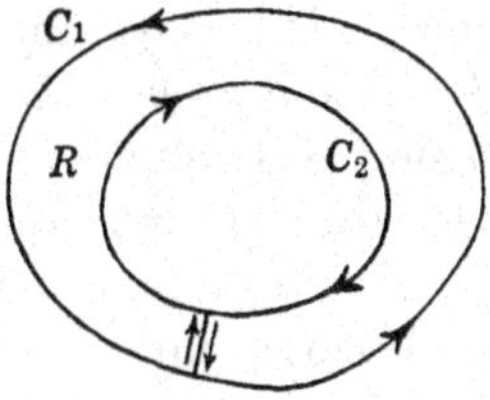

FIG. 99c

in at most two points. For in each subregion formula (1) is valid, and, when these equations are added for all the subregions, the surface integrals add up to the integral over the entire region; but the line integrals over the internal boundaries cancel, since each is traversed twice, but in opposite directions, leaving only the line integral over the external bounding traversed in a positive direction (Fig. 99b).

The boundary of R may even consist of two or more closed curves: thus in Fig. 99c the region R is interior to C_1 but exterior to C_2; we may now make a cut between C_1 and C_2 and traverse the entire boundary of R in the positive sense as shown by the arrows. The line integrals over the cut cancel, for it is traversed twice and in opposite directions; and the resultant line integral over C consists of a counterclockwise circuit of C_1 and a clockwise circuit of C_2.

Although Green's Theorem is commonly stated for scalar functions P, Q, it is evidently true when P and Q are tensors with continuous first partial derivatives; for, when P and Q are referred to a constant basis, the theorem holds for each scalar component.

100. Reduction of Surface to Line Integrals. A surface is said to be *bilateral* if it is possible to distinguish one of its sides from the other. Not all surfaces are bilateral. The simplest unilateral surface is the *Möbius strip;* this may be materialized by taking a rectangular strip of paper, giving it one twist, and pasting the ends ab and $a'b'$ together (Fig. 100a). This surface has but *one side;* if we move a point P along its median line and make a complete circuit of the strip it will arrive at the point P' directly underneath. Since we can travel on a continuous path from one side of the Möbius Strip to the opposite, these sides cannot be distinguished. This is not the case with a spherical surface, which has an *inside* and an *outside*, or any portion S of the surface bounded by a simple closed curve C; for we cannot pass from one side of S to the other without crossing C.

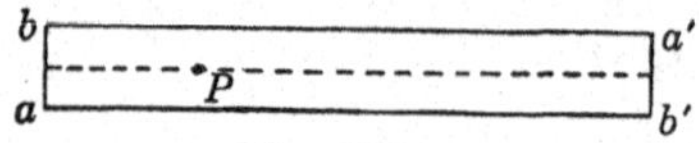

Fig. 100a

Let S be a portion of a bilateral surface $\mathbf{r} = \mathbf{r}(u, v)$ bounded by a simple closed curve C which consists of a finite number of smooth arcs. The surface itself is assumed to consist of a finite number of parts over which the unit normal $\mathbf{n}$ is continuous. The positive sense on C is such that a person, erect in the direction of $\mathbf{n}$, will have S to the left on making the circuit C. If such an oriented surface is continuously deformed into a portion of the xy-plane bounded by a curve C', $\mathbf{n}$ becomes $\mathbf{k}$ (the unit vector in the direction $+z$) and the positive circuit on C' forms with $\mathbf{n} = \mathbf{k}$ a right-handed screw.

Let $f(\mathbf{r})$ be a tensor point function (scalar, vector, dyadic, etc.) whose scalar components have continuous derivatives over S. Such a tensor is said to be *continuously differentiable.* When the foregoing conditions on S and C are fulfilled, we then have the

Basic Theorem I. *If* $\mathbf{f}(\mathbf{r})$ *is a continuously differentiable tensor point function over* S, *the surface integral of* $\mathbf{n} \times \nabla\mathbf{f}$ *over* S *is equal to the integral of* $\mathbf{T}\mathbf{f}$ *taken about its boundary* C *in the positive sense:*

$$(1) \qquad \int_S \mathbf{n} \times \nabla\mathbf{f}\, dS = \oint_C \mathbf{T}\mathbf{f}\, ds.$$

Proof. Since

$$\mathbf{n} \times \nabla \mathbf{f} = \frac{1}{H}\{(\mathbf{r}_v\mathbf{f})_u - (\mathbf{r}_u\mathbf{f})_v\}, \qquad dS = H\,du\,dv,$$

$$\int_S \mathbf{n} \times \nabla \mathbf{f}\,dS = \int_{S'} \{(\mathbf{r}_v\mathbf{f})_u - (\mathbf{r}_u\mathbf{f})_v\}du\,dv,$$

where S' is the region of the uv-plane which contains all parameter values u, v corresponding to points of S. We now may apply

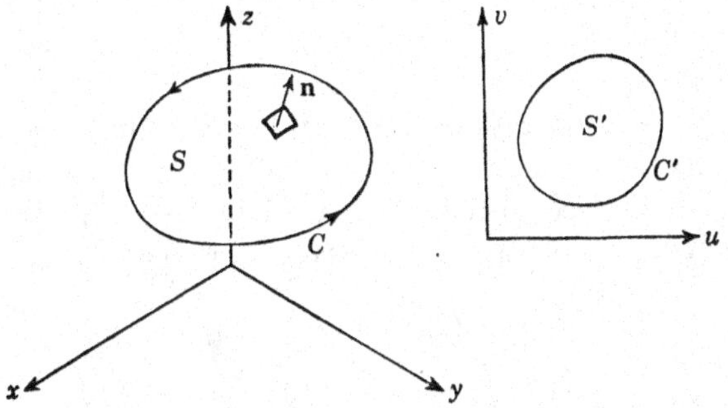

FIG. 100b

Green's Theorem in the plane to the last integral, letting u and v play the rôles of x and y; thus

$$\int_{S'} \{(\mathbf{r}_v\mathbf{f})_u - (\mathbf{r}_u\mathbf{f})_v\}du\,dv = \oint_{C'} (\mathbf{r}_u\mathbf{f}\,du + \mathbf{r}_v\mathbf{f}\,dv),$$

where the circuit integral is taken about the curve C', which forms the boundary of S', in the positive sense, that is, in the direction of a rotation of the positive u-axis into the positive v-axis. If we regard u and v as the functions of the arc s on the curve C, it becomes

$$\oint_C \left(\mathbf{r}_u\mathbf{f}\frac{du}{ds} + \mathbf{r}_v\mathbf{f}\frac{dv}{ds}\right) ds = \oint_C \mathbf{T}\mathbf{f}\,ds,$$

since

$$\mathbf{r}_u\frac{du}{ds} + \mathbf{r}_v\frac{dv}{ds} = \frac{d\mathbf{r}}{ds} = \mathbf{T},$$

the positive unit tangent along C. Formula (1) thus is established.

If we introduce the *vector* elements of surface and of arc,

$$d\mathbf{S} = \mathbf{n}\, dS, \qquad d\mathbf{r} = \boldsymbol{\tau}\, ds,$$

the basic theorem for transforming surface to line integrals becomes

$$(2) \qquad \int_S d\mathbf{S} \times \nabla \mathbf{f} = \oint_C d\mathbf{r}\, \mathbf{f},$$

in which $d\mathbf{S}$ and $d\mathbf{r}$ must be written as *prefactors.*

If $\mathbf{f}$ is a vector, dyadic, etc., we may obtain the other integral transformations from (1) by placing a dot or a cross between the first two vectors in each term (dyad, triad, etc.) of the integrands; for both of these operations are distributive with respect to addition and therefore may be carried out under the integral sign. This process applied to the tensor $\mathbf{n} \times \nabla \mathbf{f}$ gives $\mathbf{n} \cdot \nabla \times \mathbf{f}$ and $\mathbf{n} \overset{2}{\cdot} \nabla \mathbf{f} - \mathbf{n} \nabla \cdot \mathbf{f}$; we thus obtain the important formulas:

$$(3) \qquad \int \mathbf{n} \cdot \nabla \times \mathbf{f}\, dS = \oint \boldsymbol{\tau} \cdot \mathbf{f}\, ds,$$

$$(4) \qquad \int (\mathbf{n} \overset{2}{\cdot} \nabla \mathbf{f} - \mathbf{n}\, \nabla \cdot \mathbf{f})\, dS = \oint \boldsymbol{\tau} \times \mathbf{f}\, ds.$$

When $\mathbf{f}$ is a vector function, (3) becomes *Stokes' Theorem:*

$$(3)' \qquad \int \mathbf{n} \cdot \operatorname{rot} \mathbf{f}\, dS = \oint \boldsymbol{\tau} \cdot \mathbf{f}\, ds,$$

a result of first importance in differential geometry, hydrodynamics and electrodynamics.

If S is a closed surface which is divided into two parts S_1, S_2 by the curve C, we may choose $\mathbf{n}$ as the unit external normal; then, from (1),

$$\int_{S_1} \mathbf{n} \times \nabla \mathbf{f}\, dS = \oint_C \boldsymbol{\tau}_1 \mathbf{f}\, ds, \qquad \int_{S_2} \mathbf{n} \times \nabla \mathbf{f}\, dS = \oint_C \boldsymbol{\tau}_2 \mathbf{f}\, ds,$$

where $\boldsymbol{\tau}_2 = -\boldsymbol{\tau}_1$, since the positive sense on C regarded as bounding S_1 is reversed when regarded as bounding S_2. On adding these integrals, we find that the integral of $\mathbf{n} \times \nabla \mathbf{f}$ over the entire surface is zero. We indicate this by the notation,

$$(5) \qquad \oint \mathbf{n} \times \nabla \mathbf{f}\, dS = 0,$$

which denotes an integral over a *closed* surface. On taking dot and cross invariants in (5), we have also

$$\oint \mathbf{n} \cdot \nabla \times \mathbf{f}\, dS = 0, \tag{6}$$

$$\oint (\mathbf{n} \overset{2}{\cdot} \nabla \mathbf{f} - \mathbf{n}\, \nabla \cdot \mathbf{f})\, dS = 0. \tag{7}$$

Example 1. Put $f = \mathbf{r}$ in (4) and (7); then, since grad $\mathbf{r} = \mathbf{I}$, div $\mathbf{r} = 3$,

$$\int \mathbf{n}\, dS = \tfrac{1}{2} \oint \mathbf{r} \times d\mathbf{r}, \qquad \oint \mathbf{n}\, dS = 0.$$

The last result, which states that the vector area of a closed surface is zero, generalizes the polyhedron theorem of § 17.

Example 2. When div $\mathbf{f} = 0$, we have seen that $\mathbf{f} = \operatorname{rot} \mathbf{g}$ (§ 92); hence

$$\int \mathbf{n} \cdot \mathbf{f}\, dS = \int \mathbf{n} \cdot \operatorname{rot} \mathbf{g}\, dS = \oint \boldsymbol{\tau} \cdot \mathbf{g}\, ds.$$

Thus, if $\mathbf{f} = \nabla u \times \nabla v$, div $\mathbf{f} = 0$, and we may take $\mathbf{g}$ as $u \nabla v$ or $-v \nabla u$; hence

$$\int \mathbf{n} \cdot \nabla u \times \nabla v\, dS = \oint u\, dv = -\oint v\, du = \tfrac{1}{2} \oint (u\, dv - v\, du).$$

In particular, when $u = x$, $v = y$, $\mathbf{n} = \mathbf{k}$, this formula expresses an area in the xy-plane as a circuit integral $\tfrac{1}{2} \oint (x\, dy - y\, dx)$ over its boundary.

101. Alternative Form of Transformation. If we replace $\mathbf{f}$ by $\mathbf{n}\mathbf{g}$ in (100.4), where $\mathbf{g}$ is any continuously differentiable tensor, we have, from (97.13),

$$\int \{\operatorname{Grad} \mathbf{g} - (\operatorname{Div} \mathbf{n})\, \mathbf{n}\mathbf{g}\}\, dS = \oint \boldsymbol{\tau} \times \mathbf{n}\, \mathbf{g}\, ds. \tag{1}$$

This integral transformation is quite as general as that given by basic theorem I. For, if we replace $\mathbf{g}$ by $\mathbf{n}\mathbf{f}$ and then cross in the first position, the integrand on the left becomes

$$\mathbf{a} \times (\mathbf{n}_u \mathbf{f} + \mathbf{n}\, \mathbf{f}_u) + \mathbf{b} \times (\mathbf{n}_v \mathbf{f} + \mathbf{n}\, \mathbf{f}_v) = -\mathbf{n} \times \operatorname{Grad} \mathbf{f},$$

since Rot $\mathbf{n} = 0$; whereas, on the right,

$$(\boldsymbol{\tau} \times \mathbf{n}) \times \mathbf{n}\, \mathbf{f} = -\boldsymbol{\tau}\mathbf{f}.$$

We thus retrieve the basic transformation (100.1).

We now shall express (1) in slightly different notation. On the right the vector $\boldsymbol{\tau} \times \mathbf{n} = \mathbf{m}$ is the unit normal to C tangent to S and pointing *outward*. If we write $\mathbf{f}$ instead of $\mathbf{g}$ and $J = -\operatorname{Div} \mathbf{n}$ (J is not the Jacobian of § 88), (1) now becomes

(2) $$\int (\nabla_s \mathbf{f} + J\, \mathbf{n}\mathbf{f})\, dS = \oint \mathbf{m}\, \mathbf{f}\, ds.$$

On dotting and crossing in the first position of the tensor integrands, we obtain the additional formulas:

(3) $$\int (\nabla_s \cdot \mathbf{f} + J\, \mathbf{n} \cdot \mathbf{f})\, dS = \oint \mathbf{m} \cdot \mathbf{f}\, ds,$$

(4) $$\int (\nabla_s \times \mathbf{f} + J\, \mathbf{n} \times \mathbf{f})\, dS = \oint \mathbf{m} \times \mathbf{f}\, ds.$$

When $\mathbf{f}$ is a vector, $\nabla_s \cdot \mathbf{f} = \operatorname{Div} \mathbf{f}$, $\nabla_s \times \mathbf{f} = \operatorname{Rot} \mathbf{f}$.

$J = -\operatorname{Div} \mathbf{n}$ is called the *mean curvature* of the surface (§ 131). Surfaces for which $J = 0$ are called *minimal surfaces;* for such surfaces the integrands on the left reduce to the first term.

On the *plane,* $\mathbf{n} = \mathbf{k}$, a constant vector, and $\operatorname{Div} \mathbf{n} = 0$; then $J = 0$ in (2), (3), and (4). When $\mathbf{f}$ is a vector, these equations become (if we write dA for dS):

(5) $$\int \operatorname{Grad} \mathbf{f}\, dA = \oint \mathbf{m}\, \mathbf{f}\, ds,$$

(6) $$\int \operatorname{Div} \mathbf{f}\, dA = \oint \mathbf{m} \cdot \mathbf{f}\, ds,$$

(7) $$\int \operatorname{Rot} \mathbf{f}\, dA = \oint \mathbf{m} \times \mathbf{f}\, ds,$$

where $\mathbf{m}$ is the *external* unit normal to the closed plane curve forming the boundary. We may deduce (6) from (7) by replacing $\mathbf{f}$ by $\mathbf{k} \times \mathbf{f}$; for

$$\operatorname{Rot} (\mathbf{k} \times \mathbf{f}) = \mathbf{k} \operatorname{Div} \mathbf{f}, \qquad \mathbf{m} \times (\mathbf{k} \times \mathbf{f}) = \mathbf{k}\, \mathbf{m} \cdot \mathbf{f}.$$

102. Line Integrals. When $\mathbf{f}(\mathbf{r})$ is a vector, consider the *line integral,*

(1) $$\int_{\mathbf{r}_0}^{\mathbf{r}} \mathbf{f} \cdot d\mathbf{r} = \int_{t_0}^{t} \mathbf{f} \cdot \frac{d\mathbf{r}}{dt}\, dt,$$

taken over a curve $C: \mathbf{r} = \mathbf{r}(t)$ from the point P_0 to P. The value of this integral depends on the curve C and the end points P_0 and

P, but not on the parameter t. For, if we make the change of parameter $t = t(\tau)$,

$$\int_{t_0}^{t} \mathbf{f} \cdot \frac{d\mathbf{r}}{dt}\, dt = \int_{\tau_0}^{\tau} \mathbf{f} \cdot \frac{d\mathbf{r}}{dt}\frac{dt}{d\tau}\, d\tau = \int_{\tau_0}^{\tau} \mathbf{f} \cdot \frac{d\mathbf{r}}{d\tau}\, d\tau.$$

Let us consider under what conditions the line integral (1) is independent of the path C—that is, when its value is the same for all paths from P_0 to P.

We shall call a closed curve *reducible* in a region R if it can be shrunk continuously to a point without passing outside of R. Thus in the region between two concentric spheres all closed curves are reducible. However in the region composed of the points within a torus, all curves that encircle the axis of the torus are irreducible. A region in which all closed curves are reducible is called *simply connected.* Thus the region between two concentric spheres is simply connected, whereas the interior of a torus is not.

If $\mathbf{f}$ is continuously differentiable, and $\operatorname{rot} \mathbf{f} = 0$ in a region R, the line integral of $\mathbf{f}$ around any *reducible* closed curve in R is zero; for we can span a surface S over R that lies entirely within R (shrinking the curve to a point generates such a surface), and then, by Stokes' Theorem,

$$\oint \mathbf{f} \cdot d\mathbf{r} = \int_S \mathbf{n} \cdot \operatorname{rot} \mathbf{f}\, dS = 0.$$

We may now prove the

THEOREM. *If* $\mathbf{f}$ *is a continuously differentiable vector, and* $\operatorname{rot} \mathbf{f} = 0$ *in a simply connected region, the line integral* $\int \mathbf{f} \cdot d\mathbf{r}$ *between any two points of the region is the same for all paths in the region joining these points.*

Proof. If C_1 and C_2 are any two curves P_0AP, P_0BP joining P_0 to P, the line integral of f around the circuit P_0APBP_0 is zero; for all closed curves in a simply connected region are reducible. Hence

$$\int_{P_0AP} \mathbf{f} \cdot d\mathbf{r} + \int_{PBP_0} \mathbf{f} \cdot d\mathbf{r} = \int_{P_0AP} \mathbf{f} \cdot d\mathbf{r} - \int_{P_0BP} \mathbf{f} \cdot d\mathbf{r} = 0.$$

Under the conditions of this theorem, the line integral (1) from a fixed point P_0 to a variable point P of the region depends only upon P; in other words the integral defines a scalar point function,

(2) $$\varphi(\mathbf{r}) = \int_{\mathbf{r}_0}^{\mathbf{r}} \mathbf{f} \cdot d\mathbf{r}.$$

Let us compute $d\varphi/ds = \mathbf{e} \cdot \nabla\varphi$ for an arbitrary direction $\mathbf{e}$. We have

$$\varphi(\mathbf{r} + s\mathbf{e}) - \varphi(\mathbf{r}) = \int_{\mathbf{r}}^{\mathbf{r}+s\mathbf{e}} \mathbf{f} \cdot d\mathbf{r} = \int_0^s \mathbf{f} \cdot \mathbf{e}\, ds,$$

since $d\mathbf{r} = \mathbf{e}\, ds$ along the ray from $\mathbf{r}$ to $\mathbf{r} + s\mathbf{e}$. By the law of the mean for integrals,

$$\int_0^s \mathbf{f} \cdot \mathbf{e}\, ds = s\mathbf{e} \cdot \mathbf{f}(\mathbf{r} + \bar{s}\mathbf{e}), \qquad (0 < \bar{s} < s),$$

and hence

$$\frac{d\varphi}{ds} = \lim_{s \to 0} \frac{\varphi(\mathbf{r} + s\mathbf{e}) - \varphi(\mathbf{r})}{s} = \mathbf{e} \cdot \mathbf{f}(\mathbf{r}).$$

Since $\mathbf{e} \cdot \nabla\varphi = \mathbf{e} \cdot \mathbf{f}$ for any $\mathbf{e}$,

(3) $$\mathbf{f} = \nabla\varphi.$$

Thus an irrotational vector $\mathbf{f}$ may be expressed as the gradient of scalar φ, given by (2). The determination of φ given in § 91 is the line integral of $\mathbf{f}$ taken over a *step path* from (x_0, y_0, z_0) to (x, y, z). Thus in (91.9) the integrals, in reverse order, are taken over the straight segments from

$$P_0(x_0, y_0, z_0) - (x_0, y_0, z) - (x_0, y, z) - P(x, y, z).$$

There are six such step paths; for the first segment can be chosen in three ways and the second in two.

In mathematical physics it is customary to express an irrotational vector as the *negative* of a gradient:

(4) $$\mathbf{f} = -\nabla\psi; \qquad \text{then} \quad \psi(\mathbf{r}) = \int_{\mathbf{r}}^{\mathbf{r}_0} \mathbf{f} \cdot d\mathbf{r}$$

is called the *scalar potential* of $\mathbf{f}$.

103. Line Integrals on a Surface. We next consider line and circuit integrals over curves restricted to lie on a given surface, or within a certain region S of the surface. A region S of a surface is called *simply connected* if all closed surface curves in S are *reducible*—that is, can be shrunk continuously to a point without passing outside of S. In the plane, for example, the region within a circle is simply connected, but the region between two concentric circles is not.

If the vector $\mathbf{f}$ is continuously differentiable, the circuit integral $\oint \mathbf{f} \cdot d\mathbf{r}$ about any reducible curve of S is zero when rot $\mathbf{f}$ is everywhere tangential to the surface; for

$$\oint \mathbf{f} \cdot d\mathbf{r} = \int \mathbf{n} \cdot \operatorname{rot} \mathbf{f} \, dS = 0.$$

When S is a portion of the level surface $u(x, y, z) = c$ of a scalar function, ∇u on the surface is parallel to the surface normal and the condition $\mathbf{n} \cdot \operatorname{rot} \mathbf{f} = 0$ may be written

$$(1) \qquad \nabla u \cdot \operatorname{rot} \mathbf{f} = 0.$$

Just as in § 102, we may now prove the

THEOREM 1. *If* $\mathbf{f}$ *is a continuously differentiable vector, and* $\mathbf{n} \cdot \operatorname{rot} \mathbf{f} = 0$ *in a simply connected portion* S *of a surface, the line integral* $\int \mathbf{f} \cdot d\mathbf{r}$ *between any two points of* S *is the same for all surface curves joining these points.*

Under the conditions of this theorem the line integral,

$$(2) \qquad \varphi(\mathbf{r}) = \int_{\mathbf{r}_0}^{\mathbf{r}} \mathbf{f} \cdot d\mathbf{r},$$

is the same over all surface curves from P_0 to P and therefore defines a scalar point function $\varphi(\mathbf{r})$ in S. Along any definite curve $\mathbf{r} = \mathbf{r}(s)$ issuing from P_0 $(s = 0)$ φ is a function of the arc s,

$$\varphi(s) = \int_0^s \mathbf{f} \cdot \frac{d\mathbf{r}}{ds} \, ds = \int_0^s \mathbf{f} \cdot \mathbf{T} \, ds,$$

where $\mathbf{T}$ is the unit tangent vector; hence at any point P $d\varphi/ds = \mathbf{f} \cdot \mathbf{T}$. Since the curve may be varied so that $\mathbf{T}$ assumes any direction $\mathbf{e}$ at P, we have the relation,

$$(3) \qquad \frac{d\varphi}{ds} = \mathbf{e} \cdot \mathbf{f},$$

for all unit vectors $\mathbf{e}$ in the tangent plane at P. Now if $\mathbf{e}_1, \mathbf{e}_2$ are two such vectors, and $\mathbf{e}^1, \mathbf{e}^2, \mathbf{n}$ the set reciprocal to $\mathbf{e}_1, \mathbf{e}_2, \mathbf{n}$, we have, from (95.8),

$$\text{Grad}\ \varphi = \mathbf{e}^1 \frac{d\varphi}{ds_1} + \mathbf{e}^2 \frac{d\varphi}{ds_2} = \mathbf{e}^1\mathbf{e}_1 \cdot \mathbf{f} + \mathbf{e}^2\mathbf{e}_2 \cdot \mathbf{f}.$$

Since $\mathbf{f} = (\mathbf{e}^1\mathbf{e}_1 + \mathbf{e}^2\mathbf{e}_2 + \mathbf{nn}) \cdot \mathbf{f}$,

$$\text{(4)} \qquad \text{Grad}\ \varphi = \mathbf{f}_t,$$

the projection of $\mathbf{f}$ on the tangent plane at P. We have thus proved the

THEOREM 2. *If* $\mathbf{f}$ *is a continuously differentiable vector, and* $\mathbf{n} \cdot \text{rot}\ \mathbf{f} = 0$ *in a simply connected region of a surface, the tangential projection of* $\mathbf{f}$ *on the surface is*

$$\mathbf{f}_t = \text{Grad}\ \varphi \quad \text{where} \quad \varphi = \int_{\mathbf{r}_0}^{\mathbf{r}} \mathbf{f} \cdot d\mathbf{r}.$$

104. Field Lines of a Vector. When $\mathbf{f}(\mathbf{r})$ is a continuously differentiable vector function, the curves tangent to $\mathbf{f}(\mathbf{r})$ at all their points are called the *field lines* of $\mathbf{f}$. If $\mathbf{r} = \mathbf{r}(t)$ is a field line, $d\mathbf{r}/dt$ is tangent to the curve (§ 41) and consequently a multiple of $\mathbf{f}$. The field lines have therefore the differential equation:

$$\text{(1)} \qquad \mathbf{f} \times d\mathbf{r} = 0.$$

If $\mathbf{f} = f_1\mathbf{i} + f_2\mathbf{j} + f_3\mathbf{k}$, (1) is equivalent to the system,

$$\text{(2)} \qquad \frac{dx}{f_1} = \frac{dy}{f_2} = \frac{dz}{f_3}.$$

Any integral $u(x, y, z) = a$ *of this system represents a surface locus of field lines.* For, from $du = d\mathbf{r} \cdot \nabla u = 0$ and (1), we have

$$\text{(3)} \qquad \mathbf{f} \cdot \nabla u = 0;$$

and, since ∇u is normal to the surface $u = a$, the vector $\mathbf{f}$ at any point of the surface is in the tangent plane.

If $v(x, y, z) = b$ is a second integral of (2), we have also

$$\text{(4)} \qquad \mathbf{f} \cdot \nabla v = 0.$$

If v is independent of u, $\nabla u \times \nabla v \neq 0$ (§ 87, theorem 1). From (3) and (4) we conclude that $\mathbf{f}$ is parallel to $\nabla u \times \nabla v$, and hence

$$\text{(5)} \qquad \nabla u \times \nabla v = \lambda \mathbf{f}.$$

But the curve in which the surfaces $u = a$, $v = b$ intersect is everywhere tangent to $\nabla u \times \nabla v$. Hence we have the

THEOREM 1. *If $u = a$ and $v = b$ are independent integrals of the system $\mathbf{f} \times d\mathbf{r} = 0$, the surfaces they represent intersect in the field lines of* $\mathbf{f}$.

From (3) and (4), we see that u and v are independent solutions of the partial-differential equation:

$$\mathbf{f} \cdot \nabla w = f_1 \frac{\partial w}{\partial x} + f_2 \frac{\partial w}{\partial y} + f_3 \frac{\partial w}{\partial z} = 0. \tag{6}$$

In view of (5), this equation is equivalent to

$$\nabla u \times \nabla v \cdot \nabla w = 0. \tag{7}$$

From § 87, theorem 3, (7) is satisfied when and only when u, v and w are connected by a functional relation. Hence we have

THEOREM 2. *The general solution of the partial differential equation $\mathbf{f} \cdot \nabla w = 0$ is $w = \varphi(u, v)$, where φ is an arbitrary function and $u = a$, $v = b$ are independent integrals of $\mathbf{f} \times d\mathbf{r} = 0$.*

When the vector field $\mathbf{f}(\mathbf{r})$ is solenoidal, div $\mathbf{f} = 0$. In § 92 we found that we could express $\mathbf{f}$ as rot $\mathbf{g}$. We now deduce another form for $\mathbf{f}$ which gives at once its field lines.

THEOREM 3. *A solenoidal vector $\mathbf{f}$ which is continuously differentiable can be expressed in the form:*

$$\mathbf{f} = \nabla u \times \nabla v. \tag{8}$$

Proof. If we assume the relation (8), we have $\mathbf{f} \cdot \nabla u = 0$, $\mathbf{f} \cdot \nabla v = 0$; thus both u and v are solutions of $\mathbf{f} \cdot \nabla w = 0$. If we choose for u some integral of the system (2), $\mathbf{f} \cdot \nabla u = 0$. Now, if $d\mathbf{r}$ is a differential on the level surface $u = a$, we have, from (8),

$$\mathbf{f} \times d\mathbf{r} = \nabla v \, du - \nabla u \, dv = -\nabla u \, dv. \tag{9}$$

In the function $u(x, y, z)$, at least one variable is actually present. If z is present, $\partial u/\partial z = u_z$ is not identically zero; hence, on multiplying (9) by $\mathbf{k} \cdot$, we have $\mathbf{k} \cdot \mathbf{f} \times d\mathbf{r} = -u_z \, dv$ and

$$v = -\int \frac{\mathbf{k} \times \mathbf{f} \cdot d\mathbf{r}}{u_z}, \tag{10}$$

the integral being taken over a curve on the surface $u = a$. This integral is independent of the path; for, from (85.3) and (85.6),

$$\operatorname{rot}\frac{\mathbf{k}\times\mathbf{f}}{u_z} = \left(\nabla\frac{1}{u_z}\right)\times(\mathbf{k}\times\mathbf{f}) + \frac{1}{u_z}\operatorname{rot}(\mathbf{k}\times\mathbf{f})$$

$$= -\frac{1}{u_z^2}(\nabla u_z)\times(\mathbf{k}\times\mathbf{f}) - \frac{\mathbf{k}\cdot\nabla f}{u_z},$$

$$= \frac{-(\mathbf{f}\cdot\nabla u_z)\mathbf{k} + (\mathbf{k}\cdot\nabla u_z)\mathbf{f}}{u_z^2} - \frac{\mathbf{f}_z}{u_z},$$

$$\nabla u\cdot\operatorname{rot}\frac{\mathbf{k}\times\mathbf{f}}{u_z} = \frac{-\mathbf{f}\cdot\nabla u_z - \mathbf{f}_z\cdot\nabla u}{u_z} = \frac{-1}{u_z}\frac{\partial}{\partial z}(\mathbf{f}\cdot\nabla u) = 0.$$

If $u(x, y, z)$ contains x or y, we may replace $\mathbf{k}$ and u_z in (10) by $\mathbf{i}$ and u_x or $\mathbf{j}$ and u_y.

With the values of u and v thus obtained, we now have, from theorem 2 of § 103,

$$\nabla u\times\nabla v = \nabla u\times\operatorname{Grad} v = -\nabla u\times\frac{(\mathbf{k}\times\mathbf{f})_t}{u_z} = \frac{-1}{u_z}\nabla u\times(\mathbf{k}\times\mathbf{f}) = \mathbf{f},$$

where Grad v refers to the surface $u = a$, and $(\mathbf{k}\times\mathbf{f})_t$ is a tangential projection upon it. Moreover $\mathbf{f}\cdot\nabla v = 0$.

The field lines of the vector $\nabla u\times\nabla v$ are the curves in which the surfaces $u = a$, $v = b$ intersect.

Example 1. The field lines of the vector,

$$\mathbf{f} = xz\mathbf{i} + yz\mathbf{j} + xy\mathbf{k},$$

have the differential equations:

$$\text{(i)}\qquad \frac{dx}{xz} = \frac{dy}{yz} = \frac{dz}{xy}.$$

From these, we obtain the equations,

$$y\,dx - x\,dy = 0, \qquad y\,dx + x\,dy - 2z\,dz = 0,$$

which have the integrals:

$$y/x = a, \qquad xy - z^2 = b.$$

These one-parameter families of surfaces intersect in the two-parameter family of field lines.

When one integral, as $y/x = a$, is known, we may find a second by obtaining the field lines on the surface $y/x = a$. These must satisfy the equation obtained from (i) by putting $y = ax$, namely, $ax\,dx = z\,dz$. Its integral $ax^2 - z^2 = b$ gives the field lines on the surface $y/x = a$. On replacing a by y/x, we obtain a second family of surfaces $xy - z^2 = b$ which intersects the first family $y/x = a$ in the field lines.

Example 2. The field lines of the vector,

$$\mathbf{f} = x\mathbf{i} + 2x^2\mathbf{j} + (y + z)\mathbf{k},$$

have the differential equations:

$$\text{(ii)} \qquad \frac{dx}{x} = \frac{dy}{2x^2} = \frac{dz}{y + z}.$$

From $dy = 2x\,dx$, we obtain the family of surfaces,

$$y - x^2 = a.$$

As no other integrable combination is evident, we put $y = a + x^2$ in (ii); then

$$\frac{dx}{x} = \frac{dz}{a + x^2 + z}, \quad \text{or} \quad \frac{dz}{dx} - \frac{z}{x} = \frac{a}{x} + x.$$

This linear equation, with the integrating factor $1/x$, has the solution,

$$z = -a + x^2 + bx.$$

A second family of integral surfaces is therefore

$$\frac{z + y}{x} - 2x = b.$$

Example 3. The vector,

$$\mathbf{f} = x(y - z)\mathbf{i} + y(z - x)\mathbf{j} + z(x - y)\mathbf{k},$$

is solenoidal; for $\operatorname{div} \mathbf{f} = 0$. In order to express $\mathbf{f}$ as $\nabla u \times \nabla v$ we choose for u an integral of the system,

$$\frac{dx}{x(y - z)} = \frac{dy}{y(z - x)} = \frac{dz}{z(x - y)}.$$

Since the sum of the denominators is zero,

$$dx + dy + dz = 0, \qquad x + y + z = a;$$

hence we may take

$$u = x + y + z.$$

We now use (10) to compute v. On the surface $u = a$, $z = a - x - y$, and

$$\begin{aligned} -\mathbf{k} \times \mathbf{f} \cdot d\mathbf{r} &= -x(y - z)\,dy + y(z - x)\,dx \\ &= y(a - 2x - y)\,dx + x(a - x - 2y)\,dy. \end{aligned}$$

Since this is an exact differential, the method of § 91 gives

$$v = \int_0^x (ay - 2xy - y^2)\, dx = axy - x^2y - xy^2$$

$$= xy(a - x - y) = xyz.$$

With $u = x + y + z$, $v = xyz$, we may readily verify that $\mathbf{f} = \nabla u \times \nabla v$.

105. Pfaff's Problem. Since rot **f** is solenoidal, we have from theorem 3 of § 104,

$$\text{(1)} \qquad \operatorname{rot} \mathbf{f} = \nabla u \times \nabla v = \operatorname{rot} (u\, \nabla v),$$

$$\operatorname{rot} (\mathbf{f} - u\nabla v) = 0;$$

hence (§ 91) $\mathbf{f} - u\nabla v$ is the gradient ∇w of a scalar, and

$$\text{(2)} \qquad \mathbf{f} = \nabla w + u\nabla v.$$

When **f** is a continuously differentiable vector, we may find three scalars u, v, w so that (2) holds good. The determination of these scalar functions is known as *Pfaff's Problem.* We proceed to give a simple and direct solution.

Assuming the truth of (2), we at once deduce (1); hence

$$\text{(3)} \qquad \nabla u \cdot \operatorname{rot} \mathbf{f} = 0, \qquad \nabla v \cdot \operatorname{rot} \mathbf{f} = 0,$$

so that both u and v satisfy the same partial differential equation: $\nabla \varphi \cdot \operatorname{rot} \mathbf{f} = 0$. Let $v = a$ be some integral of the system $d\mathbf{r} \times \operatorname{rot} \mathbf{f} = 0$; then $\nabla v \cdot \operatorname{rot} \mathbf{f} = 0$. Now on the surface $v = a$ we have, from (2),

$$\mathbf{f} \cdot d\mathbf{r} = dw - u\, dv = dw, \qquad w = \int \mathbf{f} \cdot d\mathbf{r};$$

this integral, taken over a curve of the surface $v = a$, is independent of the path since $\nabla v \cdot \operatorname{rot} \mathbf{f} = 0$. On substituting the functions v and w thus obtained in (2), this equation uniquely determines u.

We now must show that, with u, v, w thus determined, $\nabla w + u\nabla v = \mathbf{f}$. Let Grad w and **n** denote the gradient and unit normal on a surface of the family $v =$ const; then (103.4)

$$\nabla w = \operatorname{Grad} w + \mathbf{n} \frac{dw}{dn} = \mathbf{f}_t + \mathbf{n} \frac{dw}{dn},$$

and the projection of $\nabla w + u\nabla v$ tangential to the surface is $\mathbf{f}_t$. Moreover u was chosen to make the normal projections of $\nabla w + u\nabla v$ and $\mathbf{f}$ the same, that is,

$$\mathbf{n}\frac{dw}{dn} + u\nabla v = \mathbf{f} - \mathbf{f}_t.$$

From (1) and (2),

$$\mathbf{f} \cdot \operatorname{rot} \mathbf{f} = \nabla u \times \nabla v \cdot \nabla w = \frac{\partial(u, v, w)}{\partial(x, y, z)};$$

hence $\mathbf{f} \cdot \operatorname{rot} \mathbf{f} = 0$ implies that u, v, w are functionally dependent (§ 87, theorem 3). In this case,

$$\mathbf{f} \cdot d\mathbf{r} = dw + u(v, w)\, dv = 0 \tag{4}$$

is an *ordinary* differential equation which admits solutions under general conditions—as when $u(v, w)$ and $\partial u/\partial w$ are continuous in a region R.* Hence the condition

$$\mathbf{f} \cdot \operatorname{rot} \mathbf{f} = 0 \tag{5}$$

shown in § 91, ex. 2, to be *necessary* for the integrability of $\mathbf{f} \cdot d\mathbf{r} = 0$, is also *sufficient*.

When (5) is fulfilled, let $\varphi(w, v) = C$ be the general solution of (4). Then

$$d\varphi = \frac{\partial\varphi}{\partial w} dw + \frac{\partial\varphi}{\partial v} dv = 0$$

must yield (4) upon division by $\partial\varphi/\partial w$. In other words, $\lambda = \partial\varphi/\partial w$ is an integrating factor of (4):

$$d\varphi = d\mathbf{r} \cdot \nabla\varphi = \lambda\mathbf{f} \cdot d\mathbf{r}.$$

Then $\lambda\mathbf{f} = \nabla\varphi$, and $\mathbf{f}$ is everywhere normal to the surfaces $\varphi =$ const. The field lines of $\mathbf{f}$ are then the orthogonal trajectories of these surfaces.

A family of curves is said to form a *field* in a region R if just one curve of the family passes through every point of R. The unit vectors $\mathbf{T}, \mathbf{N}, \mathbf{B}$ along the curves are then vector point functions in R; and the first Frenet Formula $d\mathbf{T}/ds = \kappa\mathbf{N}$ can be written

$$\mathbf{T} \cdot \nabla\mathbf{T} = \kappa\mathbf{N}. \tag{6}$$

* See Agnew, *Differential Equations*, New York, 1942, p. 310, *et seq.*

Now, from (85.7) and (85.9),

$$\tfrac{1}{2}\nabla(\mathbf{T}\cdot\mathbf{T}) = \mathbf{T}\cdot\nabla\mathbf{T} + \mathbf{T}\times\operatorname{rot}\mathbf{T} = 0,$$

since $\mathbf{T}\cdot\mathbf{T} = 1$; thus, from (6),

$$\mathbf{T}\times\operatorname{rot}\mathbf{T} = -\kappa\mathbf{N}. \tag{7}$$

For a *surface-normal* field

$$\mathbf{T}\cdot\operatorname{rot}\mathbf{T} = 0, \qquad (\mathbf{T}\times\operatorname{rot}\mathbf{T})\times\mathbf{T} = \operatorname{rot}\mathbf{T},$$

and, from (7), we have

$$\operatorname{rot}\mathbf{T} = \kappa\mathbf{B}, \qquad |\operatorname{rot}\mathbf{T}| = \kappa. \tag{8}$$

The Darboux vector of a surface-normal field of curves is therefore $\boldsymbol{\delta} = \tau\mathbf{T} + \operatorname{rot}\mathbf{T}$.

Example. Find u, v, w so that

$$\mathbf{f} = [2yz, zx, 3xy] = \nabla w + u\,\nabla v.$$

Solution. $\operatorname{rot}\mathbf{f} = [2x, -y, -z]$; since the system,

$$\frac{dx}{2x} = \frac{dy}{-y} = \frac{dz}{-z},$$

has the solution $z/y = a$, we take $v = z/y$. On the surface $v = a$, $z = ay$; hence

$$w = \int \mathbf{f}\cdot d\mathbf{r} = \int (2yz\,dx + zx\,dy + 3xy\,dz)$$

$$= \int (2ay^2\,dx + 4axy\,dy)$$

$$= 2axy^2 = 2xyz.$$

Now u is given by

$$[2yz, zx, 3xy] = [2yz, 2zx, 2xy] + u[0, -z/y^2, 1/y],$$

whence $u = xy^2$. Thus our solution is

$$u = xy^2, \qquad v = z/y, \qquad w = 2xyz.$$

Since $\mathbf{f}\cdot\operatorname{rot}\mathbf{f} = 0$, u, v, and w must be functionally related; in fact $w = 2uv$ The total differential equation,

$$2yz\,dx + zx\,dy + 3xy\,dz = 0, \tag{i}$$

is thus equivalent to

$$dw + u\,dv = 3u\,dv + 2v\,du = 0,$$

and has the integral $u^2v^3 = C$, or $x^2yz^3 = C$.

106. Reduction of Volume to Surface Integrals. Let S be a closed surface and V the volume it encloses. Then S has two sides, an inside and an outside, and at all regular points of S we have a definite unit normal $\mathbf{n}$ directed towards the outside. We shall suppose that S consists of a finite number of parts over which $\mathbf{n}$ is continuous.

Consider, first, a surface S which is cut in at most two points by a line parallel to the z-axis; denote them by (x, y, z_1) and (x, y, z_2), where $z_1 < z_2$. Then S has a lower portion S_1 con-

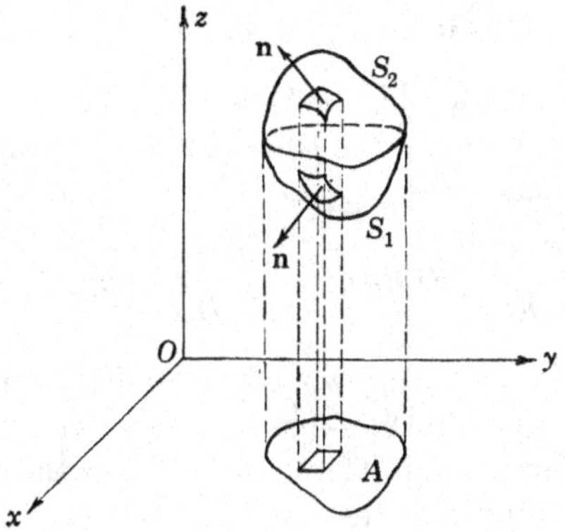

FIG. 106

sisting of all the points (x, y, z_1), and an upper portion S_2 consisting of the points (x, y, z_2). We suppose also that the points for which $z_1 = z_2$ form a closed curve separating S_1 from S_2 (Fig. 106). The equations of S_1 and S_2 may be taken as $z = z_1(x, y)$, $z = z_2(x, y)$.

Now let $\mathbf{f}(x, y, z)$ be a tensor function of valence ν whose scalar components have continuous partial derivatives throughout V. Then, if the volume integral of $\partial\mathbf{f}/\partial z$ over V is written as a triple integral with the element of volume $dV = dx\,dy\,dz$, we may effect a first integration with respect to z,

$$\text{(i)} \quad \iiint_V \frac{\partial \mathbf{f}}{\partial z}\,dx\,dy\,dz = \iint_A \mathbf{f}(x, y, z_2)\,dx\,dy - \iint_A \mathbf{f}(x, y, z_1)\,dx\,dy,$$

where the double integrals are taken over the common projection A of S_1 and S_2 on the xy-plane.

If we regard x, y as the parameters u, v on the surfaces S_1 and S_2, the vector element of surface is $\mathbf{r}_x \times \mathbf{r}_y \, dx \, dy$ (94.9). Now from

$$\mathbf{r} = \mathbf{i}x + \mathbf{j}y + \mathbf{k}z(x, y), \qquad \mathbf{r}_x = \mathbf{i} + \mathbf{k}z_x, \qquad \mathbf{r}_y = \mathbf{j} + \mathbf{k}z_y,$$

$$\mathbf{r}_x \times \mathbf{r}_y = \mathbf{k} - \mathbf{i}z_x - \mathbf{j}z_y.$$

Over S_2 ($z = z_2$) this vector has the direction of the external normal $\mathbf{n}$; but over S_1 ($z = z_1$) it has the direction of the internal normal $-\mathbf{n}$. Hence, if we denote the vector element of area *in the direction of the external normal* by $d\mathbf{S} = \mathbf{n}\, dS$,

$$\mathbf{n}\, dS = \pm(\mathbf{k} - \mathbf{i}z_x - \mathbf{j}z_y)\, dx\, dy, \qquad \mathbf{k} \cdot \mathbf{n}\, dS = \pm dx\, dy,$$

where the plus sign applies to S_2 and the minus to S_1. The two integrals over A now may be combined into a single integral over S, so that we may write

$$\text{(1)} \qquad \iiint_V \frac{\partial \mathbf{f}}{\partial z}\, dx\, dy\, dz = \iint_S \mathbf{f}(x, y, z)\, \mathbf{k} \cdot \mathbf{n}\, dS.$$

This formula is also valid when S is bounded laterally by a part of a cylinder parallel to the z-axis and separating S_1 from S_2. For (i) holds as before; and, in (1), $\mathbf{k} \cdot \mathbf{n} = 0$ over the cylinder, so that it contributes nothing to the integral over S.

We now may remove the condition that S is cut in only two points by a line parallel to the z-axis. For, if we divide V into parts bounded by surfaces which do satisfy this condition and apply formula (1) to each point and add the results, the volume integrals will combine to the left member of (1); the surface integrals over the boundaries between the parts cancel (for each appears twice but with opposed values of $\mathbf{n}$), whereas the remaining surface integrals combine to the right member of (1).

Finally we may extend (1) to regions bounded by two or more closed surfaces, that is, regions with cavities in them, by this same process of subdivision. Additional surfaces must be introduced so that the parts of V are all bounded by a single closed surface, and the surface integrals over these will cancel in pairs as before.

When x, y, z form a dextral system of axes, the same is true of y, z, x and z, x, y. Hence, if in (1) we make cyclic interchanges in x, y, z, we obtain the corresponding formulas:

$$\text{(2)} \qquad \iiint_V \frac{\partial \mathbf{f}}{\partial x}\, dx\, dy\, dz = \iint_S \mathbf{f}(x, y, z)\, \mathbf{i} \cdot \mathbf{n}\, dS,$$

$$\text{(3)} \qquad \iiint_V \frac{\partial \mathbf{f}}{\partial y}\, dx\, dy\, dz = \iint_S \mathbf{f}(x, y, z)\, \mathbf{j} \cdot \mathbf{n}\, dS.$$

If we insert the prefactors **k**, **i**, **j** in the integrands of (1), (2), and (3), respectively, add the resulting equations, and note that

$$\mathbf{i}\frac{\partial f}{\partial x} + \mathbf{j}\frac{\partial f}{\partial y} + \mathbf{k}\frac{\partial f}{\partial z} = \nabla \mathbf{f}, \qquad \mathbf{ii} \cdot \mathbf{n} + \mathbf{jj} \cdot \mathbf{n} + \mathbf{kk} \cdot \mathbf{n} = \mathbf{I} \cdot \mathbf{n} = \mathbf{n},$$

we obtain finally

$$\text{(4)} \qquad \int \nabla \mathbf{f}\, dV = \oint \mathbf{n}\, \mathbf{f}\, dS,$$

using $\oint$ to denote integration over a *closed* surface. The integrands are tensors of valence $\nu + 1$. We have thus proved

Basic Theorem II. *If* **f(r)** *is a continuously differentiable tensor point function over the region V bounded by a closed surface S, whose unit external normal* **n** *is sectionally continuous, then the integral of* ∇**f** *over the volume V is equal to the integral of* **nf** *over the surface S.*

If **f(r)** is a vector or tensor of higher valence, we may obtain from (5) other integral transformations by placing a dot or a cross between the first two vectors in each term of the integrands; for both of these operations are distributive with respect to addition and therefore may be carried out under the integral sign. We thus obtain the important formulas:

$$\text{(5)} \qquad \int \nabla \cdot \mathbf{f}\, dV = \oint \mathbf{n} \cdot \mathbf{f}\, dS,$$

$$\text{(6)} \qquad \int \nabla \times \mathbf{f}\, dV = \oint \mathbf{n} \times \mathbf{f}\, dS.$$

When **f** is a vector, (5) is known as the *divergence theorem;* the integral $\int \mathbf{n} \cdot \mathbf{f}\, dS$ then is called the *normal flux* of **f** through the surface.

Example. If rot **f** = 0 in a simply connected origin, $\mathbf{f} = \nabla\varphi$ (§ 102), and

$$\int \mathbf{f}\, dV = \int \nabla\varphi\, dV = \oint \mathbf{n}\varphi\, dS.$$

Thus the volume integral of an irrotational vector can be expressed a surface integral over the boundary.

For example, the *center of mass* of a *homogeneous* body of mass m is fixed by the position vector,

$$\mathbf{r}^* = \frac{1}{m}\int \mathbf{r}\, dm = \frac{1}{V}\int \mathbf{r}\, dV;$$

for $m = \rho V$, where ρ denotes the constant density. Since rot $\mathbf{r} = 0$, we have, from (102.2),

$$\mathbf{r} = \nabla\varphi, \qquad \text{where} \quad \varphi = \int_0^{\mathbf{r}} \mathbf{r}\cdot d\mathbf{r} = \frac{r^2}{2}.$$

Hence

$$V\mathbf{r}^* = \tfrac{1}{2}\oint \mathbf{n} r^2\, dS.$$

107. Solid Angle. The rays from a point O through the points of a closed curve generate a cone; and the surface of a unit sphere about O intercepted by this cone is called the *solid angle* Ω of the cone.

The reciprocal of the distance $r = OP$ is a harmonic function (§ 89, ex. 2); hence

$$\text{div grad}\,\frac{1}{r} = 0, \quad \text{and} \quad \text{grad}\,\frac{1}{r} = -\frac{\mathbf{R}}{r^2}$$

is a solenoidal vector.

Let us now apply the divergence theorem to the vector $\mathbf{f} = \mathbf{R}/r^2$ in the region interior to a cone of solid angle Ω and limited externally by a surface S, internally by a small sphere σ about O of radius a. Within this region $\mathbf{f}$ is continuously differentiable, div $\mathbf{f} = 0$ and $\oint \mathbf{n}\cdot\mathbf{f}\, dS = 0$. The external normal $\mathbf{n} = -\mathbf{R}$ over σ, and over the conical surface $\mathbf{n}\cdot\mathbf{R} = 0$; hence

$$\int_S \frac{\mathbf{n}\cdot\mathbf{R}}{r^2}\, dS = \int_\sigma \frac{1}{a^2}\, dS = \frac{S_a}{a^2} = \Omega, \tag{1}$$

where S_a is the area cut from σ by the cone. Note that the ratio S_a/a^2 is independent of the radius and may be computed with $a = 1$.

If S is a closed surface, we have

$$\oint \frac{\mathbf{n}\cdot\mathbf{R}}{r^2}\, dS = \begin{cases} 4\pi, & O \text{ inside of } S \\ 0, & O \text{ outside of } S. \end{cases} \tag{2}$$

When O is outside of S, $\mathbf{f}$ is continuously differentiable throughout its interior, and the foregoing result is immediate. In this case the elements of solid angle,

$$d\Omega = \frac{\mathbf{n} \cdot \mathbf{R}}{r^2}\, dS, \tag{3}$$

corresponding to the same ray cancel in pairs.

108. Green's Identities. We now apply the divergence theorem (106.5) to the vector $\mathbf{f} = \varphi \nabla \psi$, where φ and ψ are scalar functions having continuous derivatives of the first and second order, respectively, in a region R bounded by a closed surface S. Now

$$\operatorname{div}(\varphi \nabla \psi) = \nabla \varphi \cdot \nabla \psi + \varphi \operatorname{div} \nabla \psi \qquad (85.2),$$

$$\mathbf{n} \cdot (\varphi \nabla \psi) = \varphi\, d\psi/dn,$$

where the *normal derivative* $d\psi/dn$ is in the direction of the external normal to the bounding surface. Hence, on writing the operator div grad as ∇^2, we have

$$\int \nabla \varphi \cdot \nabla \psi \, dV + \int \varphi \, \nabla^2 \psi \, dV = \oint \varphi \frac{d\psi}{dn} \, dS. \tag{1}$$

In case ψ is not defined outside of S, we replace $d\psi/dn$ by the negative of the derivative along the internal normal $-\mathbf{n}$. Formula (1) is known as *Green's first identity.*

If both φ and ψ have continuous derivatives of the first and second orders, we may interchange φ and ψ in (1). On subtracting this result from (1), we obtain *Green's second identity:*

$$\int (\varphi \nabla^2 \psi - \psi \nabla^2 \varphi) \, dV = \oint \left(\varphi \frac{d\psi}{dn} - \psi \frac{d\varphi}{dn} \right) dS. \tag{2}$$

We now take $\psi = 1/r$, where r is the distance OP. If O is *interior* to S, we cannot apply (2) to the entire region enclosed since ψ becomes infinite at O. We therefore exclude O by surrounding it by a small sphere σ of radius ϵ and apply (2) to the region R' between S and σ; then

$$-\int_{R'} \frac{1}{r} \nabla^2 \varphi \, dV = \oint_S \left(\varphi \frac{d}{dn} \frac{1}{r} - \frac{1}{r} \frac{d\varphi}{dn} \right) dS + \oint_\sigma \left(\varphi \frac{d}{dn} \frac{1}{r} - \frac{1}{r} \frac{d\varphi}{dn} \right) dS.$$

On the sphere $r = \epsilon$,

$$\frac{d}{dn}\left(\frac{1}{r}\right) = -\frac{d}{dr}\left(\frac{1}{r}\right) = \frac{1}{r^2} = \frac{1}{\epsilon^2}, \qquad \frac{d\varphi}{dn} = -\frac{\partial \varphi}{\partial r}, \qquad dS = \epsilon^2 \, d\Omega,$$

$d\Omega$ denoting the solid angle subtended by dS. The integral over σ is therefore

$$\oint_\sigma \left(\frac{\varphi}{\epsilon^2} + \frac{1}{\epsilon}\frac{\partial \varphi}{\partial r}\right) dS = \oint \varphi \, d\Omega + \epsilon \oint \frac{\partial \varphi}{\partial r}\, d\Omega = 4\pi\bar{\varphi} + \epsilon \oint \frac{\partial \varphi}{\partial r}\, d\Omega,$$

where $\bar{\varphi}$ is a value of φ at some point of σ. Now, as $\epsilon \to 0$,

$$\oint_\sigma \left(\varphi \frac{d}{dn}\frac{1}{r} - \frac{1}{r}\frac{d\varphi}{dn}\right) dS \to 4\pi\varphi(O);$$

$$-\int_{R'} \frac{1}{r} \nabla^2\varphi \, dV \to -\int_R \frac{1}{r}\nabla^2\varphi \, dV$$

for the integrand remains finite if we use the spherical element of volume $dV = r^2 \sin\theta \, dr \, d\theta \, d\varphi$. We thus obtain *Green's third identity:*

$$(3) \qquad 4\pi\varphi(O) = -\int_R \frac{\nabla^2\varphi}{r}\, dV + \oint_S \left(\frac{1}{r}\frac{d\varphi}{dn} - \varphi\frac{d}{dn}\frac{1}{r}\right) dS.$$

When O is *exterior* to S we may put $\psi = 1/r$ directly in (2); in this case the right member of (3) equals zero.

We may deduce three analogous identities from (101.6),

$$\int \operatorname{Div} \mathbf{f} \, dA = \oint \mathbf{m} \cdot \mathbf{f} \, ds,$$

the divergence theorem in the *plane*. With $\mathbf{f} = \varphi \operatorname{Grad} \psi$, we find

$$(4) \qquad \int \nabla\varphi \cdot \nabla\psi \, dA + \int \varphi \nabla^2\psi \, dA = \oint \varphi \frac{d\psi}{dn}\, ds,$$

where we now interpret ∇ and ∇^2 as Grad and Div Grad and $d\psi/dn$ as a derivative in the direction of the external normal $\mathbf{m}$. By an interchange of φ and ψ, we find as before

$$(5) \qquad \int (\varphi \nabla^2\psi - \psi\nabla^2\varphi)\, dA = \oint \left(\varphi\frac{d\psi}{dn} - \psi\frac{d\varphi}{dn}\right) ds.$$

To obtain the third identity from (5), we take $\psi = \log \rho$, where ρ is the distance OP in the plane. Since $\log \rho$ is a solution of Laplace's Equation in the plane,

$$\operatorname{Div} \operatorname{Grad} \psi = 0,$$

$\nabla^2\psi = 0$ in (5). As before, we must exclude the origin from the region by surrounding it by a small circle γ of radius ϵ. Applying (5) to the region remaining, we have

$$-\int_{R'} \log\rho\cdot\nabla^2\varphi\, dA =$$
$$\oint_C\left(\varphi\frac{d}{dn}\log\rho - \log\rho\frac{d\varphi}{dn}\right)ds + \int_\gamma\left(\varphi\frac{d}{dn}\log\rho - \log\rho\frac{d\varphi}{dn}\right)ds.$$

On the circle $\rho = \epsilon$,

$$\frac{d}{dn}\log\rho = -\frac{d}{d\rho}\log\rho = -\frac{1}{\rho} = -\frac{1}{\epsilon}, \qquad \frac{d\varphi}{dn} = -\frac{\partial\varphi}{\partial\rho}, \qquad ds = \epsilon\, d\theta;$$

the integral over γ is therefore

$$\oint_\gamma\left(\log\epsilon\frac{\partial\varphi}{\partial\rho} - \frac{\varphi}{\epsilon}\right)\epsilon\, d\theta = \epsilon\log\epsilon\oint\frac{\partial\varphi}{\partial\rho}d\theta - 2\pi\,\bar\varphi,$$

where $\bar\varphi$ is a value of φ at some point of γ. When $\epsilon\to 0$, $\epsilon\log\epsilon \to 0$, and we obtain

$$(6)\quad 2\pi\varphi(O) = \int_R \log\rho\cdot\nabla^2\varphi\, dA + \oint_C\left(\varphi\frac{d}{dn}\log\rho - \log\rho\frac{d\varphi}{dn}\right)ds.$$

109. Harmonic Functions. A solution of Laplace's Equation,

$$(1)\qquad \text{div grad}\,\varphi = \nabla^2\varphi = 0,$$

is called a *harmonic function.* A function φ is said to be harmonic in a region R if it has continuous derivatives of the first and second orders and $\nabla^2\varphi = 0$ in R.

If a vector $\mathbf{f}$ is both irrotational and solenoidal, its scalar potential ψ is harmonic; for, from (102.4),

$$\mathbf{f} = -\nabla\psi, \qquad \text{div}\,\mathbf{f} = -\nabla^2\psi = 0.$$

If in Green's second identity (108.2) φ and ψ are harmonic throughout R, we have

$$(2)\qquad \oint\left(\varphi\frac{d\psi}{dn} - \psi\frac{d\varphi}{dn}\right)dS = 0.$$

Thus, if $\psi = 1$, we have, for any harmonic function φ,

$$(3)\qquad \oint\frac{d\varphi}{dn}dS = 0.$$

If φ is harmonic in the region bounded by a closed surface S, Green's third identity (108.3) gives the value $\varphi(P)$ at any *interior* point P,

$$4\pi\varphi(P) = \oint_S \left(\frac{1}{r}\frac{d\varphi}{dn} - \varphi\frac{d}{dn}\frac{1}{r}\right) dS, \tag{4}$$

where $r = PQ$, the distance from P to points Q on the surface. Thus a function φ, harmonic in the region enclosed by S, is determined completely at any interior point P by the values of φ and $d\varphi/dn$ on the boundary.

When P is *exterior* to the surface S, we have

$$0 = \oint_S \left(\frac{1}{r}\frac{d\varphi}{dn} - \varphi\frac{d}{dn}\frac{1}{r}\right) dS, \tag{5}$$

on putting $\psi = 1/r$ in (2).

If S is a sphere of radius r about P as center and lying entirely within the region R, we have, from (4),

$$4\pi\varphi(P) = \frac{1}{r}\oint \frac{d\varphi}{dn}\, dS + \frac{1}{r^2}\oint \varphi\, dS;$$

or, in view of (3),

$$\varphi(P) = \frac{1}{4\pi r^2}\oint \varphi\, dS. \tag{6}$$

Since the surface of the sphere is $4\pi r^2$, we have the

Mean Value Theorem. *The value of a function, harmonic in a region R, at any point P is equal to the mean of its values on any sphere about P as center and lying entirely within R.*

This theorem shows that a function which is harmonic in a closed bounded region R, but not constant, attains its extreme values only on the boundary. For let P be a frontier point of the set for which φ attains its minimum value m. If P were an interior point of R, there would be a sphere about P, lying within R, on which $\varphi > m$ at some points; hence the mean value of φ over the sphere would be greater than $\varphi(P) = m$.

If in Green's first identity (108.1) we take $\varphi = \psi$ and assume that φ is harmonic in R, we have

$$\int_R |\nabla\varphi|^2\, dV = \oint_S \varphi\frac{d\varphi}{dn}\, dS. \tag{7}$$

Hence, if either $\varphi = 0$ or $d\varphi/dn = 0$ on S, the volume integral in (7) vanishes; and, since $|\nabla\varphi|^2$ is continuous and never negative, we must conclude that $\nabla\varphi = 0$, and φ has a constant value throughout R. If $\varphi = 0$ on S, $\varphi = 0$ in R; and, if $d\varphi/dn = 0$ on S, $\varphi = C$ in R.

Now let φ_1 and φ_2 be two harmonic functions in R; then their difference $\varphi = \varphi_1 - \varphi_2$ is also harmonic. If $\varphi_1 = \varphi_2$ on S, $\varphi = 0$ on S and $\varphi_1 = \varphi_2$ throughout R. If $d\varphi_1/dn = d\varphi_2/dn$ on S, $d\varphi/dn = 0$ on S and $\varphi_1 = \varphi_2 + C$ throughout R.

If a harmonic function φ has a constant value C on S, $\varphi = C$ throughout R; for the harmonic function $\varphi - C$ is 0 on S.

110. Electric Point Charges. By Coulomb's Law, an electric charge e_1 at O exerts a force of

$$\mathbf{F} = \frac{e_1 e_2}{r^2}\mathbf{R}\ \dagger \tag{1}$$

upon a charge e_2 at P; $r = OP$, and $\mathbf{R}$ is a unit vector in the direction OP. Thus the charges repel when $e_1 e_1 > 0$ (charges of same sign) and attract when $e_1 e_2 < 0$. The force exerted by a charge e at O upon a unit charge at P, namely,

$$\mathbf{E} = \frac{e}{r^2}\mathbf{R}, \tag{2}$$

is called the *electric intensity* at P due to the charge e. Since $\mathbf{R} = \nabla r$,

$$\mathbf{E} = -\nabla\frac{e}{r}, \tag{3}$$

so that $\mathbf{E}$ has the *scalar potential* (102.4),

$$\varphi = \frac{e}{r}. \tag{4}$$

The vector $\mathbf{E}$ is both irrotational and solenoidal; for

$$\operatorname{rot}\mathbf{E} = 0, \qquad \operatorname{div}\mathbf{E} = -e\nabla^2\frac{1}{r} = 0. \tag{5}$$

† In a vacuum, if the charges are measured in statcoulombs and the distance in centimeters, the force is given in dynes.

If S is a closed surface, we have, from (107.2),

(6) $$\oint \mathbf{n}\cdot\mathbf{E}\,dS = e\oint \frac{\mathbf{n}\cdot\mathbf{R}}{r^2}\,dS = \begin{cases} 4\pi e & O \text{ inside of } S,\\ 0 & O \text{ outside of } S.\end{cases}$$

We assume that the intensity due to a system of point charges $e_1, e_2, \cdots, e_n$ is the sum of their separate intensities: $\mathbf{E} = \Sigma\mathbf{E}_i$. Since $\mathbf{E}_i = -\nabla e_i/r_i$, where r_i is the distance from the charge e_i to P,

$$\mathbf{E} = -\nabla\left(\frac{e_1}{r_1} + \frac{e_2}{r_2} + \cdots + \frac{e_n}{r_n}\right),$$

and the potential of the system is

(7) $$\varphi = \frac{e_1}{r_1} + \frac{e_2}{r_2} + \cdots + \frac{e_n}{r_n}\cdot$$

If this system of charges is within the closed surface S, we have, from (6),

(8) $$\oint \mathbf{n}\cdot\mathbf{E}\,dS = 4\pi\Sigma e_i.$$

The normal flux of the electric intensity through a closed surface is equal to 4π times the sum of the enclosed charges.

111. Surface Charges. If a surface S carries a distributed charge σ per unit of area, the potential at a point P due to the charged element dS at Q (regarded as a point charge $\sigma\,dS$) is $\sigma\,dS/r$, where r is the distance QP. If we assume that the *surface density* σ is continuous or piecewise continuous over S, the total potential at P is

(1) $$\varphi = \int_S \frac{\sigma\,dS}{r}\cdot$$

The electric intensity at a point P outside of S, due to the charge element $\sigma\,dS$ at Q, is $-\sigma\,dS\,\nabla_P\,1/r$; the total electric intensity at P is therefore

(2) $$\mathbf{E} = -\int_S \sigma\nabla_P\frac{1}{r}\,dS = -\nabla_P\varphi.$$

The notation ∇_P means that P must be varied in computing the gradient. Since P is outside of S, $1/r$ and its first partial derivatives are continuous for all positions of Q on S; hence in computing $\mathbf{E} = -\nabla_P \varphi$, we may differentiate (1) under the integral sign.

In differentiating functions of $r = PQ$, we may vary either P or Q, holding the other point fixed. Thus, if $\mathbf{R}$ is a unit vector in the direction $\overrightarrow{PQ}$,

$$\nabla_Q r = \mathbf{R}, \qquad \nabla_P r = -\mathbf{R},$$

when Q and P, respectively, are varied; and, in general,

$$(3) \qquad \nabla_P f(r) = f'(r)\nabla_P r = -f'(r)\nabla_Q r = -\nabla_Q f(r).$$

Consequently, we also may write (2) as

$$(2)' \qquad \mathbf{E} = \int_S \sigma \nabla_Q \frac{1}{r}\, dS.$$

In integrals, such as this, the subscript on ∇ may be omitted on the understanding that the variation is at dS.

From (1) and (2), we see that φ and $\mathbf{E}$ are continuous at all points P not on the surface. At such points φ is harmonic; for

$$\nabla_P^2 \varphi = \int_S \sigma \nabla_P^2 \frac{1}{r}\, dS = 0.$$

At a point P on the surface, the integrals for φ and $\mathbf{E}$ are improper since r passes through zero. However it can be shown ‡ that if σ is piecewise continuous on S, φ is defined on S and is everywhere continuous. Moreover, as P approaches a surface point Q from the positive side (toward which $\mathbf{n}$ points) or the negative side, under general conditions $\mathbf{E}$ approaches limiting values $\mathbf{E}_+$ and $\mathbf{E}_-$, such that

$$(4) \qquad \mathbf{E}_+ - \mathbf{E}_- = 4\pi\sigma\, \mathbf{n},$$

where σ and $\mathbf{n}$ are the surface density and unit normal at Q. Thus the normal component of $\mathbf{E}$ experiences a jump of $4\pi\sigma$ as P passes through the surface.

112. Doublets and Double Layers. The potential at P due to a charge $-e$ at Q and a charge $+e$ at Q' is (110.7)

$$\frac{-e}{QP} + \frac{e}{Q'P} = e\left(\frac{1}{r'} - \frac{1}{r}\right).$$

‡ See O. D Kellogg, *Foundations of Potential Theory*, Berlin, 1929, Chapter 3, § 5.

If Q' approaches Q and at the same time the charges increase, so that the product,

$$e\,\overrightarrow{QQ'} = \mathbf{m},$$

remains constant, the limiting result is called a *doublet of moment* **m**. The potential of this doublet is

$$\text{(1)} \qquad \varphi = \lim_{Q' \to Q} \left\{ e(QQ') \frac{\dfrac{1}{PQ'} - \dfrac{1}{PQ}}{QQ'} \right\} = \mathbf{m} \cdot \nabla_Q \frac{1}{r};$$

for the limit of the fraction in the second member is the directional derivative of $1/r$ in the direction of m.

A continuous distribution of doublets over a surface with moments everywhere in the direction of the normal **n** is called a *double layer*. If $\mu\mathbf{n}\,dS$ is the moment of the doublet at the surface element dS at Q, the potential of the double layer at an outside point P is

$$\text{(2)} \qquad \varphi = \int_S \mu\mathbf{n} \cdot \nabla_Q \frac{1}{r}\,dS,$$

where $r = PQ$. Since

$$-\mathbf{n} \cdot \nabla_Q \frac{1}{r}\,dS = \frac{\mathbf{n} \cdot \nabla_Q r}{r^2}\,dS = d\Omega$$

is the solid angle subtended by dS (107.3), we also may write

$$\text{(3)} \qquad \varphi = -\int_S \mu\,d\Omega.$$

When μ is constant, this reduces to $-\mu\Omega$, where Ω is the total solid angle subtended by S at P.

When μ, the moment density, † is piecewise continuous, φ is continuous at all points P not on the surface. At such points φ is harmonic; for

$$\nabla_P^2\varphi = \int_S \mu\mathbf{n} \cdot \nabla_Q \nabla_P^2 \frac{1}{r}\,dS = 0.$$

At a point Q where μ is continuous and the surface has continuous curvature, it can be shown * that φ has definite limits φ_+, φ_-

† In the case of a *magnetic* shell, μ is called the *density of magnetization*.

* See O. D. Kellogg, *op. cit.*, Chapter 6, § 6.

according as P approaches Q from the positive or negative side of the surface, and that

$$\varphi_+ - \varphi_1 = 4\pi\mu. \tag{4}$$

At a point P outside of S, the electric intensity $\mathbf{E} = -\nabla_P\varphi$ is continuously differentiable. Since $\mu\mathbf{n}$ is a function of Q (not P), we have, from (2),

$$\mathbf{E} = -\nabla\varphi = -\int_S \mu\mathbf{n} \cdot \nabla_Q\nabla_P \frac{1}{r}\, dS. \tag{5}$$

But, from (85.6),

$$\nabla_P \times \left(\mathbf{n} \times \nabla_Q \frac{1}{r}\right) = -\mathbf{n} \cdot \nabla_P\nabla_Q \frac{1}{r},$$

so that we also may write

$$\mathbf{E} = \nabla_P \times \int_S \mu\mathbf{n} \times \nabla_Q \frac{1}{r}\, dS. \tag{6}$$

Consequently the intensity due to a double layer has, besides the scalar potential φ, also a *vector potential* $\mathbf{A}$ (§ 92):

$$\mathbf{E} = \operatorname{rot}_P \mathbf{A}, \tag{7}$$

$$\mathbf{A} = \int_S \mu\mathbf{n} \times \nabla_Q \frac{1}{r}\, dS. \tag{8}$$

When μ is constant, $\mathbf{A}$ may be transformed into a circuit integral about the boundary of S; for, from the basic theorem (100.1),

$$\mathbf{A} = \mu \int_S \mathbf{n} \times \nabla \frac{1}{r}\, dS = \mu \oint_C \frac{\mathbf{T}}{r}\, ds. \tag{9}$$

Example. Let φ be a function harmonic in a region bounded by a closed surface S; then, at any interior point P (109.4),

$$\varphi(P) = \int_S \frac{d\varphi/dn}{4\pi r}\, dS - \int_S \frac{\varphi}{4\pi} \frac{d}{dn}\left(\frac{1}{r}\right) dS.$$

Comparison with (111.1) and (112.2) shows that φ may be regarded as the potential due to a surface change of density $\sigma = (d\varphi/dn)/4\pi$ and a double layer of moment density $\mu = -\varphi/4\pi$.

113. Space Charges. If a region V carries a distributed charge ρ per unit of volume, the potential at a point P due to the charged element dV at Q (regarded as a point charge $\rho\, dV$) is $\rho\, dV/r$, where

r is the distance QP. If we assume that *volume density* ρ is piecewise continuous over V, the total potential at a point P outside of V is

$$\varphi = \int_V \rho \frac{dV}{r}; \tag{1}$$

The electric intensity at P due to the charge element $\rho\, dV$ at Q is $-\rho\, dV\, \nabla_P\, 1/r$; the total electric intensity at P is therefore

$$\mathbf{E} = -\int_V \rho\, \nabla_P \frac{1}{r}\, dV = -\nabla_P \varphi. \tag{2}$$

Since P lies outside of V, $1/r$ and its first partial derivatives are continuous for all positions of Q within V; hence, in computing $\nabla_P \varphi$, we may differentiate (1) under the integral sign.

When P lies within the charged region V, the integrals for φ and $\mathbf{E}$ are improper since r passes through zero. But it can be shown ¶ that, when ρ is piecewise continuous, φ and $\mathbf{E}$ exist at the points of V and are continuous throughout space. Moreover the potential φ is everywhere differentiable, and $\mathbf{E} = -\nabla_P \varphi$.

The equation (110.8) also may be proved for space charges; namely,

$$\int_S \mathbf{n} \cdot \mathbf{E}\, dS = 4\pi \int \rho\, dV, \tag{3}$$

the integral on the right covering all space charges within S. The closed surface S may either completely enclose the charged region V or cut through it. If S encloses no charges, the integral (3) is zero.

When ρ is continuously differentiable, it can be shown that $\mathbf{E}$ has the same property. Now if S_1 is any closed surface enclosing a subregion V_1 of V, the divergence theorem shows that

$$\int_{S_1} \mathbf{n} \cdot \mathbf{E}\, dS = \int_{V_1} \operatorname{div} \mathbf{E}\, dV.$$

But, from (3),

$$\int_{V_1} \operatorname{div} \mathbf{E}\, dV = 4\pi \int_{V_1} \rho\, dV;$$

¶ See O. D. Kellogg, *op. cit.*, Chapter 6, § 3.

and, since this holds for any subregion V_1 of V,

(4) $$\operatorname{div} \mathbf{E} = 4\pi\rho,$$

or, if we put $\mathbf{E} = -\nabla\varphi$,

(5) $$\nabla^2\varphi = -4\pi\rho.$$

This partial-differential equation is called *Poisson's Equation.* At points outside of the charged region V, $\rho = 0$, and the potential φ satisfies *Laplace's Equation:*

(6) $$\nabla^2\varphi = 0.$$

114. Heat Conduction. In mathematical physics the integral theorems often are used in setting up differential equations. As an illustration, consider the flow of heat at a point P of a body. According to Fourier's Law, the direction of flow is normal to the isothermal surface through P; and the flow $\mathbf{F}$ (calories per second) per unit of surface is proportional to the temperature gradient at P. Thus $\mathbf{F}$ may be represented by the vector,

(1) $$\mathbf{F} = -k\,\nabla T \text{ cal./sec./cm.}^2,$$

where T is the temperature and k the thermal conductivity of the body at P. Since k is positive, the minus sign in (1) expresses the fact that heat flows in the direction of decreasing temperature.

Let ρ and c denote the density and specific heat of the body at P. Then the rate at which heat is being absorbed in a region R bounded by a closed surface S is

$$\frac{\partial}{\partial t}\int c\rho T\,dV = \int c\rho\frac{\partial T}{\partial t}\,dV.$$

If $\mathbf{n}$ is the outward unit normal to the surface S, the rate at which heat flows into R through S is

$$-\oint \mathbf{n}\cdot\mathbf{F}\,dS = \int \operatorname{div}(k\nabla T)\,dV.$$

Hence, if heat is being generated in R at the rate of h calories per unit of volume,

$$\int c\rho\frac{\partial T}{\partial t}\,dV = \int \{\operatorname{div}(k\nabla T) + h\}\,dV.$$

Since this holds for an arbitrary region R within the body,

$$c\rho \frac{\partial T}{\partial t} = \operatorname{div}(k\nabla T) + h.^* \tag{2}$$

This is the differential equation of heat conduction.

If the body is homogeneous, k is constant, and (1) becomes

$$c\rho \frac{\partial T}{\partial t} = k\,\nabla^2 T + h; \tag{3}$$

and, if there is no internal generation of heat, we have

$$\frac{\partial T}{\partial t} = \kappa\,\nabla^2 T, \tag{4}$$

where $\kappa = k/c\rho$ is called by Kelvin the "diffusivity."

When the heat flow becomes *steady* the temperature distribution is constant in time. Hence in the steady state (4) reduces to Laplace's Equation $\nabla^2 T = 0$.

Example. Find the temperature distribution in a homogeneous hollow sphere whose inner and outer surfaces are held at constant temperatures.

When the flow is steady, T is a function of r alone; hence, from (89.17),

$$\nabla^2 T = \frac{1}{r^2}\frac{d}{dr}(r^2 T_r), \qquad \frac{d}{dr}\left(r^2 \frac{dT}{dr}\right) = 0, \qquad T = \frac{A}{r} + B.$$

A and B may be determined from the conditions $T = T_1$ when $r = r_1$, $T = T_2$ when $r = r_2$.

115. Summary: Integral Transformations. If $\mathbf{f}(\mathbf{r})$ is a continuously differentiable tensor point function, the basic integral transformations are:

$$\text{(A)} \qquad \int \nabla\mathbf{f}\,dV = \oint \mathbf{n}\mathbf{f}\,dS \qquad \left(\int dV\,\nabla\mathbf{f} = \oint d\mathbf{S}\,\mathbf{f}\right);$$

$$\text{(B)} \qquad \int \mathbf{n}\times\nabla\mathbf{f}\,dS = \oint \mathbf{T}\mathbf{f}\,ds \qquad \left(\int d\mathbf{S}\times\nabla\mathbf{f} = \oint d\mathbf{r}\,\mathbf{f}\right).$$

* If $f(\mathbf{r})$ is a continuous scalar function in a region V and $\int f(\mathbf{r})\,dV = 0$ for an arbitrary subregion of V, then $f(\mathbf{r}) = 0$ throughout V. For, if $f(\mathbf{r}) \neq 0$ at a point P, we can surround P with a sphere σ so small that f does not change sign in σ, and hence $\int_\sigma f(\mathbf{r})\,dV \neq 0$, contrary to hypothesis.

In addition,

$$\text{(C)} \qquad \int_{\mathbf{r}_1}^{\mathbf{r}_2} \mathbf{T} \cdot \nabla \mathbf{f}\, ds = \int_{\mathbf{r}_1}^{\mathbf{r}_2} d\mathbf{r} \cdot \nabla \mathbf{f} = \mathbf{f}(\mathbf{r}_2) - \mathbf{f}(\mathbf{r}_1)$$

may be regarded as a third basic type. From these, other integral transformations may be deduced by dotting and crossing within tensor functions. If we dot and cross in the first position, we get, from (A):

$$\text{(A}\cdot\text{)} \qquad \int \nabla \cdot \mathbf{f}\, dV = \oint \mathbf{n} \cdot \mathbf{f}\, dS,$$

$$\text{(A}\times\text{)} \qquad \int \nabla \times \mathbf{f}\, dV = \oint \mathbf{n} \times \mathbf{f}\, dS.$$

Similarly, from (B),

$$\text{(B}\cdot\text{)} \qquad \int \mathbf{n} \cdot \nabla \times \mathbf{f}\, dS = \oint \mathbf{T} \cdot \mathbf{f}\, ds,$$

$$\text{(B}\times\text{)} \qquad \int (\mathbf{n} \overset{2}{\cdot} \nabla \mathbf{f} - \mathbf{n} \nabla \cdot \mathbf{f})\, dS = \oint \mathbf{T} \times \mathbf{f}\, ds,$$

in which ∇ may be replaced by ∇_s. In the B transformations the surface integrals vanish when taken over a *closed* surface.

When $\mathbf{f}$ is a vector function, (A ·) is known as the *divergence theorem* and (B ·) as Stokes' Theorem.

If we replace $\mathbf{f}$ by $\mathbf{nf}$ in (B ×) we obtain

$$\text{(B}'\text{)} \qquad \int (\nabla_s \mathbf{f} + J\, \mathbf{nf})\, dS = \oint \mathbf{m} \mathbf{f}\, ds,$$

a transformation of the same scope as (B); $J = -\operatorname{Div} \mathbf{n}$ is the mean curvature of the surface, and $\mathbf{m} = \mathbf{T} \times \mathbf{n}$ is the unit *external* normal to the bounding circuit. From (B′) we derive

$$\text{(B}'\cdot\text{)} \qquad \int (\nabla_s \cdot \mathbf{f} + J\, \mathbf{n} \cdot \mathbf{f})\, dS = \oint \mathbf{m} \cdot \mathbf{f}\, ds,$$

$$\text{(B}'\times\text{)} \qquad \int (\nabla_s \times \mathbf{f} + J\, \mathbf{n} \times \mathbf{f})\, dS = \oint \mathbf{m} \times \mathbf{f}\, ds.$$

On a *plane*, $\mathbf{n}$ is constant, and $J = 0$; if we write dA for dS, the B′ transformations become

$$\text{(P)} \qquad \int \nabla_s \mathbf{f}\, dA = \oint \mathbf{m} \mathbf{f}\, ds,$$

(P $\cdot$) $$\int \nabla_s \cdot \mathbf{f}\, dA = \oint \mathbf{m} \cdot \mathbf{f}\, ds,$$

(P $\times$) $$\int \nabla_s \times \mathbf{f}\, dA = \oint \mathbf{m} \times \mathbf{f}\, ds.$$

When **f** is a vector, (P $\cdot$) is the divergence theorem in the plane.

If $\mathbf{f}(\mathbf{r})$ is a continuously differentiable vector and $\operatorname{rot} \mathbf{f} = 0$ in a simply connected region of 3-space, the line integral of $\mathbf{f} \cdot d\mathbf{r}$ is independent of the path, and

$$\mathbf{f} = \nabla\varphi, \qquad \varphi(\mathbf{r}) = \int_{\mathbf{r}_0}^{\mathbf{r}} \mathbf{f} \cdot d\mathbf{r}.$$

If $\mathbf{f}(\mathbf{r})$ is a continuously differentiable vector, and $\mathbf{n} \cdot \operatorname{rot} \mathbf{f} = 0$ in a simply connected portion of a surface, the line integral of $\mathbf{f} \cdot d\mathbf{r}$ over surface curves is independent of the path, and

$$\mathbf{f}_t = \operatorname{Grad} \varphi, \qquad \varphi(\mathbf{r}) = \int_{\mathbf{r}_0}^{\mathbf{r}} \mathbf{f} \cdot d\mathbf{r} = \int_{\mathbf{r}_0}^{\mathbf{r}} \mathbf{f}_t \cdot d\mathbf{r},$$

where $\mathbf{f}_t = \mathbf{f} - (\mathbf{n} \cdot \mathbf{f})\mathbf{n}$ is the tangential projection of **f** on the surface.

PROBLEMS

1. Show that a closed curve lying in a plane with unit normal **n** encloses an area A given by

$$\mathbf{n}A = \tfrac{1}{2} \oint \mathbf{r} \times d\mathbf{r}.$$

2. Compute the integral $\int \mathbf{n} \cdot \operatorname{rot} \mathbf{f}\, dS$ over that portion of the sphere $r = a$ lying above the xy-plane when $\mathbf{f} = \varphi(r)\mathbf{c}$.

3. Prove that $\oint \mathbf{r} \times \mathbf{n}\, dS = 0$ over any closed surface S. If a body bounded by S is subjected to a uniform pressure $-p\mathbf{n}$ per unit of area, show that these forces have zero moment about any point in space.

4. If the closed curve C encloses a portion of a surface S, show that

$$\int_S \mathbf{n} \times \mathbf{r}\, dS = \tfrac{1}{2} \oint_C \mathbf{T} r^2\, ds.$$

5. If $\mathbf{f} = \mathbf{i}u(x, y) + \mathbf{j}v(x, y)$, prove that

$$\oint_C \mathbf{f} \times d\mathbf{r} = \mathbf{k} \iint_A \operatorname{div} \mathbf{f}\, dx\, dy$$

where C is closed curve in the xy-plane enclosing the region A.

6. If $\mathbf{c}$ is a constant vector, prove that

$$\oint_S \mathbf{n} \times (\mathbf{c} \times \mathbf{r})\, dS = 2V\mathbf{c}$$

where V is the volume enclosed by the surface S.

7. The closed surface S encloses a volume V. If the vector $\mathbf{f}$ is everywhere normal to S, prove that $\int_V \operatorname{rot} \mathbf{f}\, dV = 0$.

8. Show that the centroid of a volume V bounded by a closed surface S is given by

$$V\mathbf{r}^* = \tfrac{1}{2} \oint r^2 \mathbf{n}\, dS.$$

Apply this formula to find the centroid of a hemisphere of radius a, center O, lying above the xy-plane.

9. If p denotes the distance from the center of the ellipsoid $\mathbf{r} \cdot \Phi \cdot \mathbf{r} = 1$ to the tangent plane at the point $\mathbf{r}$, show that the integrals of p and $1/p$ over its surface have the values

$$\oint p\, dS = 3V, \qquad \oint dS/p = \varphi_1 V,$$

where V is the enclosed volume $[p = \mathbf{r} \cdot \mathbf{n},\ 1/p = \mathbf{r} \cdot \Phi \cdot \mathbf{n}]$. Express these results in terms of the semiaxes a, b, c of the ellipsoid.

10. Find a vector $\mathbf{v} = f(r)\mathbf{r}$ such that $\operatorname{div} \mathbf{v} = r^m$ $(m \neq -3)$. Prove that

$$\int r^m\, dV = \frac{1}{m+3} \oint r^m \mathbf{r} \cdot \mathbf{n}\, dS.$$

11. Show that

(a) $$\int r^m \mathbf{R}\, dV = \frac{1}{m+1} \oint r^{m+1} \mathbf{n}\, dS \qquad (m \neq -1);$$

(b) $$\int \frac{\mathbf{R}}{r}\, dV = \oint \log r\, \mathbf{n}\, dS.$$

12. If $\mathbf{p}$ and $\mathbf{q}$ are vector functions, prove that

(a) $$\oint \mathbf{n} \cdot \mathbf{pq}\, dS = \int (\mathbf{q} \operatorname{div} \mathbf{p} + \mathbf{p} \cdot \nabla \mathbf{q})\, dV;$$

(b) $$\oint \mathbf{n} \cdot \mathbf{rr}\, dS = 4 \int \mathbf{r}\, dV.$$

13. If $r = PQ$, the solid angle subtended at P by the surface S (over which Q ranges) is

$$\Omega_P = -\int_S \mathbf{n} \cdot \nabla_Q \frac{1}{r}\, dS \qquad (107.1),$$

where $\nabla_Q 1/r$ is computed at Q. Prove that

$$\nabla_P \Omega_P = \oint_C \mathbf{T} \times \nabla_Q \frac{1}{r}\, ds = \oint_C \frac{\mathbf{r} \times \mathbf{T}}{r^3}\, ds$$

where $\mathbf{r} = \overrightarrow{PQ}$ and the curve C is the boundary of S.

14. If C and C' are two unlinked closed curves, show that if $\mathbf{r} = \overrightarrow{PP'}$ the double-circuit integral,

$$\oint_C \oint_{C'} \frac{\mathbf{r} \cdot \mathbf{T} \times \mathbf{T}'}{r^3}\, ds\, ds' = 0.$$

If C and C' are two simple loops, linked as in a chain, show that the preceding double integral is $\pm 4\pi$, the sign depending on the sense of C'.

15. Show that tangential forces of constant magnitude σ acting along a closed plane curve C are equivalent to a couple of moment $2\sigma\mathbf{A}$, where $\mathbf{A}$ is the vector area enclosed by C. What is the moment of the couple when C is a twisted curve? [Put $\mathbf{f} = \mathbf{r}$ in (100.4).]

16. Let $\mathbf{f}(\mathbf{r})$ be a solenoidal vector function. Prove that $\mathbf{f} = \operatorname{rot} \mathbf{g}$ (§ 92) where

$$\mathbf{g} = -\mathbf{r} \times \int_0^1 \lambda \mathbf{f}(\lambda \mathbf{r})\, d\lambda \quad \text{or} \quad \mathbf{r} \times \int_1^\infty \lambda \mathbf{f}(\lambda \mathbf{r})\, d\lambda$$

provided the integrals exist. (The integrand of the former may become infinite at the lower limit.)

In particular if $\mathbf{f}(\mathbf{r})$ is homogeneous of degree $n \neq -2$, $\mathbf{f}(\lambda\mathbf{r}) = \lambda^n \mathbf{f}(\mathbf{r})$ and $\mathbf{g}$ is given by $\mathbf{g} = (\mathbf{f} \times \mathbf{r})/(n + 2)$. Prove this formula directly by making use of $\mathbf{r} \cdot \nabla \mathbf{f} = n\mathbf{f}$ and (85.6).

CHAPTER VII

HYDRODYNAMICS

116. Stress Dyadic. Let S be any closed surface inside a deformable body. It divides the body into two portions, A within S, and B without S. The forces acting on A are of two kinds: (i) *body* or *mass* forces which act on the interior particles, and (ii) *surface forces*, which act on the bounding surface S.

Let $\overline{\mathbf{R}}$ denote the average body force on the element of mass Δm; then, if Δm shrinks to zero while always enclosing a point P, the limit of $\overline{\mathbf{R}}/\Delta m$ is defined as the body force $\mathbf{R}$, per unit of mass, at P.

Let $\overline{\mathbf{F}}_n$ denote the average surface force acting on A over the vector element of surface $\mathbf{n}\Delta S$ of the boundary S; then, if ΔS shrinks to zero while always enclosing a point P, the limit of $\overline{\mathbf{F}}_n/\Delta S$ is defined as the surface force $\mathbf{F}_n$, per unit of area, acting on A at P. Here n denotes a directed line normal to S at P and directed from A to B. The notation $\mathbf{F}_n$ thus associates a surface force with a surface element of unit normal $\mathbf{n}$. The force $\mathbf{F}_n dS$ is exerted on the element $\mathbf{n}\, dS$ *by the matter toward which* $\mathbf{n}$ *points.*

The surface force acting on B on the same element $\mathbf{n}\, dS$ is, from the law of action and reaction,

$$\text{(1)} \qquad \mathbf{F}_{-n} = -\mathbf{F}_n.$$

Consider now a small tetrahedron (Fig. 116) with three faces parallel to the coordinate planes; and let outwardly directed lines normal to the faces be denoted by $-x, -y, -z, n$. If the area of the inclined face is A, the faces normal to x, y, z have areas $A \cos(n, x)$, $A \cos(n, y)$, $A \cos(n, z)$, and the volume of the tetrahedron is $\frac{1}{3}Ah$ where h is the altitude of P above the base A. If body and surface forces are replaced by averages and $\bar{\rho}$ denotes the average density, the equilibrium of the tetrahedron requires that

$$\overline{\mathbf{F}}_n A + \overline{\mathbf{F}}_{-x} A \cos(n, x) + \overline{\mathbf{F}}_{-y} A \cos(n, y) + \overline{\mathbf{F}}_{-z} A \cos(n, z) + \tfrac{1}{3}Ah\bar{\rho}\overline{\mathbf{R}} = 0.$$

If we divide out A and let $h \to 0$ so that the tetrahedron shrinks to the vertex P, we have in the limit

$$\mathbf{F}_n + \mathbf{F}_{-x} \cos(n, x) + \mathbf{F}_{-y} \cos(n, y) + \mathbf{F}_{-z} \cos(n, z) = 0.$$

If $\mathbf{n}$ is a unit vector along the line n, $\cos(n, x) = \mathbf{n} \cdot \mathbf{i}$, etc.; and, since $\mathbf{F}_{-x} = -\mathbf{F}_x$, we have

$$\mathbf{F}_n = \mathbf{n} \cdot (\mathbf{i}\,\mathbf{F}_x + \mathbf{j}\,\mathbf{F}_y + \mathbf{k}\,\mathbf{F}_z). \tag{2}$$

All surface forces in (2) now refer to the point P. Thus a plane through P normal to $\mathbf{n}$ divides the body into two parts; and the

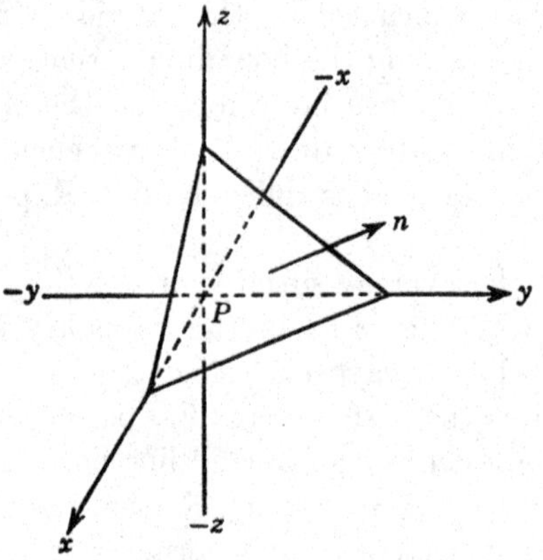

Fig. 116

part toward which $\mathbf{n}$ points acts upon the other part with the force $\mathbf{F}_n$ per unit of surface at P. The dyadic,

$$\Phi = \mathbf{i}\,\mathbf{F}_x + \mathbf{j}\,\mathbf{F}_y + \mathbf{k}\,\mathbf{F}_z, \tag{3}$$

is called the *stress dyadic* at P; it effects a synthesis of all surface forces at P by means of the relation,

$$\mathbf{F}_n = \mathbf{n} \cdot \Phi. \tag{4}$$

If the deformable body is a *fluid*, we assume as an experimental fact that *the stress across any surface element of a fluid in equilibrium is a pressure normal to the element;* hence

$$\mathbf{F}_n = -p\mathbf{n}; \qquad \mathbf{F}_x = -p_1\mathbf{i}, \qquad \mathbf{F}_y = -p_2\mathbf{j}, \qquad \mathbf{F}_z = -p_3\mathbf{k}.$$

With these values, (2) becomes

$$p\mathbf{n} = \mathbf{n} \cdot \mathbf{i}\; p_1\mathbf{i} + \mathbf{n} \cdot \mathbf{j}\; p_2\mathbf{j} + \mathbf{n} \cdot \mathbf{k}\; p_3\mathbf{k};$$

and, on multiplying by $\mathbf{i}\cdot$, $\mathbf{j}\cdot$, $\mathbf{k}\cdot$, in turn, we have

$$p = p_1 = p_2 = p_3. \tag{5}$$

At any point within a fluid the pressure is the same in all directions. At any point where the pressure is p, the stress dyadic is

$$\Phi = -p(\mathbf{ii} + \mathbf{jj} + \mathbf{kk}) = -p\mathbf{I}. \tag{6}$$

For fluids *in motion* it is no longer true that only *normal* stresses exist. *Viscous* fluids in motion do exert tangential stresses; but in many problems these tangential stresses are small and may be neglected. We shall develop the mechanics of fluids on the hypothesis of purely normal stress. We imply this assumption by speaking of *perfect* or *non-viscous* fluids; for a perfect fluid the stress dyadic is $-p\mathbf{I}$.

117. Equilibrium of a Deformable Body. If a body in a strained state is in equilibrium, any portion of it bounded by a closed surface S is in equilibrium under its body forces and the surface forces acting on S. Let the body forces be $\mathbf{R}$ per unit of mass, and let the surface forces $\mathbf{F}_n$ per unit of area be given by the stress dyadic Φ. Then, from (116.4), $\mathbf{F}_n = \mathbf{n}\cdot\Phi$, and the equations of equilibrium are

$$\int \mathbf{R}\rho\, dV + \oint \mathbf{n}\cdot\Phi\, dS = 0, \tag{1}$$

$$\int \mathbf{r}\times\mathbf{R}\rho\, dV + \oint \mathbf{r}\times(\mathbf{n}\cdot\Phi)\, dS = 0, \tag{2}$$

where (2) is the moment equation about the origin.

If we transform the surface integrals by means of (106.5),

$$\oint \mathbf{n}\cdot\Phi\, dS = \int \nabla\cdot\Phi\, dV,$$

$$\oint \mathbf{r}\times(\mathbf{n}\cdot\Phi)\, dS = -\oint \mathbf{n}\cdot\Phi\times\mathbf{r}\, dS = -\int \nabla\cdot(\Phi\times\mathbf{r})\, dV,$$

(1) and (2) become

$$\int (\rho\mathbf{R} + \nabla\cdot\Phi)\, dV = 0,$$

$$\int [\rho\mathbf{R}\times\mathbf{r} + \nabla\cdot(\Phi\times\mathbf{r})]\, dV = 0.$$

Since these integrals vanish for an arbitrary choice of S, their integrands must be identically zero. The equations of equilibrium are therefore

(3) $$\rho\mathbf{R} + \nabla \cdot \Phi = 0,$$

(4) $$\rho\mathbf{R} \times \mathbf{r} + \nabla \cdot (\Phi \times \mathbf{r}) = 0.$$

On eliminating $\rho\mathbf{R}$ from these equations, we obtain

(5) $$\nabla \cdot (\Phi \times \mathbf{r}) - (\nabla \cdot \Phi) \times \mathbf{r} = 0.$$

From (5) we conclude that $\boldsymbol{\phi}$, the vector invariant of Φ, is zero; for, if we write $\Phi = \mathbf{i}\,\mathbf{F}_x + \mathbf{j}\,\mathbf{F}_y + \mathbf{k}\,\mathbf{F}_z$,

$$\frac{\partial}{\partial x}(\mathbf{F}_x \times \mathbf{r}) + \frac{\partial}{\partial y}(\mathbf{F}_y \times \mathbf{r}) + \frac{\partial}{\partial z}(\mathbf{F}_z \times \mathbf{r}) - \left(\frac{\partial \mathbf{F}_x}{\partial x} + \frac{\partial \mathbf{F}_y}{\partial y} + \frac{\partial \mathbf{F}_z}{\partial z}\right) \times \mathbf{r} = 0,$$

$$\mathbf{F}_x \times \frac{\partial \mathbf{r}}{\partial x} + \mathbf{F}_y \times \frac{\partial \mathbf{r}}{\partial y} + \mathbf{F}_z \times \frac{\partial \mathbf{r}}{\partial z} = \mathbf{F}_x \times \mathbf{i} + \mathbf{F}_y \times \mathbf{j} + \mathbf{F}_z \times \mathbf{k} = 0,$$

and hence $\boldsymbol{\phi} = 0$. In view of the theorem of § 68, we see that the moment equation (2) requires that *the stress dyadic Φ be symmetric.*

If we write

$$\mathbf{F}_x = \mathbf{i}\,X_x + \mathbf{j}\,Y_x + \mathbf{k}\,Z_x, \quad \cdots,$$

the stress dyadic may be written in the nonion form,

(6) $$\Phi = \begin{pmatrix} X_x & Y_x & Z_x \\ X_y & Y_y & Z_y \\ X_z & Y_z & Z_z \end{pmatrix},$$

where, by virtue of its symmetry,

(7) $$X_y = Y_x, \qquad Y_z = Z_y, \qquad Z_x = X_z.$$

The *normal components* of stress X_x, Y_y, Z_z occur in the principal diagonal, whereas the *tangential components*, or components of *shear*, are equal in pairs. More specifically, *in perpendicular planes, the components of shear perpendicular to their line of intersection are equal.*

Since Φ is symmetric, we always can find three mutually orthogonal vectors **i**, **j**, **k**, so that

(8) $$\Phi = \alpha\,\mathbf{ii} + \beta\,\mathbf{jj} + \gamma\,\mathbf{kk}.$$

Then **i**, **j**, **k** give the *principal directions of stress*, and α, β, γ are the *principal stresses* at the point in question.

118. Equilibrium of a Fluid. For a fluid the stress dyadic is

$$\Phi = -p\mathbf{I} \quad \text{and} \quad \nabla \cdot \Phi = -\nabla p,$$

from (86.8); the equation of equilibrium (117.3) is therefore

$$(1) \qquad \nabla p = \rho \mathbf{R}.$$

Consider a liquid of constant density ρ in equilibrium under gravity. The body force per unit mass is then $\mathbf{g}$, the acceleration of gravity, and

$$(2) \qquad \nabla p = \rho \mathbf{g} = \rho g \mathbf{k} = \rho g \nabla z = \nabla(\rho g z),$$

when the z-axis is directed downward along a plumb line Hence

$$(3) \qquad p = \rho g z + p_0,$$

where $p = p_0$ when $z = 0$. If the origin is at a free surface, p_0 is the atmospheric pressure, and $\rho g z$ is called the *hydrostatic* pressure.

119. Floating Body. Let the surface S of a floating body V be divided by its plane section A at the water line into two parts: S_1 submerged, S_2 in air (Fig. 119). If p_0 is the atmospheric pressure, $p_1 = \rho g z$ the hydrostatic pressure at depth z, $p_0 + p_1$ acts on S_1, p_0 on S_2; or we may say that p_0 acts over S, while p_1 acts on the closed surface $S_1 + A$ ($p_1 = 0$ on A) enclosing the volume V_1 below the water line. Hence the equations of equilibrium of the body are:

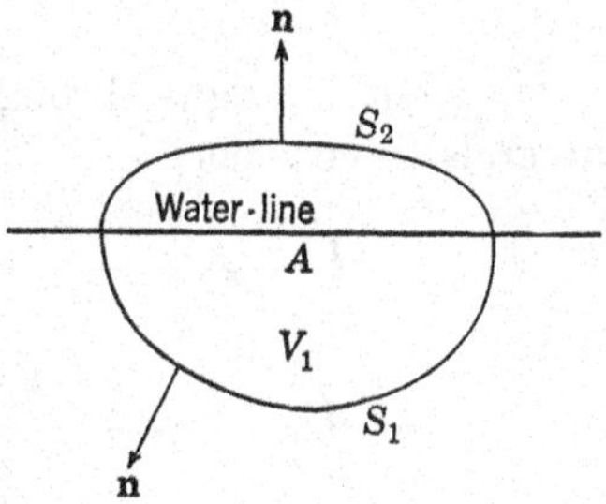

FIG. 119

$$(1) \qquad \int_V \mathbf{g}\, dm - \oint_S p_0 \mathbf{n}\, dS - \oint_{S_1+A} p_1 \mathbf{n}\, dS = 0,$$

$$(2) \qquad \int_V \mathbf{r} \times \mathbf{g}\, dm - \oint_S \mathbf{r} \times \mathbf{n}\, p_0\, dS - \oint_{S_1+A} \mathbf{r} \times \mathbf{n}\, p_1\, dS = 0.$$

We consider in turn the integrals in (1):

$$\int_V \mathbf{g}\, dm = \mathbf{W}, \quad \text{the weight of the body;}$$

$$\oint_S p_0 \mathbf{n}\, dS = p_0 \oint \mathbf{n}\, dS = 0;$$

$$\oint_{S_1+A} \mathbf{n}\, p_1\, dS = \int_{V_1} \nabla p_1\, dV = \int_{V_1} \rho \mathbf{g}\, dV = \mathbf{W}_1,$$

the weight of the displaced liquid. Therefore (1) states that $\mathbf{W} = \mathbf{W}_1$; this is the

Principle of Archimedes: *A floating body in equilibrium displaces its own weight of liquid.*

In order to interpret (2), we note that the *center of mass* of a discrete set of particles P_i of mass m_i is defined as a *centroid* (9.3); its position vector $\mathbf{r}^*$ is given by

(3) $$m\mathbf{r}^* = \Sigma m_i \mathbf{r}_i,$$

where m is the total mass. Similarly, the center of mass of a continuous body is defined by

(4) $$m\mathbf{r}^* = \int \mathbf{r}\, dm.$$

If we change signs throughout in (2), we have the following integrals to consider:

$$\int_V \mathbf{g} \times \mathbf{r}\, dm = \mathbf{g} \times \int_V \mathbf{r}\, dm = \mathbf{g} \times m\mathbf{r}^* = \mathbf{W} \times \mathbf{r}^*,$$

$$p_0 \oint_S \mathbf{n} \times \mathbf{r}\, dS = p_0 \oint_S \operatorname{rot} \mathbf{r}\, dV = 0,$$

$$\oint_{S_1+A} \mathbf{n} \times p_1 \mathbf{r}\, dS = \int_{V_1} \operatorname{rot}(p_1 \mathbf{r})\, dV = \int_{V_1} (\nabla p_1) \times \mathbf{r}\, dV$$

$$= \mathbf{g} \times \int_{V_1} \rho \mathbf{r}\, dV = \mathbf{g} \times m_1 \mathbf{r}_1^* = \mathbf{W}_1 \times \mathbf{r}_1^*.$$

Thus (2) reduces to $\mathbf{W} \times (\mathbf{r}^* - \mathbf{r}_1^*) = 0$; this states that *the center of mass of the body and of the displaced liquid are on the same vertical.*

120. Equation of Continuity. Let ρ and $\mathbf{v}$ denote the density and velocity of a fluid at the point P and at the instant t. Consider the mass of fluid $\int \rho\, dV$ within a fixed but arbitrary closed surface S. This mass is increasing at the rate,

$$\frac{\partial}{\partial t} \int \rho\, dV = \int \frac{\partial \rho}{\partial t}\, dV,$$

the time differentiation being local. This rate must equal the rate at which fluid is *entering* S, namely, $-\oint \mathbf{n}\cdot\rho\mathbf{v}\,dS$, where $\mathbf{n}$ denotes the *outward* unit normal. Hence

$$\int \frac{\partial\rho}{\partial t}\,dV = -\oint \mathbf{n}\cdot\rho\mathbf{v}\,dS = -\int \operatorname{div}(\rho\mathbf{v})\,dV,$$

when the divergence theorem (106.5) is applied. Since the integral of $\partial\rho/\partial t + \operatorname{div}(\rho\mathbf{v})$ over any closed region within the fluid is zero, we conclude that

$$\text{(1)} \qquad \frac{\partial\rho}{\partial t} + \operatorname{div}(\rho\mathbf{v}) = 0.$$

This *equation of continuity* may be put in another form by introducing the *substantial rate of change* $d\rho/dt$ instead of the *local* time rate $\partial\rho/\partial t$. Along the actual path, or *line of motion*, of a fluid particle,

$$x = x(t), \qquad y = y(t), \qquad z = z(t),$$

the density $\rho(t, x, y, z)$ becomes a function of t alone, and

$$\frac{d\rho}{dt} = \frac{\partial\rho}{\partial t} + \frac{\partial\rho}{\partial x}\frac{dx}{dt} + \frac{\partial\rho}{\partial y}\frac{dy}{dt} + \frac{\partial\rho}{\partial z}\frac{dz}{dt}$$

$$= \frac{\partial\rho}{\partial t} + \mathbf{v}\cdot\left(\mathbf{i}\frac{\partial\rho}{\partial x} + \mathbf{j}\frac{\partial\rho}{\partial y} + \mathbf{k}\frac{\partial\rho}{\partial z}\right)$$

gives the time rate at which the density of a moving fluid particle is changing. We thus obtain the important relation,

$$\text{(2)} \qquad \frac{d\rho}{dt} = \frac{\partial\rho}{\partial t} + \mathbf{v}\cdot\nabla\rho,$$

connecting the substantial and local changes of ρ.

For any tensor function $\mathbf{f}(t, \mathbf{r})$ of time and position, associated with a fluid particle moving with the velocity $\mathbf{v}$, we have, in the same way,

$$\text{(3)} \qquad \frac{d\mathbf{f}}{dt} = \frac{\partial\mathbf{f}}{\partial t} + \mathbf{v}\cdot\nabla\mathbf{f}.$$

From (85.2),

$$\operatorname{div}(\rho \mathbf{v}) = \mathbf{v} \cdot \nabla \rho + \rho \operatorname{div} \mathbf{v};$$

hence the equation of continuity (1) also may be written

$$\frac{d\rho}{dt} + \rho \operatorname{div} \mathbf{v} = 0. \tag{4}$$

Since $\operatorname{div} \mathbf{v} = -(d\rho/dt)/\rho$ we can interpret div **v** as the relative time rate of decrease of density of a fluid particle having the velocity **v**. Thus a positive value of div **v** implies a negative $d\rho/dt$ and, consequently, an attenuation of the fluid at the point considered; hence the term *divergence*.

For an incompressible fluid, $d\rho/dt = 0$, and

$$\operatorname{div} \mathbf{v} = 0. \tag{5}$$

This equation has the integral equivalent,

$$\oint \mathbf{n} \cdot \mathbf{v}\, dS = \int \operatorname{div} \mathbf{v}\, dV = 0; \tag{6}$$

the "flux" of an incompressible fluid across the boundary of a fixed closed surface within the fluid is zero. If an incompressible fluid is also homogeneous, ρ is constant.

121. Eulerian Equation for a Fluid in Motion. In the *Eulerian* or *statistical* method of treatment we aim at finding the velocity, density, and pressure ($\mathbf{v}$, ρ, p) of the fluid as functions of the time at all points of space ($\mathbf{r}$) occupied by the fluid. Consider the fluid within a fixed closed surface S at any instant t. By D'Alembert's Principle, the body and surface forces, together with the reversed inertia forces ($-m\mathbf{a}$), may be treated as a system in statical equilibrium. In a perfect fluid the surface force is a normal pressure: $\mathbf{F}_n = -p\mathbf{n}$ (§ 116). Hence, if the body force per unit mass is denoted by **R**, D'Alembert's Principle applied to the fluid enclosed by S gives the equation,

$$\int (\mathbf{R} - \mathbf{a})\rho\, dV - \oint \mathbf{n}\, p\, dS = 0, \tag{1}$$

$$\int \mathbf{r} \times (\mathbf{R} - \mathbf{a})\rho\, dV - \oint \mathbf{r} \times \mathbf{n}\, p\, dS = 0, \tag{2}$$

where **a** denotes the acceleration of the fluid particles. On transforming the surface integrals in (1) and (2), we have

$$\int \{(\mathbf{R} - \mathbf{a})\rho - \nabla p\} \, dV = 0,$$

$$\int \{\mathbf{r} \times (\mathbf{R} - \mathbf{a})\rho + \operatorname{rot}(p\mathbf{r})\} \, dV = 0.$$

Since these integrals vanish for any choice of S, the integrands are identically zero. From (85.3), rot $(p\mathbf{r}) = (\nabla p) \times \mathbf{r}$; the equations of motion are therefore

(3) $$(\mathbf{R} - \mathbf{a})\rho - \nabla p = 0,$$

(4) $$\mathbf{r} \times (\mathbf{R} - \mathbf{a})\rho - \mathbf{r} \times \nabla p = 0.$$

When (3) holds good, (4) is identically satisfied and may be omitted. The Eulerian Equation of Motion is therefore

(5) $$\mathbf{a} = \mathbf{R} - \frac{1}{\rho} \nabla p.$$

Here $\mathbf{a} = d\mathbf{v}/dt$ is the acceleration of a moving fluid particle and must be distinguished from $\partial \mathbf{v}/\partial t$, the rate of change of fluid velocity at a fixed point. These substantial and local rates of change are connected by the relation (120.3):

(6) $$\mathbf{a} = \frac{d\mathbf{v}}{dt} = \frac{\partial \mathbf{v}}{\partial t} + \mathbf{v} \cdot \nabla \mathbf{v}.$$

When the density ρ is a function of p only, we introduce the function:

(7) $$P = \int_c^p \frac{dp}{\rho}; \quad \text{then} \quad \nabla P = \frac{dP}{dp} \nabla p = \frac{1}{\rho} \nabla p.$$

Moreover, if the body force **R** has a single-valued scalar potential Q (§ 102),

(8) $$\mathbf{R} = -\nabla Q;$$

such forces are said to be *conservative*. Under these conditions (5) becomes

(9) $$\frac{d\mathbf{v}}{dt} = -\nabla(Q + P).$$

In the Eulerian method $\mathbf{r}$ and t are independent variables, so that $\partial\mathbf{r}/\partial t = 0$. Equation (120.3) applied to $\mathbf{r}$,

$$\frac{d\mathbf{r}}{dt} = \frac{\partial\mathbf{r}}{\partial t} + \mathbf{v}\cdot\nabla\mathbf{r} = \mathbf{v}\cdot I = \mathbf{v},$$

is simply an identity.

The lines of the fluid which at any instant are everywhere tangent to $\mathbf{v}$ are called its *stream-lines*. They are not the actual paths of the fluid particles except when the flow is *steady*, that is, when $\mathbf{v}$ is constant in time ($\partial\mathbf{v}/\partial t = 0$). The stream-lines have the differential equation $\mathbf{v}\times d\mathbf{r} = 0$.

Example. Revolving Fluid. If the fluid is revolving with constant angular velocity $\boldsymbol{\omega} = \omega\mathbf{k}$ about a vertical axis, its velocity distribution is that of a rigid body; hence, with the origin on the axis of revolution, we have, from (54.3), $v_P = \boldsymbol{\omega}\times\overrightarrow{OP}$, or simply $\mathbf{v} = \boldsymbol{\omega}\times\mathbf{r}$. Then $\mathbf{a} = \boldsymbol{\omega}\times\mathbf{v}$, and (9) becomes

$$\text{(i)} \qquad \boldsymbol{\omega}\times(\boldsymbol{\omega}\times\mathbf{r}) = -\nabla(Q+P).$$

Now

$$\begin{aligned}\boldsymbol{\omega}\times(\boldsymbol{\omega}\times\mathbf{r}) &= \mathbf{r}\cdot(\boldsymbol{\omega}\boldsymbol{\omega} - \omega^2\mathrm{I}) \\ &= -\tfrac{1}{2}\nabla\{\mathbf{r}\cdot(\omega^2\mathrm{I} - \boldsymbol{\omega}\boldsymbol{\omega})\cdot\mathbf{r}\} \qquad (85.14) \\ &= -\tfrac{1}{2}\omega^2\,\nabla(r^2 - z^2) \\ &= -\tfrac{1}{2}\omega^2\,\nabla(x^2 + y^2);\end{aligned}$$

hence, from (i),

$$\nabla[Q + P - \tfrac{1}{2}\omega^2(x^2+y^2)] = 0,$$

$$\text{(ii)} \qquad Q + P - \tfrac{1}{2}\omega^2(x^2+y^2) = \text{const.}$$

For gravitational body forces, $\mathbf{R} = \mathbf{g}$, and $Q = gz$, if the z-axis is directed upward; and, if the density is constant, $P = p/\rho$. In this case, (ii) becomes

$$gz + p/\rho - \tfrac{1}{2}\omega^2(x^2+y^2) = \text{const.}$$

At a free surface p is constant; a free surface is therefore a paraboloid of revolution.

122. Vorticity. Starting with the Eulerian Equation in the form,

$$\text{(1)} \qquad \frac{\partial\mathbf{v}}{\partial t} + \mathbf{v}\cdot\nabla\mathbf{v} = -\nabla(Q+P),$$

we transform $\mathbf{v}\cdot\nabla\mathbf{v}$ by means of (85.9) and (85.7):

$$\mathbf{v}\cdot\nabla\mathbf{v} = (\nabla\mathbf{v})\cdot\mathbf{v} - \mathbf{v}\times\operatorname{rot}\mathbf{v} = \tfrac{1}{2}\nabla(\mathbf{v}\cdot\mathbf{v}) - \mathbf{v}\times\operatorname{rot}\mathbf{v};$$

hence (1) may be written

$$\frac{\partial \mathbf{v}}{\partial t} - \mathbf{v} \times \operatorname{rot} \mathbf{v} = -\nabla(Q + P + \tfrac{1}{2}v^2). \tag{2}$$

Now, from (84.11),

$$\operatorname{rot}\left(\frac{\partial \mathbf{v}}{\partial t} - \mathbf{v} \times \operatorname{rot} \mathbf{v}\right) = 0,$$

$$\begin{aligned} \frac{\partial \operatorname{rot} \mathbf{v}}{\partial t} &= \operatorname{rot}(\mathbf{v} \times \operatorname{rot} \mathbf{v}) \\ &= (\operatorname{rot} \mathbf{v}) \cdot \nabla \mathbf{v} - \mathbf{v} \cdot \nabla \operatorname{rot} \mathbf{v} - (\operatorname{rot} \mathbf{v})(\operatorname{div} \mathbf{v}), \end{aligned}$$

when we make use of (85.6) and (84.13); and, since

$$\frac{\partial \operatorname{rot} \mathbf{v}}{\partial t} + \mathbf{v} \cdot \nabla \operatorname{rot} \mathbf{v} = \frac{d \operatorname{rot} \mathbf{v}}{dt},$$

$$\frac{d \operatorname{rot} \mathbf{v}}{dt} + (\operatorname{rot} \mathbf{v})(\operatorname{div} \mathbf{v}) = (\operatorname{rot} \mathbf{v}) \cdot \nabla \mathbf{v}. \tag{3}$$

When $\mathbf{v}$ gives the velocities of a rigid body having the instantaneous angular velocity $\boldsymbol{\omega}$, $\operatorname{rot} \mathbf{v} = 2\boldsymbol{\omega}$ (§ 84, ex.). For a liquid we may regard

$$\boldsymbol{\omega} = \tfrac{1}{2} \operatorname{rot} \mathbf{v} \tag{4}$$

as the molecular rotation or *vorticity* of the fluid particles.

In (3) we now replace $\operatorname{rot} \mathbf{v}$ by $2\boldsymbol{\omega}$; and, from the equation of continuity (120.4),

$$\frac{d\rho}{dt} + \rho \operatorname{div} \mathbf{v} = 0, \qquad \operatorname{div} \mathbf{v} = \rho \frac{d}{dt}\left(\frac{1}{\rho}\right). \tag{5}$$

Then (3), after division by ρ, becomes

$$\frac{1}{\rho}\frac{d\boldsymbol{\omega}}{dt} + \boldsymbol{\omega}\frac{d}{dt}\left(\frac{1}{\rho}\right) = \frac{\boldsymbol{\omega}}{\rho} \cdot \nabla \mathbf{v},$$

$$\frac{d}{dt}\left(\frac{\boldsymbol{\omega}}{\rho}\right) = \frac{\boldsymbol{\omega}}{\rho} \cdot \nabla \mathbf{v}. \tag{6}$$

This is known as *Helmholtz's Equation*.

123. Lagrangian Equation of Motion. In the *Lagrangian* or *historical* method of dealing with a moving fluid, the motion of the individual fluid particles is followed from their initial positions $\mathbf{r}_0$

to their position $\mathbf{r}$ after a time t. Thus the history of each fluid particle is traced. If the function $f(\mathbf{r})$ is associated with a fluid particle, $\mathbf{r}$ is a function of the independent variables $\mathbf{r}_0$ and t.

In space differentiation we may form gradients relative to $\mathbf{r}$ or $\mathbf{r}_0$; these are denoted by the symbols ∇ and ∇_0. Thus we may compute df (§ 90) as either

$$df = d\mathbf{r}_0 \cdot \nabla_0 f \quad \text{or} \quad df = d\mathbf{r} \cdot \nabla f;$$

and, since

(1) $$d\mathbf{r} = d\mathbf{r}_0 \cdot \nabla_0 \mathbf{r},$$

and

$$d\mathbf{r}_0 \cdot \nabla_0 f = d\mathbf{r}_0 \cdot \nabla_0 \mathbf{r} \cdot \nabla f$$

for arbitrary displacements $d\mathbf{r}_0$, we have

(2) $$\nabla_0 f = \nabla_0 \mathbf{r} \cdot \nabla f.$$

In particular, when $f = \mathbf{r}_0$, we have

(3) $$\mathbf{I} = \nabla_0 \mathbf{r} \cdot \nabla \mathbf{r}_0,$$

so that the dyadics $\nabla_0 \mathbf{r}$ and $\nabla \mathbf{r}_0$ are reciprocal. Consequently, if we multiply (2) by $\nabla \mathbf{r}_0$ as prefactor, we have

(4) $$\nabla f = \nabla \mathbf{r}_0 \cdot \nabla_0 f.$$

An element of fluid volume dV_0 is altered by the transformation (1) in the ratio $J/1$, where J denotes the third scalar invariant of $\nabla \mathbf{r}_0$ (§ 70): thus

(5) $$dV = J\, dV_0.$$

If we use rectangular coordinates,

(6) $$\nabla \mathbf{r}_0 = \begin{pmatrix} \partial x_0/\partial x & \partial y_0/\partial x & \partial z_0/\partial x \\ \partial x_0/\partial y & \partial y_0/\partial y & \partial z_0/\partial y \\ \partial x_0/\partial z & \partial y_0/\partial z & \partial z_0/\partial z \end{pmatrix},$$

and J is the determinant of this matrix, namely, the *Jacobian* $\partial(x_0, y_0, z_0)/\partial(x, y, z)$. The element of mass $\rho_0\, dV_0$ becomes $\rho\, dV = \rho J\, dV_0$; and, since mass is conserved,

(7) $$\rho J = \rho_0 \quad \text{or} \quad \frac{d}{dt}(\rho J) = 0.$$

This is the equation of continuity in the Lagrangian method. Since

$d\rho/dt = -\rho \operatorname{div} \mathbf{v}$ from (120.4),

$$\frac{d}{dt}(\rho J) = \rho\left(\frac{dJ}{dt} - J \operatorname{div} \mathbf{v}\right) = 0,$$

and (7) is equivalent to

$$\frac{dJ}{dt} = J \operatorname{div} \mathbf{v}. \tag{8}$$

The Dynamical Equation of Lagrange corresponding to the Eulerian Equation (121.9) is obtained by multiplying the latter by $\nabla_0\mathbf{r}$ as prefactor; thus

$$\nabla_0\mathbf{r} \cdot \frac{d\mathbf{v}}{dt} = -\nabla_0\mathbf{r} \cdot \nabla(Q + P) = -\nabla_0(Q + P), \tag{9}$$

from (2). Since $\mathbf{r}_0$ and t are the independent variables, the symbols ∇_0 and d/dt commute; hence the left member of (9) may be written

$$\frac{d}{dt}\{(\nabla_0\mathbf{r}) \cdot \mathbf{v}\} - (\nabla_0\mathbf{v}) \cdot \mathbf{v} = \frac{d}{dt}\{(\nabla_0\mathbf{r}) \cdot \mathbf{v}\} - \nabla_0\tfrac{1}{2}v^2,$$

and (9) becomes

$$\frac{d}{dt}\{(\nabla_0\mathbf{r}) \cdot \mathbf{v}\} = -\nabla_0(Q + P - \tfrac{1}{2}v^2). \tag{10}$$

With the aid of the dyadic $\nabla_0\mathbf{r}$ we may integrate Helmholtz's Equation (122.6). If we make use of (4), this becomes

$$\frac{d}{dt}\left(\frac{\boldsymbol{\omega}}{\rho}\right) = \frac{\boldsymbol{\omega}}{\rho} \cdot \nabla\mathbf{r}_0 \cdot \nabla_0\mathbf{v} = \frac{\boldsymbol{\omega}}{\rho} \cdot \nabla\mathbf{r}_0 \cdot \left(\frac{d}{dt}\nabla_0\mathbf{r}\right) = -\frac{\boldsymbol{\omega}}{\rho} \cdot \left(\frac{d}{dt}\nabla\mathbf{r}_0\right) \cdot \nabla_0\mathbf{r},$$

since, from (3), the time derivative of $\nabla\mathbf{r}_0 \cdot \nabla_0\mathbf{r}$ is zero. Multiplying this equation by $\nabla\mathbf{r}_0$ as postfactor now gives

$$\frac{d}{dt}\left(\frac{\boldsymbol{\omega}}{\rho}\right) \cdot \nabla\mathbf{r}_0 + \frac{\boldsymbol{\omega}}{\rho} \cdot \left(\frac{d}{dt}\nabla\mathbf{r}_0\right) = \frac{d}{dt}\left(\frac{\boldsymbol{\omega}}{\rho} \cdot \nabla\mathbf{r}_0\right) = 0,$$

$$\frac{\boldsymbol{\omega}}{\rho} \cdot \nabla\mathbf{r}_0 = \frac{\boldsymbol{\omega}_0}{\rho_0},$$

the constant $\boldsymbol{\omega}_0/\rho_0$ being the value of $(\boldsymbol{\omega}/\rho) \cdot \nabla\mathbf{r}_0$ when $t = 0$. Multiplication by $\nabla_0\mathbf{r}$ gives finally

$$\frac{\boldsymbol{\omega}}{\rho} = \frac{\boldsymbol{\omega}_0}{\rho_0} \cdot \nabla_0\mathbf{r}. \tag{11}$$

This is *Cauchy's integral* of Helmholtz's Equation.

To verify this integral, we need only differentiate (11) with respect to t: thus

$$\frac{d}{dt}\left(\frac{\boldsymbol{\omega}}{\rho}\right) = \frac{\boldsymbol{\omega}_0}{\rho_0} \cdot \left(\frac{d}{dt} \nabla_0 \mathbf{r}\right) = \frac{\boldsymbol{\omega}_0}{\rho_0} \cdot \nabla_0 \mathbf{v} = \frac{\boldsymbol{\omega}_0}{\rho_0} \cdot \nabla \mathbf{r}_0 \cdot \nabla \mathbf{v} = \frac{\boldsymbol{\omega}}{\rho} \cdot \nabla \mathbf{v}.$$

The lines of a fluid which are everywhere tangent to $\boldsymbol{\omega}$ are called its *vortex-lines*. Their differential equation is $\boldsymbol{\omega} \times d\mathbf{r} = 0$. Consider a vortex line at the instant $t = 0$, and let its differential equation be $\boldsymbol{\omega}_0 \times d\mathbf{r}_0 = 0$. After a time t, $\boldsymbol{\omega}_0$ and $d\mathbf{r}_0$ become

$$\boldsymbol{\omega} = \frac{\rho}{\rho_0} \boldsymbol{\omega}_0 \cdot \nabla_0 \mathbf{r}, \qquad d\mathbf{r} = d\mathbf{r}_0 \cdot \nabla_0 \mathbf{r}.$$

Hence, if $\boldsymbol{\omega}_0 = \lambda\, d\mathbf{r}_0$, then $\boldsymbol{\omega} = (\rho/\rho_0)\lambda\, d\mathbf{r}$; that is, $\boldsymbol{\omega}_0 \times d\mathbf{r}_0 = 0$ implies $\boldsymbol{\omega} \times d\mathbf{r} = 0$. Thus we have proved the

THEOREM. *If the body forces have a potential and* $\rho = f(p)$ *or constant, the vortex-lines move with the fluid. Vortex-lines always consist of the same fluid particles.*

124. Flow and Circulation. The tangential line integral of the velocity of a fluid along any path is called the *flow* along that path. If the path is closed, the flow is called the *circulation*.

If the path AB at time t was A_0B_0 when $t = 0$, the flow over AB is

$$\int_A^B \mathbf{v} \cdot d\mathbf{r} = \int_{A_0}^{B_0} d\mathbf{r}_0 \cdot (\nabla_0 \mathbf{r}) \cdot \mathbf{v};$$

and, since $\mathbf{r}_0$ and t are independent variables,

$$\frac{d}{dt}\int_A \mathbf{v} \cdot d\mathbf{r} = \int_{A_0}^{B_0} d\mathbf{r}_0 \cdot \frac{d}{dt}\{(\nabla_0 \mathbf{r}) \cdot \mathbf{v}\}.$$

When the body forces have a potential Q and $\rho = f(p)$, we have, from (123.10),

$$\begin{aligned}\frac{d}{dt}\int_A^B \mathbf{v} \cdot d\mathbf{r} &= -\int_{A_0}^{B_0} d\mathbf{r}_0 \cdot \nabla_0(Q + P - \tfrac{1}{2}v^2)\\ &= -\int_{A_0}^{B_0} d\mathbf{r}_0 \cdot \nabla_0 \mathbf{r} \cdot \nabla(Q + P - \tfrac{1}{2}v^2)\\ &= -\int_A^B d\mathbf{r} \cdot \nabla(Q + P - \tfrac{1}{2}v^2);\end{aligned}$$

or, since the last integrand is a perfect differential,

$$\frac{d}{dt}\int_A^B \mathbf{v}\cdot d\mathbf{r} = [-Q - P + \tfrac{1}{2}v^2]_A^B\,, \tag{1}$$

and, in particular,

$$\frac{d}{dt}\oint \mathbf{v}\cdot d\mathbf{r} = 0. \tag{2}$$

The last result is Kelvin's

Circulation Theorem. *If the body forces have a potential and $\rho = f(p)$ or constant, the circulation over any closed curve moving with the fluid does not alter with the time.*

125. Irrotational Motion. When the body forces have a potential and $\rho = f(p)$ or constant, Cauchy's Equation (123.11) shows that, if the vorticity of a fluid vanishes at any instant, it will remain zero thereafter. Then rot $\mathbf{v} = 0$ in space and time, and the motion is termed *irrotational.* In any simply connected portion of the fluid, we may express $\mathbf{v}$ as the gradient of a scalar (§ 102); thus

$$\mathbf{v} = -\nabla\varphi, \qquad \varphi = -\int_{\mathbf{r}_0}^{\mathbf{r}} \mathbf{v}\cdot d\mathbf{r}, \tag{1}$$

and φ is called the *velocity potential.* Then the velocity is everywhere normal to the equipotential surfaces $\varphi =$ const and is directed toward decreasing potentials. Hence, along any line of motion, φ continually decreases; *in a simply connected region the lines of motion cannot form closed curves.*

By Stokes' Theorem the circulation $\oint \mathbf{v}\cdot d\mathbf{r} = 0$ over any reducible curve (§ 102); and, as this curve moves with the fluid, the circulation around it remains zero (circulation theorem, § 124). From this fact we again may deduce that, if the motion is irrotational at any instant, it remains irrotational thereafter.

The equation of continuity (120.4) now becomes

$$\frac{d\rho}{dt} - \rho\nabla^2\varphi = 0. \tag{2}$$

For an incompressible fluid, $d\rho/dt = 0$, and $\nabla^2\varphi = 0$; then φ is a harmonic function.

The equation of motion (122.2) reduces to

$$\nabla\left(Q + P + \tfrac{1}{2}v^2 - \frac{\partial\varphi}{\partial t}\right) = 0, \tag{3}$$

since $\partial\mathbf{v}/\partial t = -\nabla(\partial\varphi/\partial t)$; for a *local* time differentiation, $\partial/\partial t$ ($\mathbf{r}$ constant) and a space differentiation ∇ (t constant) commute with each other. Hence

$$Q + P + \tfrac{1}{2}v^2 - \frac{\partial\varphi}{\partial t} = f(t), \tag{4}$$

an arbitrary function of the time; and, if the flow is steady, $Q + P + \frac{1}{2}v^2$ is an absolute constant.

Every harmonic function φ represents some irrotational flow of an incompressible liquid whose velocity $\mathbf{v} = -\nabla\varphi$. The problem consists in finding a solution of $\nabla^2\varphi = 0$ which conforms to the given conditions. For example, in a flow with central symmetry, φ must be a function of $\mathbf{r}$ alone; hence, from (89.17),

$$\nabla^2\varphi = \frac{1}{r^2}\frac{d}{dr}\left(r^2\frac{d\varphi}{dr}\right) = 0, \qquad r^2\frac{d\varphi}{dr} = a, \qquad \varphi = b - \frac{a}{r}.$$

Now $\mathbf{v} = a\mathbf{R}/r^2$ where $\mathbf{R}$ is a unit radial vector; and the flux per second through any sphere of radius r about the origin is constant: $4\pi r^2(a/r^2) = 4\pi a$. When this flux is given, a is determined. The value of b is immaterial, since it disappears in $\mathbf{v} = -\nabla\varphi$; as it is customary to have φ vanish at infinity, we choose $b = 0$.

126. Steady Motion. A flow is said to be *steady* when it is invariable in time. Then all local time derivatives are zero. Thus $\partial\rho/\partial t = 0$, and the equation of continuity (120.1) is simply

$$\operatorname{div}(\rho\mathbf{v}) = 0. \tag{1}$$

Also $\partial\mathbf{v}/\partial t = 0$ and the *stream-lines* are also *lines of motion*. The Eulerian Equation (122.2) now becomes

$$\mathbf{v}\times\operatorname{rot}\mathbf{v} = \nabla(Q + P + \tfrac{1}{2}v^2). \tag{2}$$

If $\mathbf{T}$ is a unit tangent along a stream-line ($\mathbf{T}\times\mathbf{v} = 0$) or a vortex-line ($\mathbf{T}\times\operatorname{rot}\mathbf{v} = 0$), $\mathbf{T}\cdot\mathbf{v}\times\operatorname{rot}\mathbf{v} = 0$; hence $\mathbf{T}\cdot\nabla(Q + P + \frac{1}{2}v^2) = 0$, or

$$\frac{d}{ds}(Q + P + \tfrac{1}{2}v^2) = 0 \text{ along a stream- or vortex-line.} \tag{3}$$

We thus have proved

Bernoulli's Theorem. *Let the body forces have a potential and* $\rho = f(p)$ *or constant; then, for a steady flow,*

$$Q + P + \tfrac{1}{2}v^2 = \text{const} \tag{3}$$

along any stream-line or vortex-line.

The constant will vary, in general, from one line to another. However, if the motion is *irrotational* as well as steady,

$$\nabla(Q + P + \tfrac{1}{2}v^2) = 0,$$

$$Q + P + \tfrac{1}{2}v^2 = C, \tag{4}$$

where C is an absolute constant—the same throughout the fluid.

Now suppose that the fluid has a constant density ρ; this is nearly fulfilled in the case of a *liquid.* The equation of continuity is $\operatorname{div} \mathbf{v} = 0$ or $\oint \mathbf{n} \cdot \mathbf{v}\, dS = 0$. Along any tube whose surface consists of stream-lines, let the normal cross section be denoted by A. If the tube is sufficiently thin, and we apply $\oint \mathbf{n} \cdot \mathbf{v}\, dS = 0$ to the portion between A_1 and A_2, we have approximately

$$v_1 A_1 - v_2 A_2 = 0, \quad \text{or} \quad vA = \text{const}$$

as the equation of continuity along a tube of flow.

If the liquid is subject only to gravitational body forces, their potential is gz, where z is measured *upward* from a horizontal reference plane. Thus with

$$P = p/\rho, \qquad Q = gz,$$

we have

$$gz + \frac{p}{\rho} + \tfrac{1}{2}v^2 = \text{const} \tag{5}$$

along a stream-line. If z, p, v are known at the section A_0 (z_0, p_0, v_0), we have, from (5),

$$\rho g(z - z_0) + (p - p_0) + \tfrac{1}{2}\rho(v^2 - v_0^2) = 0. \tag{6}$$

For a thin tube we may take $vA = v_0 A_0$, and (6) becomes

$$p - p_0 = \rho g(z_0 - z) - \tfrac{1}{2}\rho v_0^2\left(\frac{A_0^2}{A^2} - 1\right). \tag{7}$$

Thus the pressure is least at the narrowest part of the tube.

Example. Torricelli's Law. When liquid escapes from an orifice near the bottom of a vessel which is kept filled to a constant level, the flow may be regarded as steady. Consider a stream tube extending from the orifice of the area A to the upper surface where its area is A_0. At this surface p_0 is the atmospheric pressure, and $v_0 = vA/A_0$; at the orifice $p = p_0$, and the outflow speed is v. With these values we have, from (6),

$$v^2(1 - A^2/A_0^2) = 2g(z_0 - z) = 2gh,$$

where h is the distance from the free surface to the orifice. When A_0 is large compared with A, we have approximately $v^2 = 2gh$, a result known as Torricelli's Law.

127. Plane Motion. When the flow is the same in all planes parallel to a fixed plane π and the velocity has no component normal to π, the motion is said to be *plane*. Such motion is completely determined by the motion in π, which may be taken as the xy-plane. Velocity, density, and pressure are all functions of x and y alone; and, for any tensor function $\mathbf{f}(x, y)$, we have, from (97.1),

$$\text{grad}\,\mathbf{f} = \text{Grad}\,\mathbf{f} + \mathbf{k}\frac{\partial \mathbf{f}}{\partial z} = \text{Grad}\,\mathbf{f};$$

and, similarly,

$$\text{div}\,\mathbf{f} = \text{Div}\,\mathbf{f}, \qquad \text{rot}\,\mathbf{f} = \text{Rot}\,\mathbf{f}.$$

The flux across any curve C in the xy-plane is defined as the volume of liquid per second crossing a right cylindrical surface of unit height based on C. For an observer who travels the curve in the positive sense of s, the flux crossing C from (his) right to left is

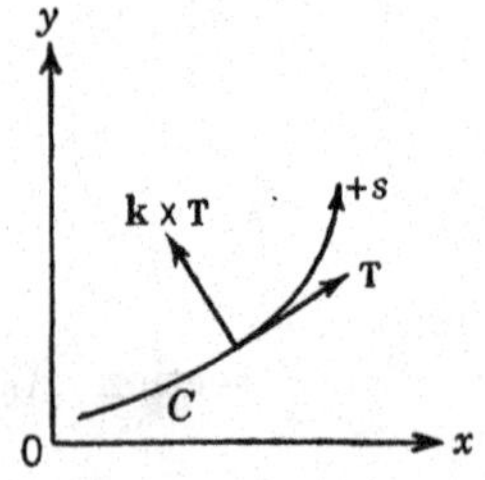

FIG. 127

$$(1) \qquad \int_C \mathbf{k} \times \mathbf{T} \cdot \mathbf{v}\, ds = \int_C \mathbf{v} \times \mathbf{k} \cdot d\mathbf{r}\cdot$$

here $\mathbf{k}$, $\mathbf{T}$, $\mathbf{k} \times \mathbf{T}$ are unit vectors along the z-axis, the tangent and normal to the curve (Fig. 127).

In any simply connected portion of the fluid the flux across a plane curve joining two points will be independent of the path, provided

$$\mathbf{k} \cdot \text{rot}\,(\mathbf{v} \times \mathbf{k}) = 0 \qquad (\text{§ 103, theorem 1}).$$

From (85.6),

$$\text{rot}\,(\mathbf{v}\times\mathbf{k}) = \mathbf{k}\cdot\nabla\mathbf{v} - \mathbf{k}\,\text{div}\,\mathbf{v} = -\mathbf{k}\,\text{div}\,\mathbf{v}.$$

For an *incompressible* fluid, div $\mathbf{v} = 0$ (120.5) and F will be independent of the path. Hence

$$\psi(\mathbf{r}) = \int_{\mathbf{r}_0}^{\mathbf{r}} \mathbf{v}\times\mathbf{k}\cdot d\mathbf{r} \tag{2}$$

defines a scalar point function; and, from § 103, theorem 2, Grad $\psi = \mathbf{v}\times\mathbf{k}$, since $\mathbf{v}\times\mathbf{k}$ lies in the xy-plane. Writing ∇ for Grad, we have

$$\nabla\psi = \mathbf{v}\times\mathbf{k}, \qquad \mathbf{v} = \mathbf{k}\times\nabla\psi. \tag{3}$$

$\nabla\psi$ is everywhere normal and $\mathbf{v}$ everywhere tangent to the curves $\psi = \text{const}$; these curves are therefore stream-lines (§ 121). Consequently, the function ψ is called the *stream function.*

If we set $\mathbf{g} = -\psi\mathbf{k}$, we have

$$\text{rot}\,\mathbf{g} = \mathbf{k}\times\nabla\psi = \mathbf{v}, \qquad \text{div}\,\mathbf{g} = 0; \tag{4}$$

thus the velocity has the vector potential $-\psi\mathbf{k}$ (§ 92).

From (85.6), we have

$$\text{rot}\,\mathbf{v} = \text{rot}\,(\mathbf{k}\times\nabla\psi) = \mathbf{k}\,\nabla^2\psi. \tag{5}$$

Therefore *the plane motion of an incompressible fluid will be irrotational when and only when the stream function is harmonic:*

$$\nabla^2\psi = 0. \tag{6}$$

In this case $\mathbf{v} = -\nabla\varphi$: the velocity has a scalar potential φ; and, since

$$\nabla^2\varphi = -\,\text{div}\,\mathbf{v} = 0, \tag{7}$$

the velocity potential is also harmonic.

From

$$\mathbf{v} = -\nabla\varphi = k\times\nabla\psi,$$

we have the relations,

$$\nabla\varphi = (\nabla\psi)\times\mathbf{k}, \qquad \nabla\psi = \mathbf{k}\times\nabla\varphi; \tag{8}$$

or, in terms of rectangular coordinates,

$$\mathbf{i}\frac{\partial\varphi}{\partial x} + \mathbf{j}\frac{\partial\varphi}{\partial y} = \mathbf{i}\frac{\partial\psi}{\partial y} - \mathbf{j}\frac{\partial\psi}{\partial x}\cdot$$

Thus (8) is equivalent to the equations:

$$\frac{\partial \varphi}{\partial x} = \frac{\partial \psi}{\partial y}, \qquad \frac{\partial \varphi}{\partial y} = -\frac{\partial \psi}{\partial x}. \tag{9}$$

But these are precisely the Cauchy–Riemann Equations which connect the real and imaginary parts of the analytic function $w = \varphi + i\psi$ of the complex variable $z = x + iy$. We thus have proved the important

THEOREM. *In any plane, irrotational motion of an incompressible fluid, the velocity potential φ and the stream function ψ are two harmonic functions which combine into an analytic function $\varphi + i\psi$ of a complex variable $x + iy$.*

Since $-\psi + i\varphi = i(\varphi + i\psi)$, we see that, if $\varphi + i\psi$ is analytic, $-\psi + i\varphi$ is also; consequently, if φ and ψ are the velocity potential and stream function for an irrotational plane flow, $-\psi$ and φ are the corresponding functions for another flow of this type.

From (8), we have $\nabla\varphi \cdot \nabla\psi = 0$: the stream-lines ($\psi$ = const) cut the equipotential lines (φ = const) at right angles.

Since the *complex potential* $w = \varphi + i\psi$ is an analytic function of z,

$$w' = \frac{dw}{dz} = \frac{\partial \varphi}{\partial x} + i\frac{\partial \psi}{\partial x} = \frac{\partial \varphi}{\partial x} - i\frac{\partial \varphi}{\partial y};$$

or, since the velocity has the components,

$$v_x = -\mathbf{i} \cdot \nabla\varphi = -\frac{\partial \varphi}{\partial x}, \qquad v_y = -\mathbf{j} \cdot \nabla\varphi = -\frac{\partial \varphi}{\partial y},$$

$$w' = -v_x + i\,v_y, \qquad -\bar{w}' = v_x + i\,v_y, \tag{10}$$

where $\bar{w}'$ denotes the conjugate of w'. Thus the velocity at any point is given by the complex vector $-\bar{w}'$; its magnitude is $|w'|$.

Example 1. Assume the complex potential $w = az^n$ (a real); then, if we write $z = re^{i\theta}$,

$$w = ar^n e^{in\theta} = ar^n(\cos n\theta + i \sin n\theta);$$

$$\varphi = ar^n \cos n\theta, \qquad \psi = ar^n \sin n\theta.$$

The stream-lines are the curves whose polar equations are $r^n \sin n\theta$ = const.

For the cases $n = 1, 2, -1$, we put $z = x + iy$, using rectangular coordinates.

(a) $\quad n = 1: \quad w = az = ax + iay.$

The stream-lines are the lines $y = \text{const}$, and the flow has the constant velocity $-\bar{w}' = -a$ in the direction of $-x$.

(b) $$n = 2: \qquad w = az^2 = a(x^2 - y^2) + i2axy.$$

The equipotential and stream-lines are the two families of equilateral hyperbolas,

$$x^2 - y^2 = \text{const}, \qquad xy = \text{const}.$$

Since the stream-line $xy = 0$ may be taken as the positive halves of the x-axis and y-axis, these may be considered as fixed boundaries and the motion regarded as a steady flow of liquid in the angle between two perpendicular walls. The velocity at any point is $-\bar{w}' = -2a\bar{z}$; its magnitude varies directly as the distance from the origin.

(c) $$n = -1: \qquad w = a/z = a(x - iy)/(x^2 + y^2).$$

The equipotential and stream-lines are two families of circles:

$$x/(x^2 + y^2) = \text{const}, \qquad y/(x^2 + y^2) = \text{const},$$

tangent to y-axis and x-axis, respectively, at the origin. The velocity $-\bar{w}' = a/\bar{z}^2$ becomes infinite at the origin.

Example 2. With the complex potential,

$$w = V(z + a^2/z) = V(re^{i\theta} + a^2r^{-1}e^{-i\theta}), \quad (V \text{ real}),$$

$$\varphi = V\left(r + \frac{a^2}{r}\right)\cos\theta, \qquad \psi = V\left(r - \frac{a^2}{r}\right)\sin\theta.$$

The stream-line $\psi = 0$ includes the circle $r = a$ and the x-axis $\sin\theta = 0$. Since $w' = V(1 - a^2/z^2)$, the complex velocity,

$$-\bar{w}' = -V\left(1 - \frac{a^2}{\bar{z}^2}\right) \to -V \quad \text{as} \quad z \to \infty.$$

Therefore we may regard the motion as a flow to the left about an infinite cylindrical obstacle of radius a. At a great distance from the obstacle, the flow has a sensibly uniform velocity $-V\mathbf{i}$.†

When body forces are neglected, the Bernoulli Equation (126.4) gives $p/\rho + \frac{1}{2}v^2 = \text{const}$. From the symmetry of the flow about the cylinder, it is clear that the total pressure exerted by the fluid on the cylinder is zero. We shall consider this matter for cylinders of arbitrary section in the next article.

128. Kutta-Joukowsky Formulas. Consider an incompressible fluid flowing past an infinite cylindrical obstacle of arbitrary cross section. The flow, in a plane perpendicular to the generators of the cylinder, is assumed to have the sensibly uniform velocity $\mathbf{v}_\infty$ at a great distance from the obstacle. Then, if the motion is

† See Lamb, *Hydrodynamics*, Cambridge, 1916, p. 75, for the stream-lines.

steady and *irrotational*, and the body forces on the fluid are neglected, we have, from (126.4),

$$(1) \qquad p = K - \frac{\rho}{2}\,\mathbf{v}\cdot\mathbf{v},$$

where K is an absolute constant.

If $\mathbf{n}$ is a unit *internal* normal to the boundary C of the obstacle (Fig. 128),

$$(2) \qquad \mathbf{F} = \oint_C \mathbf{n}\,p\,ds, \qquad \mathbf{M} = \oint_C \mathbf{r}\times\mathbf{n}\,p\,ds$$

give the resultant force and moment about the origin exerted by

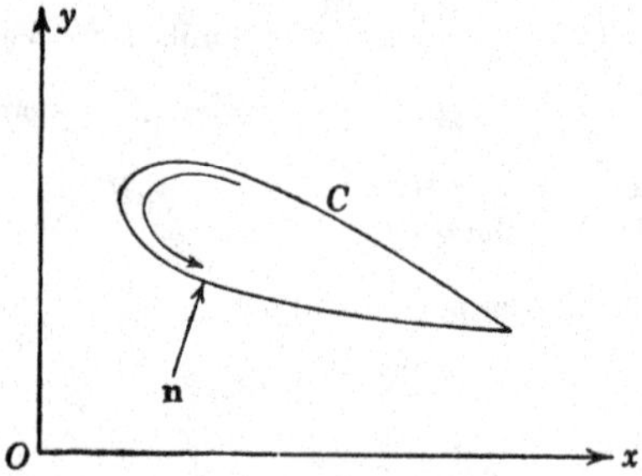

FIG. 128

fluid pressure on a unit length of cylinder. Substituting p from (1) in these integrals and noting that

$$\oint \mathbf{n}\,ds = 0, \qquad \oint \mathbf{r}\times\mathbf{n}\,ds = \int \operatorname{rot}\mathbf{r}\,dA = 0,$$

from (101.5), (101.7), we have

$$(3) \quad \mathbf{F} = -\frac{\rho}{2}\oint \mathbf{v}\cdot\mathbf{v}\,\mathbf{n}\,ds, \qquad \mathbf{M} = -\frac{\rho}{2}\oint \mathbf{v}\cdot\mathbf{v}\,\mathbf{r}\times\mathbf{n}\,ds.$$

When the flow is given by the complex potential $w = \varphi + i\psi$, we shall compute these integrals after converting them into circuit integrals in the complex z-plane.

The vectors $\mathbf{r} = x\mathbf{i} + y\mathbf{j}$ and $\bar{\mathbf{r}} = x\mathbf{i} - y\mathbf{j}$ correspond to z and its conjugate $\bar{z}$:

$$\mathbf{r} \sim z = x + iy = r(\cos\theta + i\sin\theta) = re^{i\theta},$$

$$\bar{\mathbf{r}} \sim \bar{z} = x - iy = r(\cos\theta - i\sin\theta) = re^{-i\theta}.$$

By definition,

$$\mathbf{r}_1 \cdot \mathbf{r}_2 = r_1 r_2 \cos(\theta_2 - \theta_1), \qquad \mathbf{r}_1 \times \mathbf{r}_2 = r_1 r_2 \sin(\theta_2 - \theta_1)\mathbf{k};$$

hence

$$\mathbf{r}_1 \cdot \mathbf{r}_2 + i\, \mathbf{r}_1 \times \mathbf{r}_2 \cdot \mathbf{k} = r_1 r_2 e^{i(\theta_2 - \theta_1)} = \bar{z}_1 z_2.$$

Thus $\mathbf{r}_1 \cdot \mathbf{r}_2$ and $\mathbf{r}_1 \times \mathbf{r}_2 \cdot \mathbf{k}$ correspond to the real and imaginary parts of $\bar{z}_1 z_2$:

$$\mathbf{r}_1 \cdot \mathbf{r}_2 \sim \mathcal{R}(\bar{z}_1 z_2) = \tfrac{1}{2}(\bar{z}_1 z_2 + z_1 \bar{z}_2), \tag{4}$$

$$\mathbf{r}_1 \times \mathbf{r}_2 \cdot \mathbf{k} \sim \mathcal{I}(\bar{z}_1 z_2) = \tfrac{1}{2}(\bar{z}_1 z_2 - z_1 \bar{z}_2).\dagger \tag{5}$$

Note also that $\mathbf{k} \times \mathbf{r} \sim iz$; for the multiplications by $\mathbf{k} \times$ and $i\ (= e^{i\pi/2})$ both revolve a vector in the xy-plane through $\pi/2$.

When $\mathbf{r}_1$ and $\mathbf{r}_2$ are perpendicular or parallel we have, respectively,

$$\mathbf{r}_1 \cdot \mathbf{r}_2 = 0 \sim \bar{z}_1 z_2 + z_1 \bar{z}_2 = 0, \tag{6}$$

$$\mathbf{r}_1 \times \mathbf{r}_2 = 0 \sim \bar{z}_1 z_2 - z_1 \bar{z}_2 = 0. \tag{7}$$

Turning now to the integrals (3), we have the internal normal $\mathbf{n} = \mathbf{k} \times \mathbf{T}$ for a counterclockwise circuit of C; hence

$$\mathbf{n}\, ds = \mathbf{k} \times \mathbf{T}\, ds = \mathbf{k} \times d\mathbf{r} \sim i\, dz.$$

Moreover, if $\mathbf{v} \sim v$, then $\mathbf{v} \cdot \mathbf{v} \sim v\, \bar{v}$, and

$$\mathbf{F} \sim -\frac{\rho}{2} i \oint_C v\, \bar{v}\, dz.$$

The moment $\mathbf{M}$, normal to the plane, is completely specified by the scalar,

$$M = \mathbf{k} \cdot \mathbf{M} = -\frac{\rho}{2} \oint_C \mathbf{v} \cdot \mathbf{v}\, \mathbf{k} \cdot \mathbf{r} \times \mathbf{n}\, ds,$$

and, since

$$\mathbf{k} \cdot \mathbf{r} \times (\mathbf{k} \times \mathbf{T})\, ds = (\mathbf{k} \times \mathbf{r}) \cdot (\mathbf{k} \times \mathbf{T})\, ds = \mathbf{r} \cdot d\mathbf{r} \sim \mathcal{R}(\bar{z}\, dz),$$

$$M \sim -\frac{\rho}{2} \oint_C v\, \bar{v}\, \mathcal{R}(\bar{z}\, dz) = -\frac{\rho}{2} \mathcal{R} \oint_C v\bar{v}\, \bar{z}\, dz.$$

† The real and imaginary parts of any complex number z are $\mathcal{R}(z) = \frac{1}{2}(z + \bar{z})$ and $\mathcal{I}(z) = \frac{1}{2}(z - \bar{z})$. Note also that the conjugate of a product is the product of the conjugates: $\overline{z_1 z_2} = \bar{z}_1 \bar{z}_2$.

But v and dz are parallel along C, which forms part of a stream, line; hence $\bar{v}\,dz = v\,d\bar{z}$, from (7), and we have

$$F = -\frac{\rho}{2} i \oint_C v^2\,d\bar{z}, \qquad M = -\frac{\rho}{2} \mathcal{R} \oint_C v^2 \bar{z}\,d\bar{z}$$

for the corresponding complex force and moment. Finally we form $\bar{F}$ and replace $v^2\bar{z}\,d\bar{z}$ in M by its conjugate $\bar{v}^2 z\,dz$; thus

$$\bar{F} = \frac{\rho}{2} i \oint_C \bar{v}^2\,dz = \frac{\rho}{2} i \oint_C \left(\frac{dw}{dz}\right)^2 dz, \tag{8}$$

$$M = -\frac{\rho}{2} \mathcal{R} \oint_C \bar{v}^2 z\,dz = -\frac{\rho}{2} \mathcal{R} \oint_C \left(\frac{dw}{dz}\right)^2 z\,dz, \tag{9}$$

since $\bar{v} = -dw/dz$ (127.10). These integrals exist if dw/dz remains finite over C.

Now, outside of C, $\bar{v}$ is an analytic function which approaches $\bar{v}_\infty$ at infinity. The Laurent series for $\bar{v}$ therefore has the form:

$$\bar{v} = \bar{v}_\infty + a_1/z + a_2/z^2 + a_3/z^3 + \cdots \tag{10}$$

If C_∞ is a large circle of radius R about the origin and enclosing C-we have, by Cauchy's Integral Theorem,

$$\oint_C f(z)\,dz = \oint_{C_\infty} f(z)\,dz, \quad \text{if } f(z) \text{ is analytic between } C \text{ and } C_\infty.‡$$

Thus the counterclockwise *circulation* about C is

$$\gamma = \oint_C \mathbf{v} \cdot d\mathbf{r} \sim \tfrac{1}{2} \oint_C (\bar{v}\,dz + v\,d\bar{z}) = \oint_C \bar{v}\,dz = \oint_{C_\infty} \bar{v}\,dz.$$

Moreover the *static moment of the circulation* is

$$\boldsymbol{\mu} = \oint_C \mathbf{r}\,\mathbf{v} \cdot d\mathbf{r} \sim \oint_C z\,\bar{v}\,dz = \oint_{C_\infty} z\,\bar{v}\,dz.$$

If we replace $\bar{v}$ by the series in (10), all integrals vanish except those involving

$$\oint_{C_\infty} \frac{dz}{z} = \int_0^{2\pi} \frac{Re^{i\theta}\,i\,d\theta}{Re^{i\theta}} = i \int_0^{2\pi} d\theta = 2\pi i.$$

‡ See, for example, Franklin, *A Treatise on Advanced Calculus*, New York, 1940, §§ 267, 268.

We thus find

(11) $$\gamma = 2\pi i\, a_1, \qquad \mu = 2\pi i\, a_2.$$

We now can compute the integrals in (8) and (9) in terms of v_∞, γ, and μ. From (10), we have

$$\bar{v}^2 = \bar{v}_\infty^2 + 2a_1\bar{v}_\infty/z + (2a_2\bar{v}_\infty + a_1^2)/z^2 + \cdots.$$

When this series is substituted in (8) and (9), all integrals vanish except $\oint dz/z = 2\pi i$; hence

(12) $$\bar{F} = \frac{\rho}{2} i\,(2a_1\bar{v}_\infty)2\pi i = i\rho\gamma\bar{v}_\infty, \qquad F = -i\rho\gamma v_\infty;$$

(13) $$M = -\frac{\rho}{2}\,\mathcal{R}(2a_2\bar{v}_\infty + a_1^2)2\pi i = -\rho\mathcal{R}(\mu\bar{v}_\infty) = -\rho\mathcal{R}(\bar{\mu}v_\infty),$$

since $a_1^2 = -\gamma^2/4\pi$ is real, and $\mathcal{R}(2\pi i\, a_1^2) = 0$. In vector notation these give

(14) $$\mathbf{F} = -\rho\gamma\, \mathbf{k} \times \mathbf{v}_\infty,$$

(15) $$\mathbf{M} = -\rho\boldsymbol{\mu}\cdot\mathbf{v}_\infty\,\mathbf{k} = \frac{\boldsymbol{\mu}}{\gamma}\times\mathbf{F},$$

for the resultant force and moment on a unit length of cylinder. From (15) we see that $\mathbf{F}$ acts through the point,

(16) $$\boldsymbol{\mu}/\gamma = \oint_C \mathbf{r}\,\mathbf{v}\cdot d\mathbf{r} \Big/ \oint_C \mathbf{v}\cdot d\mathbf{r},$$

which may be called the *centroid of the circulation.*

The equations (14) and (15) are called the *Kutta–Joukowsky Formulas.* From (14) we see that, for a counterclockwise circulation ($\gamma > 0$), the force $\mathbf{F}$ is *upward* if the flow is horizontal and to the left ($\mathbf{v}_\infty = -V\mathbf{i}$); we then have a *lift* of $\rho\gamma V\mathbf{j}$ per unit length of cylinder. The counterclock circulation diminishes the flow velocity below the cylinder and increases it above; by Bernoulli's Theorem this results in an excess of pressure below with a consequent upward lift.

In the absence of circulation about the cylinder ($\gamma = 0$), $\mathbf{F} = 0$, and there can be no lift. This is the case in § 127, ex. 2, where

$$w = V(z + a^2/z), \qquad \bar{v} = -dw/dz = -V(1 - a^2/z^2),$$

and $a_1 = 0$ in the Laurent series.

129. Summary: Hydrodynamics. For a perfect (non-viscous) fluid the stress dyadic is $-p\mathbf{I}$; at any point within a fluid the pressure is the same in all directions and exerted normal to a surface element.

If $\mathbf{f}(t, \mathbf{r})$ is any tensor function associated with a fluid particle moving with the velocity $\mathbf{v}$, its *substantial* rate of change is

$$\frac{d\mathbf{f}}{dt} = \frac{\partial \mathbf{f}}{\partial t} + \mathbf{v} \cdot \nabla \mathbf{f},$$

where $\partial \mathbf{f}/\partial t$ is the *local* rate of change.

In the *Eulerian method* of dealing with fluid motion, the aim is to compute $\mathbf{v}$, ρ, and p as functions of the independent variables $\mathbf{r}$, t. The (kinematic) *equation of continuity* is

$$\frac{\partial \rho}{\partial t} + \operatorname{div}(\rho \mathbf{v}) = \frac{d\rho}{dt} + \rho \operatorname{div} \mathbf{v} = 0;$$

and, for an incompressible fluid ($d\rho/dt = 0$), becomes $\operatorname{div} \mathbf{v} = 0$. The *Eulerian Equation of Motion* is

$$(E) \qquad \frac{d\mathbf{v}}{dt} = -\nabla(Q + P), \qquad P = \int \frac{dp}{\rho},$$

when the body forces have a potential Q and the density a function of p only. From this we can deduce the Differential Equation of Helmholtz for the *vorticity* $\boldsymbol{\omega} = \frac{1}{2} \operatorname{rot} \mathbf{v}$:

$$\frac{d}{dt}\left(\frac{\boldsymbol{\omega}}{\rho}\right) = \frac{\boldsymbol{\omega}}{\rho} \cdot \nabla \mathbf{v}.$$

In the *Lagrangian method* the aim is to follow the motion of the fluid particles from their initial positions $\mathbf{r}_0$ ($t = 0$) to their positions $\mathbf{r}$ after a time t. The independent variables are now $\mathbf{r}_0$ and t; the position $\mathbf{r}$ of a particle at time t is a function of $\mathbf{r}_0$ and t. We may take gradients relative to $\mathbf{r}_0(\nabla_0)$ or $\mathbf{r}(\nabla)$; and these conform to the relations,

$$\nabla_0 f = \nabla_0 \mathbf{r} \cdot \nabla f, \qquad \nabla f = \nabla \mathbf{r}_0 \cdot \nabla_0 f; \qquad \nabla_0 \mathbf{r} \cdot \nabla \mathbf{r}_0 = I.$$

If J is the third scalar of the dyadic $\nabla \mathbf{r}_0$, the *equation of continuity* becomes

$$\rho J = \rho_0, \quad \text{or} \quad dJ/dt = J \operatorname{div} \mathbf{v}.$$

The *Lagrangian Equation of Motion* corresponding to the Eulerian Equation (*E*) is

$$(L) \qquad \nabla_0 \mathbf{r} \cdot \frac{d\mathbf{v}}{dt} = -\nabla_0(Q + P).$$

From this we deduce Cauchy's integral of Helmholtz's Equation,

$$\frac{\boldsymbol{\omega}}{\rho} = \frac{\boldsymbol{\omega}_0}{\rho_0} \cdot \nabla_0 \mathbf{r};$$

and this in turn shows that vortex-lines ($\boldsymbol{\omega} \times d\mathbf{r} = 0$) move with the fluid. We find, moreover, that the circulation $\oint \mathbf{v} \cdot d\mathbf{r}$ over any closed curve moving with the fluid does not alter with the time.

When the motion is *irrotational* (rot $\mathbf{v} = 0$), $\mathbf{v} = -\nabla\varphi$, where φ is the velocity potential, and $\oint \mathbf{v} \cdot d\mathbf{r} = 0$ over any reducible curve. When the body forces are conservative and ρ is a function of p alone,

$$Q + P + \tfrac{1}{2}v^2 - \frac{\partial\varphi}{\partial t} = f(t), \text{ an arbitrary function of } t.$$

For a *steady*-motion equation (*E*) gives

$$\mathbf{v} \times \text{rot } \mathbf{v} = \nabla(Q + P + \tfrac{1}{2}v^2);$$

whence *Bernoulli's Theorem:* $Q + P + \frac{1}{2}v^2$ is constant along a stream-line or vortex-line, and this constant is absolute if the steady motion is also irrotational.

For an *incompressible fluid in plane motion*, the integral independent of the path,

$$\psi(\mathbf{r}) = \int_{\mathbf{r}_0}^{\mathbf{r}} \mathbf{v} \times \mathbf{k} \cdot d\mathbf{r} \quad (\mathbf{k} \perp \text{plane of motion}),$$

defines the *stream function.* The curves ψ = const are the stream-lines ($\mathbf{v} \times d\mathbf{r} = 0$). If the motion is also irrotational, both stream function and velocity potential are harmonic ($\nabla^2\psi = 0$, $\nabla^2\varphi = 0$), and $\nabla\psi = \mathbf{k} \times \nabla\varphi$. This is the vector equivalent of the Cauchy–Riemann Equations which guarantee that $w = \varphi + i\psi$ is an analytic function of the complex variable $z = x + iy$ in the plane of motion. The *complex potential* w therefore has a unique derivative dw/dz; and the negative of its conjugate gives the complex velocity vector: $v = -\bar{w}'$.

PROBLEMS

1. A mass of liquid is revolving about a vertical axis with the angular speed $f(r)$, where r is the perpendicular distance from the axis. With cylindrical coordinates r, θ, z (we replace ρ, φ in § 89, ex. 1 by r, θ to avoid conflict with the notation of Chapter VII), let **a**, **b**, **c** be unit vectors in the directions of $\mathbf{r}_r, \mathbf{r}_\theta, \mathbf{r}_z$ (Fig. 89a). If the angular velocity $\boldsymbol{\omega} = f(r)\mathbf{c}$, prove that

$$\mathbf{v} = rf(r)\mathbf{b}, \qquad \text{rot } \mathbf{v} = \frac{1}{r}\frac{d}{dr}[r^2 f(r)]\mathbf{c}. \qquad \text{[Cf. (89.7).]}$$

2. If the motion in Problem 1 is *irrotational*, show that

$$\boldsymbol{\omega} = \frac{\alpha}{r^2}\mathbf{c}, \qquad \mathbf{v} = \frac{\alpha}{r}\mathbf{b},$$

and that the velocity potential $\varphi = \beta - \alpha\theta$ is *not* single valued (α, β are constants).

For a liquid of constant density ρ under the action of gravity alone, show that the pressure is given by

$$gz + \frac{p}{\rho} + \frac{1}{2}\frac{\alpha^2}{r^2} = \text{const.}$$

3. If a fluid is bounded by a *fixed* surface $F(\mathbf{r}) = 0$, show that the fluid must satisfy the boundary condition $\mathbf{v} \cdot \nabla F = 0$.

More generally, if the bounding surface $F(\mathbf{r}, t) = 0$ varies with the time, show that the fluid satisfies the boundary condition

$$\frac{\partial F}{\partial t} + \mathbf{v} \cdot \nabla F = 0 \qquad \text{[Cf. (120.3).]}$$

4. A sphere of radius a is moving in a fluid with the constant velocity **u**. Show that at the surface of the sphere the velocity of the fluid satisfies the condition

$$(\mathbf{v} - \mathbf{u}) \cdot (\mathbf{r} - \mathbf{u}t) = 0.$$

5. A fluid flows through a thin tube of variable cross section A. Show that for a tube P_0P of length s

$$\frac{\partial}{\partial t}\int_0^s \rho A\, ds + \rho A v \Big|_{P_0}^{P} = 0;$$

hence deduce the equation of continuity

$$\frac{\partial}{\partial t}(\rho A) + \frac{\partial}{\partial s}(\rho A v) = 0.$$

6. From the equation of fluid equilibrium (118.1) show that when ρ is a function of p alone, rot $\mathbf{R} = 0$ and that the potential of **R** is $-P$.

If ρ is not a function of p alone, show that equilibrium is only possible when $\mathbf{R} \cdot \text{rot } \mathbf{R} = 0$. [Cf. (121.7) for P.]

7. A gas flows from a reservoir in which the pressure and density are p_0, ρ_0 into a space where the pressure is p. If the expansion takes place adiabatically, p/ρ^γ = const. (γ is the ratio of specific heats), show that

$$P = \frac{\gamma}{\gamma - 1}\frac{p}{\rho}.$$

Neglecting body forces and the velocity of the gas in the reservoir, show that the velocity $\mathbf{v}$ of efflux when the motion becomes steady is given by

$$v^2 = \frac{2\gamma}{\gamma - 1}\frac{p_0}{\rho_0}\left\{1 - (p/p_0)^{\frac{\gamma-1}{\gamma}}\right\}.$$

The velocity of sound in a gas is $c = \sqrt{\gamma p/\rho}$; hence show that

$$v^2 = \frac{2}{\gamma - 1}(c_0^2 - c^2).$$

8. If a body of liquid rotates as a whole from $r = 0$ to $r = a$ with the constant angular velocity ω_0, and rotates irrotationally with the angular velocity $\omega = \omega_0 a^2/r^2$ when $r = a$, we have the so-called "combined vortex" of Rankine. If $z = z_a$ at the free surface when $r = a$, show that the free surface is given by

$$z = z_a + \frac{\omega^2}{2g}(r^2 - a^2), \qquad r \leqq a,$$

$$z = z_a + \frac{\omega^2 a^2}{2g}\left(1 - \frac{a^2}{r^2}\right), \qquad r \geqq a.$$

Prove that the bottom of the vortex ($r = 0$) is a distance $\omega^2 a^2/g$ below the general level ($r = \infty$) of the liquid.

9. If the vorticity is constant throughout an incompressible fluid, prove that $\nabla^2 \mathbf{v} = 0$. [Cf. (84.14).]

10. When the motion of an incompressible fluid is steady, deduce from Helmholtz's Equation (122.6) that $\boldsymbol{\omega} \cdot \nabla \mathbf{v} = \mathbf{v} \cdot \nabla \boldsymbol{\omega}$.

11. If an incompressible liquid in irrotational motion occupies a simply connected region, show that

$$\int v^2\, dV = \oint \varphi \frac{d\varphi}{dn}\, dS,$$

where φ is the velocity potential and the normal derivative $d\varphi/dn$ is in the direction of the external normal to the bounding surface. [Cf. (108.1).]

12. Prove Kelvin's theorem: The irrotational motion ($\mathbf{v}$) of an incompressible fluid occupying a simply connected region S with finite boundaries has less kinetic energy ($T = \frac{1}{2}\rho \int \mathbf{v} \cdot \mathbf{v}\, dV$) than any other motion ($\mathbf{v}_1$) satisfying the same boundary conditions.

[Put $\mathbf{v}_1 = \mathbf{v} + \mathbf{v}'$; then $\operatorname{div} \mathbf{v}' = 0$ and $\mathbf{n} \cdot \mathbf{v}' = 0$, where $\mathbf{n}$ is the unit normal vector to the boundary S. Show that $T_1 = T + T'$.]

13. Under the conditions of Problem 11, show that if

(*a*) φ = const. over the boundary, or

(*b*) $d\varphi/dn = 0$ over the boundary,

then φ is constant throughout the region and $\mathbf{v} = 0$.

Hence show that the irrotational motion of a liquid occupying a simply connected region is uniquely determined when the value of either φ or $d\varphi/dn$ is specified at each point of the boundary.

14. If the body forces have a potential Q, the integrals,

$$T = \int \tfrac{1}{2}\rho v^2\,dV, \qquad U = \int \rho Q\,dV,$$

represent the kinetic and potential energy of the fluid within the region of integration. Show that for an incompressible fluid

$$\frac{d}{dt}(T + U) = -\oint p\,\mathbf{n}\cdot\mathbf{v}\,dS.$$

15. Assuming that the earth is a sphere of incompressible fluid of constant density ρ and without rotation, show that the pressure at a distance r from its center is

$$p = \tfrac{1}{2}g\rho a(1 - r^2/a^2),$$

where a is the radius of the earth.

[If γ is the constant of gravitation, the attraction on a unit mass at the distance r is

$$\tfrac{4}{3}\pi\gamma\rho r^3/r^2 = gr/a$$

where g is the attraction when $r = a$; hence the body force $-gr\mathbf{R}/a$ has the potential $\tfrac{1}{2}gr^2/a$.]

Compute the pressure at the center if ρ is taken equal to the mean density of the earth ($\rho g = 5.525 \times 62.4$ lb./ft.^{3}).

16. In example 1 of § 127, consider the motion when $n = \pi/\alpha (0 < \alpha < \pi)$. Since the lines $\theta = 0$, $\theta = \alpha$ are parts of the same stream-line $\psi = 0$, we have the steady irrotational motion of a liquid within two walls at an angle α. Find the radial and transverse components of $\mathbf{v}$ at any point (r, θ).

17. Discuss the plane, irrotational motion when the *complex potential* $w = \varphi + i\psi$ (§ 127) is given by $z = \cosh w$. Show that the equipotential lines and stream-lines are the families of confocal ellipses and hyperbolas:

$$\frac{x^2}{c^2\cosh^2\varphi} + \frac{y^2}{c^2\sinh^2\varphi} = 1,$$

$$\frac{x^2}{c^2\cos^2\psi} - \frac{y^2}{c^2\sin^2\psi} = 1,$$

with foci at $(\pm c, 0)$.

Show that the stream-lines $\psi = n\pi$, where n is any positive integer, correspond to the part of the x-axis from $x = \pm c$ to $x = \pm\infty$. If we regard this as a wall; we have the case of a liquid streaming through a slit of breadth $2c$ in an infinite plane.

CHAPTER VIII

GEOMETRY ON A SURFACE

130. Curvature of Surface Curves. Let C be any curve on the surface $\mathbf{r} = \mathbf{r}(u, v)$ with unit normal $\mathbf{n}$ (94.11). As the point P traverses C with unit speed, the Darboux vector,

$$\boldsymbol{\delta} = \tau\mathbf{T} + \kappa\mathbf{B}, \tag{1}$$

gives the angular velocity of the moving trihedral **TNB**.

Consider now the motion of the dextral trihedral **Tnp** ($\mathbf{p} = \mathbf{T} \times \mathbf{n}$) associated with the curve (Fig. 130). Since the trihedrals **Tnp** and

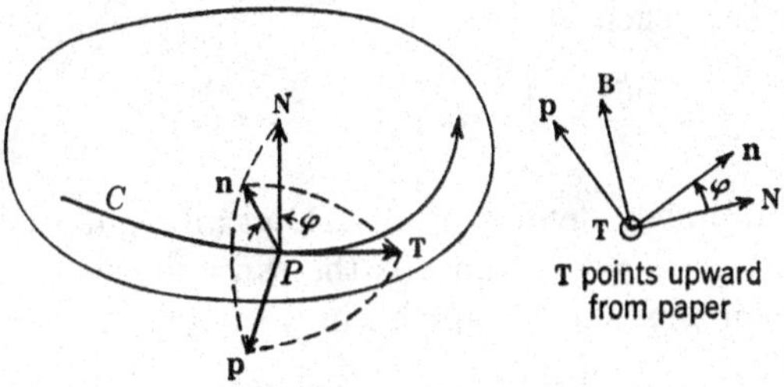

FIG. 130

TNB have **T** in common, the motion of **Tnp** relative to **TNB** is a rotation about **T**. If φ = angle ($\mathbf{N}, \mathbf{n}$), taken positive in the sense of **T**, the angular velocity of **Tnp** relative to **TNB** is $(d\varphi/ds)\mathbf{T}$. Therefore the angular velocity of **Tnp** is

$$\boldsymbol{\omega} = \boldsymbol{\delta} + \frac{d\varphi}{ds}\mathbf{T} = \left(\tau + \frac{d\varphi}{ds}\right)\mathbf{T} + \kappa\mathbf{B}. \tag{2}$$

Now $\mathbf{N} \times \mathbf{n} = \mathbf{T} \sin \varphi$ from the definition of φ and $\mathbf{n} \times (\mathbf{N} \times \mathbf{n}) = -\mathbf{p} \sin \varphi$; hence

$$\mathbf{N} = \mathbf{n} \cos \varphi - \mathbf{p} \sin \varphi, \qquad \mathbf{B} = \mathbf{T} \times \mathbf{N} = \mathbf{p} \cos \varphi + \mathbf{n} \sin \varphi.$$

Substituting this value of **B** in (2) gives

$$\boldsymbol{\omega} = (\tau + d\varphi/ds)\,\mathbf{T} + \kappa \sin \varphi\, \mathbf{n} + \kappa \cos \varphi\, \mathbf{p}. \tag{3}$$

The three scalar coefficients in (3) are written t, γ, k and are named as follows:

$$(4) \qquad t = \tau + d\varphi/ds, \quad \text{the } \textit{geodesic torsion},$$

$$(5) \qquad \gamma = \kappa \sin \varphi, \quad \text{the } \textit{geodesic curvature},$$

$$(6) \qquad k = \kappa \cos \varphi, \quad \text{the } \textit{normal curvature}.$$

With this notation,

$$(3)' \qquad \boldsymbol{\omega} = t\mathbf{T} + \gamma\mathbf{n} + k\mathbf{p}.$$

If we reverse the positive sense on the curve, we must replace $\mathbf{T}, s, \varphi$ by $-\mathbf{T}, -s, -\varphi$ (since $\mathbf{T}$ determines the sense of φ), but κ and τ are unaltered (§ 45). Therefore *a change of positive sense leaves t and k unaltered, but reverses the sign of the geodesic curvature γ.*

Since P is moving along the curve with unit speed, $ds/dt = 1$, and we take $s = t$. For any vector $\mathbf{u}$ fixed in $\mathbf{Tnb}$ we have $d\mathbf{u}/ds = \boldsymbol{\omega} \times \mathbf{u}$ (56.4); hence

$$(7) \qquad \frac{d\mathbf{n}}{ds} = \boldsymbol{\omega} \times \mathbf{n} = -k\mathbf{T} - t\,\mathbf{n} \times \mathbf{T}.$$

Now $\mathbf{n}$ is uniquely defined by (94.11) at all regular points of the surface and $d\mathbf{n}/ds = \mathbf{T} \cdot \operatorname{Grad} \mathbf{n}$ is the same for all surface curves through a point having a common unit tangent $\mathbf{T}$. From (7),

$$(8) \qquad k = -\frac{d\mathbf{n}}{ds} \cdot \mathbf{T}, \qquad t = -\frac{d\mathbf{n}}{ds} \cdot \mathbf{n} \times \mathbf{T},$$

we may therefore state

THEOREM 1. *At a given point of a surface, the normal curvature and geodesic torsion are the same for all surface curves having a common tangent there.*

Since $\cos \varphi = \mathbf{n} \cdot \mathbf{N}$, we see from (6) that the curvature $\kappa = k/\cos \varphi$ is completely determined by $\mathbf{T}$ and $\mathbf{N}$ at the point considered, provided $\varphi \neq \pi/2$. But, since $\mathbf{T}$ and $\mathbf{N}$ determine the osculating plane at the point (§ 48), we have

THEOREM 2. *All curves of a surface passing through a point and having a common osculating plane, not tangent to the surface, have the same curvature there, namely, the curvature of the plane curve cut from the surface by the osculating plane.*

In view of this theorem, we now confine our attention to the curvature of *plane* sections of the surface.

Consider now the *normal section* C_n of the surface cut by a plane through $\mathbf{n}$ and $\mathbf{T}$ at P. For this curve $\mathbf{N}_n = \pm\mathbf{n}$, $\cos\varphi = \pm 1$, and $\kappa_n = \pm k$, according as $\mathbf{N}_n$ and $\mathbf{n}$ have the same or opposite directions. If $k \neq 0$, the center of curvature of C_n at P, namely,

$$\mathbf{c}_n = \mathbf{r} + \mathbf{N}_n/\kappa_n = \mathbf{r} + \mathbf{n}/k,$$

is called the *center of normal curvature* for the direction $\mathbf{T}$. But any surface curve C tangent to $\mathbf{T}$ at P has $\mathbf{c} = \mathbf{r} + \mathbf{N}/\kappa$ for its center of curvature. Hence $\mathbf{c} - \mathbf{c}_n = \mathbf{N}/\kappa - \mathbf{n}/k$, and

$$(\mathbf{c} - \mathbf{c}_n) \cdot \mathbf{T} = 0, \qquad (\mathbf{c} - \mathbf{c}_n) \cdot \mathbf{N} = \frac{1}{\kappa} - \frac{\cos\varphi}{k} = 0,$$

from (6), so that $\mathbf{c} - \mathbf{c}_n$ is perpendicular to the osculating plane of C.

THEOREM 3 (*Meusnier*). *The center of curvature of any surface curve is the projection of the corresponding center of normal curvature upon its osculating plane if the latter is not tangent to the surface.*

Curves on a surface along which t, γ, or k vanish are named as follows:

$t = 0$: lines of curvature,

$\gamma = 0$: geodesic lines or geodesics,

$k = 0$: asymptotic lines.

When $t = 0$, $d\mathbf{n}/ds = -k\,\mathbf{T}$, from (7); hence

$$\text{(9)} \qquad \frac{d\mathbf{r}}{ds} \times \frac{d\mathbf{n}}{ds} = 0, \quad \text{along a line of curvature.}$$

When $k = 0$, $d\mathbf{n}/ds = -t\,\mathbf{n} \times \mathbf{T}$, from (7); hence

$$\text{(10)} \qquad \frac{d\mathbf{r}}{ds} \cdot \frac{d\mathbf{n}}{ds} = 0, \quad \text{along an asymptotic line.}$$

The parametric line $v = \text{const}$ is a line of curvature when $\mathbf{r}_u \times \mathbf{n}_u = 0$, and an asymptotic line when $\mathbf{r}_u \cdot \mathbf{n}_u = 0$.

131. The Dyadic $\nabla \mathbf{n}$. Let $\mathbf{e}$ and $\mathbf{e}' = \mathbf{n} \times \mathbf{e}$ be two perpendicular unit tangent vectors to the surface $\mathbf{r} = \mathbf{r}(u, v)$ at the point P_1.

If k, t and k', t' are the normal curvature and geodesic torsion associated with these directions, we have, from (130.7),

$$\frac{d\mathbf{n}}{ds} = -k\mathbf{e} - t\,\mathbf{n}\times\mathbf{e} = -k\mathbf{e} - t\mathbf{e}',$$

$$\frac{d\mathbf{n}}{ds'} = -k'\mathbf{e}' - t'\,\mathbf{n}\times\mathbf{e}' = -k'\mathbf{e}' + t'\mathbf{e};$$

hence (95.8)

(1) $$\text{Grad}\,\mathbf{n} = \mathbf{e}\frac{d\mathbf{n}}{ds} + \mathbf{e}'\frac{d\mathbf{n}}{ds'} = \mathbf{e}(-k\mathbf{e} - t\mathbf{e}') + \mathbf{e}'(-k'\mathbf{e}' + t'\mathbf{e}).$$

This dyadic is symmetric, since its vector invariant, Rot $\mathbf{n} = 0$ (96.8). We have, in fact,

(2) $$\text{Rot}\,\mathbf{n} = -(t + t')\mathbf{n} = 0, \qquad t' = -t;$$

(3) $$\nabla\mathbf{n} = -k\,\mathbf{ee} - k'\mathbf{e}'\mathbf{e}' - t(\mathbf{ee}' + \mathbf{e}'\mathbf{e}).$$

For brevity, here and elsewhere in this chapter, we use ∇ instead of Grad since no misunderstanding is possible. We state the result (2) as

THEOREM 1 (*Bonnet*). *The geodesic torsions associated with any two perpendicular tangents at a point of a surface are equal in magnitude but opposite in sign.*

From (1) we find that the first and second scalar invariants of $\nabla\mathbf{n}$ are $-(k + k')$ and $kk' + tt'$. We write

(4) $$J = k + k' = -\,\text{Div}\,\mathbf{n},$$

(5) $$K = kk' - t^2;$$

J is called the *mean curvature* and K the *total curvature* of the surface at P. Since $\nabla\mathbf{n}$ and its invariants are *point* functions over the surface, the values of J and K are independent of the choice of $\mathbf{e}$; we therefore have

THEOREM 2. *For any pair of perpendicular tangents at P, $k + k'$ and $kk' + tt'$ have the same value.*

From (3), we have

(6) $$k = -\mathbf{e}\cdot\nabla\mathbf{n}\cdot\mathbf{e}, \qquad t = -\mathbf{e}\cdot\nabla\mathbf{n}\cdot\mathbf{e}';$$

hence k and t are not altered when $\mathbf{e}$ is replaced by $-\mathbf{e}$. If we assume that k remains finite at P and is not constant, there are certain directions for which k attains its extreme values. To find these we examine the variation of k as $\mathbf{e}$ revolves about P in the tangent plane. If the angle θ between $\mathbf{e}$ and a fixed line in the tangent plane is taken positive in the sense determined by n, we have (§ 44)

$$\frac{d\mathbf{e}}{d\theta} = \mathbf{n} \times \mathbf{e} = \mathbf{e}', \qquad \frac{d\mathbf{e}'}{d\theta} = \mathbf{n} \times \mathbf{e}' = -\mathbf{e};$$

and, from (6),

$$\text{(7)} \qquad \frac{dk}{d\theta} = -\mathbf{e}' \cdot \nabla\mathbf{n} \cdot \mathbf{e} - \mathbf{e} \cdot \nabla\mathbf{n} \cdot \mathbf{e}' = 2t.$$

The extremes of k therefore occur when $t = 0$. If the direction $\mathbf{e}_1$ gives an extreme value k_1, $t_1 = 0$. Then the perpendicular direction $\mathbf{e}_2 = \mathbf{n} \times \mathbf{e}_1$, for which $t_2 = 0$ from (2), gives another extreme value k_2. But since $k + k'$ is constant, if $k = k_1$ is a maximum, $k' = k_2$ is a minimum, and vice versa.

If we take $\mathbf{e} = \mathbf{e}_1$, $\mathbf{e}' = \mathbf{e}_2$ in (3), we have

$$\text{(8)} \qquad \nabla\mathbf{n} = -k_1\mathbf{e}_1\mathbf{e}_1 - k_2\mathbf{e}_2\mathbf{e}_2,$$

and the symmetric dyadic $\nabla\mathbf{n}$ appears in the standard form (72.4). Evidently $\mathbf{e}_1$ and $\mathbf{e}_2$ are the invariant directions of ∇n with multipliers $-k_1$, $-k_2$. Moreover, if k is not constant, $\pm\mathbf{e}_1$ and $\pm\mathbf{e}_2$ are the only invariant directions at P. We state these results in

THEOREM 3. *If the normal curvature is finite and not constant at a point of the surface, it attains its maximum and minimum values for just two normal sections, at right angles to each other, and characterized by the vanishing of the geodesic torsion.*

The orthogonal directions $\mathbf{e}_1$, $\mathbf{e}_2$ are called the *principal directions* at P, and the corresponding normal curvatures k_1, k_2, the *principal curvatures*. From (4) and (5),

$$\text{(9)} \qquad J = k_1 + k_2, \qquad K = k_1k_2;$$

consequently k_1 and k_2 are roots of the quadratic,

$$\text{(10)} \qquad k^2 - Jk + K = 0,$$

the characteristic equation (§ 71) for the symmetric planar dyadic $\nabla\mathbf{n}$.

Consider finally the case when k is constant at P; from (7), $t = 0$ for all directions, and (3) becomes

$$\nabla \mathbf{n} = -k(\mathbf{ee} + \mathbf{e}'\mathbf{e}') = -k(\mathbf{I} - \mathbf{nn}). \tag{11}$$

Any direction in the tangent plane at P is an invariant direction with multiplier $-k$, and P is called an *umbilical point* or simply an *umbilic.*

If k is constant over the entire surface, we have

$$\nabla \mathbf{n} = -k\,\nabla \mathbf{r}, \qquad \nabla(\mathbf{n} + k\mathbf{r}) = 0,$$

and $\mathbf{n} + k\mathbf{r}$ is constant over the surface. If $k = 0$, $\mathbf{n}$ is a constant, and the surface is a plane. If $k \neq 0$,

$$\frac{\mathbf{n}}{k} + \mathbf{r} = \mathbf{c}, \qquad |\,\mathbf{r} - \mathbf{c}\,|^2 = 1/k^2;$$

and the surface is a sphere (center $\mathbf{c}$, radius $1/k$). Therefore *the only surfaces whose points are all umbilical are the plane and sphere.* All curves on the plane or sphere are lines of curvature ($t = 0$).

Example 1. Formulas (6) give k and t for any direction $\mathbf{e}$. If the angle $(\mathbf{e}_1\mathbf{e}) = \theta$ is reckoned positive in the sense of $\mathbf{n}$ ($\mathbf{e}_1 \times \mathbf{e} = \mathbf{n} \sin \theta$),

$$\mathbf{e} = \mathbf{e}_1 \cos \theta + \mathbf{e}_2 \sin \theta, \qquad \mathbf{e}' = -\mathbf{e}_1 \sin \theta + \mathbf{e}_2 \cos \theta.$$

Taking $\nabla \mathbf{n}$ in the form (8), we have

$$k = \mathbf{e} \cdot (k_1\mathbf{e}_1\mathbf{e}_1 + k_2\mathbf{e}_2\mathbf{e}_2) \cdot \mathbf{e} = k_1 \cos^2 \theta + k_2 \sin^2 \theta, \tag{12}$$

$$t = \mathbf{e} \cdot (k_1\mathbf{e}_1\mathbf{e}_1 + k_2\mathbf{e}_2\mathbf{e}_2) \cdot \mathbf{e}' = (k_2 - k_1) \sin \theta \cos \theta. \tag{13}$$

These equations are due, respectively, to Euler and Bonnet. Since $2t = (k_2 - k_1) \sin 2\theta$, it is clear that t attains its extreme values $\pm(k_2 - k_1)/2$ for the directions $\theta = \pm\pi/4$.

Example 2. At a surface point the directions for which $k = 0$ are called *asymptotic*. From (12), the asymptotic directions are given by $\tan^2 \theta = -k_1/k_2$ and are real and distinct only when $K = k_1k_2 < 0$. When $k_1 = 0$, $k_2 \neq 0$, both asymptotic directions coalesce with the principal direction $\mathbf{e}_1$. In an asymptotic direction, $t^2 = -K$, from (5).

Along an *asymptotic line*, $k = \kappa \cos \varphi = 0$; and, if $\kappa \neq 0$, $\cos \varphi = 0$, and $\varphi = \pm\pi/2$. Along a curved asymptotic line, the osculating plane remains tangent to the surface; moreover $\gamma = \pm\kappa$, and $t = \tau$.

Referred to the principal direction $\mathbf{e}_1$ the asymptotic directions have the slopes $\pm\sqrt{-k_1/k_2}$ and are perpendicular when and only when these slopes are ± 1; then $-k_1/k_2 = 1$, and $J = k_1 + k_2 = 0$.

Example 3. If $\mathbf{e}$ is the unit tangent vector of a surface curve, the trihedral $\mathbf{ee'n}$ has the angular velocity,

$$\boldsymbol{\omega} = t\mathbf{e} - k\mathbf{e}' + \gamma\mathbf{n} \tag{14}$$ (130.3)′,

along the curve, and

$$\frac{d\mathbf{e}}{ds} = \boldsymbol{\omega} \times \mathbf{e} = k\mathbf{n} + \gamma\mathbf{e}', \qquad \frac{d\mathbf{e}'}{ds} = \boldsymbol{\omega} \times \mathbf{e}' = t\mathbf{n} - \gamma\mathbf{e}.$$

On differentiating equations (6), we now have

$$\frac{dk}{ds} = -2(k\mathbf{n} + \gamma\mathbf{e}') \cdot \nabla\mathbf{n} \cdot \mathbf{e} - \mathbf{e} \cdot \left(\frac{d}{ds}\nabla\mathbf{n}\right) \cdot \mathbf{e},$$
$$\frac{dt}{ds} = -(k\mathbf{n} + \gamma\mathbf{e}') \cdot \nabla\mathbf{n} \cdot \mathbf{e}' - \mathbf{e} \cdot \nabla\mathbf{n} \cdot (t\mathbf{n} - \gamma\mathbf{e}) - \mathbf{e} \cdot \left(\frac{d}{ds}\nabla\mathbf{n}\right) \cdot \mathbf{e}'.$$

Since $d\nabla\mathbf{n}/ds$ is the same along all surface curves having the common tangent $\mathbf{e}$ at a point, the same is true of the expressions,

$$\frac{dk}{ds} - 2\gamma t \quad \text{and} \quad \frac{dt}{ds} + \gamma(k - k'), \tag{15}$$

respectively discovered by Laguerre and Darboux.

132. Fundamental Forms. On the surface $\mathbf{r} = \mathbf{r}(u, v)$, the gradient of a tensor function $f(u, v)$ is given by (95.5),

$$\nabla\mathbf{f} = \mathbf{a}\mathbf{f}_u + \mathbf{b}\mathbf{f}_v, \tag{1}$$

where $\mathbf{a}, \mathbf{b}, \mathbf{n}$ is the set reciprocal to $\mathbf{r}_u, \mathbf{r}_v, \mathbf{n}$:

$$\mathbf{a}\mathbf{r}_u + \mathbf{b}\mathbf{r}_v + \mathbf{nn} = \mathbf{I}. \tag{2}$$

From (1), we have, in particular,

$$\nabla u = \mathbf{a}, \qquad \nabla v = \mathbf{b}; \tag{3}$$

$$\nabla\mathbf{r} = \mathbf{a}\mathbf{r}_u + \mathbf{b}\mathbf{r}_v = \mathbf{I} - \mathbf{nn}. \tag{4}$$

The dyadic $\nabla\mathbf{r}$ is symmetric and may also be written $\mathbf{r}_u\mathbf{a} + \mathbf{r}_v\mathbf{b}$; it transforms any vector $\mathbf{f}(u, v)$ into its projection $\mathbf{f}_t$ on the tangent plane at (u, v):

$$\mathbf{f} \cdot \nabla\mathbf{r} = \nabla\mathbf{r} \cdot \mathbf{f} = (\mathbf{I} - \mathbf{nn}) \cdot \mathbf{f} = \mathbf{f}_t. \tag{5}$$

Therefore $\nabla\mathbf{r}$ acts as an idemfactor on vectors tangent to the surface at the point (u, v) and on dyadics whose vectors lie in the tangent plane.

If we form the product,

$$(\nabla\mathbf{r}) \cdot (\nabla\mathbf{r}) = (\mathbf{a}\mathbf{r}_u + \mathbf{b}\mathbf{r}_v) \cdot (\mathbf{r}_u\mathbf{a} + \mathbf{r}_v\mathbf{b}),$$

we obtain

$$\nabla \mathbf{r} = E\,\mathbf{aa} + F(\mathbf{ab} + \mathbf{ba}) + G\,\mathbf{bb}, \tag{6}$$

where

$$E = \mathbf{r}_u \cdot \mathbf{r}_u, \qquad F = \mathbf{r}_u \cdot \mathbf{r}_v, \qquad G = \mathbf{r}_v \cdot \mathbf{r}_v \tag{7}$$

are the coefficients of the *first fundamental form*,

$$d\mathbf{r} \cdot d\mathbf{r} = E\,du^2 + 2F\,du\,dv + G\,dv^2, \tag{8}$$

defined in § 94. In fact (8) follows from (6) when we form $d\mathbf{r} \cdot \nabla \mathbf{r} \cdot d\mathbf{r}$ and note that $\mathbf{a} \cdot d\mathbf{r} = du$, $\mathbf{b} \cdot d\mathbf{r} = dv$, from (3).

If we compute

$$(\nabla \mathbf{n}) \cdot (\nabla \mathbf{r}) = (\mathbf{a}\mathbf{n}_u + \mathbf{b}\mathbf{n}_v) \cdot (\mathbf{r}_u\mathbf{a} + \mathbf{r}_v\mathbf{b}),$$

we obtain, in similar fashion,

$$-\nabla \mathbf{n} = L\,\mathbf{aa} + M(\mathbf{ab} + \mathbf{ba}) + N\,\mathbf{bb}, \tag{9}$$

where

$$\begin{aligned} L &= -\mathbf{n}_u \cdot \mathbf{r}_u = \mathbf{n} \cdot \mathbf{r}_{uu}, \\ M &= -\mathbf{n}_u \cdot \mathbf{r}_v = -\mathbf{n}_v \cdot \mathbf{r}_u = \mathbf{n} \cdot \mathbf{r}_{uv}, \\ N &= -\mathbf{n}_v \cdot \mathbf{r}_v = \mathbf{n} \cdot \mathbf{r}_{vv}. \end{aligned} \tag{10}$$

The relations between scalar products in (10) are obtained by differentiating the equations,

$$\mathbf{n} \cdot \mathbf{r}_u = 0, \qquad \mathbf{n} \cdot \mathbf{r}_v = 0,$$

with respect to u and v:

$$\begin{aligned} \mathbf{n}_u \cdot \mathbf{r}_u + \mathbf{n} \cdot \mathbf{r}_{uu} &= 0, \qquad & \mathbf{n}_v \cdot \mathbf{r}_u + \mathbf{n} \cdot \mathbf{r}_{uv} &= 0, \\ \mathbf{n}_u \cdot \mathbf{r}_v + \mathbf{n} \cdot \mathbf{r}_{vv} &= 0, \qquad & \mathbf{n}_v \cdot \mathbf{r}_v + \mathbf{n} \cdot \mathbf{r}_{vv} &= 0. \end{aligned}$$

From (9) we obtain the *second fundamental form*,

$$-d\mathbf{r} \cdot d\mathbf{n} = L\,du^2 + 2M\,du\,dv + N\,dv^2, \tag{11}$$

by computing the product $d\mathbf{r} \cdot \nabla \mathbf{n} \cdot d\mathbf{r}$.

Remembering that $\nabla \mathbf{n}$ is symmetric, we next compute

$$(\nabla \mathbf{n}) \cdot (\nabla \mathbf{n}) = (\mathbf{a}\mathbf{n}_u + \mathbf{b}\mathbf{n}_v) \cdot (\mathbf{n}_u\mathbf{a} + \mathbf{n}_v\mathbf{b})$$

to obtain

$$(\nabla \mathbf{n})^2 = e\,\mathbf{aa} + f(\mathbf{ab} + \mathbf{ba}) + g\,\mathbf{bb}, \tag{12}$$

where

$$(13) \qquad e = \mathbf{n}_u \cdot \mathbf{n}_u, \qquad f = \mathbf{n}_u \cdot \mathbf{n}_v, \qquad g = \mathbf{n}_v \cdot \mathbf{n}_v.$$

From (12), we obtain the *third fundamental form*,

$$(14) \qquad d\mathbf{n} \cdot d\mathbf{n} = e\,du^2 + 2f\,du\,dv + g\,dv^2,$$

by computing the product $d\mathbf{r} \cdot (\nabla\mathbf{n})^2 \cdot d\mathbf{r}$.

The quantities (7), (10), and (13) are known, respectively, as the *fundamental quantities of the first, second, and third orders.*

If $\mathbf{e}_1$, $\mathbf{e}_2$ are unit vectors in the principal directions at P,

$$\nabla\mathbf{r} = \mathbf{e}_1\mathbf{e}_1 + \mathbf{e}_2\mathbf{e}_2,$$

$$\nabla\mathbf{n} = -k_1\,\mathbf{e}_1\mathbf{e}_1 - k_2\,\mathbf{e}_2\mathbf{e}_2,$$

$$(\nabla\mathbf{n})^2 = k_1^2\,\mathbf{e}_1\mathbf{e}_1 + k_2^2\,\mathbf{e}_2\mathbf{e}_2;$$

and, on multiplying these equations in turn by $K = k_1k_2$, $J = k_1 + k_2$, 1, and adding, we get

$$(15) \qquad (\nabla\mathbf{n})^2 + J\,\nabla\mathbf{n} + K\,\nabla\mathbf{r} = 0.$$

This is the Hamilton–Cayley Equation for the planar dyadic $\nabla\mathbf{n}$; for its first and second scalar invariants are $-J$, K, and $\nabla\mathbf{r}$ is the idemfactor in the tangent plane. Substituting from (6), (9), and (12) for the dyadics in (15), we obtain the following relations between the fundamental quantities:

$$(16) \qquad e - JL + KE = 0, \qquad f - JM + KF = 0,$$

$$g - JN + KG = 0.$$

Example 1. From the reciprocity of $\mathbf{a}$, $\mathbf{b}$, $\mathbf{n}$ and $\mathbf{r}_u$, $\mathbf{r}_v$, $\mathbf{n}$,

$$(17) \qquad \mathbf{a} = \frac{\mathbf{r}_v \times \mathbf{n}}{H}, \qquad \mathbf{b} = \frac{\mathbf{n} \times \mathbf{r}_u}{H}, \qquad \mathbf{n} = \frac{\mathbf{r}_u \times \mathbf{r}_v}{H} = H\,\mathbf{a} \times \mathbf{b},$$

where $H = \mathbf{r}_u \times \mathbf{r}_v \cdot \mathbf{n}$; hence

$$(18) \qquad \mathbf{a} \cdot \mathbf{a} = G/H^2, \qquad \mathbf{a} \cdot \mathbf{b} = -F/H^2, \qquad \mathbf{b} \cdot \mathbf{b} = E/H^2.$$

We now may compute the invariants $-J$ and K of $\nabla\mathbf{n}$:

$$\nabla\mathbf{n} = \mathbf{a}\mathbf{n}_u + \mathbf{b}\mathbf{n}_v, \qquad (\nabla\mathbf{n})_2 = \mathbf{a} \times \mathbf{b}\,\mathbf{n}_u \times \mathbf{n}_v;$$

$$-J = \mathbf{a} \cdot \mathbf{n}_u + \mathbf{b} \cdot \mathbf{n}_v = \frac{(\mathbf{n}_u \times \mathbf{r}_v + \mathbf{r}_u \times \mathbf{n}_v) \cdot \mathbf{n}}{\mathbf{r}_u \times \mathbf{r}_v \cdot \mathbf{n}},$$

$$K = (\mathbf{a} \times \mathbf{b}) \cdot (\mathbf{n}_u \times \mathbf{n}_v) = \frac{\mathbf{n}_u \times \mathbf{n}_v \cdot \mathbf{n}}{\mathbf{r}_u \times \mathbf{r}_v \cdot \mathbf{n}}.$$

Since the cross products in these equations are all parallel to **n**, these equations may be written also in the vector form:

$$(19)\qquad -J\,\mathbf{r}_u \times \mathbf{r}_v = \mathbf{n}_u \times \mathbf{r}_v + \mathbf{r}_u \times \mathbf{n}_v,$$

$$(20)\qquad K\,\mathbf{r}_u \times \mathbf{r}_v = \mathbf{n}_u \times \mathbf{n}_v.$$

If we multiply these equations by $(\mathbf{r}_u \times \mathbf{r}_v)\cdot$, apply (20.1), and introduce the fundamental quantities from (7) and (10), we have

$$(21)\qquad J(EG - F^2) = GL - 2FM + EN,$$

$$(22)\qquad K(EG - F^2) = LN - M^2.$$

These values of J and K also may be found from

$$(9)\qquad -\nabla\mathbf{n} = \mathbf{a}(L\mathbf{a} + M\mathbf{b}) + \mathbf{b}(M\mathbf{a} + N\mathbf{b}),$$

by making use of (18).

Example 2. The surface $z = z(x, y)$, with the position vector,

$$\mathbf{r} = x\mathbf{i} + y\mathbf{j} + z(x, y)\mathbf{k},$$

can be written in the parametric form

$$x = u, \quad y = v, \quad z = z(u, v).$$

If we denote the partial derivatives $z_x, z_y, z_{xx}, z_{xy} = z_{yx}, z_{yy}$ by p, q, r, s, t, respectively, we have

$$\mathbf{r}_u = \mathbf{r}_x = \mathbf{i} + p\mathbf{k}, \qquad \mathbf{r}_v = \mathbf{r}_y = \mathbf{j} + q\mathbf{k};$$

hence, from (7),

$$E = 1 + p^2, \quad F = pq, \quad G = 1 + q^2;$$

$$H^2 = EG - F^2 = 1 + p^2 + q^2;$$

$$H\mathbf{n} = \mathbf{r}_u \times \mathbf{r}_v = -p\mathbf{i} - q\mathbf{j} + \mathbf{k}.$$

Furthermore,

$$\mathbf{r}_{uu} = r\mathbf{k}, \quad \mathbf{r}_{uv} = s\mathbf{k}, \quad \mathbf{r}_{vv} = t\mathbf{k};$$

hence, from (10),

$$L = r/H, \quad M = s/H, \quad N = t/H.$$

We may now compute the mean curvature J from (21), the total curvature K from (22):

$$(23)\qquad J = \frac{(1 + q^2)r - 2pqs + (1 + p^2)t}{(1 + p^2 + q^2)^{\frac{3}{2}}},$$

$$(24)\qquad K = \frac{rt - s^2}{(1 + p^2 + q^2)^2}.$$

The principal curvatures k_1, k_2 are the roots of the quadratic (131.10):

$$(25)\qquad k^2 - Jk + K = 0.$$

133. Field of Curves. A one-parameter family of curves on a surface $\mathbf{r} = \mathbf{r}(u, v)$ is said to form a field over a portion S of the surface if one and only one curve of the family passes through every point of S. After the positive direction on one of the curves has been chosen, the positive direction on the others is taken so that the unit tangent vector to the curves is a continuous vector point function over S.

If the curves of the field have the equation $\varphi(u, v) = \text{const}$, their differential equation is

$$\varphi_u \, du + \varphi_v \, dv = 0.$$

From this we may compute $\lambda_1 = dv/du$ at all points of S. Then the vector,

$$\frac{d\mathbf{r}}{du} = \frac{\partial \mathbf{r}}{\partial u} + \frac{\partial \mathbf{r}}{\partial v}\frac{\partial v}{\partial u} = \mathbf{r}_u + \lambda_1 \mathbf{r}_v,$$

is tangent to the field curve through the point (u, v).

Similarly if $\lambda_2 = dv/du$ for a second one-parameter family of curves, the vector $\mathbf{r}_u + \lambda_2 \mathbf{r}_v$ is a tangent to the curve through (u, v). When the second family cuts the first everywhere at right angles,

$$(\mathbf{r}_u + \lambda_1 \mathbf{r}_v) \cdot (\mathbf{r}_u + \lambda_2 \mathbf{r}_v) = 0,$$

or, in terms of fundamental quantities,

$$E + (\lambda_1 + \lambda_2)F + \lambda_1 \lambda_2 \, G = 0. \tag{1}$$

Since $\lambda_1 = -\varphi_u/\varphi_v$ is known, (1) is the differential equation of the *orthogonal trajectories* of the field curves. Standard existence theorems for differential equations of the first order guarantee, under very general conditions, a solution of (1).

Since $\mathbf{r}_u, \mathbf{r}_v$ are tangent vectors to the curves $v = \text{const}$, $u = \text{const}$, the parametric curves cut at right angles when

$$\mathbf{r}_u \cdot \mathbf{r}_v = F = 0. \tag{2}$$

134. The Field Dyadic. Consider a field of curves C_1 and their orthogonal trajectories C_2 on a portion S of the surface $\mathbf{r} = \mathbf{r}(u, v)$. At every point (u, v) their unit tangents $\mathbf{e}_1$, $\mathbf{e}_2 = \mathbf{n} \times \mathbf{e}_1$, and the surface normal $\mathbf{n}$ form a dextral trihedral of unit vectors $\mathbf{e}_1\mathbf{e}_2\mathbf{n}$. As a point traverses C_1 with unit speed, $\mathbf{e}_1\mathbf{e}_2\mathbf{n}$ has the angular velocity (130.3′),

$$\boldsymbol{\omega}_1 = t_1 \mathbf{e}_1 - k_1 \mathbf{e}_2 + \gamma_1 \mathbf{n}; \tag{1}$$

since $\mathbf{p}_1 = \mathbf{e}_1 \times \mathbf{n} = -\mathbf{e}_2$. Similarly, along C_2,

(2) $$\boldsymbol{\omega}_2 = t_2\mathbf{e}_2 + k_2\mathbf{e}_1 + \gamma_2\mathbf{n},$$

since $\mathbf{p}_2 = \mathbf{e}_2 \times \mathbf{n} = \mathbf{e}_1$. Therefore

$$\frac{d\mathbf{e}_1}{ds_1} = \boldsymbol{\omega}_1 \times \mathbf{e}_1 = k_1\mathbf{n} + \gamma_1\mathbf{e}_2,$$

$$\frac{d\mathbf{e}_1}{ds_2} = \boldsymbol{\omega}_2 \times \mathbf{e}_1 = -t_2\mathbf{n} + \gamma_2\mathbf{e}_2.$$

We now have, from (95.8),

$$\begin{aligned}\operatorname{Grad} \mathbf{e}_1 &= \mathbf{e}_1 \frac{d\mathbf{e}_1}{ds_1} + \mathbf{e}_2 \frac{d\mathbf{e}_1}{ds_2} \\ &= \mathbf{e}_1(k_1\mathbf{n} + \gamma_1\mathbf{e}_2) + \mathbf{e}_2(t_1\mathbf{n} + \gamma_2\mathbf{e}_2) \\ &= (k_1\mathbf{e}_1 + t_1\mathbf{e}_2)\mathbf{n} + (\gamma_1\mathbf{e}_1 + \gamma_2\mathbf{e}_2)\mathbf{e}_2,\end{aligned}$$

since $t_2 = -t_1$ by Bonnet's Theorem (§ 131). Now, from (130.7)

$$\frac{d\mathbf{n}}{ds_1} = -k_1\mathbf{e}_1 - t_1\,\mathbf{n} \times \mathbf{e}_1;$$

and, if we write

(3) $$\mathbf{R} = \gamma_1\mathbf{e}_1 + \gamma_2\mathbf{e}_2,$$

the *field dyadic* Grad $\mathbf{e}_1$ for the curves C_1 becomes

(4) $$\nabla\mathbf{e}_1 = -\frac{d\mathbf{n}}{ds_1}\mathbf{n} + \mathbf{R}\,\mathbf{e}_2.$$

If we replace $\mathbf{e}_1$ and $\mathbf{e}_2$ in (3) by $\mathbf{n} \times \mathbf{e}_1 = \mathbf{e}_2$, $\mathbf{n} \times \mathbf{e}_2 = -\mathbf{e}_1$, $\mathbf{R}$ is unchanged, since $\gamma_2\mathbf{e}_2 + (-\gamma_1)(-\mathbf{e}_1) = \mathbf{R}$. Therefore the field dyadic for the curves C_2 is

(5) $$\nabla\mathbf{e}_2 = -\frac{d\mathbf{n}}{ds_2}\mathbf{n} - \mathbf{R}\,\mathbf{e}_1.$$

We have moreover

(6) $$\nabla\mathbf{n} = \mathbf{e}_1 \frac{d\mathbf{n}}{ds_1} + \mathbf{e}_2 \frac{d\mathbf{n}}{ds_2} = \frac{d\mathbf{n}}{ds_1}\mathbf{e}_1 + \frac{d\mathbf{n}}{ds_2}\mathbf{e}_2.$$

With the notation,

(7) $$\mathbf{P} = \frac{d\mathbf{n}}{ds_1}, \qquad \mathbf{Q} = \frac{d\mathbf{n}}{ds_2}$$

the preceding equations become

$$(4)' \qquad \nabla \mathbf{e}_1 = \mathbf{R}\mathbf{e}_2 - \mathbf{P}\mathbf{n},$$

$$(5)' \qquad \nabla \mathbf{e}_2 = -\mathbf{R}\mathbf{e}_1 - \mathbf{Q}\mathbf{n},$$

$$(6)' \qquad \nabla \mathbf{n} = \mathbf{P}\mathbf{e}_1 + \mathbf{Q}\mathbf{e}_2.$$

From (4), we have

$$\mathbf{n} \cdot \text{rot}\, \mathbf{e}_1 = \mathbf{n} \cdot (\mathbf{R} \times \mathbf{e}_2 - \mathbf{P} \times \mathbf{n}) = \mathbf{R} \cdot \mathbf{e}_2 \times \mathbf{n} = \mathbf{R} \cdot \mathbf{e}_1,$$

or, in view of (3),

$$\gamma_1 = \mathbf{n} \cdot \text{rot}\, \mathbf{e}_1.$$

The geodesic curvature is therefore a surface invariant.

THEOREM. *A field of curves having the unit tangent vector* $\mathbf{e}(u, v)$ *has at every point the geodesic curvature,*

$$(8) \qquad \gamma = \mathbf{n} \cdot \text{rot}\, \mathbf{e} = \frac{1}{H} \{(\mathbf{r}_v \cdot \mathbf{e})_u - (\mathbf{r}_u \cdot \mathbf{e})_v\}.$$

The last expression follows from (97.9).

Consider now a curve C which cuts the curves C_1 at an angle $\theta = (\mathbf{e}_1, \mathbf{e})$, reckoned positive in the sense of $\mathbf{n}$. Along C the trihedral $\mathbf{e}\mathbf{e}'\mathbf{n}$ has the angular velocity

$$(9) \qquad \boldsymbol{\omega} = t\mathbf{e} - k\mathbf{e}' + \gamma\mathbf{n} \qquad (131.14);$$

and, since the trihedral $\mathbf{e}_1\mathbf{e}_2\mathbf{n}$ has the angular velocity $-(d\theta/ds)\,\mathbf{n}$ relative to $\mathbf{e}\mathbf{e}'\mathbf{n}$, the angular velocity of $\mathbf{e}_1\mathbf{e}_2\mathbf{n}$ along C is

$$(10) \qquad \boldsymbol{\omega} - \frac{d\theta}{ds}\mathbf{n} = t\mathbf{e} - k\mathbf{e}' + \left(\gamma - \frac{d\theta}{ds}\right)\mathbf{n};$$

hence

$$\mathbf{e} \cdot \nabla \mathbf{e}_1 = \frac{d\mathbf{e}_1}{ds} = \left(\boldsymbol{\omega} - \frac{d\theta}{ds}\mathbf{n}\right) \times \mathbf{e}_1,$$

$$(11) \qquad \mathbf{e} \cdot \nabla \mathbf{e}_1 \cdot \mathbf{e}_2 = \left(\boldsymbol{\omega} - \frac{d\theta}{ds}\mathbf{n}\right) \cdot \mathbf{n} = \gamma - \frac{d\theta}{ds}\cdot$$

The left member of (11) shows that $\gamma - d\theta/ds$ is the same for all surface curves having $\mathbf{e}$ as common tangent vector.

Since $\nabla \mathbf{e}_1 \cdot \mathbf{e}_2 = \mathbf{\mathring{R}}$, from (4), we may write (11) in the form:

$$\text{(12)} \qquad \mathbf{e} \cdot \mathbf{R} = \gamma - \frac{d\theta}{ds}.$$

If C is a member of a field of curves, the angle $\theta = (\mathbf{e}_1, \mathbf{e})$ is a point function over this field, and

$$\text{(13)} \qquad \mathbf{R} = \mathbf{R} \cdot (\mathbf{ee} + \mathbf{e}'\mathbf{e}') = \gamma \mathbf{e} + \gamma' \mathbf{e}' - \nabla\theta.$$

If the field curves C_1 and C cut everywhere at the same angle, $\nabla\theta = 0$, and $\mathbf{R} = \gamma\mathbf{e} + \gamma'\mathbf{e}'$.

Example. For a field of surface curves $\varphi(u, v) = c$, the unit tangent vector is

$$\mathbf{e} = d\mathbf{r}/ds = \mathbf{r}_u \dot{u} + \mathbf{r}_v \dot{v}, \quad \text{where} \quad \varphi_u \dot{u} + \varphi_v \dot{v} = 0.$$

Therefore

$$\frac{\dot{u}}{\varphi_v} = \frac{\dot{v}}{-\varphi_u} = \frac{1}{\lambda}; \qquad \dot{u} = \frac{\varphi_v}{\lambda}, \quad \dot{v} = -\frac{\varphi_u}{\lambda};$$

and, if we substitute these values in

$$\mathbf{e} \cdot \mathbf{e} = E\dot{u}^2 + 2F\dot{u}\dot{v} + G\dot{v}^2 = 1,$$

we have

$$\lambda^2 = E\varphi_v^2 - 2F\varphi_u\varphi_v + G\varphi_u^2$$

to find λ. Thus

$$\mathbf{e} = (\mathbf{r}_u\varphi_v - \mathbf{r}_v\varphi_u)/\lambda,$$

where the sign is chosen to give the positive sense desired. The geodesic curvature now is given by (8):

$$\text{(14)} \qquad \gamma = \frac{1}{H}\left\{\frac{\partial}{\partial u}\left(\frac{F\varphi_v - G\varphi_u}{\lambda}\right) - \frac{\partial}{\partial v}\left(\frac{E\varphi_v - F\varphi_u}{\lambda}\right)\right\}.$$

This formula is due to Bonnet.

Let us apply it to find the geodesic curvature of the parametric curves in the sense of increasing u and v. For the curves $v = \text{const}$, $\varphi_u = 0$, $\varphi_v = 1$, $\lambda = \sqrt{E}$, $\mathbf{e}_1 = \mathbf{r}_u/\sqrt{E}$; for the curves $u = \text{const}$, $\varphi_u = 1$, $\varphi_v = 0$, $\lambda = -\sqrt{G}$, $\mathbf{e}_3 = \mathbf{r}_v/\sqrt{G}$; hence in the respective cases,

$$\text{(15)} \quad \gamma_1 = \frac{1}{H}\left\{\frac{\partial}{\partial u}\left(\frac{F}{\sqrt{E}}\right) - \frac{\partial}{\partial v}\sqrt{E}\right\}, \qquad \gamma_3 = \frac{1}{H}\left\{\frac{\partial}{\partial u}\sqrt{G} - \frac{\partial}{\partial v}\left(\frac{F}{\sqrt{G}}\right)\right\}.$$

These formulas also follow at once from (8) when the foregoing values of $\mathbf{e}_1$ and $\mathbf{e}_3$ are used.

The subscript 3 refers to the curves $u = \text{const}$ when they cut the curves $v = \text{const}$ at an arbitrary angle; when this angle is $\pi/2$, we use the subscript 2.

135. Geodesics. The geodesic curvature of a curve C has been defined (§ 130) as

$$(1) \qquad \gamma = \kappa \sin \varphi,$$

where φ, the angle $(\mathbf{N}, \mathbf{n})$, is taken positive in the sense of $\mathbf{T}$. Since $d\mathbf{T}/ds = \kappa\mathbf{N}$,

$$(2) \qquad \frac{d\mathbf{T}}{ds} \times \mathbf{n} = \kappa \mathbf{N} \times \mathbf{n} = \kappa \sin \varphi \, \mathbf{T} = \gamma \mathbf{T};$$

and, as $d\mathbf{T}/ds$ is unaltered by a change in positive direction, γ must change sign with $\mathbf{T}$ (§ 130).

A surface curve for which $\gamma = 0$ is called a *geodesic.* Thus, if a straight line can be drawn on a surface, it is necessarily a geodesic since $\kappa = 0$. If a geodesic is *curved* ($\kappa \neq 0$), $\gamma = 0$ implies $\sin \varphi = 0$ and $\varphi = 0$ or π. We thus have

THEOREM 1. *Along a curved geodesic the principal normal is always normal to the surface.*

Along a curved geodesic,

$$(3) \qquad k = \kappa \cos \varphi = \pm\kappa, \qquad t = \tau + \frac{d\varphi}{ds} = \tau;$$

and, if the geodesic is straight ($\kappa = \tau = 0$) and the same equations apply ($k = t = 0$).

THEOREM 2. *Along a geodesic the normal curvature is numerically the same as the curvature and the geodesic torsion equals the torsion.*

On putting $\gamma = 0$ in (2), we have

$$(4) \qquad \frac{d^2\mathbf{r}}{ds^2} \times \mathbf{n} = 0, \quad \text{or} \quad \nabla\mathbf{r} \cdot \frac{d^2\mathbf{r}}{ds^2} = 0,$$

since the tangential projection of $d^2\mathbf{r}/ds^2$ is zero (132.5). If we replace $\nabla\mathbf{r}$ by $\mathbf{a}\mathbf{r}_u + \mathbf{b}\mathbf{r}_v$, we obtain the scalar differential equations of a geodesic:

$$(5) \qquad \mathbf{r}_u \cdot \frac{d^2\mathbf{r}}{ds^2} = 0, \qquad \mathbf{r}_v \cdot \frac{d^2\mathbf{r}}{ds^2} = 0.$$

These differential equations of the *second order* show that the geodesics on a surface form a two-parameter family. In general, a geodesic may be determined by *two* conditions, by specifying (*a*) that it shall pass through a given point in a given direction, or (*b*), that it shall pass through two given points.

That a geodesic on a surface in general can be found to fulfill conditions of the type (*a*) or (*b*) is plausible in view of the following theorems from mechanics.

THEOREM 3. *A particle constrained to move on a surface and free from the action of any tangential forces will describe a geodesic with constant speed.*

Proof. In view of (52.5), the equation of motion $m\mathbf{a} = \mathbf{F}$ may be written

$$m\left(\frac{dv}{dt}\,\mathbf{T} + \kappa v^2\,\mathbf{N}\right) = p\mathbf{n},$$

where v is the speed and $p\mathbf{n}$ the normal force. Multiplying by $\mathbf{T}\cdot$ gives $dv/dt = 0$, $v = \text{const.}$ If $\kappa = 0$, the particle describes a straight line. If $\kappa \neq 0$, $\mathbf{N} = \pm\mathbf{n}$, and $\sin\varphi = 0$. Since either κ or $\sin\varphi$ must vanish, $\gamma = 0$ in both cases, and the particle describes a geodesic. The pressure $|p| = m\kappa v^2$.

THEOREM 4. *If a weightless flexible cord is stretched over a smooth surface between two of its points, its tension is constant, and the line of contact is a geodesic.*

Proof. Since the cord is in equilibrium, the vector sum of all the forces acting upon any portion of it vanishes. Let F denote

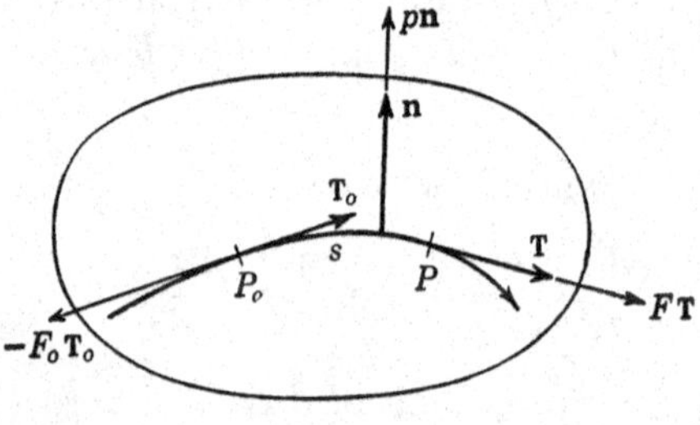

FIG. 135

the magnitude of the tension at P, distant s (arc P_0P) from the fixed end of the cord, and $p\mathbf{n}$ the normal reaction of the surface per unit length. Then from the equilibrium of the length s of the cord (Fig. 135),

$$F\mathbf{T} - F_0\mathbf{T}_0 + \int_0^s p\mathbf{n}\,ds = 0.$$

On differentiating this equation with respect to s, we have

$$\frac{dF}{ds}\mathbf{T} + F\kappa\mathbf{N} + p\mathbf{n} = 0;$$

hence $dF/ds = 0$, and F is constant. If $\kappa = 0$, the line of contact is straight; if $\kappa \neq 0$, $\mathbf{N} = \pm\mathbf{n}$, and $|p| = \kappa F$. In either case, $\gamma = 0$, and the line of contact is a geodesic.

As a cord can be stretched between any two points of a convex surface, the line of contact is a geodesic fulfilling condition (*b*). On concave surfaces we must imagine the cord replaced by a thin strip of spring steel laid flatwise.

136. Geodesic Field. Consider now a one-parameter family of geodesics that form a field of tangent vector $\mathbf{e}_1$ over a portion S of the surface. Then, since

$$\gamma_1 = \mathbf{n} \cdot \operatorname{rot} \mathbf{e}_1 = 0, \tag{1}$$

we have, from Stokes' Theorem,

$$\oint \mathbf{T} \cdot \mathbf{e}_1 \, ds = \int \mathbf{n} \cdot \operatorname{rot} \mathbf{e}_1 \, dS = 0,$$

for any sectionally smooth closed curve lying entirely within S. Writing θ = angle $(\mathbf{e}_1, \mathbf{T})$ gives

$$\oint \cos\theta \, ds = 0 \tag{2}$$

as the integral equivalent of (1). This formula leads to

THEOREM 1. *An arc of a geodesic that is one of the curves of a geodesic field is shorter than any other surface curve joining its end points and lying entirely within the field.*

Proof. Let APB be a geodesic arc of the field and AQB any other curve of surface covered by the field (Fig. 136a). Applying (2) to the circuit $APBQB$ formed by these arcs, we have

$$\int_{APB} ds + \int_{BQA} \cos\theta \, ds = 0,$$

$$\int_{APB} ds = \int_{AQB} \cos\theta \, ds \leqq \int_{AQB} |\cos\theta| \, ds < \int_{AQB} ds;$$

that is, arc $APB <$ arc AQB.

We next consider the orthogonal trajectories of a geodesic field. Their basic property is given by

THEOREM 2 (*Gauss*). *The orthogonal trajectories of a geodesic field intercept equal arcs on the geodesics.*

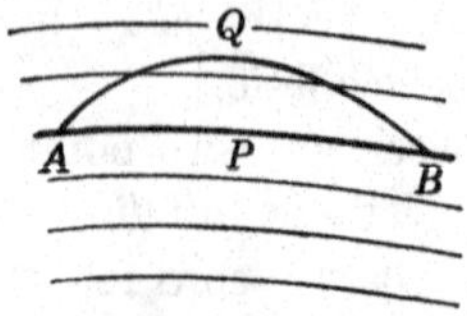

FIG. 136a

FIG. 136b

Proof. Let AB, $A'B'$ be geodesic arcs intercepted between the curves AA', BB' cutting them at right angles (Fig. 136b). Then applying (2) to the circuit $ABB'A'A$, and, noting that

$$\mathbf{T} \cdot \mathbf{e}_1 = \cos\theta = 1, 0, -1, 0$$

over AB, BB', $B'A'$, $A'A$, we have arc AB = arc $A'B'$.

137. Equations of Codazzi and Gauss. These celebrated equations in the theory of surfaces simply state that

$$\mathbf{n} \cdot \nabla \times (\nabla \mathbf{n}) = 0, \tag{1}$$

$$\mathbf{n} \cdot \nabla \times (\nabla \mathbf{e}_1) = 0. \tag{2}$$

If we remember that ∇ means ∇_s, both identities follow from (97.11).

In order to express these identities in terms of the vectors,

$$\mathbf{P} = d\mathbf{n}/ds_1, \qquad \mathbf{Q} = d\mathbf{n}/ds_2, \qquad \mathbf{R} = \gamma_1\mathbf{e}_1 + \gamma_2\mathbf{e}_2, \tag{3}$$

we make use of (97.12):

$$\mathbf{n} \cdot \nabla \times (\mathbf{pq}) = (\mathbf{n} \cdot \operatorname{rot} \mathbf{p})\mathbf{q} + \mathbf{p} \cdot \mathbf{n} \times \nabla \mathbf{q}. \tag{4}$$

On substituting

$$\nabla \mathbf{n} = \mathbf{P}\mathbf{e}_1 + \mathbf{Q}\mathbf{e}_2 \tag{5}$$

in (1), we have, from (4),

$$(\mathbf{n} \cdot \operatorname{rot} \mathbf{P})\mathbf{e}_1 + (\mathbf{n} \cdot \operatorname{rot} \mathbf{Q})\mathbf{e}_2 + \mathbf{P} \cdot \mathbf{n} \times \nabla \mathbf{e}_1 + \mathbf{Q} \cdot \mathbf{n} \times \nabla \mathbf{e}_2 = 0;$$

and, if we put (§ 134)

$$\nabla \mathbf{e}_1 = \mathbf{R}\mathbf{e}_2 - \mathbf{P}\mathbf{n}, \qquad \nabla \mathbf{e}_2 = -\mathbf{R}\mathbf{e}_1 - \mathbf{Q}\mathbf{n}, \tag{6}$$

we have

$$(\mathbf{n} \cdot \operatorname{rot} \mathbf{P} - \mathbf{Q} \cdot \mathbf{n} \times \mathbf{R})\mathbf{e}_1 + (\mathbf{n} \cdot \operatorname{rot} \mathbf{Q} + \mathbf{P} \cdot \mathbf{n} \times \mathbf{R})\mathbf{e}_2 = 0.$$

This equivalent to the two equations:

$$\mathbf{n} \cdot \operatorname{rot} \mathbf{P} = -\mathbf{n} \cdot \mathbf{Q} \times \mathbf{R}, \tag{7}$$

$$\mathbf{n} \cdot \operatorname{rot} \mathbf{Q} = -\mathbf{n} \cdot \mathbf{R} \times \mathbf{P}. \tag{8}$$

These are the *Equations of Codazzi.*

We next put $\nabla \mathbf{e}_1 = \mathbf{R}\mathbf{e}_2 - \mathbf{P}\mathbf{n}$ in (2) and obtain

$$(\mathbf{n} \cdot \operatorname{rot} \mathbf{R})\mathbf{e}_2 - (\mathbf{n} \cdot \operatorname{rot} \mathbf{P})\mathbf{n} + \mathbf{R} \cdot \mathbf{n} \times \nabla \mathbf{e}_2 - \mathbf{P} \cdot \mathbf{n} \times \nabla \mathbf{n} = 0.$$

Replacing $\nabla \mathbf{e}_2$ and $\nabla \mathbf{n}$ by the foregoing values, we have

$$(\mathbf{n} \cdot \operatorname{rot} \mathbf{R} - \mathbf{P} \cdot \mathbf{n} \times \mathbf{Q})\mathbf{e}_2 - (\mathbf{n} \cdot \operatorname{rot} \mathbf{P} + \mathbf{R} \cdot \mathbf{n} \times \mathbf{Q})\mathbf{n} = 0.$$

This is also equivalent to two equations. One of these is the same as (7); the other,

$$\mathbf{n} \cdot \operatorname{rot} \mathbf{R} = -\mathbf{n} \cdot \mathbf{P} \times \mathbf{Q}, \tag{9}$$

is the *Equation of Gauss.* In the right-hand member,

$$\mathbf{n} \cdot \mathbf{P} \times \mathbf{Q} = (\mathbf{e}_1 \times \mathbf{e}_2) \cdot (\mathbf{P} \times \mathbf{Q}) = K,$$

the second scalar invariant of $\nabla \mathbf{n}$ or the *total curvature* (§ 131). Gauss's Equation thus becomes

$$\mathbf{n} \cdot \operatorname{rot} \mathbf{R} = -K. \tag{10}$$

This is perhaps the most important result in the theory of surfaces.

The equations of Codazzi and Gauss can be expressed in terms of the quantities k_i, t_i, γ_i along the field curves.† From (130.7),

$$\frac{d\mathbf{n}}{ds_1} = -k_1\mathbf{e}_1 - t_1\mathbf{n} \times \mathbf{e}_1, \qquad \frac{d\mathbf{n}}{ds_2} = -k_2\mathbf{e}_2 - t_2\mathbf{n} \times \mathbf{e}_2;$$

and, since $t_2 = -t_1$, $\mathbf{n} \times \mathbf{e}_1 = \mathbf{e}_2$, $\mathbf{n} \times \mathbf{e}_2 = -\mathbf{e}_1$,

$$-\mathbf{P} = k_1\mathbf{e}_1 + t_1\mathbf{e}_2, \qquad -\mathbf{Q} = t_1\mathbf{e}_1 + k_2\mathbf{e}_2 \qquad (134.7). \tag{11}$$

In order to compute the left members of (7), (8), (10), we apply the identity,

$$\mathbf{n} \cdot \operatorname{rot} \lambda\mathbf{f} = \lambda\, \mathbf{n} \cdot \operatorname{rot} \mathbf{f} + \mathbf{f} \times \mathbf{n} \cdot \nabla\lambda, \tag{12}$$

† Now k_1 and k_2 denote the normal curvatures of the orthogonal field curves; only when $t_1 = t_2 = 0$ are k_1 and k_2 *principal curvatures* (§ 131).

which follows at once from (85.3), to both terms of $-\mathbf{P}$, $-\mathbf{Q}$ and $\mathbf{R}$. Remembering that $\mathbf{n} \cdot \operatorname{rot} \mathbf{e}_i = \gamma_i$, we thus obtain

$$-\mathbf{n} \cdot \operatorname{rot} \mathbf{P} = \gamma_1 k_1 + \gamma_2 t_1 - \frac{dk_1}{ds_2} + \frac{dt_1}{ds_1},$$

$$-\mathbf{n} \cdot \operatorname{rot} \mathbf{Q} = \gamma_1 t_1 + \gamma_2 k_2 - \frac{dt_1}{ds_2} + \frac{dk_2}{ds_1},$$

$$\mathbf{n} \cdot \operatorname{rot} \mathbf{R} = \gamma_1^2 + \gamma_2^2 - \frac{d\gamma_1}{ds_2} + \frac{d\gamma_2}{ds_1};$$

and, since

$$\mathbf{n} \cdot \mathbf{Q} \times \mathbf{R} = \gamma_1 k_2 - \gamma_2 t_1, \qquad \mathbf{n} \cdot \mathbf{R} \times \mathbf{P} = -\gamma_1 t_1 + \gamma_2 k_1,$$

the Equations of Codazzi and Gauss become

(13) $$\gamma_1(k_1 - k_2) + 2\gamma_2 t_1 + \frac{dt_1}{ds_1} - \frac{dk_1}{ds_2} = 0,$$

(14) $$\gamma_2(k_2 - k_1) + 2\gamma_1 t_1 + \frac{dk_2}{ds_1} - \frac{dt_1}{ds_2} = 0,$$

(15) $$\gamma_1^2 + \gamma_2^2 + \frac{d\gamma_2}{ds_1} - \frac{d\gamma_1}{ds_2} + K = 0.$$

Equation (14) states the same property for the $\mathbf{e}_2$-field that (13) states for the $\mathbf{e}_1$-field. To show this, change the subscripts 1, 2 in (13) into 2, -1 ($\mathbf{n} \times \mathbf{e}_2 = -\mathbf{e}_1$); then, since

$$k_{-1} = k_1, \qquad t_2 = -t_1, \qquad \gamma_{-1} = -\gamma_1, \qquad ds_{-1} = -ds_1,$$

we obtain (14).

138. Lines of Curvature. The curves on the surface which are everywhere tangent to the principal directions (§ 131) are called *lines of curvature.* Along a line of curvature the geodesic torsion is zero ($t = 0$), and the torsion $\tau = -d\varphi/ds$.

From (130.7), we have $d\mathbf{n}/ds = -k\mathbf{T}$ along lines of curvature. Their differential equation is therefore

(1) $$\frac{d\mathbf{r}}{ds} \times \frac{d\mathbf{n}}{ds} = 0, \quad \text{or} \quad \mathbf{n} \cdot \frac{d\mathbf{r}}{ds} \times \frac{d\mathbf{n}}{ds} = 0;$$

for $(d\mathbf{r}/ds) \times (d\mathbf{n}/ds)$ is always parallel to $\mathbf{n}$. In particular, the parametric curves are lines of curvature, if

$$\mathbf{r}_u \times \mathbf{n}_u = 0, \qquad \mathbf{r}_v \times \mathbf{n}_v = 0;$$

then, if k_1, k_2 are the *principal curvatures*,

$$\mathbf{n}_u = -k_1\mathbf{r}_u, \qquad \mathbf{n}_v = -k_2\mathbf{r}_v. \tag{2}$$

On the plane and sphere, all curves are lines of curvature (§ 131). On other surfaces there are in general two orthogonal principal directions at each point; and the lines of curvature form two fields cutting each other at right angles. If $\mathbf{e}_1$ and $\mathbf{e}_2 = \mathbf{n} \times \mathbf{e}_1$ are the corresponding unit field vectors, the Equations of Codazzi become

$$\frac{dk_1}{ds_2} = \gamma_1(k_1 - k_2), \qquad \frac{dk_2}{ds_1} = \gamma_2(k_1 - k_2). \tag{3}$$

These lead to a simple proof of

THEOREM 1. *If $K = 0$, $J \neq 0$, the lines of curvature along which the normal curvature vanishes are straight lines.*

Proof. Since $K = k_1k_2 = 0$, $J = k_1 + k_2 \neq 0$, we may suppose that $k_1 = 0$, $k_2 \neq 0$. Hence $\gamma_1 = 0$, from (3); and, since

$$k_1 = \kappa_1 \cos\varphi = 0, \qquad \gamma_1 = \kappa_1 \sin\varphi = 0,$$

we conclude that $\kappa_1 = 0$.

Along the *rulings* of the surface (which are asymptotic lines as well as lines of curvature) $t_1 = 0$, $k_1 = 0$, and $d\mathbf{n}/ds_1 = 0$ (130.7); hence $\mathbf{n}$ has a fixed direction along a ruling. This also follows from (134.1); for $\boldsymbol{\omega}_1 = 0$, and the trihedral e_1e_2n has a fixed orientation along a ruling. In general a surface has a different tangent plane at each point and is therefore the envelope of a two-parameter family of planes. However the ruled surface under consideration has the same tangent plane along an entire ruling and is therefore *the envelope of a one-parameter family of planes.* Such a surface is called *developable.*

Consider now the lines of curvature cutting the rulings orthogonally. Their tangent vector $\mathbf{e}_2 = \mathbf{n} \times \mathbf{e}_1$ is constant along a ruling. From (137.14) and (137.15),

$$\frac{dk_2}{ds_1} = -\gamma_2 k_2, \qquad \frac{d\gamma_2}{ds_1} = -\gamma_2^2; \qquad \frac{d}{ds_1}\left(\frac{\gamma_2}{k_2}\right) = 0.$$

But $\gamma_2/k_2 = \tan\varphi_2$; hence $\varphi_2 = (\mathbf{N}_2, \mathbf{n})$ is constant along a ruling, and $\mathbf{N}_2$ as well. This property lends plausibility to the fact that the developable surface may be bent into a plane (after certain cuts are made) without stretching or tearing.

We now may amplify theorem 1: *If $K = 0$, $J \neq 0$, the surface is developable.* We shall see in § 142 that developable surfaces are of three types; *cylinders* (including planes), *cones*, and *tangent surfaces* (generated by the tangents to a curve which is not a straight line).

THEOREM 2. *The normals to a surface along one of its curves have an envelope when and only when the curve is a line of curvature; in this case the envelope is the locus of the centers of normal curvature along the curve.*

Proof. Let s denote the arc measured along the curve C of the surface S, and $\mathbf{r}(s)$ and $\mathbf{n}(s)$ the position vector and unit surface normal at a point of C. The points on the family of normals have position vectors $\mathbf{r}(s) + \lambda\,\mathbf{n}(s)$, where λ is a variable scalar. If we take λ as a function of s, say $\lambda = \lambda(s)$, the curve $\bar{C}$,

$$\bar{\mathbf{r}} = \mathbf{r}(s) + \lambda(s)\mathbf{n}(s)$$

will be the envelope of the normals if

$$\frac{d\bar{\mathbf{r}}}{ds} = \mathbf{T} + \lambda\frac{d\mathbf{n}}{ds} + \frac{d\lambda}{ds}\mathbf{n} \quad \text{or} \quad \mathbf{T} + \lambda\frac{d\mathbf{n}}{ds}$$

is parallel to $\mathbf{n}$. But since both $\mathbf{T}$ and $d\mathbf{n}/ds$ are perpendicular to $\mathbf{n}$, the envelope will exist only when

$$\mathbf{T} + \lambda\frac{d\mathbf{n}}{ds} = (1 - \lambda k)\mathbf{T} - \lambda t\,\mathbf{n}\times\mathbf{T} = 0 \qquad (130.7),$$

that is, when $t = 0$ and $\lambda = 1/k$. Therefore the normals have an envelope only along lines of curvature; and then the envelope is the curve $\bar{\mathbf{r}} = \mathbf{r} + \mathbf{n}/k$, the locus of the centers of normal curvature along C.

We turn now to some theorems relative to the lines of intersection of surfaces.

THEOREM 3. *If two surfaces S_1, S_2 cut under a constant angle θ, the curve of intersection has the same geodesic torsion whether regarded as a curve of S_1 or of S_2.*

Proof. If $\mathbf{N}$ is the unit principal normal along the curve of intersection,

$$\varphi_2 - \varphi_1 = (\mathbf{N}, \mathbf{n}_2) - (\mathbf{N}, \mathbf{n}_1) = (\mathbf{n}_1, \mathbf{N}) + (\mathbf{N}, \mathbf{n}_2) = (\mathbf{n}_1, \mathbf{n}_2) = \theta,$$

and

$$t_2 - t_1 = \tau + \frac{d\varphi_2}{ds} - \tau - \frac{d\varphi_1}{ds} = \frac{d\theta}{ds} = 0.$$

From this result, we have at once

THEOREM 4 (*Joachimsthal*). *If two surfaces cut under a constant angle, their curve of intersection is a line of curvature of both or of neither; and, conversely, if the line of intersection is a line of curvature of both, the surfaces cut under a constant angle.*

We are now in position to prove the celebrated

THEOREM 5 (*Dupin*). *Two surfaces belonging to different families of a triple orthogonal system cut one another in lines of curvature of each.*

Proof. Denote the geodesic torsion at a point P on the curve of intersection of the surfaces S_i and S_j by t_{ij} or t_{ji}, according as the curve is regarded as belonging to S_i or S_j. Now at the point P we have

$$t_{ij} = t_{ji} \quad \text{(theorem 3)}, \qquad t_{ij} = -t_{ik} \quad \text{(§ 131, theorem 1)},$$

where i, j, k represents any permutation of the indices 1, 2, 3. Then

$$t_{ij} = -t_{ik} = -t_{ki} = t_{kj} = t_{jk} = -t_{ji} = -t_{ij},$$

so that $t_{ij} = 0$.

Example. A *surface of revolution* about the z-axis has the parametric equations,

$$x = u \cos v, \qquad y = u \sin v, \qquad z = z(u),$$

where u, v are the plane polar coordinates (ρ, φ). These give the vector equation,

$$\mathbf{r} = u\,\mathbf{R}(v) + \mathbf{k}\,z(u),$$

where $\mathbf{R}$ is a unit radial vector ($\mathbf{R} \cdot \mathbf{k} = 0$). Now

$$\mathbf{r}_u = \mathbf{R} + \mathbf{k}z', \qquad \mathbf{r}_v = u\mathbf{P};$$

$$\mathbf{n} = \frac{\mathbf{r}_u \times \mathbf{r}_v}{|\mathbf{r}_u \times \mathbf{r}_v|} = \frac{\mathbf{k} - z'\mathbf{R}}{(1 + z'^2)^{\frac{1}{2}}};$$

$$\mathbf{n}_u = -\frac{z''}{(1 + z'^2)^{\frac{3}{2}}}\,\mathbf{r}_u, \qquad \mathbf{n}_v = -\frac{z'}{u(1 + z'^2)^{\frac{1}{2}}}\,\mathbf{r}_v.$$

The last equations show that the parametric lines (the parallels and meridians of our surface) are the lines of curvature, and that the principal curvatures are

$$k_1 = \frac{z''}{(1 + z'^2)^{\frac{3}{2}}}, \qquad k_2 = \frac{z'}{u(1 + z'^2)^{\frac{1}{2}}}.$$

Consequently,

$$(4) \qquad K = k_1 k_2 = \frac{z' z''}{u(1 + z'^2)^2},$$

$$(5) \qquad J = k_1 + k_2 = \frac{uz'' + z'(1 + z'^2)}{u(1 + z'^2)^{\frac{3}{2}}}.$$

Let us apply these results to find K and J for the *torus* generated by revolving the circle,

$$(u - a)^2 + z^2 = b^2,$$

about the z-axis. On differentiation, we find

$$z' = \frac{a - u}{z}, \qquad 1 + z'^2 = \frac{b^2}{z^2}, \qquad z'' = -\frac{b^2}{z^3};$$

hence, from (4) and (5),

$$K = \frac{u - a}{b^2 u}, \qquad J = \frac{a - 2u}{bu}.$$

If we introduce the latitude θ on the generating circle, $u = a + b \cos \theta$, and

$$(6) \qquad K = \frac{\cos \theta}{b(a + b \cos \theta)}, \qquad -J = \frac{1}{b} + \frac{\cos \theta}{a + b \cos \theta}.$$

139. Total Curvature. The Gauss Equation (137.15),

$$(1) \qquad \gamma_1^2 + \gamma_2^2 + \frac{d\gamma_1}{ds_1} - \frac{d\gamma_1}{ds_2} + K = 0,$$

shows that, if orthogonal geodesics can be chosen as field curves, $\gamma_1 = \gamma_2 = 0$, and $K = 0$; hence

THEOREM 1. *Orthogonal geodesic fields can exist only on a surface of zero total curvature.*

We now compute K from the Gauss Equation (137.10); using (97.9), we have

$$(2) \qquad -K = \mathbf{n} \cdot \operatorname{rot} \mathbf{R} = \frac{1}{H} \{(\mathbf{r}_v \cdot \mathbf{R})_u - (\mathbf{r}_u \cdot \mathbf{R})_v\}.$$

Let the parametric curves cut at an angle $\theta = (\mathbf{e}_1, \mathbf{e}_3)$ where $\mathbf{e}_1 = \mathbf{r}_u/\sqrt{E}$, $\mathbf{e}_3 = \mathbf{r}_v/\sqrt{G}$. Then, if $\mathbf{e}_2 = \mathbf{n} \times \mathbf{e}_1$, $\mathbf{e}_4 = \mathbf{n} \times \mathbf{e}_3$, we have, from (134.13),

$$\mathbf{R} = \gamma_1 \mathbf{e}_1 + \gamma_2 \mathbf{e}_2 = \gamma_3 \mathbf{e}_3 + \gamma_4 \mathbf{e}_4 - \nabla\theta;$$

and

$$\mathbf{r}_u \cdot \mathbf{R} = \gamma_1 \sqrt{E}, \qquad \mathbf{r}_v \cdot \mathbf{R} = \gamma_3 \sqrt{G} - \frac{\partial\theta}{\partial v} \qquad (95.7).$$

Substitution in (2) now gives the elegant *Formula of Liouville* for the total curvature,

$$K = \frac{1}{H}\left\{\frac{\partial}{\partial v}(\gamma_1\sqrt{E}\,) - \frac{\partial}{\partial u}(\gamma_3\sqrt{G}\,) + \frac{\partial^2\theta}{\partial u\,\partial v}\right\}, \tag{3}$$

in which $H = \sqrt{EG - F^2}$, γ_1 and γ_3 are given by (134.15), and

$$\cos\theta = \mathbf{e}_1 \cdot \mathbf{e}_3 = \frac{\mathbf{r}_u \cdot \mathbf{r}_v}{\sqrt{EG}} = \frac{F}{\sqrt{EG}}. \tag{4}$$

When the parametric curves are orthogonal, $\theta = \pi/2$ and $F = 0$, $H = \sqrt{EG}$. Equations (134.15) now give

$$\gamma_1 = -\frac{1}{2H}\frac{\partial E/\partial v}{\sqrt{E}}, \qquad \gamma_3 = \frac{1}{2H}\frac{\partial G/\partial u}{\sqrt{G}}, \tag{5}$$

and Liouville's Formula becomes

$$K = -\frac{1}{2H}\left\{\frac{\partial}{\partial u}\left(\frac{1}{H}\frac{\partial G}{\partial u}\right) + \frac{\partial}{\partial v}\left(\frac{1}{H}\frac{\partial E}{\partial v}\right)\right\}. \tag{6}$$

From (3), we obtain an immediate proof of

THEOREM 2 (*Gauss*). *The total curvature K of a surface depends only upon the coefficients E, F, G of the first fundamental form and their first and second partial derivatives with respect to u and v.*

This fundamental theorem was first proved by Gauss after long and tedious calculations. Gauss's name of "Theorema egregium" (Latin *egregius* means literally "out of the herd") for this result shows that he was fully aware of its importance.

Example. To find a surface of revolution of constant negative total curvature, we set $K = -1/a^2$ in (138.4). The resulting differential equation,

$$\frac{2z'z''}{(1 + z'^2)^2} = -\frac{2u}{a^2},$$

has the first integral,

$$\frac{1}{1+z'^2} = \frac{u^2}{a^2} + A \quad \text{or} \quad \cos^2 \psi = \frac{u^2}{a^2} + A,$$

where $\psi = \tan^{-1} z'$ is the inclination of the tangent to the u-axis (Fig. 139). If we impose the condition $u = -a$ when $\psi = 0$, $A = 0$ and $u = -a \cos \psi$. Now

$$\frac{dz}{d\psi} = \frac{dz}{du}\frac{du}{d\psi} = \tan \psi \cdot a \sin \psi = a(\sec \psi - \cos \psi),$$

$$z = a \log (\sec \psi + \tan \psi) - a \sin \psi + B;$$

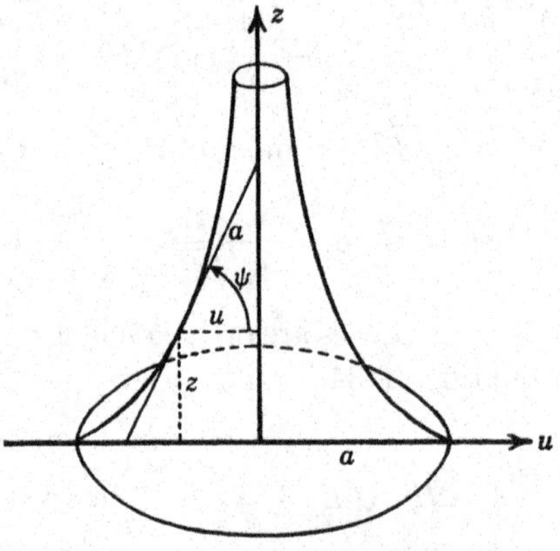

FIG. 139

and $B = 0$ if $z = 0$ when $\psi = 0$. Therefore the meridian of our surface of constant negative K has the parametric equations:

$$u = -a \cos \psi, \qquad z = a \log (\sec \psi + \tan \psi) - a \sin \psi.$$

This curve, asymptotic to the z-axis as $\psi \to \pi/2$, is called a *tractrix*. The equation $u = -a \cos \psi$ shows that the segment of any tangent between the curve and z-axis has the constant length a. *This property is characteristic of the tractrix* (cf. § 50, ex. 5).

140. Bonnet's Integral Formula. The differential formula for the total curvature,

$$(1) \qquad K = -\mathbf{n} \cdot \text{rot}\, \mathbf{R},$$

has an important integral equivalent. If we integrate K over a portion of a surface S bounded by a simple closed curve which consists of a finite number of smooth arcs, we have, by Stokes'

Theorem,

$$\int K\,dS = -\int \mathbf{n}\cdot \operatorname{rot}\mathbf{R}\,dS$$

$$= -\oint \mathbf{T}\cdot\mathbf{R}\,ds = -\oint\left(\gamma - \frac{d\theta}{ds}\right)ds,$$

from (134.12); hence

$$\text{(2)} \qquad \int K\,dS = \oint d\theta - \oint \gamma\,ds.$$

If the bounding curve has a continuously turning tangent, $\oint d\theta = 2\pi$ and (2) becomes

$$\text{(3)} \qquad \int K\,dS = 2\pi - \oint \gamma\,ds.$$

This very important formula was discovered by the French geometer, Bonnet, in 1848. The integral $\int K\,dS$ over S is called the *integral curvature* of S.

Let us first apply (2) to the figure bounded by two geodesic arcs APB and AQB meeting at A and B. Then denoting the interior angles at the corners by A and B (Fig. 140a), we have

$$\text{(4)} \quad \int K\,dS = \oint d\theta = 2\pi - (\pi - A) - (\pi - B) = A + B.$$

Since $A + B > 0$ this equation is impossible when $K = 0$; hence, on a surface of negative or zero total curvature, two geodesic arcs

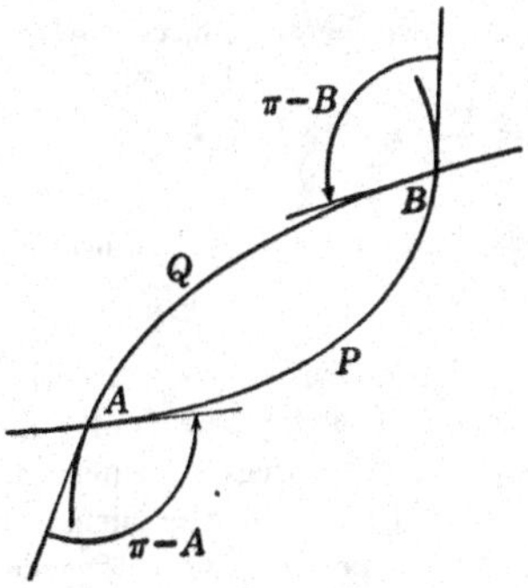

FIG. 140a

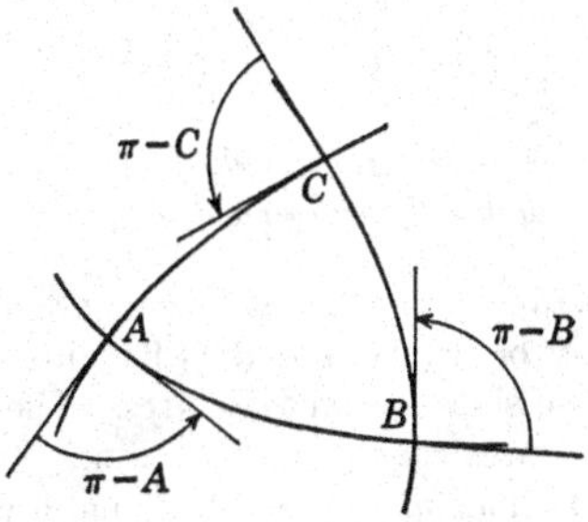

FIG. 140b

cannot meet in two points so as to enclose a simply connected area. We may state (4) as follows:

THEOREM 1. *The integral curvature of a geodesic lune is equal to the sum of its interior angles.*

We next apply (2) to a *geodesic triangle ABC* (Fig. 140b): that is, the figure enclosed by three geodesic arcs. Again denoting the interior angles at the corners by A, B, C, we have

$$\int K\,dS = \oint d\theta = 2\pi - (\pi - A) - (\pi - B) - (\pi - C),$$

$$\int K\,dS = A + B + C - \pi. \tag{5}$$

THEOREM 2 (*Gauss*). *The integral curvature of a geodesic triangle is equal to its "angular excess," that is, the excess of the sum of its angles over π.*

When K is constant, the integral curvature $\int K\,dS$ is the product of K by the area S. If K is identically zero, as on a plane, the sum of the angles of a geodesic triangle is π. If $K \neq 0$, we have

THEOREM 3 (*Gauss*). *On a surface of constant non-zero total curvature, the area of a geodesic triangle is equal to the quotient of its angular excess by the total curvature.*

Example 1. A sphere of radius r has the constant total curvature $K = 1/r^2$. For all normal sections at a point are great circles of curvature $1/r$; all points are umbilical points and $k_1k_2 = 1/r^2$. Formula (5) now states that: *The area of a spherical triangle is r^2 times its angular excess in radians.*

If ϵ denotes the angular excess in right angles, we therefore have the formula,

$$\text{Area} = r^2 \frac{\pi}{2} \epsilon = \frac{4\pi r^2}{8} \epsilon,$$

that is, *the area of a spherical triangle equals the area of a spherical octant times the angular excess in right angles.*

Example 2. Let us decompose a closed bilateral surface S, consisting entirely of regular points (§ 93), into a number f of curvilinear "polygons" S_i whose sides are analytic arcs. These sides meet in vertices at which at least three arcs come together. Since S is regular and bilateral, the surface has a continuous unit external normal **n** which defines a positive sense of circuit on each polygon by the rule of the right-handed screw. *When positive circuits* *are made about two adjoining polygons, their common side is traversed in opposite directions.* Now for each S_i we have, by Bonnet's Theorem (2),

$$\int_{S_i} K\, dS = 2\pi - \sum^{j} (\pi - \alpha_{ij}) - \oint_{S_i} \gamma\, ds,$$

where α_{ij} are the interior angles at the vertices of S_i. If these equations for all of the f polygons S_i are added, we have

$$\oint_S K\, dS = 2\pi f - \sum^{i} \sum^{j} (\pi - \alpha_{ij});$$

for the integrals $\int \gamma\, ds$ over adjoining sides cancel in pairs since γ changes sign with change of direction. In each polygon the number of interior angles equals the number of sides; hence $\Sigma\Sigma\pi = 2e\pi$, where e is the total number of sides or *edges*. Moreover the sum of the interior angles about any vertex is 2π; hence $\Sigma\Sigma\alpha_{ij} = 2v\pi$, where v is the total number of vertices. We thus obtain

$$f - e + v = \frac{1}{2\pi} \oint_S K\, dS. \tag{6}$$

From the right member of (6) we see that the number on the left is independent of the manner in which S is subdivided into polygons. From the left member we conclude that $\oint K\, dS$ is not altered when S is deformed into another completely regular surface.

If S can be continuously deformed into a sphere, we have

$$\oint K\, dS = \frac{1}{r^2} 4\pi r^2 = 4\pi \tag{7}$$

for a sphere, and

$$f - e + v = 2. \tag{8}$$

Evidently $f - e + v$ is not altered by any continuous deformation of S, even though the resulting surface is not completely regular. Thus, if S is transformed into a *polyhedron* (a closed solid with plane polygons for faces), the relation (8) still holds good and constitutes the famous *Polyhedron Formula* that Euler discovered in 1752: *Faces + vertices − edges* = 2.

Next suppose that S can be continuously deformed into a *torus*. From (138.6), for a torus,

$$\oint K\, dS = \iint K\, b(a + b \cos \theta)\, d\theta\, dv = \int_0^{2\pi} \int_0^{2\pi} \cos \theta\, d\theta\, dv = 0; \tag{9}$$

hence $f - e + v = 0$ for any surface continuously deformable into a torus. This relation holds, for example, for any ring surface with polyhedral faces.

Example 3. *Parallel Displacement.* A vector **f**, that always remains tangent to a surface S, is said to undergo a *parallel displacement* along a surface curve C when the component of $d\mathbf{f}/ds$ tangential to the surface is zero; then

$$\mathbf{n} \times \frac{d\mathbf{f}}{ds} = 0 \quad \text{or} \quad \frac{d\mathbf{f}}{ds} = \lambda\mathbf{n}. \tag{10}$$

Since

$$\frac{d}{ds}(\mathbf{f} \cdot \mathbf{f}) = 2\mathbf{f} \cdot \frac{d\mathbf{f}}{ds} = 2\lambda\mathbf{f} \cdot \mathbf{n} = 0,$$

the length of **f** *must remain constant during a parallel displacement.*

Now the trihedral **Tnp** associated with the curve C has the angular velocity,

$$\boldsymbol{\omega} = t\mathbf{T} + \gamma\mathbf{n} + k\mathbf{p} \qquad (130.3').$$

If the angle $\theta = (\mathbf{T}, \mathbf{f})$ is reckoned positive in the sense of **n**, the trihedral **fng** ($\mathbf{g} = \mathbf{f} \times \mathbf{n}$) has the angular velocity $\mathbf{n}\, d\theta/ds$ relative to **Tnp**. Hence, as **f** moves along C with unit speed, the trihedral **fng** revolves with the angular velocity,

$$\boldsymbol{\Omega} = \boldsymbol{\omega} + \mathbf{n}\frac{d\theta}{ds} = t\mathbf{T} + (\gamma + d\theta/ds)\mathbf{n} + k\mathbf{p};$$

hence $d\mathbf{f}/ds = \boldsymbol{\Omega} \times \mathbf{f}$, and

$$\frac{d\mathbf{f}}{ds} \times \mathbf{n} = (\boldsymbol{\Omega} \times \mathbf{f}) \times \mathbf{n} = (\mathbf{n} \cdot \boldsymbol{\Omega})\mathbf{f} = \left(\gamma + \frac{d\theta}{ds}\right)\mathbf{f}.$$

Hence **f** will undergo a parallel displacement along the curve C when and only when

$$\gamma + \frac{d\theta}{ds} = 0. \tag{11}$$

If **f** is tangent to C (or, more generally, when θ remains constant), **f** will undergo a parallel displacement along C only when C is a geodesic.

THEOREM. *If a tangential surface vector is given a parallel displacement about a smooth closed curve C, it will revolve through an angle,*

$$2\pi - \oint \gamma\, ds = \int K\, dS. \tag{12}$$

Proof. From (11) we see that **f** revolves through the angle,

$$\oint \frac{d\theta}{ds}\, ds = -\oint \gamma\, ds$$

relative to **T**, the unit tangent vector of C. Since **T** itself revolves through 2π in passing around C, the total rotation of **f** is $2\pi - \oint \gamma\, ds$, or $\int K\, dS$ by Bonnet's Integral Formula (3).

141. Normal Systems. At every point of the surface S, $\mathbf{r} = \mathbf{r}(u, v)$, a straight line is defined by the unit vector $\mathbf{m}(u, v)$. The points of this two-parameter family of lines are given by

$$\mathbf{r}_1 = \mathbf{r} + \lambda\mathbf{m}. \tag{1}$$

Under what circumstances do these lines admit an orthogonal surface? That is, when do they form the system of normals to a surface?

When λ is a definite continuous function $\lambda(u, v)$, $\mathbf{r}_1$ is the position vector of a surface S_1. If ∇ denotes a surface gradient relative to S, we have, from (1),

$$\nabla\mathbf{r}_1 = \nabla\mathbf{r} + (\nabla\lambda)\mathbf{m} + \lambda\nabla\mathbf{m},$$

$$(\nabla\mathbf{r}_1)\cdot\mathbf{m} = (\nabla\mathbf{r})\cdot\mathbf{m} + \nabla\lambda,$$

since $\mathbf{m}\cdot\mathbf{m} = 1$. In $\nabla\mathbf{r}_1 = \mathbf{a}\mathbf{r}_{1u} + \mathbf{b}\mathbf{r}_{1v}$, the postfactors are tangent to S_1; hence, in order that $\mathbf{m}$ be normal to S_1, it is necessary and sufficient that $(\nabla\mathbf{r}_1)\cdot\mathbf{m} = 0$, that is,

$$-\nabla\lambda = (\nabla\mathbf{r})\cdot\mathbf{m} = \mathbf{m} - (\mathbf{m}\cdot\mathbf{n})\mathbf{n} \qquad (132.5). \tag{2}$$

But (2) implies that

$$\mathbf{n}\cdot\operatorname{rot}\mathbf{m} = 0; \tag{3}$$

conversely, from § 103, theorem 2, (3) implies that

$$\mathbf{m}_t = (\nabla\mathbf{r})\cdot\mathbf{m} = -\nabla\lambda \quad \text{where} \quad \lambda = -\int_{\mathbf{r}_0}^{\mathbf{r}} \mathbf{m}\cdot d\mathbf{r}, \tag{4}$$

the integral being taken over any path from $\mathbf{r}_0$ to $\mathbf{r}(u, v)$ on S. We have thus proved

THEOREM 1. *In order that a two-parameter family of lines* $\mathbf{r}(u, v) + \lambda\mathbf{m}(u, v)$ *form a normal system, it is necessary and sufficient that* $\mathbf{n}\cdot\operatorname{rot}\mathbf{m} = 0$, *where* $\mathbf{n}$ *is the unit normal to the surface* $\mathbf{r} = \mathbf{r}(u, v)$.

When m fulfils condition (3), we can find a one-parameter family of surfaces normal to the lines $\mathbf{r} + \lambda\mathbf{m}$. We need only determine $\lambda(u, v)$ from (4) with some definite choice of $\mathbf{r}_0$. Since $\mathbf{r}_0$ may be chosen at pleasure, all possible values of λ are then given by

$\lambda + C$, where C is an arbitrary constant. We thus obtain a one-parameter family of surfaces,

$$\mathbf{r}_1 = \mathbf{r} + (\lambda + C)\mathbf{m},$$

normal to the lines.

Consider now a geodesic field over S with the unit tangent vector $\mathbf{e}$. Since $\gamma = \mathbf{n} \cdot \operatorname{rot} \mathbf{e} = 0$ at every point of S, we have

THEOREM 2 (*Bertrand*). *The tangents to the geodesics of a field form a normal system of lines.*

We conclude with an important theorem in geometrical optics.

THEOREM 3 (*Malus and Dupin*). *If a normal system of lines is reflected or refracted at any surface, it still remains a normal system.*

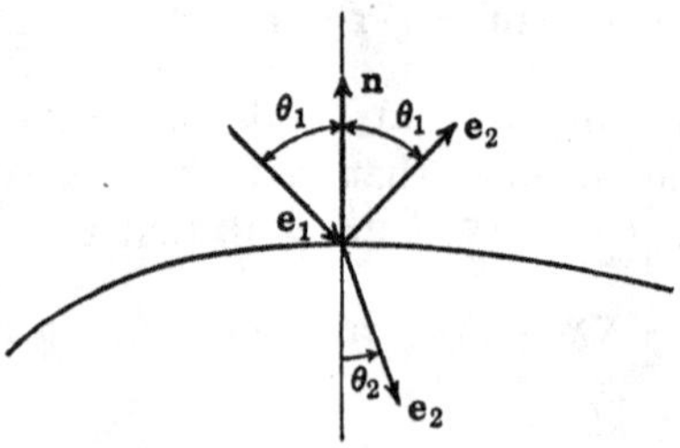

FIG. 141

Proof. Let $\mathbf{e}_1$ denote the unit vectors along the incident rays and $\mathbf{e}_2$ the unit vectors along the refracted (or reflected) ray (Fig. 141). If μ_i denotes an absolute refractive index, $\mu_1 \sin \theta_1 = \mu_2 \sin \theta_2$ (Snell's Law) and $\mathbf{e}_1$, $\mathbf{e}_2$, $\mathbf{n}$ are coplanar; hence,

$$(\mu_2\mathbf{e}_2 - \mu_1\mathbf{e}_1) \times \mathbf{n} = 0 \quad \text{and} \quad \mathbf{n} \cdot \operatorname{rot} (\mu_2\mathbf{e}_2 - \mu_1\mathbf{e}_1) = 0,$$

by the theorem following (97.10). For the reflected ray (§ 75),

$$\mathbf{e}_2 = (\mathbf{I} - 2\mathbf{nn}) \cdot \mathbf{e}_1 \quad \text{and} \quad \mathbf{n} \cdot \operatorname{rot} (\mathbf{e}_2 - \mathbf{e}_1) = 0.$$

In either case: $\mathbf{n} \cdot \operatorname{rot} \mathbf{e}_1 = 0$ implies $\mathbf{n} \cdot \operatorname{rot} \mathbf{e}_2 = 0$.

142. Developable Surfaces. A *developable surface* (§ 138) is the envelope of one-parameter family of planes. The limiting line of intersection of two neighboring tangent planes is a line, or *ruling*, on the surface. A developable is therefore a *ruled surface*. The equation of a ruled surface may be written

$$\text{(1)} \qquad \mathbf{r} = \mathbf{p}(u) + v\mathbf{e}(u),$$

where $\mathbf{p} = \mathbf{p}(u)$ is a curve C crossing the rulings and $\mathbf{e}(u)$ is a unit vector along the rulings. The surface normal is parallel to

$$\mathbf{r}_u \times \mathbf{r}_v = (\mathbf{p}_u + v\mathbf{e}_u) \times \mathbf{e}.$$

When the normal maintains the same direction along a ruling, the surface is *developable*. For the same plane is tangent to the surface along an entire ruling; and, as each value of u corresponds to a ruling and hence to a tangent plane, the surface is enveloped by a one-parameter family of planes. Consequently the surface (1) is developable when and only when $\mathbf{p}_u \times \mathbf{e}$ and $\mathbf{e}_u \times \mathbf{e}$ are parallel; that is, when $\mathbf{p}_u$, $\mathbf{e}$ and $\mathbf{e}_u$ are coplanar:

$$\mathbf{p}_u \cdot \mathbf{e} \times \mathbf{e}_u = 0. \tag{2}$$

Case 1: $\mathbf{e} \times \mathbf{e}_u = 0$. Since $\mathbf{e} \cdot \mathbf{e} = 1$, we have $\mathbf{e} \cdot \mathbf{e}_u = 0$ and hence $\mathbf{e}_u = 0$. Therefore $\mathbf{e}$ is constant, and (1) represents a general *cylinder*.

Case 2: $\mathbf{e} \times \mathbf{e}_u \neq 0$. We then may write (§ 5)

$$\mathbf{p}_u = \alpha(u)\mathbf{e} + \beta(u)\mathbf{e}_u. \tag{3}$$

With a function $\lambda(u)$, as yet undetermined, let us write (1) as

$$\mathbf{r} = (\mathbf{p} + \lambda\mathbf{e}) + (v - \lambda)\mathbf{e} = \mathbf{q} + (v - \lambda)\mathbf{e}, \tag{4}$$

where

$$\mathbf{q}(u) = \mathbf{p}(u) + \lambda(u)\mathbf{e}(u). \tag{5}$$

Then

$$\mathbf{q}_u = \mathbf{p}_u + \lambda_u\mathbf{e} + \lambda\mathbf{e}_u = (\alpha + \lambda_u)\mathbf{e} + (\beta + \lambda)\mathbf{e}_u;$$

and, if we choose $\lambda = -\beta$,

$$\mathbf{q}_u = (\alpha - \beta_u)\mathbf{e}. \tag{6}$$

If $\alpha = \beta_u$ along C, $\mathbf{q}_u = 0$ and $\mathbf{q}$ is constant. Then (4) represents a *cone* of vertex $\mathbf{q}$ if the vectors $\mathbf{e}(u)$ are not coplanar, a *plane* if they are coplanar.

In general, however, $\alpha \neq \beta_u$; then (6) shows that $\mathbf{q}(u)$ traces a curve having $\mathbf{e}$ as tangent vector. The surface (4) then is generated by the tangents of the curve $\mathbf{q} = \mathbf{q}(u)$; it is the *tangent surface* of this curve.

Developable surfaces are planes, cylinders, cones or *tangent surfaces.*

143. Minimal Surfaces. In § 131 the mean curvature of a surface was defined as $J = -\operatorname{Div} \mathbf{n} = k_1 + k_2$.

With $\mathbf{f} = 1$, the integral theorem (101.2) becomes

$$\int J\mathbf{n}\, dS = \oint \mathbf{m}\, ds. \tag{1}$$

Next put $\mathbf{f} = \mathbf{r}$ in (101.4); then, since $\nabla\mathbf{r} = \mathbf{I} - \mathbf{nn}$, Rot $\mathbf{r} = 0$, and we have

$$\int \mathbf{r} \times J\mathbf{n}\, dS = \oint \mathbf{r} \times \mathbf{m}\, ds. \tag{2}$$

These equations admit of a simple mechanical interpretation:

THEOREM 1. *The normal pressures* $J\mathbf{n}$ *(per unit of area) over any simply connected portion of surface* S *bounded by a closed curve* C *are statically equivalent to a system of unit forces (per unit of length) along the external normals to* C *and tangent to* S.

Consider, now, a soap film with a constant surface tension q per unit of length and subjected to an unbalanced normal force $p\mathbf{n}$ per unit of area. Since any portion of the film bounded by a closed curve C is in equilibrium, we must have

$$\int p\mathbf{n}\, dS + 2q \oint \mathbf{m}\, ds = \int (p + 2Jq)\mathbf{n}\, dS = 0;$$

the surface tension q is doubled, because it is exerted on *both* sides of the film. We therefore conclude that

$$p = -2Jq. \tag{3}$$

Thus, for a soap bubble of radius r, the normal curvature is everywhere

$$k = \kappa\mathbf{N} \cdot \mathbf{n} = \kappa \cos \pi = -1/r \quad \text{and} \quad J = -2/r;$$

thus the pressure inside of a soap bubble exceeds the pressure outside by an amount $p = 4q/r$.

In particular, if the film is exposed to the same pressure on both sides, $p = 0$; then $J = 0$ at all points of the film. A surface whose mean curvature J is everywhere zero is called a *minimal* surface. Thus a soap film spanned over a wire loop of any shape and with atmospheric pressure on both sides materializes a minimal surface.

Let S_0 be a minimal surface bounded by a closed curve C. Its existence is guaranteed by its physical counterpart, the soap film spanned over C. Imagine now that S_0 is embedded in a *field of minimal surfaces,* that is, a one-parameter family of surfaces in a certain region R enclosing S_0, such that through every point of R there passes one and only one surface of the family. Such a field may be generated, for example, by the parallel translation of S_0. At each point of R, the unit normal $\mathbf{n}_0$ to the minimal surfaces may be so chosen that $\mathbf{n}_0$ is a continuous vector-point function; and, from (97.2),

$$\operatorname{div} \mathbf{n}_0 = \operatorname{Div} \mathbf{n}_0 + \mathbf{n}_0 \cdot \frac{d\mathbf{n}_0}{dn} = -J + 0 = 0.$$

Now let S be any other surface in R spanning the curve C and enclosing with S_0 a volume V of unit external normal $\mathbf{n}$. From the divergence theorem,

$$\oint_{S_0+S} \mathbf{n} \cdot \mathbf{n}_0 \, dS = \int_V \operatorname{div} \mathbf{n}_0 \, dV = 0.$$

If $\mathbf{n} = -\mathbf{n}_0$ over S_0 and $\mathbf{n} \cdot \mathbf{n}_0 = \cos \psi$ over S,

$$\int_{S_0} dS = \int_S \mathbf{n} \cdot \mathbf{n}_0 \, dS \leqq \int_S \cos \psi \, dS < \int_S dS.$$

We have thus proved

THEOREM 2. *A minimal surface spanning a closed curve and belonging to a field of minimal surfaces has a smaller area than any other surface spanning the same curve and lying entirely in the field.*

Example 1. In order to find a surface of revolution which is also minimal, we set $J = 0$ in (138.5). The resulting differential equation,

$$uz'' + z'(1 + z'^2) = 0$$

may be written

$$\frac{z'z''}{z'^2} - \frac{z'z''}{1 + z'^2} + \frac{1}{u} = 0$$

and has the first integral $uz'/\sqrt{1 + z'^2} = C$. Solving for z', we have $z' = C/\sqrt{u^2 - C^2}$; hence

$$z + k = C \cosh^{-1} \frac{u}{C}, \qquad u = C \cosh \frac{z + k}{C}.$$

This represents a catenary in the uz-plane; when revolved about the z-axis it generates a surface known as the *catenoid. Catenoids are the only minimal surfaces of revolution.*

Example 2. The *right helicoid* is a surface generated by a line which always cuts a fixed axis at right angles while revolving about and sliding along the axis at uniform rates. In other words, the right helicoid is a *spiral ramp.*

A right helicoid about the z-axis has the parametric equations,

$$x = u \cos v, \qquad y = u \sin v, \qquad z = av, \tag{4}$$

where u, v are plane polar coordinates ρ, φ in the xy-plane. The lines $v =$ const are the horizontal rulings on the surface; the lines $u =$ const are circular helices on the cylinder $x^2 + y^2 = u^2$.

Equations (4) give the vector equation,

$$\mathbf{r} = u\,\mathbf{R}(v) + av\,\mathbf{k}, \tag{5}$$

where $\mathbf{R}$ is a unit radial vector ($\mathbf{R} \cdot \mathbf{k} = 0$). Now

$$\mathbf{r}_u = \mathbf{R}, \qquad \mathbf{r}_v = u\mathbf{P} + a\mathbf{k};$$

$$\mathbf{n} = \frac{\mathbf{r}_u \times \mathbf{r}_v}{|\mathbf{r}_u \times \mathbf{r}_v|} = \frac{u\mathbf{k} - a\mathbf{P}}{\sqrt{u^2 + a^2}};$$

$$\mathbf{n}_u = \frac{a}{(u^2 + a^2)^{\frac{3}{2}}}\mathbf{r}_v, \qquad \mathbf{n}_v = \frac{a}{(u^2 + a^2)^{\frac{1}{2}}}\mathbf{r}_u.$$

Since $\mathbf{r}_u \cdot \mathbf{r}_v = 0$, the parametric lines are orthogonal; and, since $\mathbf{r}_u \cdot \mathbf{n}_u = 0$, $\mathbf{r}_v \cdot \mathbf{n}_v = 0$, they are also asymptotic lines. From $\mathbf{n}_u \times \mathbf{n}_v = K\,\mathbf{r}_u \times \mathbf{r}_v$ (132.20), we have

$$K = \frac{-a^2}{(u^2 + a^2)^2};$$

and, since $\mathbf{n}_u \times \mathbf{r}_v + \mathbf{r}_u \times \mathbf{n}_v = 0$, $J = 0$ (132.19). *The right helicoid is a minimal surface.* Its negative total curvature is constant along any helix $u =$ const.

Along the lines of curvature,

$$(\mathbf{r}_u\,du + \mathbf{r}_v\,dv) \times (\mathbf{n}_u\,du + \mathbf{n}_v\,dv) = \mathbf{r}_u \times \mathbf{n}_u\,du^2 + \mathbf{r}_v \times \mathbf{n}_v\,dv^2 = 0,$$

or $du^2 - (u^2 + a^2)\,dv^2 = 0$. The two families of these lines therefore satisfy the differential equations:

$$du/\sqrt{u^2 + a^2} = \pm dv.$$

On integration, these give

$$\sinh^{-1}\frac{u}{a} = c \pm v \quad \text{or} \quad u = a \sinh (c \pm v)$$

as the finite equations of the lines of curvature.

Example 3. On a minimal surface, the asymptotic lines form an orthogonal system (§ 131, ex. 2). If we choose them as our field curves, $k_1 = k_2 = 0$, and the Codazzi Equations (137.13), (137.14) become

$$2\gamma_2 t_1 + \frac{dt_1}{ds_1} = 0, \qquad 2\gamma_1 t_1 - \frac{dt_1}{ds_2} = 0.$$

Since $K = -t_1^2$,

$$\frac{dK}{ds_1} = -2t_1 \frac{dt_1}{ds_1} = -4\gamma_2 K, \qquad \frac{dK}{ds_2} = -2t_1 \frac{dt_1}{ds_2} = 4\gamma_1 K;$$

$$\nabla K = \mathbf{e}_1 \frac{dK}{ds_1} + \mathbf{e}_2 \frac{dK}{ds_2} = 4K(\gamma_1 \mathbf{e}_2 - \gamma_2 \mathbf{e}_1) = 4K\,\mathbf{n} \times \mathbf{R} \qquad (134.3).$$

144. Summary: Surface Geometry. On a surface $\mathbf{r} = \mathbf{r}(u, v)$ of unit normal $\mathbf{n}$, the angular velocity of the trihedral $\mathbf{T n p}$ ($\mathbf{p} = \mathbf{T} \times \mathbf{n}$) along a curve is $\boldsymbol{\omega} = t\mathbf{T} + \gamma\mathbf{n} + k\mathbf{p}$. If $\varphi =$ angle $(\mathbf{N}, \mathbf{n})$, positive in the sense of $\mathbf{T}$,

$$t = \tau + d\varphi/ds, \qquad \gamma = \kappa \sin \varphi, \qquad k = \kappa \cos \varphi$$

are the *geodesic torsion*, *geodesic curvature*, and *normal curvature*, respectively. From

$$(1) \qquad \frac{d\mathbf{n}}{ds} = \boldsymbol{\omega} \times \mathbf{n} = -k\mathbf{T} - t\,\mathbf{n} \times \mathbf{T},$$

we conclude that k and t are the same for all surface curves having a common tangent at a point.

Important surface curves and their differential equations:

$$\text{Lines of curvature } (t = 0): \qquad \frac{d\mathbf{r}}{ds} \times \frac{d\mathbf{n}}{ds} = 0;$$

$$\text{Asymptotic lines } (k = 0): \qquad \frac{d\mathbf{r}}{ds} \cdot \frac{d\mathbf{n}}{ds} = 0;$$

$$\text{Geodesics } (\gamma = 0): \qquad \mathbf{n} \times \frac{d^2\mathbf{r}}{ds^2} = 0 \quad \text{or} \quad \nabla\mathbf{r} \cdot \frac{d^2\mathbf{r}}{ds^2} = 0.$$

The equations for geodesics follow from

$$\frac{d\mathbf{T}}{ds} \times \mathbf{n} = \kappa\mathbf{N} \times \mathbf{n} = \gamma\mathbf{T}.$$

If $\mathbf{e}$ and $\mathbf{e}' = \mathbf{n} \times \mathbf{e}$ are perpendicular directions in the tangent plane,

$$-\nabla\mathbf{n} = k\,\mathbf{e}\mathbf{e} + k'\,\mathbf{e}'\mathbf{e}' + t\,\mathbf{e}\mathbf{e}' - t'\,\mathbf{e}'\mathbf{e}.$$

Since $\operatorname{Rot} \mathbf{n} = 0$, $\nabla\mathbf{n}$ is symmetric, and $t + t' = 0$. The first and second scalar invariants of $-\nabla\mathbf{n}$,

$$J = -\operatorname{Div} \mathbf{n} = k + k', \quad \text{the } \textit{mean curvature;}$$

$$K = kk' + tt' = kk' - t^2, \quad \text{the } \textit{total curvature,}$$

have the same value for any pair of perpendicular directions. If the point is not an *umbilic* (k constant for all directions), k attains its extreme values in the orthogonal principal directions $\mathbf{e}_1$, $\mathbf{e}_2$ for which $t_1 = 0$, $t_2 = 0$. In terms of the *principal curvatures* k_1, k_2, $J = k_1 + k_2$, $K = k_1 k_2$.

If $\mathbf{a}, \mathbf{b}, \mathbf{n}$ are reciprocal to $\mathbf{r}_u, \mathbf{r}_v, \mathbf{n}$, the *first* and *second fundamental* forms are

$$\nabla \mathbf{r} = E\,\mathbf{aa} + F(\mathbf{ab} + \mathbf{ba}) + G\,\mathbf{bb},$$

$$-\nabla \mathbf{n} = L\,\mathbf{aa} + M(\mathbf{ab} + \mathbf{ba}) + N\,\mathbf{bb};$$

and the *fundamental quantities* are given by

$$E = \mathbf{r}_u \cdot \mathbf{r}_u, \qquad F = \mathbf{r}_u \cdot \mathbf{r}_v, \qquad G = \mathbf{r}_v \cdot \mathbf{r}_v;$$

$$L = -\mathbf{r}_u \cdot \mathbf{n}_u, \qquad M = -\mathbf{r}_u \cdot \mathbf{n}_v = -\mathbf{r}_v \cdot \mathbf{n}_u, \qquad N = -\mathbf{r}_v \cdot \mathbf{n}_v.$$

If $\mathbf{e}_1$, $\mathbf{e}_2 = \mathbf{n} \times \mathbf{e}_1$ are the unit tangent vectors along a *field* of surface curves and their orthogonal trajectories, we have

$$\nabla \mathbf{e}_1 = \mathbf{R}\mathbf{e}_2 - \mathbf{P}\mathbf{n}, \qquad \nabla \mathbf{e}_2 = -\mathbf{R}\mathbf{e}_1 - \mathbf{Q}\mathbf{n}, \qquad \nabla \mathbf{n} = \mathbf{P}\mathbf{e}_1 + \mathbf{Q}\mathbf{e}_2,$$

where

$$\mathbf{P} = d\mathbf{n}/ds_1, \qquad \mathbf{Q} = d\mathbf{n}/ds_2, \qquad \mathbf{R} = \gamma_1 \mathbf{e}_1 + \gamma_2 \mathbf{e}_2.$$

Codazzi Equations: $\mathbf{n} \cdot \nabla \times (\nabla \mathbf{n}) = 0$; or

$$\mathbf{n} \cdot \operatorname{rot} \mathbf{P} = -\mathbf{n} \cdot \mathbf{Q} \times \mathbf{R}, \qquad \mathbf{n} \cdot \operatorname{rot} \mathbf{Q} = -\mathbf{n} \cdot \mathbf{R} \times \mathbf{P};$$

Gauss Equation: $\mathbf{n} \cdot \nabla \times (\nabla \mathbf{e}_1) = 0$; or

$$\mathbf{n} \cdot \operatorname{rot} \mathbf{R} = -K.$$

For any field of curves of tangent vector $\mathbf{e}$,

$$\gamma = \mathbf{n} \cdot \operatorname{rot} \mathbf{e}.$$

The last two equations show that γ and K may be expressed in terms of E, F, G and their partial derivatives.

When embedded in a geodesic field, the arc of a geodesic is shorter than any other curve lying within the field and having the same end points.

If a field of curves of tangent vector $\mathbf{e}$ cuts the field $\mathbf{e}_1$ at the angle $\theta = (\mathbf{e}_1, \mathbf{e})$, positive in sense of $\mathbf{n}$,

$$\mathbf{R} = \gamma_1 \mathbf{e}_1 + \gamma_2 \mathbf{e}_2 = \gamma \mathbf{e} + \gamma' \mathbf{e}' - \nabla \theta.$$

Bonnet's Integral Formula,

$$\int_S K\,dS = \oint_C d\theta - \oint_C \gamma\,ds,$$

is the integral ("Stokian") equivalent of the Gauss Equation, $\mathbf{n}\cdot\text{Rot}\,\mathbf{R} = -K$. If the simple closed curve C has a continuously turning tangent $\oint d\theta = 2\pi$, but if the path has angles (*interior* angles α_i),

$$\oint d\theta = 2\pi - \Sigma(\pi - \alpha_i).$$

If C is a *geodesic triangle*, $\int K\,dS = \alpha_1 + \alpha_2 + \alpha_3 - \pi$; and, on a surface of constant $K \neq 0$, the area of a geodesic triangle is $(\alpha_1 + \alpha_2 + \alpha_3 - \pi)/K$. A sphere of radius a has the constant positive curvature $1/a^2$; a tractrix of revolution for which the tangential distance to the asymptote is a has the constant negative curvature $-1/a^2$.

A *minimal surface* ($J = 0$) embedded in a field of minimal surfaces and spanning a closed curve C has a smaller area than any other surface lying within the field and spanning C. Soap films spanned over wire framework materialize minimal surfaces. The *right helicoid* is a ruled minimal surface. The *catenoid* is the only minimal surface of revolution.

PROBLEMS

1. If u, v, are plane polar coordinates r, θ, prove that the surface obtained by revolving the curve $z = f(x)$ about the z-axis has the parametric equations:

$$x = u\cos v, \quad y = u\sin v, \quad z = z(u).$$

2. A straight line, which always cuts the z-axis at right angles, is revolved about and moved along this axis. The surface thus generated is called a *conoid.* If u and v are plane polar coordinates in the xy-plane, show that the conoid has the parametric equations:

$$x = u\cos v, \quad y = u\sin v, \quad z = z(v).$$

In particular when dz/dv is constant the conoid is a *right helicoid:*

$$x = u\cos v, \quad y = u\sin v, \quad z = av.$$

3. The central quadric surface,

$$\frac{x^2}{a^2} \pm \frac{y^2}{b^2} \pm \frac{z^2}{c^2} = 1,$$

is an *ellipsoid*, a *hyperboloid of one sheet*, or a *hyperboloid of two sheets* according as the terms on the left have the signs $(+, +, +)$, $(+, +, -)$, $(+, -, -)$. Show that the corresponding parametric equations are

$$x = a \sin u \cos v, \quad y = b \sin u \sin v, \quad z = c \cos u;$$

$$x = a \cosh u \cos v, \quad y = b \cosh u \sin v, \quad z = c \sinh u;$$

$$x = a \cosh u \cosh v, \quad y = b \cosh u \sinh v, \quad z = c \sinh u.$$

4. The *paraboloid*,

$$\frac{x^2}{a^2} \pm \frac{y^2}{b^2} = \frac{2z}{c},$$

is *elliptic* or *hyperbolic*, according as the terms on the left have the signs $(+, +)$ or $(+, -)$. Show that the corresponding parametric equations are

$$x = au \cos v, \quad y = bu \sin v, \quad z = \tfrac{1}{2}cu^2;$$

$$x = au \cosh v, \quad y = bu \sinh v, \quad z = \tfrac{1}{2}cu^2.$$

5. The position vector,

$$\mathbf{r} = \mathbf{f}(u) + \mathbf{g}(v),$$

traces a *surface of translation*. Show that any curve $u = a$ (const.) may be obtained by giving the curve $\mathbf{r}_1 = \mathbf{g}(v)$ a translation $\mathbf{f}(a)$; and that any curve $v = b$ may be obtained by giving the curve $\mathbf{r}_2 = \mathbf{f}(u)$ a translation $\mathbf{g}(b)$.

What are the surfaces,

$$\mathbf{r} = \mathbf{f}(u) + \mathbf{b}v, \qquad \mathbf{r} = \mathbf{a}u + \mathbf{b}v + \mathbf{c}?$$

6. Show that a right circular cylinder of radius a has the constant mean curvature $J = 1/a$.

7. For the surface $xyz = a^3$ show that

$$K = \frac{3a^6}{(x^2y^2 + y^2z^2 + z^2x^2)^2}, \qquad J = \frac{2a^3(x^2 + y^2 + z^2)}{(x^2y^2 + y^2z^2 + z^2x^2)^{\frac{3}{2}}}.$$

[Cf. § 132, ex. 2.]

8. For the elliptic paraboloid,

$$x^2/a^2 + y^2/b^2 = 2z/c$$

prove that the total curvature $K \leq c^2/a^2b^2$.

9. For the *surface of revolution* about the z-axis,

$$\mathbf{r} = \mathbf{i}u \cos v + \mathbf{j}u \sin v + \mathbf{k}z(u),$$

show that

$$E = 1 + z'^2 \quad F = 0, \quad G = u^2;$$

$$L = z''/(1 + z'^2)^{\frac{1}{2}}, \quad M = 0, \quad N = uz'/(1 + z'^2)^{\frac{1}{2}}. \qquad \text{[Cf. § 132.]}$$

From (132.21) and (132.22) deduce the values of K and J and from (132.25) find the principal curvatures. Check with (138.4) and (138.5).

10. For the *conoid*,

$$\mathbf{r} = \mathbf{i}u \cos v + \mathbf{j}u \sin v + \mathbf{k}z(v),$$

show that

$$E = 1, \quad F = 0, \quad G = u^2 + z'^2 = H^2;$$

$$L = 0, \quad M = -z'/H, \quad N = uz''/H;$$

$$K = -z'^2/H^4, \quad J = uz''/H^3. \qquad \text{[Cf. § 132.]}$$

When $z = av$, show that the resulting *right helicoid* is *minimal* (§ 143); and that its principal curvatures are $\pm a/(u^2 + a^2)$.

11. Show that the curvatures J and K have the dimensions of $(\text{length})^{-1}$ and $(\text{length})^{-2}$. Verify this in Problems 6, 7, 8, 9, 10.

12. The position vector of a twisted curve Γ is given as a function $\mathbf{r}_0(s)$ of the arc. The surface generated by its tangents is

$$\mathbf{r} = \mathbf{r}_0(v) + u\mathbf{T}(v)$$

where $v = s$ and u is the distance along a tangent measured from Γ. Show that this *tangent surface* has the curvatures $K = 0$, $J = \tau/|u|\kappa$. What are the principal curvatures?

13. If a parabola is revolved about its directrix, show that the principal curvatures of the surface of revolution satisfy the relation $2k_1 + k_2 = 0$.

[With the notation of § 138, ex., the parabola has the equation $z^2 = 4a(u - a)$ when the directrix is the z-axis and $u = 2a$ at the focus.]

14. The vectors $\mathbf{a}$, $\mathbf{b}$, $\mathbf{n}$ and $\mathbf{r}_u$, $\mathbf{r}_v$, $\mathbf{n}$ form reciprocal sets of vectors over the surface $\mathbf{r} = \mathbf{r}(u, v)$ [Cf. § 95]. Hence show that any vector $\mathbf{f}(u, v)$, defined over the surface, may be written

$$\mathbf{f} = (\mathbf{f} \cdot \mathbf{a})\mathbf{r}_u + (\mathbf{f} \cdot \mathbf{b})\mathbf{r}_v + (\mathbf{f} \cdot \mathbf{n})\mathbf{n}.$$

15. Prove that

$$H^2\mathbf{a} = G\mathbf{r}_u - F\mathbf{r}_v, \qquad H^2\mathbf{b} = -F\mathbf{r}_u + E\mathbf{r}_v.$$

[Use Prob. 14 and (132.18).]

16. Prove that

$$\mathbf{n}_u \cdot \mathbf{a} = \frac{FM - GL}{H^2}, \qquad \mathbf{n}_u \cdot \mathbf{b} = \frac{FL - EM}{H^2};$$

$$\mathbf{n}_v \cdot \mathbf{a} = \frac{FN - GM}{H^2}, \qquad \mathbf{n}_v \cdot \mathbf{b} = \frac{FM - EN}{H^2}.$$

[Use Prob. 15.]

17. Prove the *derivative formulas of Weingarten:*

$$H^2\mathbf{n}_u = (FM - GL)\mathbf{r}_u + (FL - EM)\mathbf{r}_v,$$

$$H^2\mathbf{n}_v = (FN - GM)\mathbf{r}_u + (FM - EN)\mathbf{r}_v.$$

[Use Prob. 14 and Prob. 16.]

18. Show that the *asymptotic lines* of the surface $\mathbf{r} = \mathbf{r}(u, v)$ have the differential equation,

$$(\mathbf{r}_u \, du + \mathbf{r}_v \, dv) \cdot (\mathbf{n}_u \, du + \mathbf{n}_v \, dv) = 0$$

[Cf. (130.10)]; or, in terms of fundamental quantities,

$$L \, du^2 + 2M \, du \, dv + N \, dv^2 = 0.$$

19. Prove that the asymptotic lines on a right helicoid (Prob. 10) are the parametric curves and form an orthogonal net.

20. For the ruled surface,

$$\mathbf{r} = \mathbf{p}(u) + v\mathbf{e}(u), \qquad (142.1),$$

prove that $M = \mathbf{p}_u \cdot \mathbf{e} \times \mathbf{e}_u / H$, $N = 0$.

21. Prove that a ruled surface is developable when, and only when, $M = 0$. [Cf. (142.2).]

22. Prove that the asymptotic lines of a developable surface, not a plane, are the generating lines (counted twice). [Differential equation: $du^2 = 0$.]

23. Prove that, along a *curved* asymptotic line,

(*a*) The osculating plane is tangent to the surface.

(*b*) The geodesic torsion equals the torsion: $t = \tau$.

(*c*) The geodesic curvature is numerically equal to the curvature: $\gamma = \pm\kappa$.

(*d*) The square of the torsion is equal to the negative of the total curvature. $\tau^2 = -K$.

24. Prove that

(*a*) If an asymptotic line is a plane curve other than a straight line, it is also a line of curvature.

(*b*) An asymptotic line of curvature is plane.

25. Prove that the following conditions are necessary and sufficient in order that the surface be

(i) A plane: $L = M = N = 0$.

(ii) A sphere: $E/L = F/M = G/N$.

26. Show that the lines of curvature of the surface $\mathbf{r} = \mathbf{r}(u, v)$ have the differential equation,

$$\mathbf{n} \cdot (\mathbf{r}_u \, du + \mathbf{r}_v \, dv) \times (\mathbf{n}_u \, du + \mathbf{n}_v \, dv) = 0$$

[Cf. (138.1)]; or, in terms of fundamental quantities,

$$\begin{vmatrix} dv^2 & -du \, dv & du^2 \\ E & F & G \\ L & M & N \end{vmatrix} = 0.$$

27. Show that the parametric curves are lines of curvature when, and only when, $F = 0$, $M = 0$.

28. In the notation of § 132, ex. 2, prove that the lines of curvature on the surface $z = z(x, y)$ have the differential equation,

$$\begin{vmatrix} dy^2 & -dx \, dy & dx^2 \\ 1 + p^2 & pq & 1 + q^2 \\ r & s & t \end{vmatrix} = 0.$$

29. Show that the hyperbolic paraboloid of Prob. 4 has the parametric equations:

$$x = \tfrac{1}{2}a(u + v), \quad y = \tfrac{1}{2}b(u - v), \quad z = \tfrac{1}{2}cuv.$$

Prove that the differential equations of its asymptotic lines and lines of curvature are, respectively,

$$du\,dv = 0, \qquad \left(\frac{dv}{du}\right)^2 = \frac{a^2 + b^2 + c^2v^2}{a^2 + b^2 + c^2u^2};$$

and find the uv-equations of the asymptotic lines and lines of curvature.

30. If the parametric curves on a surface are its lines of curvature, show that

$$\mathbf{n}_u = -k_1\mathbf{r}_u, \qquad \mathbf{n}_v = -k_2\mathbf{r}_v.$$

31. Two surfaces that have the same normal lines are called *parallel.* Surface points on the same normal are said to *correspond;* thus

$$\bar{\mathbf{r}}(u, v) = \mathbf{r}(u, v) + \lambda\mathbf{n}(u, v)$$

determines corresponding points on the parallel surfaces S and $\bar{S}$. Prove that the distance λ between the parallel surfaces is constant.

[If we take the lines of curvature of S as parametric curves,

$$\bar{\mathbf{r}}_u = (1 - \lambda k_1)\mathbf{r}_u + \lambda_u\mathbf{n}, \qquad \bar{\mathbf{r}}_v = (1 - \lambda k_2)\mathbf{r}_v + \lambda_v\mathbf{n}$$

(Prob. 30); and, on cross multiplication,

$$\bar{H}\bar{\mathbf{n}} = (1 - \lambda k_1)(1 - \lambda k_2)H\mathbf{n} + \lambda_u(1 - \lambda k_2)\mathbf{n}\times\mathbf{r}_v + \lambda_v(1 - \lambda k_1)\mathbf{r}_u\times\mathbf{n}.$$

Now $\bar{\mathbf{n}} = \epsilon\mathbf{n}$ where $\epsilon = \pm 1$; and as $\mathbf{n}, \mathbf{r}_u, \mathbf{r}_v$ are mutually orthogonal,

$$\bar{H} = \epsilon(1 - \lambda k_1)(1 - \lambda k_2)H, \qquad \lambda_u = \lambda_v = 0.]$$

32. Prove that the lines of curvature on parallel surfaces correspond and that the corresponding principal curvatures satisfy the equations,

$$\bar{k}_1 = \frac{\epsilon k_1}{1 - \lambda k_1}, \qquad \bar{k}_2 = \frac{\epsilon k_2}{1 - \lambda k_2}.$$

Hence the mean and total curvature of $\bar{S}$ are given by

$$\bar{J} = \frac{\epsilon(J - 2\lambda K)}{1 - \lambda J + \lambda^2 K}, \qquad \bar{K} = \frac{K}{1 - \lambda J + \lambda^2 K}.$$

33. Show that surfaces parallel to a surface of revolution are also surfaces of revolution.

34. Prove the "*Theorema egregium*" of Gauss by establishing Baltzer's formula for the total curvature K:

$$H^4K = \begin{vmatrix} F_{uv} - \tfrac{1}{2}G_{uu} - \tfrac{1}{2}E_{vv} & \tfrac{1}{2}E_u & F_u - \tfrac{1}{2}E_v \\ F_v - \tfrac{1}{2}G_u & E & F \\ \tfrac{1}{2}G_v & F & G \end{vmatrix} - \begin{vmatrix} 0 & \tfrac{1}{2}E_v & \tfrac{1}{2}G_u \\ \tfrac{1}{2}E_v & E & F \\ \tfrac{1}{2}G_u & F & G \end{vmatrix}.$$

Method. From (132.10) and (132.22) show that

$$H^4K = [\mathbf{r}_{uu}\mathbf{r}_u\mathbf{r}_v][\mathbf{r}_{vv}\mathbf{r}_u\mathbf{r}_v] - [\mathbf{r}_{uv}\mathbf{r}_u\mathbf{r}_v]^2.$$

Using (24.14), the right member becomes

$$\begin{vmatrix} \mathbf{r}_{uu}\cdot\mathbf{r}_{vv} & \mathbf{r}_{uu}\cdot\mathbf{r}_u & \mathbf{r}_{uu}\cdot\mathbf{r}_v \\ \mathbf{r}_u\cdot\mathbf{r}_{vv} & E & F \\ \mathbf{r}_v\cdot\mathbf{r}_{vv} & F & G \end{vmatrix} - \begin{vmatrix} \mathbf{r}_{uv}\cdot\mathbf{r}_{uv} & \mathbf{r}_{uv}\cdot\mathbf{r}_u & \mathbf{r}_{uv}\cdot\mathbf{r}_v \\ \mathbf{r}_{uv}\cdot\mathbf{r}_u & E & F \\ \mathbf{r}_{uv}\cdot\mathbf{r}_v & F & G \end{vmatrix}$$

In both determinants the upper left element has the cofactor $EG - F^2$; hence we may replace these elements by $\mathbf{r}_{uu}\cdot\mathbf{r}_{vv} - \mathbf{r}_{uv}\cdot\mathbf{r}_{uv}$ and 0, respectively. Now, by differentiating

$$\mathbf{r}_u\cdot\mathbf{r}_u = E,\quad \mathbf{r}_u\cdot\mathbf{r}_v = F,\quad \mathbf{r}_v\cdot\mathbf{r}_v = G,$$

we may show that

$$\mathbf{r}_{uu}\cdot\mathbf{r}_u = \tfrac{1}{2}E_u,\quad \mathbf{r}_{uv}\cdot\mathbf{r}_u = \tfrac{1}{2}E_v,\quad \mathbf{r}_{vv}\cdot\mathbf{r}_u = F_v - \tfrac{1}{2}G_u,$$

$$\mathbf{r}_{vv}\cdot\mathbf{r}_v = \tfrac{1}{2}G_v,\quad \mathbf{r}_{uv}\cdot\mathbf{r}_v = \tfrac{1}{2}G_u,\quad \mathbf{r}_{uu}\cdot\mathbf{r}_v = F_u - \tfrac{1}{2}E_v;$$

$$\mathbf{r}_{uu}\cdot\mathbf{r}_{vv} - \mathbf{r}_{uv}\cdot\mathbf{r}_{uv} = F_{uv} - \tfrac{1}{2}G_{uu} - \tfrac{1}{2}E_{vv}.$$

35. On the surface $\mathbf{r} = \mathbf{r}(u, v)$ let us write

$$u^1 = u,\quad u^2 = v;\qquad \mathrm{D}_1 = \frac{\partial}{\partial u},\quad \mathrm{D}_2 = \frac{\partial}{\partial v};$$

$$\mathbf{e}_1 = \mathbf{r}_u,\quad \mathbf{e}_2 = \mathbf{r}_v;\qquad \mathbf{e}^1 = \mathbf{a},\quad \mathbf{e}^2 = \mathbf{b};\qquad \text{then}\quad \mathbf{e}_\alpha\cdot\mathbf{e}^\beta = \delta_\alpha^\beta,$$

where the Greek indices have the range 1, 2. Moreover let

$$g_{\alpha\beta} = \mathbf{e}_\alpha\cdot\mathbf{e}_\beta,\quad g^{\alpha\beta} = \mathbf{e}^\alpha\cdot\mathbf{e}^\beta,\quad g = \det g_{\alpha\beta} = \begin{vmatrix} g_{11} & g_{12} \\ g_{21} & g_{22} \end{vmatrix};$$

$$h_{\alpha\beta} = \mathbf{n}\cdot\mathrm{D}_\alpha\mathrm{D}_\beta\mathbf{r} = h_{\beta\alpha};\qquad h = \det h_{\alpha\beta}.$$

Prove that

(*a*) $g_{11} = E$, $g_{12} = F$, $g_{22} = G$; $g = EG - F^2$.

(*b*) $g^{11} = \dfrac{g_{22}}{g}$, $g^{12} = \dfrac{-g_{12}}{g}$, $g^{22} = \dfrac{g_{22}}{g}$;

that is, $g^{\alpha\beta}$ is the reduced cofactor of $g_{\alpha\beta}$ in det $g_{\alpha\beta}$.

(*c*) $h_{11} = L$, $h_{12} = M$, $h_{22} = N$; $h = LN - M^2$;

(*d*) $\mathrm{D}_\alpha\mathbf{e}_\beta = \mathrm{D}_\beta\mathbf{e}_\alpha$;

(*e*) $\mathrm{D}_\alpha g_{\beta\gamma} + \mathrm{D}_\beta g_{\gamma\alpha} - \mathrm{D}_\gamma g_{\alpha\beta} = 2\mathbf{e}_\gamma\cdot\mathrm{D}_\alpha\mathbf{e}_\beta$.

(*f*) $\mathrm{D}_\alpha\mathbf{e}_\beta = \tfrac{1}{2}\mathbf{e}^\gamma(\mathrm{D}_\alpha g_{\beta\gamma} + \mathrm{D}_\beta g_{\gamma\alpha} - \mathrm{D}_\gamma g_{\alpha\beta}) + \mathbf{n}h_{\alpha\beta}$ (summed over $\gamma = 1, 2$).

(*g*) $\mathbf{e}^\alpha = g^{\alpha\beta}\mathbf{e}_\beta$ (summed over $\beta = 1, 2$). [Cf. § 145.]

36. The *Christoffel symbols* $\Gamma_{\alpha\beta}^\gamma(\alpha, \beta, \gamma = 1, 2)$ for a surface $\mathbf{r} = \mathbf{r}(u^1, u^2)$ having $g_{\alpha\beta}\,du^\alpha\,du^\beta$ as first fundamental form (§ 132) are computed from the equations:

$$\Gamma_{\alpha\beta}^\gamma = \mathbf{e}^\gamma\cdot\mathrm{D}_\alpha\mathbf{e}_\beta.$$

With reference to Prob. 35 prove that

(a) $\Gamma^{\gamma}_{\alpha\beta} = \Gamma^{\gamma}_{\beta\alpha}$

(b) $\Gamma^{\rho}_{\alpha\beta} = \frac{1}{2}g^{\rho\gamma}(\mathrm{D}_{\alpha}g_{\beta\gamma} + \mathrm{D}_{\beta}g_{\gamma\alpha} - \mathrm{D}_{\gamma}g_{\alpha\beta})$ (summed over $\rho = 1, 2$).

(c) $2g\Gamma^{1}_{11} = g_{22}\,\mathrm{D}_1 g_{11} - 2g_{12}\,\mathrm{D}_1 g_{12} + g_{12}\,\mathrm{D}_2 g_{11}$,
$2g\Gamma^{1}_{12} = g_{22}\,\mathrm{D}_2 g_{11} - g_{12}\,\mathrm{D}_1 g_{22}$,
$2g\Gamma^{1}_{22} = -g_{12}\,\mathrm{D}_2 g_{22} + 2g_{22}\,\mathrm{D}_2 g_{12} - g_{22}\,\mathrm{D}_1 g_{22}$.

[Use the formula of part b; for example,

$$2\Gamma^{1}_{11} = g^{11}(\mathrm{D}_1 g_{11} + \mathrm{D}_1 g_{11} - \mathrm{D}_1 g_{11}) + g^{12}(\mathrm{D}_1 g_{12} + \mathrm{D}_1 g_{21} - \mathrm{D}_2 g_{11}),$$

$$2g\Gamma^{1}_{11} = g_{22}\,\mathrm{D}_1 g_{11} - g_{12}(2\mathrm{D}_1 g_{12} - \mathrm{D}_2 g_{11}).]$$

(d) From the formulas for $\Gamma^{1}_{\alpha\beta}$ obtain the formulas for $\Gamma^{2}_{\alpha\beta}$.

(e) From Prob. 35 (*f*) prove the *derivative formulas of Gauss:*

$$\mathrm{D}_{\alpha}\mathbf{e}_{\beta} = \Gamma^{\lambda}_{\alpha\beta}\mathbf{e}_{\lambda} + h_{\alpha\beta}\mathbf{n} \quad \text{(summed over } \lambda = 1, 2).$$

37. From $\mathbf{n}\cdot\mathbf{e}_{\beta} = 0$ show that $\mathbf{e}_{\beta}\cdot\mathrm{D}_{\alpha}\mathbf{n} = -h_{\alpha\beta}$. Hence prove the *derivative formulas of Weingarten;*

$$\mathrm{D}_{\alpha}\mathbf{n} = -h_{\alpha\beta}\mathbf{e}^{\beta} \quad \text{(summed over } \beta = 1, 2).$$

Check these results with Prob. 17.

38. With the notations of Prob. 35 show that equations (132.21) and (132.22) may be written

$$J = g^{\alpha\beta}h_{\alpha\beta}, \qquad K = h/g \quad (\alpha,\ \beta \text{ summed over } 1,\ 2).$$

39. Show that the tangential projection of the vector curvature $d\mathbf{T}/ds = \kappa\mathbf{N}$ of a surface curve is

$$\nabla\mathbf{r}\cdot\frac{d^2\mathbf{r}}{ds^2} = \mathbf{e}_{\gamma}\left(\frac{d^2u^{\gamma}}{ds^2} + \Gamma^{\gamma}_{\alpha\beta}\frac{du^{\alpha}}{ds}\frac{du^{\beta}}{ds}\right)$$

summed over $\alpha, \beta, \gamma = 1, 2$. Hence prove that the equations of a geodesic on a surface are

$$\frac{d^2u^{\gamma}}{ds^2} + \Gamma^{\gamma}_{\alpha\beta}\frac{du^{\alpha}}{ds}\frac{du^{\beta}}{ds} = 0 \qquad (\gamma = 1, 2).$$

$$\left[\mathbf{T} = \mathbf{e}_{\beta}\frac{du^{\beta}}{ds},\quad \frac{d\mathbf{T}}{ds} = \mathbf{e}_{\beta}\frac{d^2u^{\beta}}{ds^2} + \mathrm{D}_{\alpha}\mathbf{e}_{\beta}\frac{du^{\alpha}}{ds}\frac{du^{\beta}}{ds},\quad \nabla\mathbf{r} = \mathbf{e}_{\gamma}\mathbf{e}^{\gamma}.\quad \text{Cf. Prob. 36.}\right]$$

CHAPTER IX

TENSOR ANALYSIS

145. The Summation Convention. Expressions which consist of a sum of similar terms may be condensed greatly, without any essential loss of clarity, by indicating summation by means of repeated indices. This usage, originally due to Einstein, is stated precisely in the following

SUMMATION CONVENTION. *Any term, in which the same index (subscript or superscript) appears twice, shall stand for the sum of all such terms obtained by giving this index its complete range of values.* This range of values, if not understood, must be specified in advance.

By way of illustration, we repeat some of the formulas of § 24, using the summation convention. The index range is 1, 2, 3.

The two forms of a vector (24.1) and (24.2) become

(1) $$\mathbf{u} = u^i\mathbf{e}_i, \qquad \mathbf{u} = u_j\mathbf{e}^j.$$

In these equations i and j are *summation* (*or dummy*) *indices. But any other letter will do as well;* thus

$$\mathbf{u} = u^r\mathbf{e}_r = u^1\mathbf{e}_1 + u^2\mathbf{e}_2 + u^3\mathbf{e}_3.$$

In order to compute $\mathbf{u} \cdot \mathbf{v}$, we first recall the defining equation for reciprocal sets,

(2) $$\mathbf{e}_i \cdot \mathbf{e}^j = \delta_i^j \qquad (23.3).$$

Now

(3) $$\mathbf{u} \cdot \mathbf{v} = (u^i\mathbf{e}_i) \cdot (v_j\mathbf{e}^j) = u^iv_j\,\delta_i^j = u^iv_i;$$

we here use different indices in expressing $\mathbf{u}$ and $\mathbf{v}$ in order to get the $3^2 = 9$ terms in the expanded product (six are zero). Similarly,

(4) $$\mathbf{u} \cdot \mathbf{v} = (u_i\mathbf{e}^i) \cdot (v^j\mathbf{e}_j) = u_iv^j\,\delta_j^i = u_iv^i.$$

Note that the summation indices in the preceding examples appear once as subscript and once as superscript. The significance of this arrangement (*not required* by the summation convention) soon will be apparent.

146. Determinants. A permutation of the first n natural numbers is said to be *even* or *odd*, according as it can be formed from $123 \cdots n$ by an even or odd number of interchanges of adjacent numbers. The total number of permutations of n different numbers is $n!$; one half of these are even and one half odd. For example, when $n = 3$, 123, 231, 312 are even permutations, 213, 132, 321 are odd.

Consider now a permutation $ijk \cdots r$ of the numbers $123 \cdots n$. We then define the permutation symbols $\epsilon_{ijk\cdots r}$ and $\epsilon^{ijk\cdots r}$ as equal to 1 or -1, according as $ijk \cdots r$ is an even or an odd permutation of $123 \cdots n$; but, if any index is repeated, the epsilon is zero. For example, when $n = 3$,

$$\epsilon^{123} = \epsilon^{231} = \epsilon^{312} = 1, \qquad \epsilon^{213} = \epsilon^{132} = \epsilon^{321} = -1,$$

$$\epsilon^{112} = \epsilon^{122} = \epsilon^{222} = 0.$$

The epsilons just defined are useful in dealing with determinants. To be specific, we shall take $n = 3$; but all the formulas apply without change of form for any value of n.

From the definition of a determinant:

$$(1) \qquad a = \det a_i^j = \begin{vmatrix} a_1^1 & a_1^2 & a_1^3 \\ a_2^1 & a_2^2 & a_2^3 \\ a_3^1 & a_3^2 & a_3^3 \end{vmatrix} = \begin{cases} \epsilon_{ijk} a_1^i a_2^j a_3^k \\ \epsilon^{ijk} a_i^1 a_j^2 a_k^3. \end{cases}$$

The implied summations on i, j, k produce $3^3 = 27$ terms; of these, 21 involve repeated indices and therefore give a zero epsilon, whereas 6 involve permutations of 123 and give precisely the $3! = 6$ terms of the determinant. (Write them out!)

The theorems relative to an interchange of rows and columns are given by

$$(2) \qquad a\epsilon_{rst} = \epsilon_{ijk} a_r^i a_s^j a_t^k, \qquad a\epsilon^{rst} = \epsilon^{ijk} a_i^r a_j^s a_k^t.$$

Proof. When r, s, t are 1, 2, 3, equations (2) reduce to (1). Since ϵ_{rst} changes sign when two adjacent indices are interchanged,

(2) will be established when the right members are shown to have this same property. Now

$$\epsilon_{ijk}a^i_r a^j_s a^k_t = \epsilon_{ijk}a^j_s a^i_r a^k_t = -\epsilon_{jik}a^j_s a^i_r a^k_t;$$

and, if we interchange the summation indices i, j, the last expression becomes $-\epsilon_{ijk}a^i_s a^j_r a^k_t$.

If we multiply the first equation of (2) by ϵ^{rst} and sum on rst, we obtain

$$3!\,a = \epsilon_{ijk}\epsilon^{rst}a^i_r a^j_s a^k_t. \tag{3}$$

On the left $\epsilon^{rst}\epsilon_{rst}$, summed over the 3! permutations of 123, equals 3!; on the right, the summation extends over all six indices and produces $3^6 = 729$ terms. Many with zero epsilons vanish; and each non-vanishing term such as $a^1_1 a^2_2 a^3_3$ appears six times, corresponding to the 3! permutations of its factors. Equation (3) thus is not useful for computation; its importance rests on the information it gives about the determinant when its elements are transformed.

The *cofactor* of an element a^j_i in the determinant (1) is defined as its coefficient in the expansion of the determinant. The cofactor of a^j_i is denoted by A^i_j. To find A^i_j, strike out the row and column in which a^j_i stands; then A^i_j equals the resulting minor taken with the sign $(-1)^{i+j}$. If the elements of any row or column are multiplied by their respective cofactors and added, we obtain the determinant—its *Laplace expansion;* but, if the elements of any row (column) are multiplied by the corresponding cofactors of another row (column) and added, the sum is zero. These important properties both are included in

$$a^r_i A^j_r = a^j_r A^r_i = a\,\delta^j_i. \tag{4}$$

Here r is the summation index. If in (4) we put $j = i$, *both* r and i are summation indices, and we get

$$a^r_i A^i_r = a^i_r A^r_i = a\,\delta^i_i, \tag{5}$$

where $\delta^i_i = \delta^1_1 + \delta^2_2 + \delta^3_3 = 3$ $(n = 3)$. But, if we put $j = i$ in (4) and *suspend summation on* i, we obtain the Laplace expansions,

$$a^r_i A^i_r = a^i_r A^r_i = a \quad (i \text{ fixed}), \tag{6}$$

for $\delta^i_i = 1$ for a fixed i.

A cofactor divided by the value of the determinant is called a *reduced cofactor*. The reduced cofactor of a_i^j is therefore

(7) $$\alpha_j^i = A_j^i/a \qquad (a \neq 0).$$

In terms of reduced cofactors, equations (4) become

(8) $$a_i^r \alpha_r^j = a_r^j \alpha_i^r = \delta_i^j.$$

When the elements of a determinant are written a_{ij}, its definition becomes

(9) $$a = \epsilon^{ijk} a_{1i} a_{2j} a_{3k} = \epsilon^{ijk} a_{i1} a_{j2} a_{k3}.$$

The reduced cofactor of a_{ij} then is written a^{ij}; and equations (8) become

(10) $$a_{ir} a^{jr} = a_{ri} a^{rj} = \delta_i^j.$$

If the elements a_{ij} are functions of a variable x, we have, from (9),

$$\frac{da}{dx} = \epsilon^{ijk} \left(\frac{da_{1i}}{dx} a_{2j} a_{3k} + a_{1i} \frac{da_{2j}}{dx} a_{3k} + a_{1i} a_{2j} \frac{da_{3k}}{dx} \right)$$

$$= \frac{da_{1i}}{dx} A^{1i} + \frac{da_{2j}}{dx} A^{2j} + \frac{da_{3k}}{dx} A^{3k},$$

where A^{ij} denotes the cofactor of a_{ij}. Summing on *two* indices, we may write

(11) $$\frac{da}{dx} = \frac{da_{ij}}{dx} A^{ij} \qquad (i, j = 1, 2, 3).$$

The derivative of a determinant is the sum of the products formed by differentiating each element and multiplying by the cofactor of the element.

The properties (8) of reduced cofactors enable us to solve a system of linear equations when the determinant of the coefficients is not zero. Consider, for example, the equations:

(12) $$a_j^i x^j = y^i \qquad (i, j = 1, 2, 3; \quad a \neq 0).$$

To solve these for x^j, multiply (12) by the reduced cofactor α_i^k and sum on i: the left member becomes $\delta_j^k x^j = x^k$, and we obtain $x^k = \alpha_i^k y^i$, or, on replacing k by j,

(13) $$x^j = \alpha_i^j y^i \qquad (i, j = 1, 2, 3).$$

To find the product of the determinants $a = \det a_i^j$, $b = \det b_i^j$, we have, from (1) and (2),

$$
\begin{aligned}
(14) \qquad ab &= \epsilon_{ijk} a_1^i a_2^j a_3^k b \\
&= a_1^i a_2^j a_3^k \epsilon_{rst} b_i^r b_j^s b_k^t \\
&= \epsilon_{rst} (a_1^i b_i^r)(a_2^j b_j^s)(a_3^k b_k^t) \\
&= \epsilon_{rst} c_1^r c_2^s c_3^t,
\end{aligned}
$$

where

$$(15) \qquad c_i^j = a_i^r b_r^j = a_i^1 b_1^j + a_i^2 b_2^j + a_i^3 b_3^j.$$

From (15), we see that the element in the ith row and jth column of the product ab is given by the sum of the products of the corresponding terms in the ith row of a and the jth column of b—the so-called "row–column" rule.

Since the value of a determinant is not altered by an interchange of rows and columns, we also can compute ab by "row–row," "column–row," and "column–column" rules. For example, the product

$$(16) \qquad \begin{vmatrix} a_1^1 & a_1^2 & a_1^3 \\ a_2^1 & a_2^2 & a_2^3 \\ a_3^1 & a_3^2 & a_3^3 \end{vmatrix} \begin{vmatrix} \alpha_1^1 & \alpha_2^1 & \alpha_3^1 \\ \alpha_1^2 & \alpha_2^2 & \alpha_3^2 \\ \alpha_1^3 & \alpha_2^3 & \alpha_3^3 \end{vmatrix} = \begin{vmatrix} 1 & 0 & 0 \\ 0 & 1 & 0 \\ 0 & 0 & 1 \end{vmatrix} = 1.$$

by use of the row–row rule.

THEOREM. *If a determinant $a \neq 0$, the determinant formed by replacing each element by its reduced cofactor is $1/a$.*

If we solve equations (13) for y^i by using the reduced cofactors in $\det \alpha_i^j$, we will obtain (12); in other words, the reduced cofactor of α_i^j in its determinant is a_j^i. The two determinants in (16) thus are reciprocally related: each is formed from the reduced cofactors of the other.

147. Contragredient Transformations. Let us now introduce a new basis $\bar{\mathbf{e}}_1, \bar{\mathbf{e}}_2, \bar{\mathbf{e}}_3$ by means of the linear transformation,

$$
(1) \qquad
\begin{aligned}
\bar{\mathbf{e}}_1 &= c_1^1 \mathbf{e}_1 + c_1^2 \mathbf{e}_2 + c_1^3 \mathbf{e}_3, \\
\bar{\mathbf{e}}_2 &= c_2^1 \mathbf{e}_1 + c_2^2 \mathbf{e}_2 + c_2^3 \mathbf{e}_3, \\
\bar{\mathbf{e}}_3 &= c_3^1 \mathbf{e}_1 + c_3^2 \mathbf{e}_2 + c_3^3 \mathbf{e}_3,
\end{aligned}
$$

where the coefficients c_i^j are real constants whose determinant $c \neq 0$. In brief,

$$\bar{\mathbf{e}}_i = c_i^j \mathbf{e}_j, \qquad c = \det c_i^j \neq 0. \tag{1}$$

The condition $c \neq 0$ ensures that $\bar{\mathbf{e}}_1$, $\bar{\mathbf{e}}_2$, $\bar{\mathbf{e}}_3$ are linearly independent. For, if there were three constants λ^i such that

$$\lambda^i \bar{\mathbf{e}}_i = 0, \quad \text{then} \quad \lambda^i c_i^j \mathbf{e}_j = 0;$$

the linear independence of the vectors $\mathbf{e}_j$ requires that $c_i^j \lambda^i = 0$, and, since $c \neq 0$, these equations only admit the solution $\lambda^i = 0$. This argument applies without change to space of n dimensions, in which a basis consists of n linearly independent vectors.

In the present case ($n = 3$), we also may argue as follows. From (1), we have

$$\bar{\mathbf{e}}_1 \cdot \bar{\mathbf{e}}_2 \times \bar{\mathbf{e}}_3 = c_1^i c_2^j c_3^k \, \mathbf{e}_i \cdot \mathbf{e}_j \times \mathbf{e}_k = c_1^i c_2^j c_3^k \epsilon_{ijk} \, \mathbf{e}_1 \cdot \mathbf{e}_2 \times \mathbf{e}_3;$$

hence, on writing $E = \mathbf{e}_1 \cdot \mathbf{e}_2 \times \mathbf{e}_3$, we have

$$\bar{E} = (\det c_i^j) E = cE. \tag{2}$$

Since the vectors $\mathbf{e}_i$ form a basis $E \neq 0$, hence $c \neq 0$ implies $\bar{E} \neq 0$ and the linear independence of the vectors $\bar{\mathbf{e}}_i$.

For any vector $\mathbf{u}$,

$$\mathbf{u} = \bar{u}^i \bar{\mathbf{e}}_i = u^j \mathbf{e}_j; \tag{3}$$

hence, on substitution from (1),

$$\bar{u}^i c_i^j \mathbf{e}_j = u^j \mathbf{e}_j,$$

or, since the vectors $\mathbf{e}_j$ are linearly independent,

$$u^j = c_i^j \bar{u}^i. \tag{4}$$

If γ_j^i is the reduced cofactor of c_i^j in the determinant c, we have on solving equations (4),

$$\bar{u}^i = \gamma_j^i u^j, \tag{5}$$

or, written out in full,

$$\begin{aligned} \bar{u}^1 &= \gamma_1^1 u^1 + \gamma_2^1 u^2 + \gamma_3^1 u^3, \\ \bar{u}^2 &= \gamma_1^2 u^1 + \gamma_2^2 u^2 + \gamma_3^2 u^3, \\ \bar{u}^3 &= \gamma_1^3 u^1 + \gamma_2^3 u^2 + \gamma_3^3 u^3. \end{aligned} \tag{5}$$

The linear transformations (1) and (5) are said to be *contragredient.* Their matrices,

$$C = \begin{pmatrix} c_1^1 & c_1^2 & c_1^3 \\ c_2^1 & c_2^2 & c_2^3 \\ c_3^1 & c_3^2 & c_3^3 \end{pmatrix}, \qquad \Gamma = \begin{pmatrix} \gamma_1^1 & \gamma_2^1 & \gamma_3^1 \\ \gamma_1^2 & \gamma_2^2 & \gamma_3^2 \\ \gamma_1^3 & \gamma_2^3 & \gamma_3^3 \end{pmatrix},$$

are so related that any element of Γ is the reduced cofactor of the corresponding element of C in its determinant c; and any element of C is the reduced cofactor of the corresponding element of Γ in its determinant γ.

In order to state this definition analytically, we remind the reader of the following definitions from matrix algebra. The *product* AB of two square matrices A and B is the matrix whose element in the ith row and jth column is the sum of the corresponding products of the elements in the ith row of A and jth column of B (row–column rule). The *transpose* of a matrix A is a matrix A' obtained from A by interchanging rows and columns. The *unit matrix* I has the elements δ_i^j (ones in the principal diagonal, zeros elsewhere).

If we compute $C\Gamma'$ or $\Gamma C'$ and make use of the equations,

$$(6) \qquad c_i^r \gamma_r^j = \delta_i^j, \qquad \gamma_r^i c_j^r = \delta_j^i,$$

we find that

$$(7) \qquad C\Gamma' = \Gamma C' = I.$$

These equations characterize contragredient transformations.

Two matrices whose product is I are said to be *reciprocal.* Thus C and Γ' or C' and Γ are reciprocal matrices. From (7), we see that *contragredient matrices are so related that the transpose of either is the reciprocal of the other.*

148. Covariance and Contravariance. When new base vectors are introduced by means of the transformation,

$$(1) \qquad \bar{\mathbf{e}}_i = c_i^j \mathbf{e}_j, \qquad \det c_i^j \neq 0,$$

the components of any vector $\mathbf{u}$ are subjected to the contragredient transformation,

$$(2) \qquad \bar{u}^i = \gamma_j^i u^j.$$

In view of (24.3), this may be written

$$\mathbf{u} \cdot \bar{\mathbf{e}}^i = \gamma_j^i \mathbf{u} \cdot \mathbf{e}^j;$$

and, as this holds for every vector $\mathbf{u}$,

(3) $$\bar{\mathbf{e}}^i = \gamma^i_j \mathbf{e}^j.$$

The basis $\mathbf{e}^i$, reciprocal to $\mathbf{e}_i$, thus is transformed in the same way as u^i. To find the transformation for the components u_i, multiply (1) by $\mathbf{u} \cdot$; then, from (24.4),

(4) $$\bar{u}_i = c^j_i u_j,$$

a transformation *cogredient* with (1). Thus quantities written with subscripts transform in the same way (cogrediently); and quantities written with superscripts also transform in the same way—*but the latter transformations are contragredient to the former. Thus the position of the index indicates the character of the transformation.*

A vector $\mathbf{u}$ is said to have the *contravariant* components u^i, the *covariant* components u_i. These terms suggest variation *unlike* and *like* that of the base vectors $\mathbf{e}_i$; in other words, the transformation (1) is regarded as a standard. Thus $[u^1, u^2, u^3]$ and $[u_1, u_2, u_3]$ are two ways of representing the same vector $\mathbf{u}$; in the first it is referred to the basis $\mathbf{e}_i$, in the second to the reciprocal basis $\mathbf{e}^i$. $[u^1, u^2, u^3]$ often is called a *contravariant vector*, $[u_1, u_2, u_3]$ a *covariant vector*. Actually, the vector $\mathbf{u}$ is neither contravariant or covariant, but *invariant*.

By using the properties of the reduced cofactors,

(5) $$c^r_i \gamma^j_r = c^j_r \gamma^r_i = \delta^j_i,$$

we may solve equations (1) to (4) for the original base vectors and components (§ 146). Thus the equations,

(6) $$\bar{\mathbf{e}}_i = c^j_i \mathbf{e}_j, \qquad \bar{u}_i = c^j_i u_j; \qquad \bar{\mathbf{e}}^i = \gamma^i_j \mathbf{e}^j, \qquad \bar{u}^i = \gamma^i_j u^j,$$

have the solutions,

(7) $$\mathbf{e}_i = \gamma^j_i \bar{\mathbf{e}}_j, \qquad u_i = \gamma^j_i \bar{u}_j; \qquad \mathbf{e}^i = c^i_j \bar{\mathbf{e}}^j, \qquad u^i = c^i_j \bar{u}^j.$$

Note that the matrices c^i_j, γ^j_i in (7) are the matrices c^j_i, γ^i_j of (6) *transposed.* To be quite clear on this point the reader should write out equations $\bar{\mathbf{e}}_i = c^j_i \mathbf{e}_j$ and $\mathbf{e}^i = c^i_j \bar{\mathbf{e}}^j$ in full.

From (1) and (3), we find that

$$\bar{\mathbf{e}}_i \cdot \bar{\mathbf{e}}^j = c^r_i \mathbf{e}_r \cdot \gamma^j_s \mathbf{e}^s = c^r_i \gamma^j_s \delta^s_r = c^r_i \gamma^j_r = \delta^j_i,$$

which shows that the new bases have the fundamental property of reciprocal sets.

An expression, such as $u_i v^i$, $u_i \mathbf{e}^i$, $\mathbf{e}_i \cdot \mathbf{e}^i$, summed over the same upper and lower indices, maintains its form under the transformation. Take, for example, the scalar product $u_i v^i$: making use of (2) and (4), we have

$$\bar{u}_i \bar{v}^i = c_i^r u_r \gamma_s^i v^s = \delta_s^r u_r v^s = u_r v^r.$$

The index notation thus automatically indicates quantities that are *invariant* to affine transformations of the base vectors.

149. Orthogonal Transformations. When the basis $\mathbf{e}_i$ and the new basis $\bar{\mathbf{e}}_i$ both consist of mutually orthogonal triples of unit vectors, the transformation,

$$\bar{\mathbf{e}}_i = c_i^j \mathbf{e}_j, \tag{1}$$

is called *orthogonal*. Since both bases are self-reciprocal (§ 23), the corresponding transformation (148.3) between the reciprocal bases,

$$\bar{\mathbf{e}}^i = \gamma_j^i \mathbf{e}^j, \quad \text{now becomes} \quad \bar{\mathbf{e}}_i = \gamma_j^i \mathbf{e}_j.$$

Since this transformation must be the same as (1),

$$C = \begin{pmatrix} c_1^1 & c_1^2 & c_1^3 \\ c_2^1 & c_2^2 & c_2^3 \\ c_3^1 & c_3^2 & c_3^3 \end{pmatrix} = \begin{pmatrix} \gamma_1^1 & \gamma_2^1 & \gamma_3^1 \\ \gamma_1^2 & \gamma_2^2 & \gamma_3^2 \\ \gamma_1^3 & \gamma_2^3 & \gamma_3^3 \end{pmatrix} = \Gamma. \tag{2}$$

The matrix of an orthogonal transformation is called an *orthogonal matrix*. From (2), we have the

THEOREM. *Every element of an orthogonal matrix is its own reduced cofactor in the determinant of the matrix.*

In view of the properties of reduced cofactors, the coefficients of an orthogonal transformation satisfy the equations:

$$c_i^r c_j^r = c_r^i c_r^j = \delta_i^j. \tag{3}$$

The matric equation (147.7) characterizing contragredient transformations becomes

$$CC' = I \tag{4}$$

for orthogonal transformations. Thus *a real matrix is orthogonal when its transpose equals its reciprocal.*

If c denotes the determinant of C, we have, from (4),

$$c^2 = 1, \qquad c = \pm 1. \tag{5}$$

If the bases $\mathbf{e}_i$ and $\bar{\mathbf{e}}_i$ are both dextral or both sinistral, we can bring the trihedral $\mathbf{e}_1\mathbf{e}_2\mathbf{e}_3$ into coincidence with $\bar{\mathbf{e}}_1\bar{\mathbf{e}}_2\bar{\mathbf{e}}_3$ by continuous motion—in fact, by a rotation about an axis through the origin. At any stage of the motion, the base vectors are related by equations of the form (1); and, as the motion progresses, the coefficients c_i^j change continuously from their initial to their final values δ_i^j, and $c = \det c_i^j$ becomes $\det \delta_i^j = 1$. But, since this determinant equals ± 1 at all stages of the motion and must change continuously if at all, $c = 1$ when the bases have the same orientation.

If one basis is dextral, the other sinistral, $\mathbf{e}_1\mathbf{e}_2\mathbf{e}_3$ and $\bar{\mathbf{e}}_1\bar{\mathbf{e}}_2(-\bar{\mathbf{e}}_3)$ have the same orientation. Hence the determinant of the transformation,

$$\bar{\mathbf{e}}_1 = c_1^r\mathbf{e}_r, \qquad \bar{\mathbf{e}}_2 = c_2^r\mathbf{e}_r, \qquad -\bar{\mathbf{e}}_3 = -c_3^r\mathbf{e}_r,$$

namely, $-c$, must equal 1; thus $c = -1$ when the bases have different orientations.

150. Quadratic Forms. A real quadratic polynomial,

$$(1) \qquad A(x, x) = a_{ij}x^ix^j \qquad (i, j = 1, 2, \cdots, n),$$

for which $a_{ij} = a_{ji}$ is called a *real quadratic form* in the variables $x^1, x^2, \cdots, x^n$. The symmetry requirement $a_{ij} = a_{ji}$ entails no loss in generality; for, if $a_{ij} \neq a_{ji}$, the form is not altered if we replace a_{ij} and a_{ji} by $\frac{1}{2}(a_{ij} + a_{ji})$.

The determinant of the coefficients, $a = \det a_{ij}$, is called the *discriminant* of the form. The form is said to be *singular* if $\det a_{ij} = 0$, *non-singular* if $\det a_{ij} \neq 0$.

Associated with the quadratic form $A(x, x)$ is the bilinear form,

$$(2) \qquad A(x, y) = a_{ij}x^iy^j = A(y, x),$$

known as its *polar form.*

The expansion of $a_{ij}(x^i + \lambda y^i)(x^j + \lambda y^j)$ shows that a quadratic form and its polar are related by the identity,

$$(3) \quad A(x + \lambda y, x + \lambda y) = A(x, x) + 2\lambda A(x, y) + \lambda^2 A(y, y).$$

If we make the linear transformation,

$$(4) \qquad x^i = c_r^iy^r, \qquad c = \det c_r^i \neq 0,$$

the form (1) becomes

$$(5) \qquad B(y, y) = b_{rs}y^ry^s \quad \text{where} \quad b_{rs} = a_{ij}c_r^ic_s^j.$$

From the rule for multiplying determinants,

$$\det b_{rs} = \det (a_{ij}c_r^i) \det c_s^j = \det a_{ij} \det c_r^i \det c_s^j;$$

the discriminant of $B(y, y)$ is therefore

$$(6) \qquad b = \det b_{rs} = c^2 a.$$

Linear transformations of non-zero determinant do not alter the singular or non-singular character of a quadratic form.

If the form $A(x, x)$ is singular, it can be expressed in terms of fewer than n variables which are linear functions of x. For since $\det a_{ij} = 0$, the system of n linear equations $a_{ij}x^i = 0$ has a solution $(x_0^1, x_0^2, \cdots, x_0^n)$ which does not consist entirely of zeros; thus we may assume that $x_0^1 \neq 0$. Now $A(x_0, y) = a_{ij}x_0^i y^j = 0$, and, from (3),

$$A(x + \lambda x_0, x + \lambda x_0) = A(x, x),$$

irrespective of the value of λ. If we now choose $\lambda = -x^1/x_0^1$ and write

$$y^i = x^i + \lambda x_0^i = x^i - x_0^i x^1/x_0^1,$$

we have $A(y, y) = A(x, x)$. Since $y^1 = 0$, $A(y, y)$ is expressed in terms of the $n - 1$ variables $y^2, y^3, \cdots, y^n$.

Henceforth we shall consider only non-singular forms. A non-singular form is said to be *definite* when it vanishes only for $x^1 = x^2 = \cdots = x^n = 0$. For all other sets of values, the sign of a definite quadratic form is always the same. To prove this, suppose that $A(x, x) > 0$, for the set $x^1, x^2, \cdots, x^n$ and $A(y, y) < 0$ for the set $y^1, y^2, \cdots, y^n$. Then, from (3),

$$(7) \qquad A(x + \lambda y, x + \lambda y) = 0$$

is a quadratic equation in λ having two distinct real roots λ_1, λ_2; for

$$\{A(x, y)\}^2 - A(x, x)A(y, y) > 0.$$

Consequently, the form vanishes for *two* different sets of values, $x^i + \lambda_1 y^i$ and $x^i + \lambda_2 y^i$ and therefore cannot be definite.

For a definite quadratic form, the quadratic equation (7) in λ must have a pair of either complex roots or equal roots; hence, for a definite form,

$$(8) \qquad \{A(x, y)\}^2 - A(x, x)A(y, y) \leq 0,$$

the equal sign corresponding to the case $x^i + \lambda y^i = 0$, $(i = 1, \cdots, n)$, in which the sets x^i and y^i are proportional.

A definite quadratic form is called *positive definite* or *negative definite* according as its sign (for non-zero sets) is positive or negative. A non-singular form which is not definite is called *indefinite:* such a form may vanish for values x^i other than zero.

When $A(x, x)$ is positive definite, it is well known that we can find a real linear transformation (4) which will reduce $A(x, x)$ to a sum of squares: *

$$B(x, x) = \delta_{ij}^{ij} y^i y^j = y^1 y^1 + y^2 y^2 + \cdots + y^n y^n. \tag{9}$$

The discriminant of a positive definite form is therefore positive; for, from (6), $b = 1 = c^2 a$, and $a = 1/c^2 > 0$.

151. The Metric. Using the notation,

$$g_{ij} = \mathbf{e}_i \cdot \mathbf{e}_j = g_{ji}, \tag{1}$$

for the nine scalar products of the base vectors, we have

$$\mathbf{u} \cdot \mathbf{v} = (u^i \mathbf{e}_i) \cdot (v^j \mathbf{e}_j) = g_{ij} u^i v^j, \qquad \mathbf{u} \cdot \mathbf{u} = g_{ij} u^i u^j. \tag{2}$$

Thus $\mathbf{u} \cdot \mathbf{u}$ is a real quadratic form,

$$g_{11} u^1 u^1 + g_{22} u^2 u^2 + g_{33} u^3 u^3 + 2g_{12} u^1 u^2 + 2g_{23} u^2 u^3 + 2g_{31} u^3 u^1,$$

in the variables u^1, u^2, u^3; and, since $\mathbf{u} \cdot \mathbf{u} > 0$ when $\mathbf{u} \neq 0$, this form is *positive definite.* Moreover, $\mathbf{u} \cdot \mathbf{v}$ is given by the associated bilinear (polar) form in u^i, v^j.

From (2), we have

$$|\mathbf{u}| = \sqrt{g_{ij} u^i u^j}, \tag{3}$$

$$\cos(\mathbf{u}, \mathbf{v}) = \frac{\mathbf{u} \cdot \mathbf{v}}{|\mathbf{u}| |\mathbf{v}|} = \frac{g_{ij} u^i v^j}{\sqrt{g_{ij} u^i u^j} \sqrt{g_{ij} v^i v^j}}. \tag{4}$$

Thus, when the six constants g_{ij} are known, we can find the lengths of vectors and the angles between them when a fixed unit of length is adopted. The quantities g_{ij}, since they make *measurements* possible, are said to determine the *metric* of our 3-space; and $g_{ij} u^i u^j$ is called the *metric* (or *fundamental*) *quadratic form.*

152. Relations between Reciprocal Bases. The discriminant of the metric form is

$$g = \begin{vmatrix} g_{11} & g_{12} & g_{13} \\ g_{21} & g_{22} & g_{23} \\ g_{31} & g_{32} & g_{33} \end{vmatrix}. \tag{1}$$

* Bôcher, *Introduction to Higher Algebra,* New York, 1907, p. 150.

The *reduced* cofactor of g_{ij} in this determinant is denoted by g^{ij}; and we have, from (146.10), the relations:

(2) $$g_{ir}g^{jr} = g_{ri}g^{rj} = \delta_i^j.$$

Since any vector may be written

$$\mathbf{u} = u_j\mathbf{e}^j = \mathbf{u} \cdot \mathbf{e}_j\mathbf{e}^j,$$

we have, on taking $\mathbf{u} = \mathbf{e}_i$,

(3) $$\mathbf{e}_i = g_{ij}\mathbf{e}^j.$$

We may solve these equations for $\mathbf{e}^j$ by multiplying by g^{ik} and summing on i; thus

$$g^{ik}\mathbf{e}_i = g^{ik}g_{ij}\mathbf{e}^j = \delta_j^k\mathbf{e}^j = \mathbf{e}^k,$$

or, if we interchange i and k,

(4) $$\mathbf{e}^i = g^{ki}\mathbf{e}_k.$$

From (4), we have also

$$\mathbf{e}^i \cdot \mathbf{e}^j = g^{ki}\mathbf{e}_k \cdot \mathbf{e}^j = g^{ki}\delta_k^j = g^{ji},$$

(5) $$g^{ij} = \mathbf{e}^i \cdot \mathbf{e}^j = g^{ji}.$$

Making use of (3) and (4), we now have, for any vector $\mathbf{u}$,

(6) $$u_i = \mathbf{e}_i \cdot \mathbf{u} = g_{ij}\mathbf{e}^j \cdot \mathbf{u} = g_{ij}u^j,$$

(7) $$u^i = \mathbf{e}^i \cdot \mathbf{u} = g^{ij}\mathbf{e}_j \cdot \mathbf{u} = g^{ij}u_j.$$

The equations enable us to convert contravariant components of a vector to covariant and vice versa.

Finally, from (24.14), we have

(8) $$g = \det g_{ij} = \det(\mathbf{e}_i \cdot \mathbf{e}_j) = [\mathbf{e}_1\mathbf{e}_2\mathbf{e}_3]^2 = E^2,$$

(9) $$\det g^{ij} = \det(\mathbf{e}^i \cdot \mathbf{e}^j) = [\mathbf{e}^1\mathbf{e}^2\mathbf{e}^3]^2 = E^{-2} = \frac{1}{g}\cdot$$

153. The Affine Group. When an origin O is given, the constant basis $\mathbf{e}_1, \mathbf{e}_2, \mathbf{e}_3$ defines a Cartesian coordinate system x^1, x^2, x^3; for any point P is determined by the components of its position vector:

$$\overrightarrow{OP} = x^1\mathbf{e}_1 + x^2\mathbf{e}_2 + x^3\mathbf{e}_3.$$

The components x^i of $\overrightarrow{OP}$ are called the *Cartesian coordinates* of P relative to the basis $\mathbf{e}_i$.

When the base vectors $\mathbf{e}_i$ are subjected to the transformation,

$$(1) \qquad \bar{\mathbf{e}}_i = c_i^j \mathbf{e}_j, \qquad c = \det c_i^j \neq 0,$$

the invariance of $\overrightarrow{OP}$, namely,

$$(2) \qquad x^j \mathbf{e}_j = \bar{x}^i \bar{\mathbf{e}}_i,$$

induces a contragredient transformation on the coordinates. For, if we substitute from (1) in (2), we find that $x^j = c_i^j \bar{x}^i$; that is,

$$(3) \qquad x^i = c_j^i \bar{x}^j.$$

The matrix c_i^j in (1) is *transposed* in (3); and, while (1) expresses the new base vectors in terms of the old, (3) expresses the old coordinates in terms of the new. We now can solve (3) for the new coordinates by multiplying by the reduced cofactors γ_i^k and summing; thus we find $x^i \gamma_i^k = \bar{x}^k$, or

$$(4) \qquad \bar{x}^j = \gamma_i^j x^i, \qquad \gamma = \det \gamma_i^j = 1/c.$$

The transformation (3) is called *affine*, or, more specifically, *centered affine*, since the origin O has not been altered. The centered-affine transformations with non-zero determinant form a *group*, that is,

(*a*) the set includes the identity transformation:

$$x^i = \delta_j^i \bar{x}^j = \bar{x}^i, \qquad \det \delta_j^i = 1;$$

(*b*) each transformation of the set has an inverse;
(*c*) the succession of two transformations of the set is equivalent to a single transformation of the set.

Thus the transformations,

$$x^i = a_j^i \bar{x}^j, \qquad \bar{x}^j = b_k^j \bar{\bar{x}}^k, \quad \text{give} \quad x^i = c_k^i \bar{\bar{x}}^k,$$

where

$$c_k^i = a_j^i b_k^j \quad \text{and} \quad \det c_k^i = (\det a_j^i)(\det b_k^j) \neq 0.$$

Our present point of view is that a transformation of the base vectors induces a definite transformation of coordinates. But the reverse point of view is adopted when transformation groups more general than the affine are under consideration: *the coordinates are transformed, and we then inquire as to corresponding transformation of the base vectors.*

In the case of the affine group the transformation (3) of coordinates obviously entails the transformation (1) of base vectors. For, if we put $x^j = \bar{x}^i c_i^j$ in (2), we have

$$\bar{x}^i(c_i^j \mathbf{e}_j - \bar{\mathbf{e}}_i) = 0$$

for all $\bar{x}^i$, that is, (1) must hold.

154. Dyadics. Under the affine transformation,

$$(1) \qquad \bar{\mathbf{e}}_i = c_i^j \mathbf{e}_j, \qquad \bar{x}^i = \gamma_j^i x^j \qquad (c \neq 0),$$

we have also

$$(2) \qquad \bar{\mathbf{e}}^i = \gamma_j^i \mathbf{e}^j, \qquad \bar{x}_i = c_i^j x_j \qquad (148.6).$$

Contravariant and covariant components of a vector, u^i and u_i, transform like x^i and x_i, respectively. The sets of components u^i and u_i often are called contravariant and covariant vectors. Actually they are different representations of the same *invariant* vector,

$$(3) \qquad \mathbf{u} = u^i \mathbf{e}_i = u_i \mathbf{e}^i.$$

Nine (3×3) numbers T^{ij} which transform like the nine products of vector components $u^i v^j$, namely,

$$(4) \qquad \bar{T}^{ij} = \gamma_r^i \gamma_s^j T^{rs},$$

are called *contravariant* components of a dyadic. Similarly, nine numbers T_{ij} which transform like $u_i v_j$, namely,

$$(5) \qquad \bar{T}_{ij} = c_i^r c_j^s T_{rs},$$

are called *covariant* components of a dyadic. Finally, nine numbers $T^i_{\cdot j}$ or $T_i^{\cdot j}$ which transform like $u^i v_j$ or $u_i v^j$, respectively, namely,

$$(6) \qquad \bar{T}^i_{\cdot j} = \gamma_r^i c_j^s T^r_{\cdot s}, \qquad \bar{T}_i^{\cdot j} = c_i^r \gamma_s^j T_r^{\cdot s},$$

are called *mixed* components of a dyadic. Note that in $T^i_{\cdot j}$ the upper index comes first, in $T_i^{\cdot j}$ the lower index.

Just as u^i and u_i are two representations of a single invariant vector $\mathbf{u}$, the components T^{ij}, T_{ij}, $T^i_{\cdot j}$, $T_i^{\cdot j}$ are four representations of one and the same dyadic,

$$(7) \qquad \mathbf{T} = T^{ij}\mathbf{e}_i\mathbf{e}_j = T_{ij}\mathbf{e}^i\mathbf{e}^j = T^i_{\cdot j}\mathbf{e}_i\mathbf{e}^j = T_i^{\cdot j}\mathbf{e}^i\mathbf{e}_j.$$

Note that the indices on components and base vectors always are placed in the same order from left to right. All forms of $\mathbf{T}$ given in (7) are invariant; for example, on using (1) and (4), we have

$$\bar{T}^{ij}\bar{\mathbf{e}}_i\bar{\mathbf{e}}_j = \gamma^i_r\gamma^j_s T^{rs} c^h_i \mathbf{e}_h c^k_j \mathbf{e}_k = \delta^h_r \delta^k_s T^{rs}\mathbf{e}_h\mathbf{e}_k = T^{hk}\mathbf{e}_h\mathbf{e}_k,$$

from the properties of reduced cofactors (146.8).

The coefficients g_{ij} of the metric quadratic form are tensor components; for, from (151.1),

$$\bar{g}_{ij} = \bar{\mathbf{e}}_i \cdot \bar{\mathbf{e}}_j = (c^r_i \mathbf{e}_r) \cdot (c^s_j \mathbf{e}_s) = c^r_i c^s_j g_{rs}, \tag{8}$$

in agreement with (5). The same is true of g^{ij}; for, from (152.5),

$$\bar{g}^{ij} = \bar{\mathbf{e}}^i \cdot \bar{\mathbf{e}}^j = (\gamma^i_r \mathbf{e}^r) \cdot (\gamma^j_s \mathbf{e}^s) = \gamma^i_r \gamma^j_s g^{rs}, \tag{9}$$

in agreement with (4). Again, if we *define* $\bar{\delta}^i_j = \delta^i_j$, δ^i_j transforms as a mixed tensor,

$$\bar{\delta}^i_j = \bar{\mathbf{e}}^i \cdot \bar{\mathbf{e}}_j = (\gamma^i_r \mathbf{e}^r) \cdot (c^s_j \mathbf{e}_s) = \gamma^i_r c^s_j \delta^r_s. \tag{10}$$

Indeed, g_{ij}, g^{ij} and δ^i_j are all components of the same tensor,

$$g_{ij}\mathbf{e}^i\mathbf{e}^j = g^{ij}\mathbf{e}_i\mathbf{e}_j = \delta^i_j\mathbf{e}_i\mathbf{e}^j = \delta^j_i\mathbf{e}^i\mathbf{e}_j = \mathbf{e}_i\mathbf{e}^i = \mathbf{e}^i\mathbf{e}_i, \tag{11}$$

namely, the *idemfactor* (§ 66). In fact, if we use the formulas,

$$\mathbf{e}_i = g_{ir}\mathbf{e}^r, \qquad \mathbf{e}^i = g^{ir}\mathbf{e}_r \qquad (\S\ 152), \tag{12}$$

and remember the properties of g_{ij}, g^{ij} as reduced cofactors, any two members of (11) can be shown to be equal; for example,

$$g_{ij}\mathbf{e}^i\mathbf{e}^j = g_{ij}g^{ir}\mathbf{e}_r\mathbf{e}^j = \delta^r_j\mathbf{e}_r\mathbf{e}^j = \mathbf{e}_j\mathbf{e}^j,$$

$$g_{ij}\mathbf{e}^i\mathbf{e}^j = g_{ij}\mathbf{e}^i g^{jr}\mathbf{e}_r = \delta^r_i\mathbf{e}^i\mathbf{e}_r = \mathbf{e}^i\mathbf{e}_i.$$

Since $\mathbf{e}_i\mathbf{e}^i$ and $\mathbf{e}^i\mathbf{e}_i$ represent the same tensor, the *order* of the indices in its mixed components, the Kronecker deltas, is immaterial.

Any component of $\mathbf{T}$ may be expressed in terms of components of another type by means of the metric form. For example, we have, from (7) and (12),

$$\begin{aligned} T^{ij} = \mathbf{e}^i \cdot \mathbf{T} \cdot \mathbf{e}^j &= g^{ir}\, \mathbf{e}_r \cdot \mathbf{T} \cdot \mathbf{e}^j = g^{ir}T_r{}^{\cdot j}, \\ &= g^{jr}\, \mathbf{e}^i \cdot \mathbf{T} \cdot \mathbf{e}_r = g^{jr}T^i{}_{\cdot r}, \\ &= g^{ir}g^{js}\, \mathbf{e}_r \cdot \mathbf{T} \cdot \mathbf{e}_s = g^{ir}g^{js}T_{rs}. \end{aligned}$$

155. Absolute Tensors. We next define *absolute tensors* with respect to the centered-affine group of transformations: $x^i = c^i_j\bar{x}^j$ $(\bar{x}^j = \gamma^j_i x^i)$.

Scalars $\varphi(x^1, x^2, x^3)$ which have one component in each coordinate system given by

$$(1) \qquad \bar{\varphi}(\bar{x}^1, \bar{x}^2, \bar{x}^3) = \varphi(x^1, x^2, x^3)$$

are said to be *absolute tensors of valence zero.*

Vectors have three components, and these may be of two types u_i, u^i. The laws of transformation,

$$(2) \qquad \bar{u}_i = c_i^j u_j, \qquad \bar{u}^i = \gamma_j^i u^j,$$

characterize *absolute tensors of valence one.* The vector itself,

$$\mathbf{u} = u_i \mathbf{e}^i = u^i \mathbf{e}_i,$$

is *invariant* to the transformation: $\bar{\mathbf{u}} = \mathbf{u}$.

Dyadics have $3^2 = 9$ components, and these may be of $2^2 = 4$ types: $T_{ij}, T_i^{\cdot j}, T^i_{\cdot j}, T^{ij}$. If these components transform, respectively, like the products $u_i v_j, u_i v^j, u^i v_j, u^i v^j$ of components of two absolute vectors, they are said to form *absolute tensors of valence two.* The dyadic itself,

$$\mathbf{T} = T_{ij}\mathbf{e}^i\mathbf{e}^j = T_i^{\cdot j}\mathbf{e}^i\mathbf{e}_j = T^i_{\cdot j}\mathbf{e}_i\mathbf{e}^j = T^{ij}\mathbf{e}_i\mathbf{e}_j,$$

is *invariant* to the transformation: $\bar{\mathbf{T}} = \mathbf{T}$.

In general, an *absolute tensor of valence m* is a set of 3^m components that transform like the product of m absolute-vector components; and, since each of these may be covariant or contravariant, the components are of 2^m types. One of these types is purely contravariant ($T^{ij\cdots k}$), another purely covariant ($T_{ij\cdots k}$); the remaining types are mixed and have both upper and lower indices.

Consider, for example, a tensor $\mathbf{T}$ of valence three (a triadic); its $3^3 = 27$ components may be of $2^3 = 8$ types. Its covariant components T_{ijk} will transform like $u_i v_j w_k$, namely,

$$(3) \qquad \bar{T}_{ijk} = c_i^r c_j^s c_k^t T_{rst};$$

and the mixed components $T_i^{\cdot jk}$ like $u_i v^j w^k$, namely,

$$(4) \qquad \bar{T}_i^{\cdot jk} = c_i^r \gamma_s^j \gamma_t^k T_r^{\cdot st}.$$

The tensor itself,

$$\mathbf{T} = T_{ijk}\mathbf{e}^i\mathbf{e}^j\mathbf{e}^k = T_i^{\cdot jk}\mathbf{e}^i\mathbf{e}_j\mathbf{e}_k = \cdots,$$

is *invariant* to the transformation; for example,

$$\begin{aligned}\bar{\mathbf{T}} &= \bar{T}_{ijk}\bar{\mathbf{e}}^i\bar{\mathbf{e}}^j\bar{\mathbf{e}}^k \\ &= (c_i^r c_j^s c_k^t T_{rst})(\gamma_a^i \mathbf{e}^a)(\gamma_b^j \mathbf{e}^b)(\gamma_c^k \mathbf{e}^c) \\ &= \delta_a^r \delta_b^s \delta_c^t T_{rst}\mathbf{e}^a\mathbf{e}^b\mathbf{e}^c \\ &= T_{rst}\mathbf{e}^r\mathbf{e}^s\mathbf{e}^t = \mathbf{T}.\end{aligned}$$

Note that the indices on the base vectors in $\mathbf{T}$ have the same order as the indices on the component but occupy opposed positions.

We may solve the 27 equations (3) for the original components T_{rst} by multiplying by $\gamma_a^i\gamma_b^j\gamma_c^k$ and summing; we thus find

$$\gamma_a^i\gamma_b^j\gamma_c^k\bar{T}_{ijk} = \delta_a^r\delta_b^s\delta_c^t T_{rst} = T_{abc}. \tag{5}$$

In similar fashion we find, from (4),

$$\gamma_a^i c_j^b c_k^c \bar{T}_i^{\cdot jk} = \delta_a^r \delta_s^b \delta_t^c T_r^{\cdot st} = T_a^{\cdot bc}. \tag{6}$$

156. Relative Tensors. When the coordinates are subjected to an affine transformation,

$$x^i = c_r^i \bar{x}^r, \qquad \bar{x}^i = \gamma_r^i x^r \qquad (c = \det c_j^i \neq 0), \tag{1}$$

the transformation equations for all tensor components thus far considered contain only the coefficients c_j^i, γ_j^i. More generally, we may have equations of transformation, such as

$$\bar{T}_i^{\cdot jk} = c^N c_i^r \gamma_s^j \gamma_t^k T_r^{\cdot st}, \tag{2}$$

which involve the Nth power of the determinant c. In this case the quantities in question are said to be components of a *relative tensor of weight* N. When $N = 0$, equations (2) reduce to (155.4) for an *absolute* tensor. As a consequence of (2), the relative tensor $\mathbf{T} = T_r^{\cdot st}\mathbf{e}^r\mathbf{e}_s\mathbf{e}_t$ becomes

$$\bar{\mathbf{T}} = c^N\mathbf{T} \tag{3}$$

when the transformation (1) is effected.

The law of transformation of a tensor component is determined by

(i) its *valence:* the number of its indices;
(ii) its *type:* the position of its indices from left to right;
(iii) its *weight:* the power of c that enters into the transformation equations.

If the weight of a tensor is not specified, it is assumed to be zero; the tensor is then absolute.

A *relative scalar* φ of weight N (valence zero) is transformed according to

$$\bar{\varphi}(\bar{x}^1, \bar{x}^2, \bar{x}^3) = c^N \varphi(x^1, x^2, x^3). \tag{4}$$

The box product of the base vectors, $E = \mathbf{e}_1 \cdot \mathbf{e}_2 \times \mathbf{e}_3$, is a relative scalar of weight 1; for, from (147.2),

$$\bar{E} = cE. \tag{5}$$

The box product of the reciprocal base vectors, $\mathbf{e}^1 \cdot \mathbf{e}^2 \times \mathbf{e}^3 = E^{-1}$, is a relative scalar of weight -1; for $\bar{E}^{-1} = c^{-1}E^{-1}$, from (5).

The determinant,

$$g = \det g_{ij} = E^2 \qquad (152.8), \tag{6}$$

is a relative scalar of weight 2; for $\bar{E}^2 = c^2E^2$.

THEOREM. *The permutation symbols ϵ_{ijk} and ϵ^{ijk}, regarded as the same set of numbers in all coordinate systems, are components of relative tensors of weight -1 and 1, respectively.*

Proof. If the formulas (146.2) are applied to the determinants,

$$c = \det c_i^j, \qquad \gamma = \det \gamma_i^j = 1/c,$$

we have

$$c\epsilon_{ijk} = c_i^r c_j^s c_k^t \epsilon_{rst}, \qquad \gamma\epsilon^{ijk} = \gamma_r^i \gamma_s^j \gamma_t^k \epsilon^{rst}.$$

Since $\bar{\epsilon}_{ijk} = \epsilon_{ijk}$, $\bar{\epsilon}^{ijk} = \epsilon^{ijk}$ by definition, these equations assume the form,

$$\bar{\epsilon}_{ijk} = c^{-1} c_i^r c_j^s c_k^t \epsilon_{rst}, \qquad \bar{\epsilon}^{ijk} = c\gamma_r^i \gamma_s^j \gamma_t^k \epsilon^{rst},$$

required by relative tensors of weight -1 and 1.

157. General Transformations. Three equations,

$$\bar{x}^i = f^i(x^1, x^2, x^3), \tag{1}$$

in which the functions f^i are single-valued for all points of a region R and which can be solved reciprocally to give the three equations,

$$x^i = g^i(\bar{x}^1, \bar{x}^2, \bar{x}^3), \tag{2}$$

in which the functions g^i also are single valued, determine a one-to-one correspondence between the sets of numbers x^i and $\bar{x}^i$. If we regard both sets of numbers as coordinates of the same point,

(1) defines a *transformation of coordinates.* The affine transformation is the particular case of (1) in which the functions f^i are linear and homogeneous in x^1, x^2, x^3.

Consider now the totality of such transformations in which the functions $f^i(x^1, x^2, x^3)$ are *analytic functions having a non-vanishing Jacobian* in R:

$$\left| \frac{\partial \bar{x}}{\partial x} \right| = \det \frac{\partial \bar{x}^i}{\partial x^j} \neq 0. \tag{3}$$

The implicit-function theorem ensures the existence of the solutions (2) of equations (1) in a sufficiently restricted neighborhood of any point. In order to have a transformation of coordinates in R, we must assume that the solution (2) exists and is single valued throughout R.

The coordinate transformations (1) thus defined form a *group;* for

(*a*) they include the identity transformation $\bar{x}^i = x^i$ whose Jacobian is 1;

(*b*) each transformation (1) has an inverse (2) whose Jacobian $|\partial x/\partial \bar{x}|$ is the reciprocal of (3);

(*c*) the succession of two transformations of the set is equivalent to another transformation of the set.

As to (*c*), the two transformations,

$$\bar{x}^i = f^i(x^1, x^2, x^3), \qquad \bar{\bar{x}}^i = h^i(\bar{x}^1, \bar{x}^2, \bar{x}^3),$$

are equivalent to

$$\bar{\bar{x}}^i = h^i[f^1(x), f^2(x), f^3(x)] = j^i(x^1, x^2, x^3),$$

for which the Jacobian,

$$\det \frac{\partial \bar{\bar{x}}^i}{\partial x^j} = \det \left(\frac{\partial \bar{\bar{x}}^i}{\partial \bar{x}^r} \frac{\partial \bar{x}^r}{\partial x^j} \right) = \left(\det \frac{\partial \bar{\bar{x}}^i}{\partial \bar{x}^r} \right) \left(\det \frac{\partial \bar{x}^r}{\partial x^j} \right) \neq 0.$$

Moreover, since f^i and h^i are analytic functions, the functions j^i are likewise analytic.

In the affine transformation,

$$\bar{x}^i = \gamma^i_r x^r, \qquad x^i = c^i_r \bar{x}^r, \tag{4}$$

we have

$$\frac{\partial \bar{x}^i}{\partial x^r} = \gamma^i_r, \qquad \frac{\partial x^i}{\partial \bar{x}^r} = c^i_r; \tag{5}$$

the Jacobians,

$$\text{(6)} \qquad \det \frac{\partial \bar{x}^i}{\partial x^r} = \gamma, \qquad \det \frac{\partial x^i}{\partial \bar{x}^r} = c, \qquad \text{and} \quad \gamma c = 1.$$

A tensor of weight N, with respect to the affine group, transforms according to the pattern,

$$\text{(7)} \qquad \bar{T}^{ij}_{\cdot\cdot k} = c^N \gamma^i_r \gamma^j_s c^t_k T^{rs}_{\cdot\cdot t}.$$

In view of (5) and (6), this equation may be written

$$\text{(8)} \qquad \bar{T}^{ij}_{\cdot\cdot k} = \left| \frac{\partial x}{\partial \bar{x}} \right|^N \frac{\partial \bar{x}^i}{\partial x^r} \frac{\partial \bar{x}^j}{\partial x^s} \frac{\partial x^t}{\partial \bar{x}^k} T^{rs}_{\cdot\cdot t}$$

Now, *by definition*, a set of 3^3 functions $T^{rs}_{\cdot\cdot t}$ is said to form a tensor of valence 3 and weight N with respect to the general transformations (1) provided the components transform according to the pattern (8). This equation becomes (7) when the transformation is affine and constitutes a natural generalization of (7). The corresponding equation for a tensor of any valence or type is now obvious. In particular, for contravariant and covariant components of an absolute vector,

$$\text{(9), (10)} \qquad \bar{u}^i = \frac{\partial \bar{x}^i}{\partial x^r} u^r, \qquad \bar{u}_i = \frac{\partial x^r}{\partial \bar{x}^i} u_r.$$

Since the new tensor components are linear and homogeneous in terms of the old components, we have the important result: *If the components of a tensor vanish in one coordinate system, they vanish in all coordinate systems.*

More generally, *tensor equations maintain their form in all coordinate systems.* If any geometrical or physical property is expressed by means of an equation between tensors, this equation in an arbitrary coordinate system expresses the same property.

Although the coordinates themselves are not vector components in the general group (1), their *differentials* dx^i are the components of a contravariant vector; for, from (1),

$$\text{(11)} \qquad d\bar{x}^i = \frac{\partial \bar{x}^i}{\partial x^1} dx^1 + \frac{\partial \bar{x}^i}{\partial x^2} dx^2 + \frac{\partial \bar{x}^i}{\partial x^3} dx^3 = \frac{\partial \bar{x}^i}{\partial x^r} dx^r.$$

We regard the differentials dx^r as the prototype of contravariant vectors; and the rule of total differentiation gives the correct pattern for their transformation.

The partial derivatives $\partial\varphi/\partial x^i$ of an absolute scalar $\varphi(x^1, x^2, x^3)$ are the components of a covariant vector; for, since

$$\bar{\varphi}(\bar{x}^1, \bar{x}^2, \bar{x}^3) = \varphi(x^1, x^2, x^3),$$

when we replace x^i on the right by the values (2),

$$\frac{\partial\bar{\varphi}}{\partial\bar{x}^i} = \frac{\partial\varphi}{\partial x^1}\frac{\partial x^1}{\partial\bar{x}^i} + \frac{\partial\varphi}{\partial x^2}\frac{\partial x^2}{\partial\bar{x}^i} + \frac{\partial\varphi}{\partial x^3}\frac{\partial x^3}{\partial\bar{x}^i} = \frac{\partial x^r}{\partial\bar{x}^i}\frac{\partial\varphi}{\partial x^r}. \tag{12}$$

We regard the derivatives $\partial\varphi/\partial x^r$ as the prototype of covariant vectors; and the chain rule for partial differentiation gives the correct pattern for their transformation.

In the affine transformation, the position vector $\mathbf{r} = x^i\mathbf{e}_i$, and

$$\mathbf{e}_i = \partial\mathbf{r}/\partial x^i. \tag{13}$$

We now adopt (13) as the definition of $\mathbf{e}_i$ in any coordinate system x^i; then

$$\bar{\mathbf{e}}_i = \frac{\partial\mathbf{r}}{\partial\bar{x}^i} = \frac{\partial\mathbf{r}}{\partial x^r}\frac{\partial x^r}{\partial\bar{x}^i} = \frac{\partial x^r}{\partial\bar{x}^i}\mathbf{e}_r. \tag{14}$$

The base vectors $\mathbf{e}_i$ thus transform after the pattern in (10) and therefore merit a subscript.

The base vectors $\mathbf{e}^i$ must transform after the pattern in (9), namely,

$$\bar{\mathbf{e}}^i = \frac{\partial\bar{x}^i}{\partial x^r}\mathbf{e}^r; \tag{15}$$

for, since $\mathbf{e}_r \cdot \mathbf{e}^s = \delta_r^s$,

$$\bar{\mathbf{e}}_i \cdot \bar{\mathbf{e}}^j = \frac{\partial x^r}{\partial\bar{x}^i}\frac{\partial\bar{x}^j}{\partial x^s}\mathbf{e}_r \cdot \mathbf{e}^s = \frac{\partial x^r}{\partial\bar{x}^i}\frac{\partial\bar{x}^j}{\partial x^r} = \frac{\partial\bar{x}^j}{\partial\bar{x}^i} = \delta_i^j,$$

and the sets $\bar{\mathbf{e}}_i$, $\bar{\mathbf{e}}^j$ are also reciprocal.

We now can show that

$$g_{ij} = \mathbf{e}_i \cdot \mathbf{e}_j, \qquad g^{ij} = \mathbf{e}^i \cdot \mathbf{e}^j, \qquad \delta_j^i = \mathbf{e}^i \cdot \mathbf{e}_j \tag{16}$$

transform as tensor components; we need only make the replacements indicated by (5) in the proofs of § 154. In fact g_{ij}, g^{ij}, δ_j^i are all components of the idemfactor $\mathbf{e}_r\mathbf{e}^r = \mathbf{e}^r\mathbf{e}_r$; thus δ_j^i sometimes is written g_j^i. Moreover,

$$\mathbf{e}_i = \mathbf{e}_i \cdot \mathbf{e}_r\mathbf{e}^r = g_{ir}\mathbf{e}^r, \qquad \mathbf{e}^i = \mathbf{e}^i \cdot \mathbf{e}^r\mathbf{e}_r = g^{ir}\mathbf{e}_r. \tag{17}$$

158. Permutation Tensor. The box product $E = \mathbf{e}_1 \cdot \mathbf{e}_2 \times \mathbf{e}_3$ is a relative scalar of weight 1:

$$\bar{E} = \left| \frac{\partial x}{\partial \bar{x}} \right| E; \tag{1}$$

for the proof of (147.2) applies when we replace c^i_j by $\partial x^i / \partial \bar{x}^j$. Moreover, $E^{-1} = \mathbf{e}^1 \cdot \mathbf{e}^2 \times \mathbf{e}^3$ is a relative scalar of weight -1.

The *permutation tensor* is defined by

$$\mathbf{e}_i \cdot \mathbf{e}_j \times \mathbf{e}_k \; \mathbf{e}^i\mathbf{e}^j\mathbf{e}^k = \mathbf{e}^i \cdot \mathbf{e}^j \times \mathbf{e}^k \; \mathbf{e}_i\mathbf{e}_j\mathbf{e}_k. \tag{2}$$

These triadics are equal absolute tensors. For, if we put

$$\mathbf{e}^i = g^{ir}\mathbf{e}_r, \qquad \mathbf{e}^j = g^{js}\mathbf{e}_s, \qquad \mathbf{e}^k = g^{kt}\mathbf{e}_t,$$

the left member becomes $\mathbf{e}^r \cdot \mathbf{e}^s \times \mathbf{e}^t \; \mathbf{e}_r\mathbf{e}_s\mathbf{e}_t$. Moreover, (157.14) and (157.15) show that

$$\bar{\mathbf{e}}_i \cdot \bar{\mathbf{e}}_j \times \bar{\mathbf{e}}_k \; \bar{\mathbf{e}}^i\bar{\mathbf{e}}^j\bar{\mathbf{e}}^k = \mathbf{e}_i \cdot \mathbf{e}_j \times \mathbf{e}_k \; \mathbf{e}^i\mathbf{e}^j\mathbf{e}^k. \tag{3}$$

From (2), we see that the covariant and contravariant components of the permutation tensor are

$$\mathbf{e}_i \cdot \mathbf{e}_j \times \mathbf{e}_k = \epsilon_{ijk}E, \qquad \mathbf{e}^i \cdot \mathbf{e}^j \times \mathbf{e}^k = \epsilon^{ijk}E^{-1}. \tag{4}$$

Since E and E^{-1} are relative scalars of weight 1 and -1, ϵ_{ijk} and ϵ^{ijk} are components of relative tensors of weight -1 and 1. This is the theorem of § 156; the former proof still holds when obvious changes are made.

159. Operations with Tensors. The three basic operations on tensors are *addition*, *multiplication*, and *contraction*.

1. *Addition* of tensor components of the *same valence, weight, and type* generates a tensor component of precisely the same characters.

Example. If P_{ij}, Q_{ij} are both of weight N,

$$\bar{P}_{ij} + \bar{Q}_{ij} = \left| \frac{\partial x}{\partial \bar{x}} \right|^N \frac{\partial x^r}{\partial \bar{x}^i} \frac{\partial x^s}{\partial \bar{x}^j} (P_{rs} + Q_{rs}).$$

Hence $T_{ij} = P_{ij} + Q_{ij}$ transforms in exactly the same manner as P_{ij} and Q_{ij}.

2. *Multiplication* of tensor components of valence m_1, m_2, of weight N_1, N_2, and of arbitrary type generates a tensor component of valence $m_1 + m_2$, of weight $N_1 + N_2$, and of a type which is defined by the position of the indices in the factors.

Example. Let P^{ij} and u_k have the weights 2 and 1; then

$$\bar{P}^{ij}\bar{u}_k = \left|\frac{\partial x}{\partial \bar{x}}\right|^2 \frac{\partial \bar{x}^i}{\partial x^r}\frac{\partial \bar{x}^j}{\partial x^s} P^{rs} \left|\frac{\partial x}{\partial \bar{x}}\right| \frac{\partial x^t}{\partial \bar{x}^k} u_t = \left|\frac{\partial x}{\partial \bar{x}}\right|^3 \frac{\partial \bar{x}^i}{\partial x^r}\frac{\partial \bar{x}^j}{\partial x^s}\frac{\partial x^t}{\partial \bar{x}^k} P^{rs} u_t.$$

Hence $T^{ij}_{\cdot\cdot k} = P^{ij}u_k$ transforms as a tensor of valence 3, of weight 3, and of the type indicated by its indices. Even though $P^{ij}u_k = u_kP^{ij}$, the product *tensors,*

$$\mathbf{Pu} = P^{ij}u_k\mathbf{e}_i\mathbf{e}_j\mathbf{e}^k, \qquad \mathbf{uP} = u_kP^{ij}\mathbf{e}^k\mathbf{e}_i\mathbf{e}_j,$$

are not the same.

3. *Contraction.* In any *mixed* tensor component an upper and lower index may be set equal and summed over the index range: this generates a tensor component of the same weight and of valence two less. Its type is determined by the remaining indices, not involved in the summation.

Example 1. In the absolute component $T^{ij}_{\cdot\cdot k}$ set $j = k$; then

$$u^i = T^{ij}_{\cdot\cdot j} = T^{i1}_{\cdot\cdot 1} + T^{i2}_{\cdot\cdot 2} + T^{i3}_{\cdot\cdot 3}$$

is a contravariant vector, for

$$\bar{u}^i = \bar{T}^{ij}_{\cdot\cdot j} = \frac{\partial \bar{x}^i}{\partial x^r}\frac{\partial \bar{x}^j}{\partial x^s}\frac{\partial x^t}{\partial \bar{x}^j} T^{rs}_{\cdot\cdot t} = \frac{\partial \bar{x}^i}{\partial x^r}\frac{\partial x^t}{\partial x^s} T^{rs}_{\cdot\cdot t}$$

$$= \frac{\partial \bar{x}^i}{\partial x^r}\delta^t_s T^{rs}_{\cdot\cdot t} = \frac{\partial \bar{x}^i}{\partial x^r} T^{rt}_{\cdot\cdot t} = \frac{\partial \bar{x}^i}{\partial x^r} u^r.$$

A tensor component may also be contracted with respect to two indices on the same level; we need only raise or lower one index and then contract as previously.

Example 2. To contract the absolute component $T^{ij}_{\cdot\cdot k}$ on the indices i, j we first lower the index j

$$T^i_{\cdot jk} = T^{ir}_{\cdot\cdot k}g_{jr};$$

setting $i = j$ now gives the covariant vector

$$v_k = T^{ir}_{\cdot\cdot k}g_{ir} = T^{ij}_{\cdot\cdot k}g_{ij}.$$

We may also obtain this vector by lowering the index i and setting $i = j$:

$$T_{i\cdot k}^{\cdot j} = T^{rj}_{\cdot\cdot k}g_{ri}, \qquad T_{i\cdot k}^{\cdot i} = T^{ri}_{\cdot\cdot k}g_{ri}.$$

Contraction on a pair of indices is equivalent to the scalar multiplication of the corresponding base vectors in the complete tensor.

In the preceding examples, $\mathbf{T} = T^{ij}_{\cdot\cdot k}\mathbf{e}_i\mathbf{e}_j\mathbf{e}^k$,

$$\mathbf{u} = T^{ij}_{\cdot\cdot k}\mathbf{e}_i\mathbf{e}_j \cdot \mathbf{e}^k = T^{ij}_{\cdot\cdot k}\delta^k_j\mathbf{e}_i = T^{ij}_{\cdot\cdot j}\mathbf{e}_i,$$

$$\mathbf{v} = T^{ij}_{\cdot\cdot k}\mathbf{e}_i \cdot \mathbf{e}_j\mathbf{e}^k = T^{ij}_{\cdot\cdot k}g_{ij}\mathbf{e}^k.$$

The proof that contraction produces *tensors* follows from the equation:

$$\bar{T}^{ij}_{\cdot\cdot k}\bar{\mathbf{e}}_i\bar{\mathbf{e}}_j\bar{\mathbf{e}}^k = T^{ij}_{\cdot\cdot k}\mathbf{e}_i\mathbf{e}_j\mathbf{e}^k.$$

When the new base vectors on the left are expressed in terms of the old, this becomes an identity; and, since scalar multiplication is distributive with respect to addition, it remains an identity when base vectors in the same position are multiplied on both sides. Thus, in ex. 2,

$$\bar{\mathbf{v}} = \bar{T}^{ij}_{\cdot\cdot k}\bar{g}_{ij}\bar{\mathbf{e}}^k = T^{ij}_{\cdot\cdot k}g_{ij}\mathbf{e}^k = \mathbf{v}.$$

As long as a contracted tensor has two or more indices, the foregoing process may be repeated, each contraction reducing the valence by two but leaving the weight unaltered. Thus the tensor $T^{ij}_{\cdot\cdot kh}$ may be contracted twice to yield two different scalars $T^{ij}_{\cdot\cdot ij}$ and $T^{ij}_{\cdot\cdot ji}$, each consisting of nine terms. These are obtained from the invariant tensor $T^{ij}_{\cdot\cdot kh}\mathbf{e}_i\mathbf{e}_j\mathbf{e}^k\mathbf{e}^h$ by the formation of $\mathbf{e}_i \cdot \mathbf{e}^k$, $\mathbf{e}_j \cdot \mathbf{e}^h$ and $\mathbf{e}_i \cdot \mathbf{e}^h$, $\mathbf{e}_j \cdot \mathbf{e}^k$, respectively.

Contraction often is combined with multiplication. For example, if we multiply the vectors u^i, v_j and then contract, we obtain their scalar product $u^i v_i$. Again the product $\mathbf{A} \cdot \mathbf{B}$ of two dyadics defined in § 65 is equivalent to tensor multiplication followed by contraction on the two inner indices. Thus, if $\mathbf{A} = A_{ij}\mathbf{e}^i\mathbf{e}^j$, $\mathbf{B} = B^{kh}\mathbf{e}_k\mathbf{e}_h$,

$$\mathbf{A} \cdot \mathbf{B} = A_{ij}B^{kh}\mathbf{e}^i\mathbf{e}^j \cdot \mathbf{e}_k\mathbf{e}_h = A_{ij}B^{jh}\mathbf{e}^i\mathbf{e}_h.$$

The product of the idemfactor $\mathbf{I} = \mathbf{e}_r\mathbf{e}^r$ and a tensor $\mathbf{T}$ of valence m is a tensor $\mathbf{IT}$ of valence $m + 2$. In general $\mathbf{IT}$ differs from $\mathbf{TI}$; but the *contracted* products reproduce $\mathbf{T}$:

$$\mathbf{I} \cdot \mathbf{T} = \mathbf{T} \cdot \mathbf{I} = \mathbf{T}.$$

160. Symmetry and Antisymmetry. A tensor is said to be *symmetric* in two indices *of the same type* (both covariant or both contravariant) if the value of any component is not changed by permuting them. It is *antisymmetric* or *alternating* in two indices of the same type if permuting them in any component merely changes its sign. Thus, if

$$T_{abc} = T_{bac}, \qquad T_{abc} = -T_{cba},$$

the tensor T is symmetric in a, b, alternating in a, c.

Symmetry or antisymmetry are properties that subsist after a general transformation of coordinates. Thus, for the preceding example, we have

$$\overline{T}_{ijk} = \frac{\partial x^a}{\partial \bar{x}^i}\frac{\partial x^b}{\partial \bar{x}^j}\frac{\partial x^c}{\partial \bar{x}^k}T_{abc} = \frac{\partial x^b}{\partial \bar{x}^j}\frac{\partial x^a}{\partial \bar{x}^i}\frac{\partial x^c}{\partial \bar{x}^k}T_{bac} = \overline{T}_{jik}$$

$$= \frac{\partial x^c}{\partial \bar{x}^k}\frac{\partial x^b}{\partial \bar{x}^j}\frac{\partial x^a}{\partial \bar{x}^i}(-T_{cba}) = -\overline{T}_{kji}.$$

A tensor cannot be symmetric or alternating in two indices of different types; for a property such as $T^i_{\cdot j} = T^j_{\cdot i}$ does not subsist after a transformation of coordinates.

A tensor is said to be *symmetric* in any set of upper or lower indices if its components are not altered in value by any permutation of the set. The subsistence of this property in one coordinate system ensures it in all.

A tensor is said to be *alternating* (or *antisymmetric*) in any set of upper or lower indices if its components are not altered in value by any *even* permutation of the indices and are merely changed in sign by an *odd* permutation of these indices.

A tensor T_{ij} or T_{ijk} (T^{ij}, T^{ijk}) which is alternating in all indices is called a *bivector* or *trivector*, respectively. In such *alternating tensors,* all components having two equal indices are zero. In a trivector T_{ijk}, the non-zero components can have but two values, $\pm T_{123}$; moreover, the contracted product,

$$\epsilon^{ijk}T_{ijk} = 3!T_{123}.$$

In this connection we remind the reader that a given permutation of n indices from some standard order can be accomplished by a succession of transpositions of two adjacent indices and that the total number of transpositions required to bring about a definite permutation is always even or always odd. The permutation in question is said to be *even* or *odd* in the respective cases.

161. Kronecker Deltas. We now generalize the simple Kronecker delta δ^i_j by introducing two others, defined as follows:

(1) $$\delta^{ijk}_{abc} = \epsilon_{abc}\epsilon^{ijk},$$

(2) $$\delta^{ij}_{ab} = \epsilon_{abr}\epsilon^{ijr} = \delta^{ijr}_{abr}.$$

Since the epsilons have weights of 1 and -1, δ^{ijk}_{abc} is an absolute tensor of valence six; hence δ^{ij}_{ab}, formed by contracting δ^{ijk}_{abc}, is an absolute tensor of valence four.

Those definitions show that δ^{ij}_{ab} and δ^{ijk}_{abc} can only assume the values 0, 1, -1; they are evidently alternating in both upper and lower indices. Their precise values in any case are as follows:

If both upper and lower indices of a generalized delta consist of the same *distinct* numbers chosen from 1, 2, 3, the delta is 1 or -1 according as the upper indices form an even or odd permutation of the lower; in all other cases the delta is zero.

This rule is a direct consequence of the properties of the epsilons. We have, for example,

$$\delta^{12}_{12} = 1, \qquad \delta^{32}_{23} = -1, \qquad \delta^{23}_{11} = \delta^{13}_{21} = 0;$$

$$\delta^{123}_{123} = \delta^{231}_{123} = 1, \qquad \delta^{213}_{123} = \delta^{321}_{123} = -1, \qquad \delta^{322}_{123} = 0.$$

162. Vector Algebra in Index Notation. The three operations on tensors enable us to give a succinct account of vector algebra. Vectors are to be regarded as absolute unless stated to be relative; and we shall often speak of *components* as vectors.

If $\mathbf{w} = \mathbf{u} + \mathbf{v}$, we have

$$w^i = u^i + v^i, \quad \text{or} \quad w_i = u_i + v_i;$$

obviously vector addition is commutative and associative.

The tensor product of two vectors $\mathbf{u}$, $\mathbf{v}$ is the dyad $\mathbf{uv}$. On contraction, $\mathbf{uv}$ yields the scalar product,

$$(1) \qquad \mathbf{u} \cdot \mathbf{v} = u^i v_i = u_i v^i = g_{ij} u^i v^j = g^{ij} u_i v_j.$$

From (1), we have

$$\mathbf{u} \cdot \mathbf{v} = \mathbf{v} \cdot \mathbf{u}, \qquad \mathbf{u} \cdot (\mathbf{v} + \mathbf{w}) = \mathbf{u} \cdot \mathbf{v} + \mathbf{u} \cdot \mathbf{w}.$$

The antisymmetric dyadic (bivector) $P = \mathbf{uv} - \mathbf{vu}$ is called the *outer product* of $\mathbf{u}$ and $\mathbf{v}$; its covariant components are

$$(2) \qquad P_{jk} = u_j v_k - u_k v_j = \delta^{ab}_{jk} u_a v_b.$$

The *dual* of P_{jk} (cf. § 170) is the contravariant vector of weight 1:

$$(3) \qquad p^i = \tfrac{1}{2}\epsilon^{ijk} P_{jk} = \epsilon^{ijk} u_j v_k;$$

its components,

$$u_2 v_3 - u_3 v_2, \qquad u_3 v_1 - u_1 v_3, \qquad u_1 v_2 - u_2 v_1,$$

are the same as the non-zero components P_{jk}. Since E^{-1} is a scalar of weight -1, the components $E^{-1} p^i$ are absolute. The corresponding vector,

$$(4)\ E^{-1}p^i\mathbf{e}_i = E^{-1}\epsilon^{ijk}\mathbf{e}_i u_j v_k = E^{-1}\begin{vmatrix} \mathbf{e}_1 & \mathbf{e}_2 & \mathbf{e}_3 \\ u_1 & u_2 & u_3 \\ v_1 & v_2 & v_3 \end{vmatrix} = \mathbf{u}\times\mathbf{v} \quad (24.9).$$

Similarly, from the outer product,

$$(2)' \qquad P^{jk} = \delta^{jk}_{ab}u^a v^b = u^j v^k - u^k v^j,$$

we obtain as dual the covariant vector of weight -1:

$$(3)' \qquad p_i = \tfrac{1}{2}\epsilon_{ijk}P^{jk} = \epsilon_{ijk}u^j v^k.$$

The absolute components Ep_i again give $\mathbf{u}\times\mathbf{v}$:

$$(4)' \qquad Ep_i\mathbf{e}^i = E\epsilon_{ijk}\mathbf{e}^i u^j v^k = E\begin{vmatrix} \mathbf{e}^1 & \mathbf{e}^2 & \mathbf{e}^3 \\ u^1 & u^2 & u^3 \\ v^1 & v^2 & v^3 \end{vmatrix} = \mathbf{u}\times\mathbf{v} \qquad (24.10).$$

From (4) or (4)′, we have

$$\mathbf{u}\times\mathbf{v} = -\mathbf{v}\times\mathbf{u}, \qquad \mathbf{u}\times(\mathbf{v}+\mathbf{w}) = \mathbf{u}\times\mathbf{v} + \mathbf{u}\times\mathbf{w}.$$

The components of the triple product $\mathbf{u}\times(\mathbf{v}\times\mathbf{w})$ are

$$\begin{aligned} (\mathbf{u}\times(\mathbf{v}\times\mathbf{w}))^i &= E^{-1}\epsilon^{ijk}u_j(\mathbf{v}\times\mathbf{w})_k && (4) \\ &= E^{-1}\epsilon^{ijk}u_j E\epsilon_{kab}v^a w^b && (4)' \\ &= \epsilon^{ijk}\epsilon_{abk}u_j v^a w^b \\ &= \delta^{ij}_{ab}u_j v^a w^b && (161.2) \\ &= u_j(v^i w^j - v^j w^i) \\ &= (\mathbf{u}\cdot\mathbf{w})v^i - (\mathbf{u}\cdot\mathbf{v})w^i; \end{aligned}$$

hence

$$\mathbf{u}\times(\mathbf{v}\times\mathbf{w}) = (\mathbf{u}\cdot\mathbf{w})\mathbf{v} - (\mathbf{u}\cdot\mathbf{v})\mathbf{w}.$$

The box product $\mathbf{u}\times\mathbf{v}\cdot\mathbf{w}$ is given by either of the absolute scalars,

$$(5) \qquad (\mathbf{u}\times\mathbf{v})^i w_i = E^{-1}\epsilon^{ijk}u_j v_k w_i = E^{-1}\begin{vmatrix} u_1 & u_2 & u_3 \\ v_1 & v_2 & v_3 \\ w_1 & w_2 & w_3 \end{vmatrix},$$

$$(5)' \qquad (\mathbf{u}\times\mathbf{v})_i w^i = E\epsilon_{ijk}u^j v^k w^i = E\begin{vmatrix} u^1 & u^2 & u^3 \\ v^1 & v^2 & v^3 \\ w^1 & w^2 & w^3 \end{vmatrix},$$

in agreement with (24.12).

163. The Affine Connection. Any vector can be expressed as a linear combination of the base vectors. When the 3^2 derivatives of the base vectors $\partial \mathbf{e}_j/\partial x^i = \mathrm{D}_i\mathbf{e}_j$ are thus expressed,

$$\mathrm{D}_i\mathbf{e}_j = \Gamma_{ij}^1\mathbf{e}_1 + \Gamma_{ij}^2\mathbf{e}_2 + \Gamma_{ij}^3\mathbf{e}_3 = \Gamma_{ij}^r\mathbf{e}_r, \tag{1}$$

the 3^3 coefficients Γ_{ij}^k are called *components of the affine connection.* If we multiply (1) by $\mathbf{e}^k \cdot$, the right-hand member becomes $\Gamma_{ij}^r\delta_r^k = \Gamma_{ij}^k$; hence

$$\Gamma_{ij}^k = \mathbf{e}^k \cdot \mathrm{D}_i\mathbf{e}_j. \tag{2}$$

The law of transformation for Γ_{ij}^k is given by

$$\bar{\Gamma}_{ij}^k = \bar{\mathbf{e}}^k \cdot \bar{\mathrm{D}}_i\bar{\mathbf{e}}_j = \left(\frac{\partial \bar{x}^k}{\partial x^c}\mathbf{e}^c\right) \cdot \left(\frac{\partial x^a}{\partial \bar{x}^i}\mathrm{D}_a\right)\left(\frac{\partial x^b}{\partial \bar{x}^j}\mathbf{e}_b\right), \tag{3}$$

where

$$\bar{\mathrm{D}}_i = \frac{\partial}{\partial \bar{x}^i} = \frac{\partial x^a}{\partial \bar{x}^i}\frac{\partial}{\partial x_a} = \frac{\partial x^a}{\partial \bar{x}^i}\mathrm{D}_a; \tag{4}$$

The differential operator D_i *transforms like a covariant vector* (hence the subscript). In (3), D_a acts on both scalar and vector factors following; hence, from (1),

$$\bar{\Gamma}_{ij}^k = \frac{\partial \bar{x}^k}{\partial x^c}\mathbf{e}^c \cdot \left(\frac{\partial^2 x^b}{\partial \bar{x}^i \partial \bar{x}^j}\mathbf{e}_b + \frac{\partial x^a}{\partial \bar{x}^i}\frac{\partial x^b}{\partial \bar{x}^j}\Gamma_{ab}^r\mathbf{e}_r\right).$$

Since $\mathbf{e}^c \cdot \mathbf{e}_b = \delta_b^c$, $\mathbf{e}^c \cdot \mathbf{e}_r = \delta_r^c$, this gives

$$\bar{\Gamma}_{ij}^k = \frac{\partial^2 x^b}{\partial \bar{x}^i \partial \bar{x}^j}\frac{\partial \bar{x}^k}{\partial x^b} + \frac{\partial x^a}{\partial \bar{x}^i}\frac{\partial x^b}{\partial \bar{x}^j}\frac{\partial \bar{x}^k}{\partial x^c}\Gamma_{ab}^c \tag{5}$$

for the desired law of transformation. Therefore Γ_{ij}^k is a tensor component when and only when

$$\frac{\partial^2 x^b}{\partial \bar{x}^i\,\partial \bar{x}^j} = 0, \qquad \frac{\partial x^b}{\partial \bar{x}^j} = c_j^b \text{ (const)},$$

and hence $x^b = c_j^b\bar{x}^j$, if the coordinates x^i and $\bar{x}^i$ have a common origin.

The components Γ_{ij}^k *of the affine connection* (between the derivatives of the base vectors and the vectors themselves) *are tensor components only with respect to affine transformations.*

Although the Γ_{ij}^k are not tensor components for general transformations, we shall see that the index notation still serves a useful purpose.†

Since $\mathbf{e}_j = D_j\mathbf{r}$ (157.13),

$$\Gamma_{ij}^k = \mathbf{e}^k \cdot D_iD_j\mathbf{r} = \mathbf{e}^k \cdot D_jD_i\mathbf{r} = \Gamma_{ji}^k; \tag{6}$$

and, from (5), we see that the symmetry of Γ_{ij}^k in the subscripts persists after a transformation of coordinates.

If we transform coordinates from $\bar{x}$ to $\tilde{x}$, we have, from (3),

$$\tilde{\Gamma}_{pq}^r = \left(\frac{\partial \tilde{x}^r}{\partial \bar{x}^k}\bar{\mathbf{e}}^k\right) \cdot \left(\frac{\partial \bar{x}^i}{\partial \tilde{x}^p}\bar{D}_i\right)\left(\frac{\partial \bar{x}^j}{\partial \tilde{x}^q}\bar{\mathbf{e}}_j\right), \tag{7}$$

or, on making the replacements,

$$\bar{\mathbf{e}}^k = \frac{\partial \bar{x}^k}{\partial x^c}\mathbf{e}_c, \qquad \bar{D}_i = \frac{\partial x^a}{\partial \bar{x}^i}D_a, \qquad \bar{\mathbf{e}}_j = \frac{\partial x^b}{\partial \bar{x}^j}\mathbf{e}_b,$$

$$\tilde{\Gamma}_{pq}^r = \left(\frac{\partial \tilde{x}^r}{\partial x^c}\mathbf{e}_c\right) \cdot \left(\frac{\partial x^a}{\partial \tilde{x}^p}D_a\right)\left(\frac{\partial x^b}{\partial \tilde{x}^q}\mathbf{e}_b\right), \tag{8}$$

a relation of the same form as (3). Consequently, the succession of transformations $\Gamma \to \bar{\Gamma} \to \tilde{\Gamma}$ produces a transformation $\Gamma \to \tilde{\Gamma}$ of the same form as $\Gamma \to \bar{\Gamma}$. We express this property by saying that the transformation (3), or its equivalent (5), is *transitive*.

If, in particular, $\tilde{x}^i = x^i$, we may delete all the tildas (~) in (7); this equation then gives, on expansion,

$$\Gamma_{pq}^r = \frac{\partial^2 \bar{x}^j}{\partial x^p\,\partial x^q}\frac{\partial x^r}{\partial \bar{x}^j} + \frac{\partial \bar{x}^i}{\partial x^p}\frac{\partial \bar{x}^j}{\partial x^q}\frac{\partial x^r}{\partial \bar{x}^k}\bar{\Gamma}_{ij}^k. \tag{9}$$

These equations constitute the transformation inverse to (5) and also may be derived by solving these equations for Γ_{ab}^c. This solution may be effected by the multiplication of (5) by
$\dfrac{\partial \bar{x}^i}{\partial x^p}\dfrac{\partial \bar{x}^j}{\partial x^q}\dfrac{\partial x^r}{\partial \bar{x}^k}$.

† The coordinates x^i present a similar situation. For affine transformations they are components of a vector; for general transformations this is not the case, but the indices still serve to indicate that their *differentials* dx^i are vector components.

On writing $y = \bar{x}$, the equation of transformation (5) becomes:

$$(10) \qquad \bar{\Gamma}^k_{ij} = \left\{\frac{\partial^2 x^c}{\partial y^i\, \partial y^j} + \frac{\partial x^a}{\partial y^i}\frac{\partial x^b}{\partial y^j}\Gamma^c_{ab}\right\}\frac{\partial y^k}{\partial x^c}.$$

If we now make the change of coordinates,

$$(11) \qquad x^r = x^r_0 + y^r - \tfrac{1}{2}(\Gamma^r_{pq})_0 y^p y^q,$$

where the gammas are computed for $x = x_0$, the point $x^r = x^r_0$ corresponds to $y^r = 0$. Since $\partial y^r/\partial y^i = \delta^r_i$,

$$\frac{\partial x^r}{\partial y^i} = \delta^r_i - (\Gamma^r_{iq})_0 y^q, \qquad \frac{\partial^2 x^r}{\partial y^i\, \partial y^j} = -(\Gamma^r_{ij})_0;$$

hence, at the point $x^r = x^r_0$ $(y^r = 0)$, the brace in (10) becomes

$$-(\Gamma^c_{ij})_0 + \delta^a_i \delta^b_j (\Gamma^c_{ab})_0 = 0.$$

Consequently, all the gammas $\bar{\Gamma}^k_{ij} = 0$ vanish at the origin $y^r = 0$ of the new coordinates. Such a system of coordinates is termed *geodesic*. Since the gammas are *not tensors*, the equations $\bar{\Gamma}^k_{ij} = 0$ $(y = 0)$ do not imply $\Gamma^k_{ij} = 0$ $(x = x_0)$.

Example 1. For cylindrical coordinates ρ, φ, z, we have (§ 89, ex. 1)

$$\mathbf{r} = [x, y, z] = [\rho \cos \varphi, \rho \sin \varphi, z].$$

If we put $x^1 = \rho$, $x^2 = \varphi$, $x^3 = z$,

$$\mathbf{e}_1 = D_1\mathbf{r} = [\cos \varphi, \sin \varphi, 0],$$

$$\mathbf{e}_2 = D_2\mathbf{r} = \rho[-\sin \varphi, \cos \varphi, 0],$$

$$\mathbf{e}_3 = D_3\mathbf{r} = [0, 0, 1];$$

$$D_1\mathbf{e}_1 = 0, \qquad D_2\mathbf{e}_1 = [-\sin \varphi, \cos \varphi, 0] = \frac{1}{\rho}\mathbf{e}_2, \qquad D_3\mathbf{e}_1 = 0,$$

$$D_2\mathbf{e}_2 = -\rho[\cos \varphi, \sin \varphi, 0] = -\rho\mathbf{e}_1, \qquad D_3\mathbf{e}_2 = 0,$$

$$D_3\mathbf{e}_3 = 0.$$

Hence all gammas are zero except

$$\Gamma^2_{12} = \Gamma^2_{21} = 1/\rho, \qquad \Gamma^1_{22} = -\rho.$$

Example 2. For spherical coordinates r, φ, θ, we have (§ 89, ex. 2)

$$\mathbf{r} = [x, y, z] = r[\sin \theta \cos \varphi, \sin \theta \sin \varphi, \cos \theta].$$

If we put $x^1 = r$, $x^2 = \varphi$, $x^3 = \theta$,*

$$\mathbf{e}_1 = D_1\mathbf{r} = [\sin\theta\cos\varphi,\ \sin\theta\sin\varphi,\ \cos\theta],$$

$$\mathbf{e}_2 = D_2\mathbf{r} = r\sin\theta[-\sin\varphi,\ \cos\varphi,\ 0],$$

$$\mathbf{e}_3 = D_3\mathbf{r} = r[\cos\theta\cos\varphi,\ \cos\theta\sin\varphi,\ -\sin\theta];$$

$$D_1\mathbf{e}_1 = 0, \qquad D_2\mathbf{e}_1 = \frac{1}{r}\,\mathbf{e}_2, \qquad D_3\mathbf{e}_1 = \frac{1}{r}\,\mathbf{e}_3,$$

$$D_2\mathbf{e}_2 = -r\sin^2\theta\,\mathbf{e}_1 - \sin\theta\cos\theta\,\mathbf{e}_3, \qquad D_2\mathbf{e}_3 = \cot\theta\,\mathbf{e}_2,$$

$$D_3\mathbf{e}_3 = -r\mathbf{e}_1.$$

The non-zero gammas are therefore

$$\Gamma_{12}^2 = \Gamma_{21}^2 = 1/r, \qquad \Gamma_{13}^3 = \Gamma_{31}^3 = 1/r, \qquad \Gamma_{22}^1 = -r\sin^2\theta,$$

$$\Gamma_{22}^3 = -\sin\theta\cos\theta, \qquad \Gamma_{23}^2 = \Gamma_{32}^2 = \cot\theta, \qquad \Gamma_{33}^1 = -r.$$

164. Kinematics of a Particle. We now may find velocity and acceleration of a particle in general coordinates:

$$\mathbf{v} = v^k\mathbf{e}_k, \qquad \mathbf{a} = d\mathbf{v}/dt = a^k\mathbf{e}_k. \tag{1}$$

In rectangular coordinates $x^1 = x$, $x^2 = y$, $x^3 = z$ we have $v^k = dx^k/dt$ (52.7). Hence, in general coordinates $\bar{x}^i$, these components become

$$\bar{v}^k = \frac{\partial\bar{x}^k}{\partial x^i}\frac{dx^i}{dt} = \frac{d\bar{x}^k}{dt}. \tag{2}$$

The time derivatives of the coordinates, which we write $\dot{x}^k$, thus transform as contravariant vector components.

For the acceleration we have, from (1),

$$\mathbf{a} = \frac{dv^k}{dt}\,\mathbf{e}_k + v^k\frac{\partial\mathbf{e}_k}{\partial x^i}\frac{dx^i}{dt} = \dot{v}^k\mathbf{e}_k + v^iv^k\Gamma_{ik}^j\mathbf{e}_j;$$

and, on interchanging the summation indices j, k in the last term,

$$\mathbf{a} = (\dot{v}^k + v^iv^j\Gamma_{ij}^k)\mathbf{e}_k. \tag{3}$$

Therefore, in any coordinate system the velocity and acceleration components are

$$v^k = \dot{x}^k, \qquad a^k = \ddot{x}^k + \dot{x}^i\dot{x}^j\Gamma_{ij}^k. \tag{4}$$

Since the base vectors $\mathbf{e}_i$ are not, in general, *unit* vectors, we must distinguish between the components v^k and a^k and the numerical values of the terms in $v^k\mathbf{e}_k$, $a^k\mathbf{e}_k$. Thus the terms $v^1\mathbf{e}_1$, $a^1\mathbf{e}_1$ have the numerical values $v^1|\mathbf{e}_1|$, $a^1|\mathbf{e}_1|$.

* The order r, φ, θ is *sinistral* and $[\mathbf{e}_1\mathbf{e}_2\mathbf{e}_3] < 0$; cf. p. 197.

Example 1. *Cylindrical Coordinates* ρ, φ, z. From § 163, ex. 1, $\mathbf{e}_1, \mathbf{e}_2, \mathbf{e}_3$ have the lengths $1, \rho, 1$; hence the values of the velocity terms in (1) are $\dot\rho, \rho\dot\varphi, \dot z$.

From the non-zero gammas $\Gamma^1_{22} = -\rho$, $\Gamma^2_{12} = 1/\rho$, we have

$$a^1 = \ddot\rho - \rho\dot\varphi^2, \qquad a^2 = \ddot\varphi + 2\dot\rho\dot\varphi/\rho, \qquad a^3 = \ddot z;$$

and the numerical values of the acceleration terms are

$$\ddot\rho - \rho\dot\varphi^2, \qquad \rho\ddot\varphi + 2\dot\rho\dot\varphi, \qquad \ddot z.$$

Example 2. *Spherical Coordinates* r, φ, θ. From § 163, ex. 2, $\mathbf{e}_1, \mathbf{e}_2, \mathbf{e}_3$ have the lengths $1, r\sin\theta, r$; hence the velocity terms in (1) have the values,

$$\dot r, \; r\dot\varphi\sin\theta, \; r\dot\theta.$$

With the non-zero gammas $\Gamma^1_{22} = -r\sin^2\theta$, $\Gamma^1_{33} = -r$ we have, from (4),

$$a^1 = \ddot r - \dot\varphi^2\Gamma^1_{22} + \dot\theta^2\Gamma^1_{33} = \ddot r - r\dot\varphi^2\sin^2\theta - r\dot\theta^2;$$

with $\Gamma^2_{12} = 1/r$, $\Gamma^2_{23} = \cot\theta$,

$$a^2 = \ddot\varphi + 2\dot r\dot\varphi\Gamma^2_{12} + 2\dot\varphi\dot\theta\Gamma^2_{23} = \ddot\varphi + 2\dot r\dot\varphi/r + 2\dot\varphi\dot\theta\cot\theta;$$

and, with $\Gamma^3_{13} = 1/r$, $\Gamma^3_{22} = -\sin\theta\cos\theta$,

$$a^3 = \ddot\theta + 2\dot r\dot\theta\Gamma^3_{13} + \dot\varphi^2\Gamma^3_{22} = \ddot\theta + 2\dot r\dot\theta/r - \dot\varphi^2\sin\theta\cos\theta.$$

Hence the values of the acceleration terms are

$$\ddot r - r\dot\varphi^2\sin^2\theta - r\dot\theta^2, \qquad r\ddot\varphi\sin\theta + 2\dot r\dot\varphi\sin\theta + 2r\dot\varphi\dot\theta\cos\theta,$$

$$r\ddot\theta + 2\dot r\dot\theta - r\dot\varphi^2\sin\theta\cos\theta.$$

165. Derivatives of $\mathbf{e}^i$ and E. From the relation $\mathbf{e}^j \cdot \mathbf{e}_r = \delta^j_r$, we have, on differentiation,

$$(D_i\mathbf{e}^j)\cdot\mathbf{e}_r = -\mathbf{e}^j\cdot D_i\mathbf{e}_r = -\Gamma^j_{ir},$$

and hence

$$D_i\mathbf{e}^j = -\Gamma^j_{ir}\mathbf{e}^r. \tag{1}$$

From $D_i\mathbf{e}_j = \Gamma^r_{ij}\mathbf{e}_r$ and (1), we have

$$D_i(\mathbf{e}_j\mathbf{e}^j) = \Gamma^r_{ij}\mathbf{e}_r\mathbf{e}^j - \Gamma^j_{ir}\mathbf{e}_j\mathbf{e}^r = 0, \tag{2}$$

on interchanging the summation indices r, j in the second term.

Since the product $E = \mathbf{e}_1\cdot\mathbf{e}_2\times\mathbf{e}_3$ is distributive with respect to addition, its partial derivatives may be found by the familiar rule for differentiating a product,

$$\begin{aligned} D_iE &= (D_i\mathbf{e}_1)\cdot\mathbf{e}_2\times\mathbf{e}_3 + \mathbf{e}_1\cdot(D_i\mathbf{e}_2)\times\mathbf{e}_3 + \mathbf{e}_1\cdot\mathbf{e}_2\times(D_i\mathbf{e}_3) \\ &= \Gamma^r_{i1}\mathbf{e}_r\cdot\mathbf{e}_2\times\mathbf{e}_3 + \Gamma^r_{i2}\mathbf{e}_1\cdot\mathbf{e}_r\times\mathbf{e}_3 + \Gamma^r_{i3}\mathbf{e}_1\cdot\mathbf{e}_2\times\mathbf{e}_r \\ &= (\Gamma^1_{i1} + \Gamma^2_{i2} + \Gamma^3_{i3})\mathbf{e}_1\cdot\mathbf{e}_2\times\mathbf{e}_3, \end{aligned}$$

or, if we apply the summation convention,

$$\text{(3)} \qquad \mathrm{D}_i E = \Gamma^r_{ir} E.$$

The derivative of $E^{-1} = \mathbf{e}^1 \cdot \mathbf{e}^2 \times \mathbf{e}^3$ is therefore

$$\text{(4)} \qquad \mathrm{D}_i E^{-1} = -E^{-2}\mathrm{D}_i E = -\Gamma^r_{ir} E^{-1}.$$

From (3), we have

$$\mathrm{D}_j \mathrm{D}_i E = (\mathrm{D}_j \Gamma^r_{ir} + \Gamma^r_{ir}\Gamma^s_{js})E;$$

and, since $\mathrm{D}_j \mathrm{D}_i E = \mathrm{D}_i \mathrm{D}_j E$,

$$\text{(5)} \qquad \mathrm{D}_j \Gamma^r_{ir} = \mathrm{D}_i \Gamma^r_{jr}.$$

166. Relation between Affine Connection and Metric Tensor. On differentiating $g_{ij} = \mathbf{e}_i \cdot \mathbf{e}_j$, we have

$$\text{(1)} \qquad \mathrm{D}_k g_{ij} = \Gamma^r_{ki}\mathbf{e}_r \cdot \mathbf{e}_j + \Gamma^r_{kj}\mathbf{e}_i \cdot \mathbf{e}_r = \Gamma^r_{ki} g_{rj} + \Gamma^r_{kj} g_{ir}.$$

Since $D_k g_{ij}$ and Γ^k_{ij} are both symmetric in the indices ij, there are $3 \times 6 = 18$ quantities in each set. Equations (1), 18 in number and linear in the 18 gammas, may be solved for the latter.

We first introduce the notation,

$$\text{(2)} \qquad \Gamma^r_{ij} g_{rk} = \Gamma_{ij,k},$$

just as if Γ^k_{ij} were a tensor whose upper index was lowered. Then we have also

$$\text{(3)} \qquad \Gamma_{ij,r} g^{rk} = \Gamma^k_{ij};$$

for the left member equals

$$\Gamma^s_{ij} g_{sr} g^{rk} = \Gamma^s_{ij} \delta^k_s = \Gamma^k_{ij}.$$

We note that $\Gamma_{ij,\,k}$ is also symmetric in the indices i, j. Moreover

$$\text{(4)} \qquad \mathrm{D}_i \mathbf{e}_j = \Gamma^r_{ij}\mathbf{e}_r = \Gamma_{ij,s}\mathbf{e}^s,$$

if we put $\mathbf{e}_r = g_{rs}\mathbf{e}^s$.

We may now write (1) in the form:

$$\text{(5)} \qquad \Gamma_{ki,j} + \Gamma_{kj,i} = \mathrm{D}_k g_{ij}.$$

If we permute i, j, k cyclically in (5), we obtain the equations,

$$\Gamma_{ij,k} + \Gamma_{ik,j} = \mathrm{D}_i g_{jk},$$

$$\Gamma_{jk,i} + \Gamma_{ji,k} = \mathrm{D}_j g_{ki};$$

and, upon subtracting (5) from their sum, we obtain

$$\Gamma_{ij,k} = \tfrac{1}{2}(\mathrm{D}_i g_{jk} + \mathrm{D}_j g_{ki} - \mathrm{D}_k g_{ij}). \tag{6}$$

We may now compute Γ_{ij}^k from (3).

In the older literature, the components of the affine connection are denoted by

$$[ij,k] = \Gamma_{ij,k}, \qquad \left\{ {k \atop ij} \right\} = \Gamma_{ij}^k.$$

These *Christoffel symbols* of the first and second kind, respectively, therefore are defined by the equations:

$$[ij, k] = \tfrac{1}{2}(\mathrm{D}_i g_{jk} + \mathrm{D}_j g_{ki} - \mathrm{D}_k g_{ij}), \qquad \left\{ {k \atop ij} \right\} = g^{kr}[ij, r]. \tag{7}$$

167. Covariant Derivative. The *gradient* of an absolute tensor **T** is defined as

$$\nabla \mathbf{T} = \mathbf{e}^h \frac{\partial \mathbf{T}}{\partial x^h}. \tag{1}$$

Since $\overline{\mathbf{T}} = \mathbf{T}$,

$$\bar{\mathbf{e}}^h \frac{\partial \overline{\mathbf{T}}}{\partial \bar{x}^h} = \frac{\partial \bar{x}^h}{\partial x^r} \mathbf{e}^r \frac{\partial \mathbf{T}}{\partial x^s} \frac{\partial x^s}{\partial \bar{x}^h} = \delta_r^s \mathbf{e}^r \frac{\partial \mathbf{T}}{\partial x^s} = \mathbf{e}^r \frac{\partial \mathbf{T}}{\partial x^r};$$

hence $\nabla \mathbf{T}$ is a tensor of valence one greater than **T**.

If the components of **T** are $T_{ij \cdots k}^{ab \cdots c}$ (the *order* of the indices is not specified), the components of $\nabla \mathbf{T}$ are written

$$\nabla_h T_{ij \cdots k}^{ab \cdots c}, \tag{2}$$

the index h on ∇ corresponding to the differentiation $\partial / \partial x_h$. This is a covariant index, for the operator $\mathrm{D}_h = \partial / \partial x^h$ transforms like a covariant vector:

$$\overline{\mathrm{D}}_h = \frac{\partial}{\partial \bar{x}^h} = \frac{\partial x^r}{\partial \bar{x}^h} \frac{\partial}{\partial x^r} = \frac{\partial x^r}{\partial \bar{x}^h} \mathrm{D}_r. \tag{3}$$

For this reason the components (2) are called *covariant derivatives* of $T_{ij \cdots k}^{ab \cdots c}$.

If **T** is a relative tensor of weight N and valence m, $E^{-N}\mathbf{T}$ is an absolute tensor, whose gradient,

$$\mathbf{e}^h \mathrm{D}_h (E^{-N} \mathbf{T}),$$

is an absolute tensor of valence $m + 1$. If this is multiplied by E^N, we again obtain a relative tensor of weight N; this tensor is defined to be the gradient of **T** and written

$$\nabla \mathbf{T} = E^N \mathbf{e}^h \mathrm{D}_h (E^{-N} \mathbf{T}). \tag{4}$$

When $N = 0$, (4) reduces to (1). The components of $\nabla\mathbf{T}$, denoted by prefixing ∇_h to the components of $\mathbf{T}$ as in (2), are called *covariant derivatives.*

From (165.3), we have

$$\mathrm{D}_h E^{-N} = -NE^{-N-1}\mathrm{D}_h E = -N\Gamma^r_{hr}E^{-N},$$

and hence

$$\mathrm{D}_h(E^{-N}\mathbf{T}) = E^{-N}(\mathrm{D}_h\mathbf{T} - N\Gamma^r_{hr}\mathbf{T}),$$

$$\nabla\mathbf{T} = \mathbf{e}^h(\mathrm{D}_h - N\Gamma^r_{hr})\mathbf{T}. \tag{5}$$

Thus the operator,

$$\nabla = \mathbf{e}^h(\mathrm{D}_h - N\Gamma^r_{hr}),$$

applied to any invariant tensor $\mathbf{T}$ (with its complement of base vectors) generates another tensor $\nabla\mathbf{T}$ of the same weight and valence one greater.

We next compute the components of $\nabla\mathbf{T}$ explicitly, where

$$\mathbf{T} = T^{ab\cdots c}_{ij\cdots k}\mathbf{e}_a\mathbf{e}_b\cdots\mathbf{e}_c\mathbf{e}^i\mathbf{e}^j\cdots\mathbf{e}^k. \tag{6}$$

If $\mathbf{T}$ is a relative tensor of weight N, $\nabla\mathbf{T}$ contains the term,

$$-N\Gamma^r_{hr}T^{ab\cdots c}_{ij\cdots k}\mathbf{e}^h\mathbf{e}_a\mathbf{e}_b\cdots\mathbf{e}_c\mathbf{e}^i\mathbf{e}^j\cdots\mathbf{e}^k.$$

It remains to compute the part of $\nabla\mathbf{T}$ due to the operator $\mathbf{e}^h\mathrm{D}_h$. Now D_h acts on the "product" of $T^{ab\cdots c}_{ij\cdots k}$ and a series of base vectors. Since this product is distributive with respect to addition, $\mathrm{D}_h\mathbf{T}$ may be computed by the usual rule for a product; hence

$$\begin{aligned}\mathrm{D}_h\mathbf{T} = {} & (\mathrm{D}_h T^{ab\cdots c}_{ij\cdots k})\mathbf{e}_a\mathbf{e}_b\cdots\mathbf{e}_c\mathbf{e}^i\mathbf{e}^j\cdots\mathbf{e}^k \\ & + T^{ab\cdots c}_{ij\cdots k}(\mathrm{D}_h\mathbf{e}_a)\mathbf{e}_b\cdots\mathbf{e}_c\mathbf{e}^i\mathbf{e}^j\cdots\mathbf{e}^k + \cdots \\ & + T^{ab\cdots c}_{ij\cdots k}\mathbf{e}_a\mathbf{e}_b\cdots\mathbf{e}_c(\mathrm{D}_h\mathbf{e}^i)\mathbf{e}^j\cdots\mathbf{e}^k + \cdots.\end{aligned}$$

In the second line put

$$\mathrm{D}_h\mathbf{e}_a = \Gamma^r_{ha}\mathbf{e}_r, \cdots, \qquad \mathrm{D}_h\mathbf{e}_c = \Gamma^r_{hc}\mathbf{e}_r,$$

and in the successive terms interchange the summation indices r and a, r and b, $\cdots$, r and c. In the third line put

$$\mathrm{D}_h\mathbf{e}^i = -\Gamma^i_{hr}\mathbf{e}^r, \cdots, \qquad \mathrm{D}_h\mathbf{e}^k = -\Gamma^k_{hr}\mathbf{e}^r,$$

and in the successive terms interchange the summation indices r and i, r and j, $\cdots$, r and k. When this is done, each term of $\nabla\mathbf{T}$ contains the same complex of base vectors,

$$\mathbf{e}^h\mathbf{e}_a\mathbf{e}_b\cdots\mathbf{e}_c\mathbf{e}^i\mathbf{e}^j\cdots\mathbf{e}^k,$$

and the component $\nabla_h T^{ab\cdots c}_{ij\cdots k}$ of $\nabla\mathbf{T}$ equals the sum of their scalar coefficients:

$$(7)\qquad \begin{aligned}\nabla_h T^{ab\cdots c}_{ij\cdots k} = \mathrm{D}_h T^{ab\cdots c}_{ij\cdots k} &+ T^{rb\cdots c}_{ij\cdots k}\Gamma^a_{hr} + \cdots + T^{ab\cdots r}_{ij\cdots k}\Gamma^c_{hr}\\ &- T^{ab\cdots c}_{rj\cdots k}\Gamma^r_{hi} - \cdots - T^{ab\cdots c}_{ij\cdots r}\Gamma^r_{hk}\\ &- N\Gamma^r_{hr}T^{ab\cdots c}_{ij\cdots k}.\end{aligned}$$

This is the general formula for the covariant derivative of any tensor component, absolute or relative. The last term is absent when T is absolute ($N = 0$). For every upper index *,

$$\nabla_h T^{a\cdot\cdot *\cdot\cdot c}_{i\cdots\cdots k} \quad \text{contains a term} \quad T^{a\cdot\cdot r\cdot\cdot c}_{i\cdots\cdots k}\Gamma^*_{hr};$$

and, for every lower index *,

$$\nabla_h T^{a\cdots\cdots c}_{i\cdot\cdot *\cdot\cdot k} \quad \text{contains a term} \quad -T^{a\cdots\cdots c}_{i\cdot\cdot r\cdot\cdot k}\Gamma^r_{h*}.$$

We consider now some important special cases.

If φ is a relative scalar of weight N,

$$(8)\qquad \nabla_h\varphi = \mathrm{D}_h\varphi - N\Gamma^r_{hr}\varphi.$$

Since E and E^{-1} are scalars of weight 1 and -1, we have, from (165.3) and (165.4),

$$(9)\qquad \nabla_h E = \mathrm{D}_h E - \Gamma^r_{hr}E = 0,$$

$$(10)\qquad \nabla_h E^{-1} = \mathrm{D}_h E^{-1} + \Gamma^r_{hr}E^{-1} = 0.$$

Again, since $g = \det g_{ij} = E^2$ is a relative scalar of weight 2,

$$(11)\qquad \nabla_h g = \mathrm{D}_h E^2 - 2\Gamma^r_{hr}E^2 = 0.$$

If $\mathbf{v} = v^i\mathbf{e}_i = v_i\mathbf{e}^i$ is an absolute vector,

$$(12)\qquad \nabla_h v^i = \mathrm{D}_h v^i + v^r\Gamma^i_{hr},$$

$$(13)\qquad \nabla_h v_i = \mathrm{D}_h v_i - v_r\Gamma^r_{hi}.$$

These expressions are the mixed and covariant components of one and the same dyadic $\nabla\mathbf{v}$.

For an absolute dyadic $\mathbf{T}$, the components of $\nabla\mathbf{T}$ may have the forms:

$$(14)\qquad \nabla_h T^{ij} = \mathrm{D}_h T^{ij} + T^{rj}\Gamma^i_{hr} + T^{ir}\Gamma^j_{hr},$$

$$(15)\qquad \nabla_h T^i_{\cdot j} = \mathrm{D}_h T^i_{\cdot j} + T^r_{\cdot j}\Gamma^i_{hr} - T^i_{\cdot r}\Gamma^r_{hj},$$

$$(16)\qquad \nabla_h T_{ij} = \mathrm{D}_h T_{ij} - T_{rj}\Gamma^r_{hi} - T_{ir}\Gamma^r_{hj}.$$

For the metric tensor,

$$\mathbf{G} = g_{ij}\mathbf{e}^i\mathbf{e}^j = g^{ij}\mathbf{e}_i\mathbf{e}_j = \mathbf{e}_i\mathbf{e}^i = \mathbf{I}, \tag{17}$$

we have, from (165.2),

$$\nabla\mathbf{G} = \mathbf{e}^h\mathrm{D}_h\mathbf{G} = \mathbf{e}^h\mathrm{D}_h(\mathbf{e}_i\mathbf{e}^i) = 0. \tag{18}$$

The components of $\nabla\mathbf{G}$ therefore vanish:

$$\nabla_h g_{ij} = 0, \qquad \nabla_h g^{ij} = 0, \qquad \nabla_h \delta_i^j = 0. \tag{19}$$

Note that (166.1) is equivalent to $\nabla_k g_{ij} = 0$.

Example. We can write

$$\nabla\mathbf{T} = \mathbf{e}^h(\mathrm{D}_h - N\Gamma_{hr}^r)\mathbf{T} = \mathbf{e}^h\nabla_h\mathbf{T}, \tag{20}$$

if we regard ∇_h *as an operator that acts only on scalars.* Covariant differentiation then is defined by this operational equation. With this convention, we have also

$$\nabla\nabla\mathbf{T} = \mathbf{e}^i(\mathrm{D}_i - N\Gamma_{ir}^r)\mathbf{e}^j(\mathrm{D}_j - N\Gamma_{jr}^r)\mathbf{T} = \mathbf{e}^i\mathbf{e}^j\nabla_i\nabla_j\mathbf{T}.$$

The second member may be written

$$\mathbf{e}^i\mathbf{e}^j(\mathrm{D}_i - N\Gamma_{ir}^r)(\mathrm{D}_j - N\Gamma_{jr}^r)\mathbf{T} - \mathbf{e}^i\Gamma_{is}^j\mathbf{e}^s(\mathrm{D}_j - N\Gamma_{jr}^r)\mathbf{T}$$
$$= \mathbf{e}^i\mathbf{e}^j\{(\mathrm{D}_i - N\Gamma_{ir}^r)(\mathrm{D}_j - N\Gamma_{jr}^r) - \Gamma_{ij}^s(\mathrm{D}_s - N\Gamma_{sr}^r)\}\mathbf{T}$$

when indices j and s are interchanged; hence

$$\nabla_i\nabla_j\mathbf{T} = (\mathrm{D}_i - N\Gamma_{ir}^r)(\mathrm{D}_j - N\Gamma_{jr}^r)\mathbf{T} - \Gamma_{ij}^s\nabla_s\mathbf{T},$$
$$\nabla_j\nabla_i\mathbf{T} = (\mathrm{D}_j - N\Gamma_{jr}^r)(\mathrm{D}_i - N\Gamma_{ir}^r)\mathbf{T} - \Gamma_{ji}^s\nabla_s\mathbf{T},$$

and, on subtraction,

$$(\nabla_i\nabla_j - \nabla_j\nabla_i)\mathbf{T} = (\mathrm{D}_i\mathrm{D}_j - \mathrm{D}_j\mathrm{D}_i)\mathbf{T}, \tag{21}$$

in view of (165.5). On the left the operators ∇_r act only upon the scalar components of $\mathbf{T}$. We note that (21) applies to relative as well as absolute tensors.

168. Rules of Covariant Differentiation.

1. *The covariant derivative of the sum or product of two tensors may be computed by the rules for ordinary differentiation.*

If we introduce a system of geodesic coordinates y^i (§ 163), the corresponding gammas Γ_{ij}^k will all vanish at the point $y^i = 0$; hence, from (167.7),

$$\nabla_h T_{ij\cdots k}^{ab\cdots c} = \mathrm{D}_h T_{ij\cdots k}^{ab\cdots c}, \tag{1}$$

at the origin of geodesic coordinates, which moreover can be chosen at pleasure. For example, we have the *tensor equation,*

$$\nabla_h(T^{ij}u_k) = (\nabla_h T^{ij})u_k + T^{ij}\nabla_h u_k,$$

in geodesic coordinates and therefore in any coordinates (§ 157). Consequently, the sum and product rules of ordinary differentiation also apply in covariant differentiation.

2. *The covariant derivatives of the epsilons and Kronecker deltas are zero.*

Since these tensors have *constant components*, their covariant derivatives vanish at the origin of a system of geodesic coordinates; hence they vanish in all coordinate systems.

3. *The operation of contraction is commutative with covariant differentiation.* For example, if we contract $T^i_{\cdot jk}$ on the indices i, j to form

$$T^i_{\cdot ik} = \delta^j_i T^i_{\cdot jk},$$

we have

$$\nabla_h T^i_{\cdot ik} = \delta^j_i \nabla_h T^i_{\cdot jk}.$$

4. *The components of the metric tensor (g_{ij}, g^{ij}, δ^j_i) may be treated as constants in covariant differentiation.*

This follows at once from (167.19). For example,

$$\nabla_i v_j = \nabla_i(g_{jr}v^r) = g_{jr}\nabla_i v^r.$$

Thus we may find $\nabla_i v_j$ by lowering the index j in $\nabla_i v^j$; in other words, $\nabla_i v_j$ and $\nabla_i v^j$ are components of one and the same tensor:

$$\nabla v = \nabla_i v_j \mathbf{e}^i\mathbf{e}^j = \nabla_i v^j \mathbf{e}^i\mathbf{e}_j.$$

Evidently *the operation of raising or lowering an index is commutative with covariant differentiation.*

5. *The relative scalars E, E^{-1} and g may be treated as constants in covariant differentiation.*

Since E, E^{-1} and $g = E^2$ are relative scalars of weight 1, -1, 2, respectively, we have, from (167.4),

$$\nabla E = 0, \qquad \nabla E^{-1} = 0, \qquad \nabla g = 0;$$

for in each case $\mathrm{D}_h(E^{-N}\mathbf{T}) = \mathrm{D}_h 1 = 0$.

169. Riemannian Geometry. Any set of objects which can be placed in one-to-one correspondence with the totality of ordered sets of real numbers $(x^1, x^2, \cdots, x^n)$ satisfying certain inequalities,

$$|x^i - a^i| < k^i \qquad (a^i \text{ and } k^i > 0 \text{ are constants}),$$

is said to form a region of *space of n dimensions.*‡ We speak of $(x^1, x^2, \cdots, x^n)$ as a *point;* but the actual objects may be very

‡ Veblen, *Invariants of Quadratic Differential Forms*, Cambridge, 1933, p. 13. In some applications the numbers x^i may be complex.

diverse. Thus an event in the space–time of relativity may be pictured as a point in four-space; and the configurations of a dynamical system, determined by n *generalized coordinates*, often are regarded as points in n-space.

If we associate the space $(x^1, x^2, \cdots, x^n)$ with an arbitrary non-singular quadratic form,

$$(1) \qquad g_{ij}\, dx^i\, dx^j, \qquad (g_{ij} = g_{ji}, \qquad g = \det g_{ij} \neq 0),$$

we have a *Riemannian space* with a definite system of measurement prescribed by this form (§ 151); and the geometry of this metric space is called *Riemannian geometry.* We assume that the coefficients g_{ij} are continuous twice-differentiable functions of x^i. The base vectors $\mathbf{e}_i$ are not specified; but their lengths and the angles between them may be found from the relations:

$$(2) \qquad g_{ij} = \mathbf{e}_i \cdot \mathbf{e}_j.$$

The reciprocal base vectors now are given by

$$(3) \qquad \mathbf{e}^i = g^{ir}\mathbf{e}_r,$$

where g^{ij} is the reduced cofactor of g_{ij} in $\det g_{ij}$; for

$$(4) \qquad \mathbf{e}^i \cdot \mathbf{e}_j = g^{ir}g_{jr} = \delta^i_j \qquad (152.2).$$

Moreover

$$(5) \qquad \mathbf{e}^i \cdot \mathbf{e}^j = g^{ir}\delta^j_r = g^{ij}.$$

If we now transform coordinates from x^i to $\bar{x}^i$, we assume that the new base vectors are given by

$$(6) \qquad \bar{\mathbf{e}}_i = \frac{\partial x^r}{\partial \bar{x}^i}\mathbf{e}_r, \qquad \bar{\mathbf{e}}^i = \frac{\partial \bar{x}^i}{\partial x^s}\mathbf{e}^s;$$

then $\bar{\mathbf{e}}_i$ and $\bar{\mathbf{e}}^j$ still form reciprocal sets, for

$$\bar{\mathbf{e}}_i \cdot \bar{\mathbf{e}}^j = \frac{\partial x^r}{\partial \bar{x}^i}\frac{\partial \bar{x}^j}{\partial x^s}\,\delta^s_r = \delta^j_i.$$

In view of (2) and (5), equations (6) show that g_{ij} and g^{ij} transform like absolute dyadics. These tensors often are called the fundamental covariant and contravariant tensors of Riemannian geometry. By use of the equations,

$$(7) \qquad \mathbf{e}^i = g^{ir}\mathbf{e}_r, \qquad \mathbf{e}_i = g_{ir}\mathbf{e}^r,$$

they permit us to raise and lower indices (§ 154) and thus represent any tensor of valence m by 2^m types of components.

At a given point, g_{ij} gives the orientation of the base vectors $\mathbf{e}_i$ relative to each other. In order to determine the relative orientation of sets of base vectors at *different* points, we must know the components Γ_{ij}^k of the affine connection, defined by

$$\text{(8)} \qquad \mathrm{D}_i \mathbf{e}_j = \Gamma_{ij}^r \mathbf{e}_r.$$

Then, just as in § 165, we have also

$$\text{(9)} \qquad \mathrm{D}_i \mathbf{e}^j = -\Gamma_{ir}^j \mathbf{e}^r \qquad (165.1).$$

We now assume that the affine connection is symmetric ($\Gamma_{ij}^k = \Gamma_{ji}^k$). The gammas then are determined by the metric tensor (§ 166):

$$\text{(10)} \qquad \Gamma_{ij}^k = \tfrac{1}{2} g^{kr}(\mathrm{D}_i g_{jr} + \mathrm{D}_j g_{ri} - \mathrm{D}_r g_{ij}).\dagger$$

The epsilons in n-space, defined in § 146, have n subscripts or superscripts. The equations,

$$\text{(11)} \qquad \left|\frac{\partial \bar{x}}{\partial x}\right| \epsilon^{ij\cdots k} = \frac{\partial \bar{x}^i}{\partial x^a}\frac{\partial \bar{x}^j}{\partial x^b}\cdots\frac{\partial \bar{x}^k}{\partial x^c}\,\epsilon^{ab\cdots c},$$

$$\text{(12)} \qquad \left|\frac{\partial x}{\partial \bar{x}}\right| \epsilon_{ij\cdots k} = \frac{\partial x^a}{\partial \bar{x}^i}\frac{\partial x^b}{\partial \bar{x}^j}\cdots\frac{\partial x^c}{\partial \bar{x}^k}\,\epsilon_{ab\cdots c},$$

† We shall call the base vectors *constant* if $\mathrm{D}_i \mathbf{e}_j = 0$ $(i, j = 1, 2, \cdots, n)$; then $\Gamma_{ij}^k = 0$ and the functions $g_{ij} = \mathbf{e}_i \cdot \mathbf{e}_j$ are also constant. Conversely, when g_{ij} are constants, (10) shows that $\Gamma_{ij}^k = 0$, and hence $\mathrm{D}_i \mathbf{e}_j = 0$.

In Riemannian space it is not, in general, possible to introduce coordinates x^i for which the base vectors $\mathbf{e}_i$ are constant. Only *flat space* (§ 175) is compatible with such *Cartesian coordinates;* then each point has the position vector $\mathbf{r} = x^i \mathbf{e}_i$ and $\mathbf{e}_i = \partial \mathbf{r}/\partial x^i$. Moreover, for any coordinates $\bar{x}^i$ in flat space, we have (cf. § 163, ex. 1, 2)

$$\text{(i)} \qquad \bar{\mathbf{e}}_i = \frac{\partial x^j}{\partial \bar{x}^i}\mathbf{e}_j = \frac{\partial \mathbf{r}}{\partial x^j}\frac{\partial x^j}{\partial \bar{x}^i} = \frac{\partial \mathbf{r}}{\partial \bar{x}^i}.$$

The geometry on a surface with the metric tensor g_{ij} is Riemannian ($n = 2$). Unless the surface is flat (a plane, for example), constant base vectors cannot be introduced. If we regard the surface as immersed in Euclidean 3-space each surface point has the position vector $\mathbf{r} = x\mathbf{i} + y\mathbf{j} + z\mathbf{k}$, where

$$x = x(u, v), \qquad y = y(u, v), \qquad z = z(u, v)$$

are the Cartesian equations of the surface. If we write $x^1 = u$, $x^2 = v$, the base vectors on the surface may be taken as $\mathbf{e}_i = \partial \mathbf{r}/\partial x^i$; for equation (i) shows that these vectors transform in the manner prescribed in (6). Here $\mathbf{r}$ is a position vector in Euclidean 3-space; but, in general, the surface points have no position vector in the Riemannian 2-space they define.

Any Riemannian space of n dimensions may be regarded as immersed in a Euclidean space (§ 178) of $n(n+1)/2$ dimensions. This theorem has not as yet been rigorously proved; see Veblen, *op. cit.*, p. 69.

generalized from (146.2), show that $\epsilon^{ij\cdots k}$ and $\epsilon_{ij\cdots k}$ are relative tensors of weight 1 and -1; for these are the powers of the Jacobian $|\partial x/\partial \bar{x}|$ when it is transferred to the right-hand side.

There are n types of Kronecker deltas in n-space: δ^i_a, δ^{ij}_{ab}, δ^{ijk}_{abc}, up to $\delta^{ijk\cdots m}_{abc\cdots f}$ with $2n$ indices. As in § 161, they are defined in terms of the epsilons. In the case $n = 4$, for example,

$$\delta^i_a = \frac{1}{3!}\epsilon_{abcd}\epsilon^{ibcd}, \qquad \delta^{ij}_{ab} = \frac{1}{2!}\epsilon_{abcd}\epsilon^{ijcd},$$

$$\delta^{ijk}_{abc} = \frac{1}{1!}\epsilon_{abcd}\epsilon^{ijkd}, \qquad \delta^{ijkh}_{abcd} = \epsilon_{abcd}\epsilon^{ijkh}.$$

The rule given in § 161 for the value (0, 1, or -1) of a generalized delta still applies. Moreover the preceding definitions show that all the deltas are absolute tensors, alternating in both upper and lower indices. See Prob. 42.

The n-rowed determinant,

$$(13) \qquad g = \frac{1}{n!}\epsilon^{ij\cdots k}\epsilon^{rs\cdots t}g_{ir}g_{js}\cdots g_{kt} \qquad \text{[cf. (146.3)]},$$

generalized from (152.1), is the contracted product of two epsilons of weight 1 and n absolute dyadics. Hence *the discriminant g of the fundamental quadratic form is a relative scalar of weight* 2.

The cofactor of g_{ij} in g is gg^{ij}; hence, from (146.11) and (166.5), we have

$$D_h g = gg^{ij}D_h g_{ij} = gg^{ij}(\Gamma_{hi,\,j} + \Gamma_{hj,\,i}),$$

or, in view of (166.3),

$$(14) \qquad D_h g = g\Gamma^i_{hi} + g\Gamma^j_{hj} = 2g\Gamma^r_{hr}.$$

From (14), we have also

$$(15) \qquad D_h\sqrt{g} = \sqrt{g}\,\Gamma^r_{hr},$$

a result of the same form as (165.3) with E replaced by $\sqrt{g}$.

In defining the gradient of a tensor, we replace E by $\sqrt{g}$ in (167.4); thus, in Riemannian n-space,

$$(16) \qquad \nabla\mathbf{T} = g^{\frac{N}{2}}\mathbf{e}^h D_h(g^{-\frac{N}{2}}\mathbf{T}).$$

The components of $\nabla\mathbf{T}$ again are given by (167.7). Hence this formula for the covariant derivative is still valid in Riemannian geometry.

Since the metric tensor $\mathbf{G} = g_{jk}\mathbf{e}^j\mathbf{e}^k = \mathbf{e}_k\mathbf{e}^k$ still has the property $\mathrm{D}_h\mathbf{G} = 0$, $\nabla\mathbf{G}$ and its components vanish as before:

$$(17) \qquad \nabla_h g_{ij} = 0, \qquad \nabla_h g^{ij} = 0, \qquad \nabla_h \delta_i^j = 0.$$

Moreover, from (16), $\nabla g = g\mathbf{e}^h\mathrm{D}_h 1 = 0$, and

$$(18) \qquad \nabla_h g = 0.$$

170. Dual of a Tensor. An *m-vector* is a tensor of valence m which is alternating in all indices (cf. § 160). For convenience in wording, we also regard scalars ($m = 0$) and vectors ($m = 1$) as m-vectors. In n-space we can associate with any m-vector ($0 \leq m \leq n$) an $(n - m)$-vector, its *dual*, defined as follows:

If $P_{ij\cdots k}$ and $Q^{ij\cdots k}$ are m-vectors of weight N, their duals are the $(n - m)$-vectors,

$$(1) \qquad p^{ab\cdots c} = \frac{1}{m!}\epsilon^{ab\cdots c\, ij\cdots k}P_{ij\cdots k},$$

$$(2) \qquad q_{ab\cdots c} = \frac{1}{m!}Q^{ij\cdots k}\epsilon_{ij\cdots k\, ab\cdots c},$$

of weights $N + 1$ and $N - 1$, respectively. Note that the epsilons have n indices in n-space; and, in the contracted products, *the contravariant tensor is written first, and the summation indices are adjacent.*

A scalar φ has two duals, $\epsilon^{ab\cdots h}\varphi$, $\epsilon_{ab\cdots h}\varphi$, according as we use (1) or (2); they have the same numerical components but different weights. The dual of an n-vector $T_{ijk\cdots h}$ is the scalar:

$$\frac{1}{n!}\epsilon^{ijk\cdots h}T_{ijk\cdots h} = T_{123\cdots n}.$$

THEOREM. *If* $\mathbf{T}$ *is an m-vector* ($0 \leqq m \leqq n$),

$$(3) \qquad \text{dual dual } \mathbf{T} = \mathbf{T}.$$

Proof. Write $\mathbf{T} = \mathbf{P}$, dual $\mathbf{T} = \mathbf{p}$. Using (2) and (1), we have

$$\begin{aligned}
(\text{dual } \mathbf{p})_{rs\cdots t} &= \frac{1}{(n - m)!}p^{ab\cdots c}\epsilon_{ab\cdots c\, rs\cdots t} \\
&= \frac{1}{(n - m)!m!}\epsilon^{ab\cdots c\, ij\cdots k}P_{ij\cdots k}\epsilon_{ab\cdots c\, rs\cdots t} \\
&= \frac{1}{m!}\delta^{ij\cdots k}_{rs\cdots t}P_{ij\cdots k} \qquad \text{[Cf. Prob. 42.]} \\
&= P_{rs\cdots t},
\end{aligned}$$

since $P_{ij\cdots k}$ is alternating in all indices. Hence dual $\mathbf{p} = \mathbf{P}$; and, similarly, dual $\mathbf{q} = \mathbf{Q}$.

When $\mathbf{T}$ is an n-vector, say $T_{ijk\cdots h}$, (1) gives the dual $T_{123\cdots n}$; now (2) gives

$$\text{dual dual } \mathbf{T} = \epsilon_{ijk\cdots h} T_{123\cdots n} = T_{ijk\cdots h}.$$

Thus the theorem holds in this case also, provided *both* dualizing equations (1), (2) are used.

171. Divergence. The gradient of $\mathbf{v} = v^i \mathbf{e}_i$ is

$$\nabla \mathbf{v} = \nabla_h v^i \mathbf{e}^h \mathbf{e}_i.$$

The first scalar of this dyadic is

$$\text{(1)} \qquad \operatorname{div} \mathbf{v} = \nabla_h v^i \delta_i^h = \nabla_i v^i.$$

If v^i is an absolute vector, the divergence is the absolute scalar $\nabla_i v^i$; this definition applies in n-space and for any coordinate system.

When v^i is a relative vector of weight 1,

$$\nabla_h v^i = \mathrm{D}_h v^i + v^r \Gamma_{hr}^i - v^i \Gamma_{hr}^r \qquad (167.7).$$

On contracting with $h = i$, the second and third terms cancel; for $v^r \Gamma_{ir}^i = v^i \Gamma_{ir}^r$, since r and i are summation indices. Hence

$$\text{(2)} \qquad \operatorname{div} \mathbf{v} = \mathrm{D}_i v^i \qquad (\text{wt. } \mathbf{v} = 1).$$

If v^i is absolute, $\sqrt{g}\, v^i$ has the weight 1 imparted by the scalar $\sqrt{g}$; hence

$$\nabla_i v^i = \nabla_i (g^{-\frac{1}{2}} g^{\frac{1}{2}} v^i) = g^{-\frac{1}{2}} \nabla_i (g^{\frac{1}{2}} v^i) = g^{-\frac{1}{2}} \mathrm{D}_i (g^{\frac{1}{2}} v^i),$$

in view of (2). Thus

$$\text{(3)} \qquad \operatorname{div} \mathbf{v} = \frac{1}{\sqrt{g}} \mathrm{D}_i (\sqrt{g}\, v^i) \qquad (\text{wt. } \mathbf{v} = 0).$$

The Laplacian $\nabla^2 \varphi$ of the scalar φ is defined as div $\nabla \varphi$. If φ is absolute, $\mathbf{v} = \nabla \varphi$ is an absolute vector; then

$$v_r = \mathrm{D}_r \varphi, \qquad v^i = g^{ir} \mathrm{D}_r \varphi,$$

and, from (3),

$$\text{(4)} \qquad \nabla^2 \varphi = \frac{1}{\sqrt{g}} \mathrm{D}_i (\sqrt{g}\, g^{ir} \mathrm{D}_r \varphi).$$

The divergence of any tensor $\mathbf{T}$ is defined as the gradient $\nabla \mathbf{T}$ contracted on the first and last indices. Thus if $\mathbf{T}$ has the components $T^{ij\cdots kh}$ of valence m and weight N,

$$(5) \qquad (\text{div } \mathbf{T})^{ij\cdots k} = \nabla_h T^{ij\cdots kh}$$

is a tensor of valence $m - 1$ and weight N.

THEOREM. *When $T^{ij\cdots kh}$ is an m-vector of weight* 1, div **T** *is the $(m - 1)$-vector,*

$$(6) \qquad (\text{div } \mathbf{T})^{ij\cdots k} = \mathrm{D}_h T^{ij\cdots kh},$$

where ∇ *in the defining equation is replaced by* D. *Moreover,*

$$(7) \qquad \text{div div } \mathbf{T} = 0 \qquad (m > 1).$$

Proof. From (167.7),

$$\nabla_h T^{ij\cdots kh} = \mathrm{D}_h T^{ij\cdots kh} + \Gamma^i_{hr} T^{rj\cdots kh} + \cdots + \Gamma^k_{hr} T^{ij\cdots rh} + \Gamma^h_{hr} T^{ij\cdots kr} - \Gamma^r_{hr} T^{ij\cdots kh}.$$

The two final terms cancel, as we see on interchanging the summation indices h, r in the last term. The remaining terms containing Γ^*_{hr} vanish separately on summing over h and r; for Γ^*_{hr} is symmetric and **T** antisymmetric in these indices. Thus (6) is proved.

From (6) and the alternating character of **T**,

$$(\text{div div } \mathbf{T})^{ij\cdots} = \mathrm{D}_k \mathrm{D}_h T^{ij\cdots kh} = 0.$$

172. Stokes Tensor. The gradient of the covariant vector v_k is the dyadic $\nabla_h v_k$. From this we form the antisymmetric dyadic,

$$(1) \qquad P_{ij} = \delta^{hk}_{ij} \nabla_h v_k,$$

known as the *Stokes tensor.*† When v_k is absolute,

$$\nabla_h v_k = \mathrm{D}_h v_k - \Gamma^r_{hk} v_r;$$

and, since $\delta^{hk}_{ij} \Gamma^r_{hk} = 0$ owing to the symmetry of Γ^r_{hk}, we have

$$(2) \qquad P_{ij} = \delta^{hk}_{ij} \mathrm{D}_h v_k = \mathrm{D}_i v_j - \mathrm{D}_j v_i \qquad (\text{wt. } \mathbf{v} = 0).$$

In 2-space we can form from P_{ij} the relative scalar of weight 1,

$$(3) \qquad \tfrac{1}{2}\epsilon^{ij} P_{ij} = P_{12} = \mathrm{D}_1 v_2 - \mathrm{D}_2 v_1 \qquad (\text{wt. } \mathbf{v} = 0),$$

† Veblen, *op. cit.*, p. 64.

and from this the absolute scalar,

(4) $$\frac{1}{\sqrt{g}}(\mathrm{D}_1 v_2 - \mathrm{D}_2 v_1) \qquad (\text{wt. } \mathbf{v} = 0).$$

This is the absolute invariant on the surface with the fundamenta form $g_{ij}\, dx^i\, dx^j$, written $\mathbf{n} \cdot \text{rot } \mathbf{v}$ or $\mathbf{n} \cdot \nabla \times \mathbf{v}$ ‡ in § 97.

In 3-space we can form from P_{ij} the relative vector of weight 1,

(5) $$w^i = \tfrac{1}{2}\epsilon^{ijk} P_{jk} = \epsilon^{ijk}\mathrm{D}_j v_k,$$

having the components,

$$\mathrm{D}_2 v_3 - \mathrm{D}_3 v_2, \qquad \mathrm{D}_3 v_1 - \mathrm{D}_1 v_3, \qquad \mathrm{D}_1 v_2 - \mathrm{D}_2 v_2.$$

The absolute vector,

(6) $$\text{rot } \mathbf{v} = \frac{1}{\sqrt{g}} w^i \mathbf{e}_i = \frac{1}{\sqrt{g}} \epsilon^{ijk} \mathbf{e}_i \mathrm{D}_j v_k,$$

may be written as a symbolic determinant (cf. § 146):

(7) $$\text{rot } \mathbf{v} = \frac{1}{\sqrt{g}} \begin{vmatrix} \mathbf{e}_1 & \mathbf{e}_2 & \mathbf{e}_3 \\ \mathrm{D}_1 & \mathrm{D}_2 & \mathrm{D}_3 \\ v_1 & v_2 & v_3 \end{vmatrix}.$$

Comparing this with (88.19) now shows that rot $\mathbf{v}$ is in fact the *rotation* of $\mathbf{v}$ previously defined.*

In § 91 we proved that a vector $\mathbf{v}$ is the gradient of a scalar in 3-space when and only when rot $\mathbf{v} = 0$. In n-space we have the corresponding

THEOREM. *Let* $\mathbf{v}$ *be a continuously differentiable vector in a region* R. *Then, in order that* $\mathbf{v}$ *be a gradient vector,*

(8) $$\mathbf{v} = \nabla\varphi, \qquad v_i = \mathrm{D}_i\varphi,$$

it is necessary and sufficient that the Stokes tensor vanish in R:

(9) $$P_{ij} = \mathrm{D}_i v_j - \mathrm{D}_j v_i = 0.$$

Proof. The condition is *necessary;* for, if $v_i = \mathrm{D}_i\varphi$, P_{ij} vanishes identically, owing to the continuity of the second derivatives of φ.

‡ In (97.9), $H = \sqrt{g}$, $u = x^1$, $v = x^2$, $\mathbf{r}_u = \mathbf{e}_1$, $\mathbf{r}_v = \mathbf{e}_2$, $\mathbf{r}_u \cdot \mathbf{f} = f_1$, $\mathbf{r}_v \cdot \mathbf{f} = f_2$ in our present notation.

* In (88.19), $J = \sqrt{g}$, $u = x^1$, $\mathbf{r}_u = \mathbf{e}_1$, $\mathbf{r}_u \cdot \mathbf{f} = f_1$, etc., in our present notation.

The condition is *sufficient*. For the method of § 91, extended to case of n coordinates, leads to the function,

$$(10) \qquad \varphi = \int_{a^1}^{x^1} v_1(x^1, x^2, \cdots, x^n)\, dx^1$$
$$+ \int_{a^2}^{x^2} v_2(a^1, x^2, \cdots, x^n)\, dx^2$$
$$+ \int_{a^3}^{x^3} v_3(a^1, a^2, x^3, \cdots, x^n)\, dx^3 + \cdots$$
$$+ \int_{a^n}^{x^n} v_n(a^1, a^2, \cdots, a^{n-1}, x^n)\, dx^n,$$

where $a^1, a^2, \cdots, a^n$ are the coordinates of a fixed but arbitrary point. In the ith integral x^i is the variable of integration while $x^{i+1}, \cdots, x^n$ are regarded as constant parameters. We now can show from (9) and (10) that $\mathrm{D}_i\varphi = v_i$. Let us compute, for example, $\mathrm{D}_3\varphi$. Only the first three integrals in (10) contain x^3; their derivatives with respect to x^3 are, respectively,

$$v_3(x^1, x^2, \cdots, x^n) - v_3(a^1, x^2, \cdots, x^n),$$
$$v_3(a^1, x^2, \cdots, x^n) - v_3(a^1, a^2, \cdots, x^n), \qquad v_3(a^1, a^2, \cdots, x^n),$$

when we make use of $\mathrm{D}_3v_1 = \mathrm{D}_1v_3$, $\mathrm{D}_3v_2 = \mathrm{D}_2v_3$ in the first and second; hence $\mathrm{D}_3\varphi = v_3(x^1, x^2, \cdots, x^n)$.

173. Curl. We define the *curl* of a covariant tensor $T_{bc\cdots d}$ of valence m $(m < n)$ as the $(m+1)$-vector:

$$(1) \qquad (\mathrm{curl}\ \mathbf{T})_{hij\cdots k} = \frac{1}{m!}\, \delta^{abc\cdots d}_{hij\cdots k} \nabla_a T_{bc\cdots d}.$$

When $m = 0$, $T = \varphi$ (a scalar), $0! = 1$, and

$$\mathrm{curl}\ \varphi = \delta^a_h \nabla_a \varphi = \nabla_h \varphi$$

is the gradient of φ.

When $m = 1$, $\mathbf{T} = \mathbf{v}$ (a vector), curl $\mathbf{v}$ is the Stokes tensor (172.1).

In general, we have the

THEOREM. *When $T_{ab\cdots d}$ is an absolute tensor of valence $m < n$,*

$$(2) \qquad (\mathrm{curl}\ \mathbf{T})_{hij\cdots k} = \frac{1}{m!}\, \delta^{abc\cdots d}_{hij\cdots k} \mathrm{D}_a T_{bc\cdots d},$$

where ∇ *in the defining equation is replaced by* D. *Moreover,*

(3) $$\text{curl curl } \mathbf{T} = 0 \qquad (m < n - 1).$$

Proof. From (167.7), we have

$$\nabla_a T_{bc\cdots d} = \mathrm{D}_a T_{bc\cdots d} - \Gamma^r_{ab} T_{rc\cdots d} - \cdots \Gamma^r_{ad} T_{bc\cdots r}.$$

Hence in the right member of (1) there are m terms of the type,

$$-\frac{1}{m!}\,\delta^{abc\cdots d}_{hij\cdots k}\Gamma^r_{ab}T_{rc\cdots d};$$

these all vanish separately when we sum over the subscripts of Γ^r_{a*}. Thus (2) is proved.

From (2) we have

$$(\text{curl curl } \mathbf{T})_{rst\cdots v} = \frac{1}{(m+1)!}\,\delta^{ghi\cdots k}_{rst\cdots v}\mathrm{D}_g\left(\frac{1}{m!}\,\delta^{ab\cdots d}_{hij\cdots k}\mathrm{D}_a T_{b\cdots d}\right)$$

$$= \frac{1}{m!}\,\delta^{gab\cdots d}_{rst\cdots v}\mathrm{D}_g\mathrm{D}_a T_{b\cdots d}$$

which vanishes when we sum over g and a.

When $T_{bc\cdots d}$ is an m-vector, the summation in (1) or (2) may be carried out in $m + 1$ stages by setting $a = h, i, j, \cdots, k$ in turn and summing over the other m indices. Thus, from (1), we have

(4) $$(\text{curl } T)_{hij\cdots k} = \nabla_h T_{ij\cdots k} + (-1)^m \nabla_i T_{j\cdots kh} + \cdots,$$

taking the $m + 1$ cyclical permutations of the subscripts $hij \cdots k$ in order and placing $(-1)^m$ before the second, fourth, $\cdots$ terms. In the cases $m = 1, 2, 3$, we thus obtain

(5) $$(\text{curl } \mathbf{T})_{ij} = \nabla_i T_j - \nabla_j T_i,$$

(6) $$(\text{curl } \mathbf{T})_{ijk} = \nabla_i T_{jk} + \nabla_j T_{ki} + \nabla_k T_{ij},$$

(7) $$(\text{curl } \mathbf{T})_{hijk} = \nabla_h T_{ijk} - \nabla_i T_{jkh} + \nabla_j T_{khi} - \nabla_k T_{hij}.$$

When $T_{bc\cdots d}$ is an *absolute* m-vector, we obtain, from (2), an equation of the form (4) with ∇ replaced by D.

When **S** is an absolute tensor of valence $m - 1$, $\mathbf{T} = \text{curl } \mathbf{S}$ is an absolute m-vector and curl $\mathbf{T} = 0$. Then (4) becomes

(8) $$0 = \mathrm{D}_h T_{ij\cdots k} + (-1)^m \mathrm{D}_i T_{jk\cdots h} + \cdots.$$

Thus if v_i is absolute, $T_{ij} = \delta_{ij}^{ab} \mathrm{D}_a v_b$ is an absolute bivector, and

(9) $$\mathrm{D}_i T_{jk} + \mathrm{D}_j T_{ki} + \mathrm{D}_k T_{ij} = 0.$$

174. Relation between Divergence and Curl. For alternating tensors, we have the

THEOREM. *If* $T_{bc\cdots d}$ *is an m-vector* $(m < n)$,

(1) $$\text{dual curl } \mathbf{T} = \text{div dual } \mathbf{T},$$

provided dual **T** *is taken contravariant when* **T** *is a scalar.*

Proof. From § 170,

$$\begin{aligned}(\text{dual curl } T)^{pq\cdots r} &= \frac{1}{(m+1)!}\,\epsilon^{pq\cdots r\, ij\cdots k}(\text{curl } \mathbf{T})_{ij\cdots k} \\ &= \frac{1}{(m+1)!m!}\,\epsilon^{pq\cdots r\, ij\cdots k}\delta_{ij\cdots k}^{ab\cdots d}\nabla_a T_{b\cdots d} \\ &= \frac{1}{m!}\,\epsilon^{pq\cdots r\, ab\cdots d}\nabla_a T_{b\cdots d} \\ &= \nabla_a\left(\frac{1}{m!}\,\epsilon^{pq\cdots r\, ab\cdots d}T_{b\cdots d}\right) \\ &= \nabla_a(\text{dual } \mathbf{T})^{pq\cdots ra} \\ &= (\text{div dual } \mathbf{T})^{pq\cdots r}.\end{aligned}$$

On taking duals of both members of (1), we have

(2) $$\text{curl } \mathbf{T} = \text{dual div dual } \mathbf{T}.$$

Moreover, if we replace **T** in (1) by dual **T**, we have also

(3) $$\text{div } \mathbf{T} = \text{dual curl dual } \mathbf{T}.$$

175. Parallel Displacement. A vector **p** is said to undergo a parallel displacement along a curve $x^i = \varphi^i(t)$ when $d\mathbf{p}/dt = 0$ along this curve. If $\mathbf{p} = p^k \mathbf{e}_k$, we have

$$\frac{d\mathbf{p}}{dt} = \frac{dp^k}{dt}\mathbf{e}_k + p^k \frac{\partial \mathbf{e}_k}{\partial x^i}\frac{dx^i}{dt} = \frac{dp^k}{dt}\mathbf{e}_k + p^k\frac{dx^i}{dt}\Gamma_{ik}^j \mathbf{e}_j,$$

or, on interchanging summation indices j, k in the last term,

$$\frac{d\mathbf{p}}{dt} = \left(\frac{dp^k}{dt} + p^j\frac{dx^i}{dt}\Gamma_{ij}^k\right)\mathbf{e}_k.$$

Hence, if $d\mathbf{p}/dt = 0$, the components p^k satisfy the differential equations:

$$\frac{dp^k}{dt} + p^j \frac{dx^i}{dt} \Gamma^k_{ij} = 0 \qquad (k = 1, 2, \cdots, n). \tag{1}$$

A solution $p^k(t)$ of this system, satisfying the arbitrary initial conditions $p^k(0) = a^k$, defines a vector at each point t of the curve. The vector a^k at the point P_0 $(t = 0)$ is said to undergo a parallel displacement along the curve into the vector $p^k(t)$ at the point P. In (1) dx^i/dt depends upon the functions $\varphi^i(t)$ defining the curve; hence, in general, the solutions $p^k(t)$ will change when the curve is altered. In other words, the vector p^k obtained by a parallel displacement of a^k from P_0 to P depends, in general, upon the path connecting these points.

The length of a vector $\mathbf{p}$ and the angle between two vectors $\mathbf{p}$, $\mathbf{q}$ remain constant during a parallel displacement; for, if $d\mathbf{p}/dt$ and $d\mathbf{q}/dt$ vanish along a curve, we have also

$$\frac{d}{dt}(\mathbf{p} \cdot \mathbf{p}) = 0, \qquad \frac{d}{dt}(\mathbf{p} \cdot \mathbf{q}) = 0.$$

We shall say that a vector remains *constant* during a parallel displacement.

If s is the arc along the curve,

$$ds^2 = g_{ij}\, dx^i\, dx^j = g_{ij}\dot{x}^i \dot{x}^j\, dt^2.$$

If we choose the arc as parameter $(t = s)$, we have $g_{ij}\dot{x}^i\dot{x}^j = 1$, an equation which states that the tangent vector dx^i/ds to the curve is of unit length. If a curve has the property that its unit tangent vectors dx^i/ds are parallel with respect to the curve, it is said to be a *path curve* for the parallel displacement. With $t = s$, $p^k = dx^k/ds$, (1) gives the differential equations of the path curves,

$$\frac{d^2x^k}{ds^2} + \Gamma^k_{ij} \frac{dx^i}{ds} \frac{dx^j}{ds} = 0. \tag{2}$$

The path curves are the "straightest" curves in our Riemannian space—the analogues of straight lines in Euclidean geometry.

When equations (1) can be satisfied by functions $p^k(x^1, \cdots, x^n)$ of the coordinates alone, the parallel displacement is *independent of the path*, and the space is said to be *flat*. Then

$$\frac{dp^k}{dt} = \frac{\partial p^k}{\partial x^i} \frac{dx^i}{dt},$$

and the ordinary differential equations (1) are replaced by the partial-differential equations:

$$\frac{\partial p^k}{\partial x^i} + p^j \Gamma_{ij}^k = \nabla_i p^k = 0. \tag{3}$$

Since $\nabla_i p^k$ are the components of $\nabla \mathbf{p} = \mathbf{e}^i \mathrm{D}_i \mathbf{p}$ (§ 167), we see that a flat space contains vectors $\mathbf{p}(x^1, \cdots, x^n)$, such that $\nabla \mathbf{p} = 0$, or

$$\mathrm{D}_i \mathbf{p} = 0 \qquad (i = 1, 2, \cdots, n). \tag{4}$$

Since $p_k = \mathbf{p} \cdot \mathbf{e}_k$, we see that (4) is equivalent to

$$\mathrm{D}_i p_k = \mathbf{p} \cdot \mathrm{D}_i \mathbf{e}_k. \tag{5}$$

For any fixed value of k, the n functions $\mathbf{p} \cdot \mathrm{D}_i \mathbf{e}_k$ are components of a gradient vector, and for this it is necessary and sufficient that

$$\mathrm{D}_i(\mathbf{p} \cdot \mathrm{D}_j \mathbf{e}_k) - \mathrm{D}_j(\mathbf{p} \cdot \mathrm{D}_i \mathbf{e}_k) = 0 \quad (\text{§ 172, theorem});$$

or, since $\mathrm{D}_i \mathbf{p} = 0$,

$$\mathbf{p} \cdot (\mathrm{D}_i \mathrm{D}_j - \mathrm{D}_j \mathrm{D}_i) \mathbf{e}_k = 0.$$

These equations must hold independently of the initial conditions imposed upon $\mathbf{p}$ and are therefore equivalent to

$$(\mathrm{D}_i \mathrm{D}_j - \mathrm{D}_j \mathrm{D}_i) \mathbf{e}_k = 0. \tag{6}$$

Since D_i transforms like a covariant vector (167.3), the operator

$$\mathrm{D}_{ij} = \mathrm{D}_i \mathrm{D}_j - \mathrm{D}_j \mathrm{D}_i = \delta_{ij}^{ab} \mathrm{D}_a \mathrm{D}_b, \tag{7}$$

transforms like a covariant dyadic; for

$$\overline{\mathrm{D}}_i \overline{\mathrm{D}}_j = \left(\frac{\partial x^a}{\partial \bar{x}^i} \mathrm{D}_a\right)\left(\frac{\partial x^b}{\partial \bar{x}^j} \mathrm{D}_b\right) = \frac{\partial^2 x^b}{\partial \bar{x}^i \partial \bar{x}^j} \mathrm{D}_b + \frac{\partial x^a}{\partial \bar{x}^i} \frac{\partial x^b}{\partial \bar{x}^j} \mathrm{D}_a \mathrm{D}_b,$$

$$\overline{\mathrm{D}}_{ij} = \overline{\mathrm{D}}_i \overline{\mathrm{D}}_j - \overline{\mathrm{D}}_j \overline{\mathrm{D}}_i = \frac{\partial x^a}{\partial \bar{x}^i} \frac{\partial x^b}{\partial \bar{x}^j} \mathrm{D}_{ab}.$$

Moreover,

$$\overline{\mathrm{D}}_{ij} \bar{\mathbf{e}}_k = \left(\frac{\partial x^a}{\partial \bar{x}^i} \frac{\partial x^b}{\partial \bar{x}^j} \mathrm{D}_{ab}\right)\left(\frac{\partial x^c}{\partial \bar{x}^k} \mathbf{e}_c\right) = \frac{\partial x^a}{\partial \bar{x}^i} \frac{\partial x^b}{\partial \bar{x}^j} \frac{\partial x^c}{\partial \bar{x}^k} \mathrm{D}_{ab} \mathbf{e}_c,$$

since $\overline{\mathrm{D}}_{ij} \overline{\mathrm{D}}_k x^c = 0$; hence

$$\bar{\mathbf{e}}^h \cdot \overline{\mathrm{D}}_{ij} \bar{\mathbf{e}}_k = \frac{\partial x^a}{\partial \bar{x}^i} \frac{\partial x^b}{\partial \bar{x}^j} \frac{\partial x^c}{\partial \bar{x}^k} \frac{\partial \bar{x}^h}{\partial x^d} \mathbf{e}^d \cdot \mathrm{D}_{ab} \mathbf{e}_c.$$

This equation shows that $\mathbf{e}^h \cdot \mathrm{D}_{ij}\mathbf{e}_k$ is an absolute mixed tensor of valence four, say

(8) $$R_{ijk}^{\cdot\cdot\cdot h} = \mathbf{e}^h \cdot \mathrm{D}_{ij}\mathbf{e}_k.$$

The components of this *curvature tensor* **R** are thus the coefficients in the equation,

(9) $$\mathrm{D}_{ij}\mathbf{e}_k = R_{ijk}^{\cdot\cdot\cdot h}\mathbf{e}_h;$$

the condition (6), necessary for flat space, now assumes the tensor form,

(10) $$R_{ijk}^{\cdot\cdot\cdot h} = 0.$$

If it holds in one coordinate system, it holds in all.

When the space is flat, we can choose n linearly independent vectors $\mathbf{a}_i$ at a point P and, by giving them parallel displacements to neighboring points obtain a set of *constant* base vectors $\bar{\mathbf{e}}_i = \mathbf{a}_i$ in a region about P. For these base vectors, we have

$$\bar{g}_{ij} = \mathbf{a}_i \cdot \mathbf{a}_j = \text{const}, \qquad \bar{\Gamma}_{ij}^k = 0 \qquad (166.6),$$

and the corresponding coordinate system $\bar{x}$ is called *Cartesian*.

To determine the Cartesian coordinates $y = \bar{x}$ corresponding to the base vectors $\mathbf{a}_i$, we have

$$\mathbf{e}_k = \frac{\partial y^r}{\partial x^k}\mathbf{a}_r,$$

(11) $$\frac{\partial \mathbf{e}_k}{\partial x^j} = \frac{\partial^2 y^r}{\partial x^j\,\partial x^k}\mathbf{a}_r,$$

and, on dot-multiplying (11), member for member, by

$$\frac{\partial y^s}{\partial x^r}\mathbf{e}^r = \mathbf{a}^s,$$

(12) $$\frac{\partial y^s}{\partial x^r}\Gamma_{jk}^r = \frac{\partial^2 y^s}{\partial x^j\,\partial x^k}.\dagger$$

From (11), we have the necessary conditions for the integrability of equations (12):

$$\mathrm{D}_j\mathbf{e}_k = \mathrm{D}_k\mathbf{e}_j, \qquad \mathrm{D}_{ij}\mathbf{e}_k = 0;$$

† This equation also follows from (163.9) with $\bar{\Gamma}_{ij}^k = 0$.

that is,

$$\Gamma^r_{jk} = \Gamma^r_{kj}, \qquad R_{ijk}^{\cdot\cdot\cdot h} = 0.$$

These conditions are also *sufficient* for the complete integrability of equations (12).‡ When these conditions are fulfilled, (12) admits solutions $y(x^1, x^2, \cdots, x^n)$ which with $\partial y/\partial x^i$ take on arbitrary values at a given point. If we place the origin of the Cartesian coordinate system at the point x_0, we have $y = 0$ when $x = x_0$; and, if we choose n linearly independent sets of initial values,

$$\frac{\partial y}{\partial x^1} = p^i_1, \quad \frac{\partial y}{\partial x^2} = p^i_2, \quad \cdots, \quad \frac{\partial y}{\partial x^n} = p^i_n \qquad (i = 1, 2, \cdots, n),$$

when $x = x_0$, we obtain n corresponding solutions $y^i(x)$ whose Jacobian $|\partial y/\partial x| = \det p^i_j \neq 0$ when $x = x_0$. In the region about $x = x_0$ for which $|\partial y/\partial x| \neq 0$, the n functions $y_i(x)$ thus obtained define a Cartesian coordinate system. In brief, we have the important

THEOREM. *A necessary and sufficient condition that a Riemann space, with symmetric affine connection, be flat is that its curvature tensor vanish identically.*

We may readily compute the components $R_{ijk}^{\cdot\cdot\cdot h}$ from (8):

$$\begin{aligned} R_{ijk}^{\cdot\cdot\cdot h} &= \mathbf{e}^h \cdot (\mathrm{D}_i\mathrm{D}_j\mathbf{e}_k - \mathrm{D}_j\mathrm{D}_i\mathbf{e}_k) \\ &= \mathbf{e}^h \cdot \{\mathrm{D}_i(\Gamma^r_{jk}\mathbf{e}_r) - \mathrm{D}_j(\Gamma^r_{ik}\mathbf{e}_r)\} \\ &= \mathbf{e}^h \cdot \{(\mathrm{D}_i\Gamma^r_{jk})\mathbf{e}_r - (\mathrm{D}_j\Gamma^r_{ik})\mathbf{e}_r\} + \mathbf{e}^h \cdot \{\Gamma^r_{jk}\Gamma^s_{ir}\mathbf{e}_s - \Gamma^r_{ik}\Gamma^s_{jr}\mathbf{e}_s\}, \end{aligned}$$

and, on putting $\mathbf{e}^h \cdot \mathbf{e}_r = \delta^h_r$, $\mathbf{e}^h \cdot \mathbf{e}_s = \delta^h_s$, we have

$$(13) \qquad R_{ijk}^{\cdot\cdot\cdot h} = \mathrm{D}_i\Gamma^h_{jk} - \mathrm{D}_j\Gamma^h_{ik} + \Gamma^r_{jk}\Gamma^h_{ir} - \Gamma^r_{ik}\Gamma^h_{jr}.$$

176. Curvature Tensor. From the defining equation for the curvature tensor,

$$\mathrm{D}_{ij}\mathbf{e}_k = R_{ijk}^{\cdot\cdot\cdot r}\mathbf{e}_r,$$

we obtain the covariant components,

$$(1) \qquad \mathbf{e}_h \cdot \mathrm{D}_{ij}\mathbf{e}_k = R_{ijk}^{\cdot\cdot\cdot r}g_{rh} = R_{ijkh}.$$

‡ Cf. Veblen, *op. cit.*, p. 70–1.

We now can express the *covariant curvature tensor* R_{ijkh} in terms of the gammas:

$$R_{ijkh} = \mathbf{e}_h \cdot (\mathrm{D}_i\mathrm{D}_j\mathbf{e}_k - \mathrm{D}_j\mathrm{D}_i\mathbf{e}_k)$$
$$= \mathbf{e}_h \cdot \{\mathrm{D}_i(\Gamma_{jk,r}\mathbf{e}^r) - \mathrm{D}_j(\Gamma_{ik,r}\mathbf{e}^r)\} \qquad (166.4)$$
$$= \delta^r_h\{\mathrm{D}_i\Gamma_{jk,r} - \mathrm{D}_j\Gamma_{ik,r}\} - \delta^s_h\{\Gamma_{jk,r}\Gamma^r_{is} - \Gamma_{ik,r}\Gamma^r_{js}\};$$

(2) $$R_{ijkh} = \mathrm{D}_i\Gamma_{jk,h} - \mathrm{D}_j\Gamma_{ik,h} - \Gamma_{jk,r}\Gamma^r_{ih} + \Gamma_{ik,r}\Gamma^r_{jh}.$$

Since

$$\Gamma_{jk,r}\Gamma^r_{ih} = g_{rs}\Gamma^s_{jk}\Gamma^r_{ih} = \Gamma^s_{jk}\Gamma_{ih,s},$$

we also may write (2) in the form:

(3) $$R_{ijkh} = \mathrm{D}_i\Gamma_{jk,h} - \mathrm{D}_j\Gamma_{ik,h} - \Gamma^r_{jk}\Gamma_{ih,r} + \Gamma^r_{ik}\Gamma_{jh,r}.$$

R_{ijkh} has the following types of symmetry:

(I) $$R_{ijkh} + R_{jikh} = 0;$$

(II) $$R_{ijkh} + R_{ijhk} = 0;$$

(III) $$R_{ijkh} + R_{jkih} + R_{kijh} = 0;$$

(IV) $$R_{ijkh} - R_{khij} = 0.$$

Proofs. (I) follows from $\mathrm{D}_{ij} = -\mathrm{D}_{ji}$. Since the scalars $\mathbf{e}_k \cdot \mathbf{e}_h = g_{kh}$ have continuous second derivatives (§ 169),

$$\mathrm{D}_{ij}\,(\mathbf{e}_k \cdot \mathbf{e}_h) = \mathbf{e}_h \cdot \mathrm{D}_{ij}\,\mathbf{e}_k + \mathbf{e}_k \cdot \mathrm{D}_{ij}\,\mathbf{e}_h = 0$$

gives (II). Thus R_{ijkh} is alternating in its first two and last two indices.

Since the affine connection is symmetric (§ 169), we have

(4) $$\mathrm{D}_i\mathbf{e}_j = \Gamma^r_{ij}\mathbf{e}_r = \Gamma^r_{ji}\mathbf{e}_r = \mathrm{D}_j\mathbf{e}_i.$$

Hence, on adding the identities,

$$\mathrm{D}_k(\mathrm{D}_i\mathbf{e}_j - \mathrm{D}_j\mathbf{e}_i) = 0,$$
$$\mathrm{D}_i(\mathrm{D}_j\mathbf{e}_k - \mathrm{D}_k\mathbf{e}_j) = 0,$$
$$\mathrm{D}_j(\mathrm{D}_k\mathbf{e}_i - \mathrm{D}_i\mathbf{e}_k) = 0,$$

we obtain

$$\mathrm{D}_{ij}\mathbf{e}_k + \mathrm{D}_{jk}\mathbf{e}_i + \mathrm{D}_{ki}\mathbf{e}_j = 0,$$

which, on multiplication by $\mathbf{e}_h\,\cdot$, gives (III).

Now (IV) is a consequence of (I), (II), and (III). From (III), we have

$$\underline{R_{ijkh}} + R_{jkih} + R_{kijh} = 0,$$

$$R_{jkhi} + \underline{R_{khji}} + R_{hjki} = 0,$$

$$-\underline{R_{khij}} - R_{hikj} - R_{ikhj} = 0,$$

$$-R_{hijk} - \underline{R_{ijhk}} - R_{jhik} = 0.$$

When we add these equations and make use of (I) and (II), only the underlined terms survive, namely, $2R_{ijkh} - 2R_{khij}$, and we obtain (IV).

The symmetry relations (I) to (IV) reduce the number of independent components of R_{ijkh} to $\frac{1}{12}n^2(n^2 - 1)$.

Proof. $R_{ijkh} = 0$ when $i = j$ or $k = h$ (I, II); hence, in general, the number of non-zero components is $({}_nC_2)^2 = n^2(n - 1)^2/4$. But, when $ij \neq kh$, these are paired, because $R_{ijkh} = R_{khij}$ (III); hence, if we add the number ${}_nC_2$ of unpaired components R_{ijij} to the preceding total, we obtain double the number of components with distinct values. The number of components with distinct values thus is reduced to

$$\frac{1}{2}\left\{\frac{n^2(n-1)^2}{4} + \frac{n(n-1)}{2}\right\} = \frac{1}{8}\,n(n-1)(n^2 - n + 2).$$

These are still further reduced by the ${}_nC_4$ relations (III); for i, j, k, h must all be different in order to get a new relation. If, for example, $i = j$,

$$R_{iikh} + R_{ikih} + R_{kiih} = R_{ikih} + R_{kiih} = 0$$

is already included in (I). The number of linearly independent components is therefore

$$\frac{1}{8}\,n(n-1)(n^2 - n + 2) - \frac{n(n-1)(n-2)(n-3)}{24} = \frac{1}{12}\,n^2(n^2 - 1).$$

For $n = 2, 3, 4$ this gives 1, 6, 20 linearly independent components R_{ijkh}, respectively.

Example 1. When $n = 2$, the contracted product,

$$\epsilon^{ij}\epsilon^{kh}R_{ijkh} = 4R_{1212},$$

is a relative scalar of weight 2 (§ 169); hence R_{1212}/g is an absolute scalar. Now, from (2),

$$(5) \qquad R_{1212} = D_1\Gamma_{21,2} - D_2\Gamma_{11,2} - \Gamma_{21,r}\Gamma^r_{12} + \Gamma_{11,r}\Gamma^r_{22}.$$

We shall compute this expression when the base vectors are orthogonal $g_{12} = 0$. From (166.6),

$$\Gamma_{11,1} = \tfrac{1}{2}D_1 g_{11}, \qquad \Gamma_{11,2} = -\tfrac{1}{2}D_2 g_{11},$$

$$\Gamma_{12,1} = \tfrac{1}{2}D_2 g_{11}, \qquad \Gamma_{12,2} = \tfrac{1}{2}D_1 g_{22},$$

$$\Gamma_{22,1} = -\tfrac{1}{2}D_1 g_{22}, \qquad \Gamma_{22,2} = \tfrac{1}{2}D_2 g_{22}.$$

Moreover, since $g = g_{11}g_{22}$, $g^{11} = 1/g_{11}$, $g^{22} = 1/g_{22}$; hence

$$\begin{aligned}\Gamma_{21,r}\Gamma^r_{12} &= \Gamma_{21,1}\Gamma^1_{12} + \Gamma_{21,2}\Gamma^2_{12}\\ &= \Gamma_{21,1}\Gamma_{12,1}\, g^{11} + \Gamma_{21,2}\Gamma_{12,2}\, g^{22}\\ &= \frac{(D_2 g_{11})^2}{4g_{11}} + \frac{(D_1 g_{22})^2}{4g_{22}};\\ \Gamma_{11,r}\Gamma^r_{22} &= \Gamma_{11,1}\Gamma^1_{22} + \Gamma_{11,2}\Gamma^2_{22}\\ &= \Gamma_{11,1}\Gamma_{22,1}\, g^{11} + \Gamma_{11,2}\Gamma_{22,2}\, g^{22}\\ &= -\frac{(D_1 g_{11})(D_1 g_{22})}{4g_{11}} - \frac{(D_2 g_{11})(D_2 g_{22})}{4g_{22}}.\end{aligned}$$

Substituting these results in (5) gives

$$\begin{aligned}R_{1212} &= \tfrac{1}{2}D_1D_1 g_{22} + \tfrac{1}{2}D_2D_2 g_{11}\\ &\quad - \tfrac{1}{4}\left\{\left(\frac{D_1 g_{11}}{g_{11}} + \frac{D_1 g_{22}}{g_{22}}\right)D_1 g_{22} + \left(\frac{D_2 g_{11}}{g_{11}} + \frac{D_2 g_{22}}{g_{22}}\right)D_2 g_{11}\right\}\\ &= \frac{\sqrt{g}}{2}\left\{D_1\left(\frac{1}{\sqrt{g}}D_1 g_{22}\right) + D_2\left(\frac{1}{\sqrt{g}}D_2 g_{11}\right)\right\}\end{aligned}$$

The absolute scalar,

$$(6) \qquad K = -\frac{R_{1212}}{g} = -\frac{1}{2\sqrt{g}}\left\{D_1\left(\frac{1}{\sqrt{g}}D_1 g_{22}\right) + D_2\left(\frac{1}{\sqrt{g}}D_2\, g_{11}\right)\right\},$$

is precisely the total curvature of the surface whose fundamental differential form is

$$ds^2 = g_{11}\, dx^1\, dx^1 + g_{22}\, dx^2\, dx^2;$$

for, if we put $x^1 = u$, $x^2 = v$, $g_{11} = E$, $g_{22} = G$, $H = \sqrt{g}$, (6) agrees with (139.6).

Example 2. We may contract the tensor,

$$(7) \qquad R_{ijk}^{\cdot\cdot\cdot h} = g^{hr}R_{ijkr},$$

in essentially two different ways.

With $h = k$, we have

(8) $$R_{ijk}^{\cdot\cdot\cdot k} = g^{kr}R_{ijkr} = 0,$$

since g^{kr} and R_{ijkr} are, respectively, symmetric and antisymmetric in k, r. From (175.13), we see that (8) is equivalent to the identity:

(9) $$\mathrm{D}_i\Gamma_{jr}^r = \mathrm{D}_j\Gamma_{ir}^r.$$

With $h = i$, we obtain the *Ricci tensor*,

(10) $$R_{jk} = R_{ijk}^{\cdot\cdot\cdot i} = g^{ih}R_{ijkh};$$

this is a symmetric dyadic; for, from (IV),

(11) $$R_{kj} = g^{ih}R_{ikjh} = g^{ih}R_{jhik} = g^{hi}R_{hjki} = R_{jk}.$$

The first scalar of this dyadic,

(12) $$R = g^{jk}R_{jk} = g^{jk}g^{ih}R_{ijkh},$$

is an absolute invariant. In the case $n = 2$, $g_{12} = 0$ considered in ex. 1, we have

(13) $$R = 2g^{11}g^{22}R_{2112} = -2R_{1212}/g = 2K.$$

177. Identities of Ricci and Bianchi.

Ricci Identity. In analogy with

$$\mathrm{D}_{ij} = \mathrm{D}_i\mathrm{D}_j - \mathrm{D}_j\mathrm{D}_i, \quad \text{we also write} \quad \nabla_{ij} = \nabla_i\nabla_j - \nabla_j\nabla_i.$$

With this notation, (167.21) becomes

(1) $$\nabla_{ij}\mathbf{T} = \mathrm{D}_{ij}\mathbf{T}$$

for any tensor, absolute or relative, with its base vectors. On the left ∇_{ij} acts only on scalars (cf. § 167, ex.); on the right D_{ij} acts only on base vectors, for $\mathrm{D}_{ij}\varphi = 0$ when φ is a scalar. Equation (1) yields the *Ricci identity* when we evaluate $\mathrm{D}_{ij}\mathbf{T}$ by making use of the formulas:

(2) $$\mathrm{D}_{ij}\mathbf{e}_k = R_{ijk}^{\cdot\cdot\cdot h}\mathbf{e}_h \qquad (175.9),$$

(3) $$\mathrm{D}_{ij}\mathbf{e}^k = -R_{ijh}^{\cdot\cdot\cdot k}\mathbf{e}^h.$$

Equation (3) is proved as follows:

$$\mathrm{D}_{ij}\mathbf{e}^k = \mathrm{D}_{ij}g^{kr}\mathbf{e}_r = g^{kr}R_{ijr}^{\cdot\cdot\cdot s}\mathbf{e}_s = g^{kr}g^{hs}R_{ijrh}\mathbf{e}_s$$
$$= -g^{kr}g^{hs}R_{ijhr}\mathbf{e}_s = -R_{ijh}^{\cdot\cdot\cdot k}\mathbf{e}^h.$$

For the vector $v^k\mathbf{e}_k$, we have

$$\nabla_{ij}v^k\mathbf{e}_k = v^k\mathrm{D}_{ij}\mathbf{e}_k = v^kR_{ijk}^{\cdot\cdot\cdot s}\mathbf{e}_s = v^sR_{ijs}^{\cdot\cdot\cdot k}\mathbf{e}_k,$$

(4) $$\nabla_{ij}v^k = v^sR_{ijs}^{\cdot\cdot\cdot k}.$$

Similarly, for $v_k \mathbf{e}^k$,

$$\nabla_{ij} v_k \mathbf{e}^k = v_k \mathrm{D}_{ij} \mathbf{e}^k = -v_k R_{ijs}^{\cdots k} \mathbf{e}^s = -v_s R_{ijk}^{\cdots s} \mathbf{e}^k,$$

(5) $$\nabla_{ij} v_k = -v_s R_{ijk}^{\cdots s}.$$

The general Ricci identity now is readily established. Thus, if the components of $\mathbf{T}$ are $T_{hk\cdots m}^{ab\cdots c}$, for every upper index * in $\mathbf{T}$,

$$\nabla_{ij} T_{h\cdots\cdots m}^{a\cdot\cdot*\cdot\cdot c} \quad \text{contains a term,} \quad T_{h\cdots\cdots m}^{a\cdot\cdot r\cdot\cdot c} R_{ijr}^{\cdots *};$$

and, for every lower index * in $\mathbf{T}$,

$$\nabla_{ij} T_{h\cdot\cdot*\cdot\cdot m}^{a\cdots\cdots c} \quad \text{contains a term,} \quad -T_{h\cdot\cdot r\cdot\cdot m}^{a\cdots\cdots c} R_{ij*}^{\cdots r}.$$

Bianchi Identity. At the origin of geodesic coordinates (§ 163), we have, from (175.13),

$$\nabla_i R_{jkh}^{\cdots m} = \mathrm{D}_i R_{jkh}^{\cdots m} = \mathrm{D}_i \mathrm{D}_j \Gamma_{kh}^m - \mathrm{D}_i \mathrm{D}_k \Gamma_{jh}^m.$$

By permuting ijk cyclically in this equation, we obtain two others. On adding the three equations, we find that the right members cancel; we thus obtain the *Bianchi identity:*

(6) $$\nabla_i R_{jkh}^{\cdots m} + \nabla_j R_{kih}^{\cdots m} + \nabla_k R_{ijh}^{\cdots m} = 0.$$

Since this tensor equation holds at any point, it is also true for general coordinates.

178. Euclidean Geometry. When the space is *flat*, we can determine a Cartesian coordinate system x^i (§ 175). The corresponding metric tensor g_{ij} has then constant components. If in addition the metric form $g_{ij}x^i x^j$ is *positive definite,* the space and its geometry are termed *Euclidean.* We can then always make a real linear transformation to coordinates y^i for which $g_{ij} = \delta_j^i$, the Kronecker delta, and the metric form becomes a sum of squares (§ 150):

$$\delta_j^i y^i y^j = y^1 y^1 + y^2 y^2 + \cdots + y^n y^n.$$

The corresponding base vectors $\mathbf{a}_i$ then form an orthogonal set ($\mathbf{a}_i \cdot \mathbf{a}_j = \delta_j^i$), and the coordinates y^i are said to form an orthogonal system.

Let x^i denote a Cartesian coordinate system with the base vectors $\mathbf{a}_i$. If we transform to another Cartesian system $\bar{x}^i$ with the base vectors $\bar{\mathbf{a}}_i$, we have

$$\bar{\mathbf{a}}_j = \frac{\partial x^i}{\partial \bar{x}^j} \mathbf{a}_i;$$

and, since both $\bar{\mathbf{a}}_j$ and $\mathbf{a}_i$ are constant,

$$\frac{\partial \bar{\mathbf{a}}_j}{\partial \bar{x}^k} = \frac{\partial^2 x^i}{\partial \bar{x}^k \, \partial \bar{x}^j} \mathbf{a}_i = 0, \qquad \frac{\partial x^i}{\partial \bar{x}^k \, \partial \bar{x}^j} = 0.$$

On integrating this equation twice, we have

$$x^i = c^i_j \bar{x}^j + C^i, \tag{1}$$

where c^i_j and C^i are sets of constants. The transformation between any two Cartesian coordinate systems is therefore linear with constant coefficients. As in the general transformations of § 157, we require that the Jacobian,

$$\det \frac{\partial x^i}{\partial \bar{x}^j} = \det c^i_j \neq 0. \tag{2}$$

The transformations (1) with non-vanishing determinant form a group—the *affine group*.

When the Cartesian coordinate systems y and $\bar{y}$ are both orthogonal, the law of transformation,

$$\bar{g}_{ij} = \frac{\partial y^a}{\partial \bar{y}^i} \frac{\partial y^b}{\partial \bar{y}^j} g_{ab}, \quad \text{becomes} \quad \frac{\partial y^a}{\partial \bar{y}^i} \frac{\partial y^a}{\partial \bar{y}^j} = \delta^i_j.$$

If we multiply this equation by $\partial \bar{y}^i / \partial y^k$ and sum with respect to i, we obtain

$$\frac{\partial y^k}{\partial \bar{y}^j} = \frac{\partial \bar{y}^j}{\partial y^k}. \tag{3}$$

Since orthogonal coordinate systems are also Cartesian, the transformation between two orthogonal systems has the form:

$$y^k = c^k_j \bar{y}^j + C^k. \tag{4}$$

The inverse transformation is

$$\bar{y}^j = \gamma^j_k (y^k - C^k), \tag{5}$$

where γ^j_k is the reduced cofactor of c^k_j in $\det c^k_j$. Equation (3) thus becomes

$$c^k_j = \gamma^j_k. \tag{6}$$

This is precisely the condition that the matrix c^k_j be orthogonal (§ 149, theorem); thus *a transformation between orthogonal coordi-*

nate systems is orthogonal (has an orthogonal matrix). Condition (6), which characterizes an orthogonal matrix, also implies that

(7) $$c_i^r c_j^r = c_r^i c_r^j = \delta_j^i,$$

in view of the relations (146.8) between reduced cofactors. Conditions (7), in turn, imply (6); either (6) or (7) is a necessary and sufficient condition that the matrix c_j^k be orthogonal.

Two orthogonal transformations,

$$y^i = a_j^i \bar{y}^j + A^i, \qquad \bar{y}^j = b_k^j y^k + B^j,$$

have an orthogonal resultant,

$$y^i = c_k^i y^k + C^i, \qquad c_k^i = a_j^i b_k^j;$$

we have, for example,

$$c_i^r c_j^r = (a_s^r b_i^s)(a_t^r b_j^t) = \delta_t^s b_i^s b_j^t = b_i^s b_j^s = \delta_j^i.$$

Moreover (5), the inverse of (4), and the identity transformation $y^i = \delta_j^i \bar{y}^j$ are orthogonal. Consequently, *the orthogonal transformations form a group.*

In view of (3), the equations,

$$\bar{v}_j = \frac{\partial y^k}{\partial \bar{y}^j} v_k, \qquad \bar{v}^j = \frac{\partial \bar{y}^j}{\partial y^k} v^k,$$

show that covariant and contravariant vectors transform alike under orthogonal transformations. *Within the orthogonal group, the distinction between covariance and contravariance vanishes,* and tensor components may be written indifferently with upper or lower indices. For example, we may write δ_{ij} or δ^{ij} for the Kronecker delta.

The orthogonal group of transformations admits as a subgroup those transformations for which

(8) $$\det c_j^i = \left| \frac{\partial y}{\partial \bar{y}} \right| = 1.$$

If we regard (4) as a transformation between the points y and $\bar{y}$ in the same Cartesian coordinate system, the transformation is called a *displacement* or *rigid motion.* In fact in 3-space the transformation (5) may be written

(9) $$\bar{\mathbf{r}} = \Phi \cdot \mathbf{s}, \qquad \mathbf{s} = \mathbf{r} - \mathbf{a},$$

where the dyadic Φ in nonion form is given by the matrix (c^i_k). Since this matrix is orthogonal and its determinant is 1, the transformation (9) is a translation followed by a rotation (§ 75, theorem), in brief, a *displacement.* The subgroup characterized by $\det c^i_j = 1$ is therefore called the *displacement group.* In view of (8), we see that, *within the displacement group, the distinction between absolute and relative tensors also vanishes.*

A displacement which leaves the origin invariant is called a *rotation.* Thus the transformation $y^i = c^i_j \bar{y}^j$ is a rotation if the matrix (c^i_j) is orthogonal and has the determinant 1. Rotations form a subgroup of the displacement group.

179. Surface Geometry in Tensor Notation. The equations,

$$x^i = x^i(u^1, u^2) \qquad (i = 1, 2, 3),$$

define a surface embedded in Euclidean 3-space. The space coordinates x^i are rectangular Cartesian and are designated by *italic* indices (range 1, 2, 3); the surface coordinates u^α are curvilinear and are designated by Greek indices (range 1, 2). If we write $\mathbf{a}_1 = \mathbf{i}$, $\mathbf{a}_2 = \mathbf{j}$, $\mathbf{a}_3 = \mathbf{k}$, the position vector to the surface is $\mathbf{r} = x^i \mathbf{a}_i$. The metric tensor in space is then

$$\delta_{ij} = \mathbf{a}_i \cdot \mathbf{a}_j. \tag{1}$$

If we limit the coordinate transformations $x \to \bar{x}$ to the displacement group (§ 178) the distinction between covariance and contravariance as well as the distinction between absolute and relative tensors does not exist in 3-space.

First fundamental form. The base vectors on the surface are

$$\mathbf{e}_\alpha = \frac{\partial \mathbf{r}}{\partial u^\alpha} = \frac{\partial \mathbf{r}}{\partial x^i}\frac{\partial x^i}{\partial u^\alpha} = x^i_\alpha\, \mathbf{a}_i, \tag{2}$$

where $x^i_\alpha = \partial x^i/\partial u^\alpha$. Since

$$\mathbf{e}_\alpha \cdot \mathbf{e}_\beta = (x^i_\alpha \mathbf{a}_i) \cdot (x^j_\beta \mathbf{a}_j) = x^i_\alpha x^j_\beta\, \delta_{ij} = x^i_\alpha x^i_\beta,$$

the metric tensor for the surface is

$$g_{\alpha\beta} = \mathbf{e}_\alpha \cdot \mathbf{e}_\beta = x^i_\alpha x^i_\beta. \tag{3}$$

This tensor defines the *first fundamental form* on the surface $ds^2 = g_{\alpha\beta}\, du^\alpha du^\beta$.

Note that x^i_α is a covariant surface vector; for if we make the transformation $u \to \bar{u}$, we have

$$\frac{\partial x^i}{\partial \bar{u}^\alpha} = \frac{\partial x^i}{\partial u^\beta}\frac{\partial u^\beta}{\partial \bar{u}^\alpha}.$$

If a vector $\mathbf{v}$ has the "surface components" v^α ($\alpha = 1, 2$),

$$\mathbf{v} = v^\alpha \mathbf{e}_\alpha = v^\alpha x^i_\alpha \mathbf{a}_i$$

and $v^i = v^\alpha x^i_\alpha$ ($i = 1, 2, 3$) are the "space components" of $\mathbf{v}$.

Unit surface normal. The space vector,

$$\mathbf{N} = \mathbf{e}_1 \times \mathbf{e}_2 = \epsilon_{ijk}\mathbf{a}^i x^j_1 x^k_2,$$

has the components $N_i = \epsilon_{ijk}\, x^j_1 x^k_2$; moreover

$$\begin{aligned} N^2 &= \epsilon_{ijk}\epsilon_{ist}\, x^j_1 x^k_2\, x^s_1 x^t_2 = \delta^{st}_{jk}\, x^j_1 x^k_2\, x^s_1 x^t_2 \\ &= x^j_1 x^k_2\, x^j_1 x^k_2 - x^j_1 x^k_2\, x^k_1 x^j_2 = g_{11}\, g_{22} - g^2_{12}. \end{aligned}$$

Thus $N^2 = \det g_{\alpha\beta} = g$ †; hence the *unit normal* $\mathbf{n}$ to the surface has the components $N_i/\sqrt{g}$:

$$n_i = n^i = \frac{1}{\sqrt{g}}\,\epsilon_{ijk}\, x^j_1 x^k_2. \tag{4}$$

Second fundamental form. On the surface with metric tensor $g_{\alpha\beta}$, the Christoffel symbols are given by (cf. § 166)

$$\Gamma^\lambda_{\alpha\beta} = \tfrac{1}{2}g^{\lambda\gamma}(D_\alpha g_{\beta\gamma} + D_\beta g_{\gamma\alpha} - D_\gamma g_{\alpha\beta}). \tag{5}$$

Covariant derivatives are then computed from the formulas of § 167. In particular we have, from (167.13),

$$\nabla_\alpha x^k_\beta = D_\alpha D_\beta x^k - \Gamma^\lambda_{\alpha\beta} x^k_\lambda = \nabla_\beta x^k_\alpha. \tag{6}$$

Since the covariant derivative of the metric tensor g_{ij} is zero (167.19),

$$\nabla_\alpha g_{\beta\gamma} = \nabla_\alpha\, x^k_\beta x^k_\gamma = 0. \tag{7}$$

If the product rule (§ 168, 1) is used, this equation and its cyclical permutations give

$$x^k_\gamma \nabla_\alpha x^k_\beta + x^k_\beta \nabla_\alpha x^k_\gamma = 0, \tag{7a}$$

$$x^k_\alpha \nabla_\beta x^k_\gamma + x^k_\gamma \nabla_\beta x^k_\alpha = 0, \tag{7b}$$

$$x^k_\beta \nabla_\gamma x^k_\alpha + x^k_\alpha \nabla_\gamma x^k_\beta = 0. \tag{7c}$$

† This also follows from the expansion of $(\mathbf{e}_1 \times \mathbf{e}_2) \cdot (\mathbf{e}_1 \times \mathbf{e}_2)$ given in (20.1).

Subtract the third equation from the sum of the first two; then, in view of (6), we obtain

$$x_\gamma^k \nabla_\alpha x_\beta^k = 0.$$

This equation states that the space vector $\nabla_\alpha x_\beta^k$ is perpendicular to both $\mathbf{e}_1$ and $\mathbf{e}_2$ (whose space components are x_1^k, x_2^k); that is $\nabla_\alpha x_\beta^k$ is a multiple of the unit normal n^k, say

$$\nabla_\alpha x_\beta^k = h_{\alpha\beta} n^k. \tag{8}$$

The symmetric covariant tensor,

$$h_{\alpha\beta} = n^k \nabla_\alpha x_\beta^k = h_{\beta\alpha}, \tag{9}$$

defines the *second fundamental form* on the surface: $h_{\alpha\beta} du^\alpha du^\beta$.

Derivative formulas. From (6) and (8) we have

$$\mathrm{D}_\alpha x_\beta^k = \Gamma_{\alpha\beta}^\lambda x_\lambda^k + h_{\alpha\beta} n^k; \tag{10}$$

these equations are the *derivative formulas of Gauss.* If we adjoin the (constant) base vector $\mathbf{a}_k$ to each term they become

$$\mathrm{D}_\alpha \mathbf{e}_\beta = \Gamma_{\alpha\beta}^\lambda \mathbf{e}_\lambda + h_{\alpha\beta} \mathbf{n}. \dagger \tag{10)'$$

On multiplying (10) by x_γ^k and summing on k, we have also

$$x_\gamma^k \mathrm{D}_\alpha x_\beta^k = \Gamma_{\alpha\beta}^\lambda g_{\lambda\gamma} = \Gamma_{\alpha\beta,\gamma}. \tag{11}$$

Since n^k is a space vector with no components along $\mathbf{e}_1$ or $\mathbf{e}_2$, $\nabla_\alpha n^k = \mathrm{D}_\alpha n^k$. On differentiating $n^k n^k = 1$, we obtain

$$n^k \nabla_\alpha n^k = n^k \mathrm{D}_\alpha n^k = 0; \tag{12}$$

hence $\mathrm{D}_\alpha n^k$ ($\perp n^k$) is a tangential surface vector. Similarly, from $n^k x_\beta^k = 0$, we obtain

$$n^k \nabla_\alpha x_\beta^k + x_\beta^k \nabla_\alpha n^k = 0;$$

or, in view of (9) and the symmetry of $h_{\alpha\beta}$,

$$h_{\alpha\beta} = -x_\beta^k \nabla_\alpha n^k = -x_\alpha^k \nabla_\beta n^k. \tag{13}$$

Hence

$$h_{\alpha\beta} x_\lambda^k = -g_{\alpha\lambda} \nabla_\beta n^k,$$

$$g^{\beta\lambda} h_{\alpha\beta} x_\lambda^k = -\delta_\alpha^\beta \nabla_\beta n^k = -\nabla_\alpha n^k,$$

$$\nabla_\alpha n^k = \mathrm{D}_\alpha n^k = -h_\alpha^{\cdot\lambda} x_\lambda^k. \tag{14}$$

† Note that the term $h_{\alpha\beta}\mathbf{n}$ in this equation is in apparent disagreement with (163.1); this is due to the fact that our 2-dimensional geometry is not *intrinsic* but that of a 2-space embedded in a 3-space.

These equations are the *derivative formulas of Weingarten.* If we adjoin the (constant) base vector $\mathbf{a}_k$ to each term, they become

(14)′ $$\mathrm{D}_\alpha \mathbf{n} = -h_\alpha^{\cdot\lambda} \mathbf{e}_\lambda.$$

Equations of Codazzi and Gauss. From (8) we have

$$\nabla_\beta x_\gamma^k = h_{\beta\gamma} n^k,$$
$$\nabla_\alpha \nabla_\beta x_\gamma^k = (\nabla_\alpha h_{\beta\gamma}) n^k + h_{\beta\gamma} \nabla_\alpha n^k,$$
$$= (\nabla_\alpha h_{\beta\gamma}) n^k - h_{\beta\gamma} h_\alpha^{\cdot\lambda} x_\lambda^k$$

in view of (14). Now form $\nabla_\beta \nabla_\alpha x_\lambda^k$ and subtract it from the last equation; writing $\nabla_{\alpha\beta}$ for the operator $\nabla_\alpha \nabla_\beta - \nabla_\beta \nabla_\alpha$, we thus obtain

(15) $$\nabla_{\alpha\beta} x_\gamma^k = (\nabla_\alpha h_{\beta\gamma} - \nabla_\beta h_{\alpha\gamma}) n^k + (h_{\alpha\gamma} h_\beta^{\cdot\lambda} - h_{\beta\gamma} h_\alpha^{\cdot\lambda}) x_\lambda^k.$$

If we replace $\nabla_{\alpha\beta} x_\lambda^k$ by the value,

$$\nabla_{\alpha\beta} x_\lambda^k = -R_{\alpha\beta\gamma}^{\cdot\cdot\cdot\lambda}\, x_\lambda^k,$$

given by Ricci's Identity (177.5) and adjoin $\mathbf{a}_k$ to each term, (15) becomes

(16) $$-R_{\alpha\beta\gamma}^{\cdot\cdot\cdot\lambda}\, \mathbf{e}_\lambda = (\nabla_\alpha h_{\beta\gamma} - \nabla_\beta h_{\alpha\gamma}) \mathbf{n} + (h_{\alpha\gamma} h_\beta^{\cdot\lambda} - h_{\beta\gamma} h_\alpha^{\cdot\lambda}) \mathbf{e}_\lambda.$$

This vector equation is equivalent to the scalar equations:

(17) $$0 = \nabla_\alpha h_{\beta\gamma} - \nabla_\beta h_{\alpha\gamma},$$

(18) $$R_{\alpha\beta\gamma}^{\cdot\cdot\cdot\lambda} = h_{\beta\gamma} h_\alpha^{\cdot\lambda} - h_{\alpha\gamma} h_\beta^{\cdot\lambda}.$$

Equations (17) are the *Equations of Codazzi.* When $\alpha = \beta$ the right member vanishes identically; and an interchange of α and β repeats the same equation. Hence there are but *two* independent Codazzi equations; these may be written with $\alpha = 1$, $\beta = 2$, $\gamma = 1, 2$;

(19) $$\nabla_1 h_{21} - \nabla_2 h_{11} = 0, \qquad \nabla_1 h_{22} - \nabla_2 h_{12} = 0.$$

On multiplying (18) by $g_{\lambda\delta}$ and summing on λ, we obtain the covariant curvature tensor:

(20) $$R_{\alpha\beta\gamma\delta} = h_{\beta\gamma} h_{\alpha\delta} - h_{\alpha\gamma} h_{\beta\delta}.$$

From the symmetry $h_{\alpha\beta} = h_{\beta\alpha}$ we may verify at once the four symmetry relations of $R_{\alpha\beta\gamma\delta}$:

$$R_{\alpha\beta\gamma\delta} = -R_{\beta\alpha\gamma\delta}, \qquad R_{\alpha\beta\gamma\delta} = -R_{\alpha\beta\delta\gamma},$$
$$R_{\alpha\beta\gamma\delta} + R_{\beta\gamma\alpha\delta} + R_{\gamma\alpha\beta\delta} = 0, \qquad R_{\alpha\beta\gamma\delta} = R_{\gamma\delta\alpha\beta}.$$

As Greek indices range over 1, 2, these relations show that there is but one independent equation (20). This *Equation of Gauss* may be taken as

$$(21) \qquad -R_{1212} = h_{11}h_{22} - h_{12}^2 = h,$$

where $h = \det h_{\alpha\beta}$.

Total and mean curvature. The contracted product,

$$\epsilon^{\alpha\beta}\,\epsilon^{\gamma\delta}\,R_{\alpha\beta\gamma\delta} = 4\,R_{1212} \quad \text{and} \quad g = g_{11}g_{22} - g_{12}^2$$

are both relative scalars of weight 2; hence,

$$(22) \qquad K = -R_{1212}/g = h/g$$

is an absolute scalar, namely the *total curvature* of the surface (§ 176, Ex. 1).

The *mean curvature* of the surface is defined as the absolute scalar $J = g^{\alpha\beta}h_{\alpha\beta}$.

Since $g^{\alpha\beta}$ is the cofactor of $g_{\alpha\beta}$ in $\det g_{\alpha\beta}$,

$$(23) \qquad g^{11} = g_{22}/g,\ g^{22} = g_{11}/g,\ g^{12} = g^{21} = -g_{12}/g;$$

hence

$$(24) \qquad J = g^{\alpha\beta}h_{\alpha\beta} = (g_{22}h_{11} - 2g_{12}h_{12} + g_{11}h_{22})/g.$$

180. Summary: Tensor Analysis. Under general transformations,

$$\bar{x}^i = f^i(x^1, x^2, \cdots, x^n), \qquad \left|\frac{\partial \bar{x}}{\partial x}\right| \neq 0,$$

the component of a relative tensor of weight N transforms according to the pattern:

$$\bar{T}^{ij}_{\cdot\cdot k} = \left|\frac{\partial x}{\partial \bar{x}}\right|^N \frac{\partial \bar{x}^i}{\partial x^a}\frac{\partial \bar{x}^j}{\partial x^b}\frac{\partial x^c}{\partial \bar{x}^k}\,T^{ab}_{\cdot\cdot c}.$$

When $N = 0$, the tensor is *absolute*. For brevity, *components* of tensors often are called tensors.

The number of indices on a tensor component is called its *valence*. In n-space a tensor of valence m has n^m components.

A tensor of valence zero is a *scalar*. A scalar $\varphi(x)$ has one component in each coordinate system given by

$$\bar{\varphi}(\bar{x}) = \left|\frac{\partial x}{\partial \bar{x}}\right|^N \varphi(x).$$

A tensor of valence one is a *vector*. The *differentials of the coordinates* and the *gradients of an absolute scalar* are the prototypes of absolute contravariant and covariant vectors:

$$d\bar{x}^i = \frac{\partial \bar{x}^i}{\partial x^r} dx^r, \qquad \frac{\partial \bar{\varphi}}{\partial \bar{x}^i} = \frac{\partial \bar{\varphi}}{\partial x^r} \frac{\partial x^r}{\partial \bar{x}^i}.$$

Measurement is introduced into Riemannian geometry by the non-singular quadratic form:

$$ds^2 = g_{ij}\, dx^i\, dx^j \qquad (g_{ij} = g_{ji}, \quad g = \det g_{ij} \neq 0).$$

The character of the geometry depends upon the choice of the $n(n + 1)/2$ functions $g_{ij}(x)$ of the coordinates. The relations $\mathbf{e}_i \cdot \mathbf{e}_j = g_{ij}$ determine the lengths of the *base vectors* $\mathbf{e}_i$ and the angles between them.

Any vector in n-space at the point x is linearly dependent upon the n base vectors $\mathbf{e}_i$ at this point. The reciprocal base vectors $\mathbf{e}^i$ are defined by $\mathbf{e}^i \cdot \mathbf{e}_j = \delta^i_j$. If $g = \det g_{ij}$ and g^{ij} is the reduced cofactor of g_{ij} in g,

(1) $$\mathbf{e}^i = g^{ir}\mathbf{e}_r, \qquad \mathbf{e}_i = g_{ir}\mathbf{e}^r;$$

(2) $$\mathbf{e}_i \cdot \mathbf{e}_j = g_{ij}, \qquad \mathbf{e}_i \cdot \mathbf{e}^j = \delta^j_i, \qquad \mathbf{e}^i \cdot \mathbf{e}^j = g^{ij}.$$

In passing from coordinates x to $\bar{x}$, the transformation of the base vectors is prescribed by

(3) $$\bar{\mathbf{e}}_i = \frac{\partial x^r}{\partial \bar{x}^i} \mathbf{e}_r, \qquad \bar{\mathbf{e}}^i = \frac{\partial \bar{x}^i}{\partial x^r} \mathbf{e}^r.$$

These equations show that g_{ij}, δ^j_i, g^{ij} are components of an absolute dyadic, the *metric tensor* $\mathbf{G} = g_{ij}\mathbf{e}^i\mathbf{e}^j = \mathbf{e}_j\mathbf{e}^j$.

Use of equations (1) permits indices to be raised or lowered on tensor components:

$$T^{\cdot i \cdot} = g^{ir} T^{\cdot}{}_{r}{}^{\cdot}, \qquad T_{\cdot i \cdot} = g_{ir} T^{\cdot r \cdot}.$$

If $\mathbf{T}$ is a tensor of weight N (say $\mathbf{T} = T^{ij}{}_{\cdot\cdot k}\mathbf{e}_i\mathbf{e}_j\mathbf{e}^k$),

$$\bar{\mathbf{T}} = \left| \frac{\partial x}{\partial \bar{x}} \right|^N \mathbf{T}.$$

Addition of tensor components of the same valence, weight, and type produces a component of this same character.

Multiplication of tensor components of valence m_1, m_2, of weight N_1, N_2, and of arbitrary type produces a component of valence $m_1 + m_2$ and of weight $N_1 + N_2$.

Contraction of a tensor of valence $m > 1$ results on forming the dot product of any two of its base vectors. If the vectors in question are $\mathbf{e}_i \cdot \mathbf{e}^j = \delta_i^j$, the components of the contracted tensor are obtained from the original components by putting $i = j$ and performing the implied summation.

The components of the *affine connection* Γ_{ij}^k are functions of the coordinates defined by

$$\text{(4)} \qquad \mathrm{D}_i\mathbf{e}_j = \Gamma_{ij}^r\mathbf{e}_r \qquad (\Gamma_{ij}^k = \mathbf{e}^k \cdot \mathrm{D}_i\mathbf{e}_j).$$

Then also

$$\text{(5)} \qquad \mathrm{D}_i\mathbf{e}^j = -\Gamma_{ir}^j\mathbf{e}^r;$$

$$\text{(6)} \qquad \mathrm{D}_i\mathbf{G} = 0 \qquad (\mathbf{G} = \mathbf{e}_j\mathbf{e}^j);$$

$$\text{(7)} \qquad \mathrm{D}_i g = 2g\Gamma_{ir}^r.$$

By definition,

$$\Gamma_{ij,k} = g_{kr}\Gamma_{ij}^r; \qquad \text{then} \quad \Gamma_{ij}^k = g^{kr}\Gamma_{ij,r}.$$

When Γ_{ij}^k is symmetric in ij,

$$\Gamma_{ij,k} = \tfrac{1}{2}(\mathrm{D}_i g_{jk} + \mathrm{D}_j g_{ki} - \mathrm{D}_k g_{ij}).$$

The gradient of a tensor of weight N and valence m is defined as the tensor,

$$\text{(9)} \qquad \nabla\mathbf{T} = g^{\frac{N}{2}}\mathbf{e}^h\mathrm{D}_h(g^{-\frac{N}{2}}\mathbf{T})$$

of weight N and valence $m + 1$. When $\mathbf{T}$ is absolute, $\nabla\mathbf{T} = \mathbf{e}^h\mathrm{D}_h\mathbf{T}$.

The components of $\nabla\mathbf{T}$, denoted by prefixing ∇_h to the components of $\mathbf{T}$, are called *covariant derivatives*. For any tensor $T_{\cdot j \cdot}^{\cdot i \cdot}$, we have

$$\text{(10)} \quad \nabla_h T_{\cdot j \cdot}^{\cdot i \cdot} = \mathrm{D}_h T_{\cdot j \cdot}^{\cdot i \cdot} - N\Gamma_{hr}^r T_{\cdot j \cdot}^{\cdot i \cdot} + (T_{\cdot j \cdot}^{\cdot r \cdot}\Gamma_{hr}^i + \cdots) - (T_{\cdot r \cdot}^{\cdot i \cdot}\Gamma_{hj}^r + \cdots);$$

the first parenthesis contains one term for every upper index, the second contains one term for every lower index.

The metric tensor $\mathbf{G}$ is absolute and $\nabla\mathbf{G} = 0$, from (6); hence

$$\nabla_h g_{ij} = 0, \qquad \nabla_h \delta_i^j = 0, \qquad \nabla_h g^{ij} = 0.$$

Since g is a relative scalar of weight 2, $\nabla g = 0$ from (9), and $\nabla_h g = 0$.

The covariant derivatives of the epsilons and Kronecker deltas are zero.

The *divergence* of a tensor $\mathbf{T}$ is defined as the gradient $\nabla\mathbf{T}$ contracted on the first and last indices: thus

$$(\operatorname{div} \mathbf{T})^{ij\cdots k} = \nabla_h T^{ij\cdots kh};$$

when $\mathbf{T}$ is an m-vector of weight 1, we may replace ∇_h by D_h.

The *curl* of a covariant tensor $T_{bc\cdots d}$ of valence $m < n$ is defined as the $(m+1)$-vector:

$$(\operatorname{curl} \mathbf{T})_{hij\cdots k} = \frac{1}{m!}\,\delta^{abc\cdots d}_{hij\cdots k}\nabla_a T_{bc\cdots d}.$$

When $\mathbf{T}$ is absolute, we may replace ∇_a by D_a.

A Riemannian n-space x^i with the metric tensor g_{ij} and base vectors $\mathbf{e}_i$ has the associated *curvature tensor*,

$$R_{ijk}^{\cdot\cdot\cdot h} = \mathbf{e}^h \cdot (\mathrm{D}_i\mathrm{D}_j - \mathrm{D}_j\mathrm{D}_i)\mathbf{e}_k.$$

Its covariant components,

$$R_{ijkh} = \mathbf{e}_h \cdot (\mathrm{D}_i\mathrm{D}_j - \mathrm{D}_j\mathrm{D}_i)\mathbf{e}_k,$$

have four types of symmetry:

$$R_{ijkh} + R_{jikh} = 0, \qquad R_{ijkh} + R_{ijhk} = 0,$$

$$R_{ijkh} + R_{jkih} + R_{kijh} = 0, \qquad R_{ijkh} - R_{khij} = 0.$$

These relations reduce the number of linearly independent components R_{ijkh} to $n^2(n^2 - 1)/12$. When $n = 2$, there is but one independent component, say R_{1212}; and the absolute scalar $-R_{1212}/g = K$, the total curvature of a surface whose fundamental form is $g_{ij}\,dx^i\,dx^j$.

A *Cartesian coordinate system* y^i is one in which the components g_{ij} of the metric tensor are constants; then all $\Gamma^k_{ij} = 0$, and the base vectors $\mathbf{a}_i$ remain invariable in space $(\partial\mathbf{a}_j/\partial y^i = 0)$.

The Riemannian space x^i with metric tensor g_{ij} is said to be *flat* if it is possible to transform to a Cartesian coordinate system. When the affine connection is symmetric, a necessary and sufficient condition for a flat space is that the curvature tensor vanish.

If the space is flat and the metric form $g_{ij}x^ix^j$ is positive definite, the space and its geometry are termed *Euclidean*. We then can make a real linear transformation to orthogonal coordinates y^i for which $g_{ij} = \delta^i_j$; the metric form then becomes a sum of squares. A transformation $y^i = c^i_j \bar{y}^j + C^i$ between orthogonal coordinate systems is characterized by the relations:

$$\frac{\partial y^i}{\partial \bar{y}^j} = \frac{\partial \bar{y}^j}{\partial y^i} \qquad \text{or} \qquad c^i_j = \gamma^j_i \text{ (orthogonal matrix)};$$

then

$$c^r_i c^r_j = c^i_r c^j_r = \delta^i_j, \quad \text{and} \quad \det c^i_j = \pm 1.$$

Within this *orthogonal group* of transformations, the distinction between covariance and contravariance vanishes. Orthogonal transformations for which $\det c^i_j = 1$ form the *displacement subgroup* in which the distinction between absolute and relative tensors vanishes.

When y^i and $\bar{y}^j$ are regarded as points in the same coordinate system, the transformation $y^i = c^i_j \bar{y}^j$ is a *rotation* when its matrix is orthogonal and its determinant $+1$.

PROBLEMS

Summation Convention. Index range is 1, 2, 3 unless otherwise stated.

1. Prove the following:

(a) $$\epsilon_{ijk}\, \epsilon^{rjk} = 2!\, \delta^r_i;$$

(b) $$\epsilon_{ijk}\, \delta^{ijk}_{rst} = 3!\, \epsilon_{rst};$$

(c) $$\delta^i_i = 3, \quad \delta^{ij}_{ij} = 3 \cdot 2, \quad \delta^{ijk}_{ijk} = 3!$$ [§ 161.]

2. Show that

$$\begin{vmatrix} u^1 & u^2 & u^3 \\ v^1 & v^2 & v^3 \\ w^1 & w^2 & w^3 \end{vmatrix} = \epsilon_{ijk}\, u^i v^j w^k.$$

3. Show that the two-rowed determinant formed by columns i, j of the matrix,

$$\begin{pmatrix} x^1\, x^2 \cdots x^n \\ y^1\, y^2 \cdots y^n \end{pmatrix} \quad \text{is} \quad \delta^{ij}_{rs}\, x^r y^s.$$

4. For the dyadic φ_{ij} in (77.1) show that the scalar invariants are

$$\varphi_1 = \varphi_{ii}, \quad \varphi_2 = \Phi^{ii}, \quad \varphi_3 = \frac{1}{3!}\, \epsilon^{ijk}\, \epsilon^{rst}\, \varphi_{ir}\varphi_{js}\varphi_{kt};$$

and that the vector invariant has the components $\frac{1}{2}\epsilon^{ijk}\, \varphi_{jk}$.

5. Show that the general solution of the equations:

$$a_i x^i = 0, \quad b_i x^i = 0 \quad \text{is} \quad x^i = \lambda\, \epsilon^{ijk}\, a_j b_k.$$

6. Show that the cofactor A^i_r of the element a^r_i in $a = \det a^j_i$ is

$$A^i_r = \frac{1}{2!}\epsilon^{ijk}\,\epsilon_{rst}\,a^s_j a^t_k.$$

$$[A^i_r a^p_i = \tfrac{1}{2}\epsilon_{rst}\,\epsilon^{ijk}\,a^p_i a^s_j a^t_k = \tfrac{1}{2}\epsilon_{rst}\,\epsilon^{pst}\,a = \delta^p_r\,a.]$$

7. If y^1, y^2, y^3 are functions of x^1, x^2, x^3, $\det(\partial y^i/\partial x^j)$, is called their *Jacobian* and written

$$\frac{\partial(y^1, y^2, y^3)}{\partial(x^1, x^2, x^3)}, \quad \text{or, more briefly,} \quad \left|\frac{\partial y}{\partial x}\right|.$$

If z^1, z^2, z^3 are functions of y^1, y^2, y^3, prove that

$$\left|\frac{\partial z}{\partial x}\right| = \left|\frac{\partial z}{\partial y}\right|\left|\frac{\partial y}{\partial x}\right|.$$

If, in particular, the functions

$$z^i(y^1, y^2, y^3) = x^i,$$

show that

$$\left|\frac{\partial y}{\partial x}\right|\left|\frac{\partial x}{\partial y}\right| = 1.$$

8. Prove that

$$\frac{\partial(y^i, y^j, y^k)}{\partial(x^r, x^s, x^t)} = \delta^{ijk}_{rst}\,\frac{\partial(y^1, y^2, y^3)}{\partial(x^1, x^2, x^3)}.$$

9. Prove that the

$$\text{Cofactor of } \frac{\partial y^i}{\partial x^j} \text{ in } \left|\frac{\partial y}{\partial x}\right| \text{ is } \frac{\partial x^j}{\partial y^i}\left|\frac{\partial y}{\partial x}\right|.$$

$$\left[\frac{\partial y^i}{\partial x^j}\frac{\partial x^j}{\partial y^k} = \frac{\partial y^i}{\partial y^k} = \delta^i_k.\right]$$

10. If the elements of $a = \det a_{ij}$ are functions of x^1, x^2, x^3, prove that

$$\frac{\partial a}{\partial x^r} = \frac{\partial a_{st}}{\partial x^r}A^{st}.$$

[Cf. (146.11).]

11. Prove that

$$\frac{\partial}{\partial x_r}\left|\frac{\partial y}{\partial x}\right| = \frac{\partial^2 y^i}{\partial x_r\,\partial x_j}\frac{\partial x^j}{\partial y^i}\left|\frac{\partial y}{\partial x}\right|.$$

[Apply Probs. 9 and 10.]

12. Prove that the *bordered determinant:*

$$\begin{vmatrix} v_1 & v_2 & v_3 & 0 \\ a_{11} & a_{12} & a_{13} & u_1 \\ a_{21} & a_{22} & a_{23} & u_2 \\ a_{31} & a_{32} & a_{33} & u_3 \end{vmatrix} = u_i v_j A^{ij}.$$

This determinant is formed by bordering $\det a_{ij}$ with the vectors v_i and u_j. If a_{ij} is symmetric it also equals $v_i u_j A^{ij}$.

13. When the index range is from 1 to n show that the bordered determinant (written compactly),

$$\begin{vmatrix} v_i & 0 \\ a_{ij} & u_j \end{vmatrix} = (-1)^{n+1} u_i v_j A^{ij} \qquad (i, j = 1, 2, \cdots, n).$$

14. If A^{ij} is the cofactor of a_{ij} in $a = \det a_{ij}$, and $A = \det A^{ij}$, show that $aA = a^3$ (and hence $A = a^2$ when $a \neq 0$). State the corresponding theorem when the index range is from 1 to n.

15. Show that the n-rowed determinant,

$$a = \det a_{pq} = \epsilon^{ij\cdots k} a_{1i} a_{2j} \cdots a_{nk}$$

$$= \frac{1}{n!} \epsilon^{rs\cdots t} \epsilon^{ij\cdots k} a_{ri} a_{sj} \cdots a_{tk},$$

and that the cofactor of a_{ri} is

$$A^{ri} = \frac{1}{(n-1)!} \epsilon^{rs\cdots t} \epsilon^{ij\cdots k} a_{sj} \cdots a_{tk}.$$

16. If $\det a_{ij}$ is symmetric ($a_{ij} = a_{ji}$), show that $\det A^{ij}$ is also ($A^{ij} = A^{ji}$).

17. If the n-rowed determinant $a = \det a_{ji}$ is antisymmetric ($a_{ij} = -a_{ji}$), show that $a = 0$ when n is odd.

18. Show that the linear equations,

$$a_{ij} x^j = 0, \qquad (i, j = 1, 2, \cdots, n),$$

for which $\det a_{ij} = 0$ and not *all* the cofactors A^{ij} vanish has a non-zero solution of the form $x^j = A^{kj}$ for some value of k.

19. If a_{ij} and g_{ij} are symmetric dyadics and $g_{ij}x^ix^j$ is a positive definite quadratic form (§ 150), prove that the roots of the cubic equation,

$$(1) \qquad \det (a_{ij} - \lambda g_{ij}) = 0, \qquad (i, j = 1, 2, \cdots, n),$$

are all real.

[The system of n linear equations,

$$(a_{ij} - \lambda g_{ij}) z^j = 0,$$

must have a solution $z^j = x^i + iy^j$ other than $(0, 0, \cdots, 0)$ when $\lambda = \alpha + i\beta$ is a root of (1). Hence

$$[a_{ij} - (\alpha + i\beta) g_{ij}](x^j + iy^j) = 0;$$

and, on equating real and imaginary parts to zero, we have

$$(2) \qquad a_{ij} x^j - \alpha g_{ij} x^j + \beta g_{ij} y^j = 0,$$

$$(3) \qquad a_{ij} y^j - \alpha g_{ij} y^j - \beta g_{ij} x^j = 0.$$

On multiplying (2) by y^i and (3) by x^i, and subtracting, we find that

$$\beta(g_{ij} x^i x^j + g_{ij} y^i y^j) = 0, \quad \text{and hence} \quad \beta = 0;$$

for if $g_{ij} x^i x^j = g_{ij} y^i y^j = 0$, $x^j = y^j = 0$ and consequently $z^j = 0$, contrary to hypothesis.]

Tensor Character.

20. Prove the theorem: If a_{ij}, b^{ij}, c^i_j are absolute dyadics,

$$a = \det a_{ij}, \quad b = \det b^{ij}, \quad c = \det c^i_j$$

are scalars of weight 2, −2, 0; the cofactors A^{ij}, B_{ij}, C^j_i in these determinants are dyadics of weight 2, −2, 0; and the reduced cofactors A^{ij}/a, B_{ij}/b, C^j_i/c are all absolute dyadics. [Cf. Prob. 15.]

21. Show that, if **u**, **v**, **w** are absolute vectors,

$$\begin{vmatrix} u_1 & u_2 & u_3 \\ v_1 & v_2 & v_3 \\ w_1 & w_2 & w_3 \end{vmatrix} \quad \text{and} \quad \begin{vmatrix} u^1 & u^2 & u^3 \\ v^1 & v^2 & v^3 \\ w^1 & w^2 & w^3 \end{vmatrix}$$

are relative scalars of weight 1 and −1.

22. If a_{ij} and b^{ij} are absolute antisymmetric dyadics in 3-space, show that
(*a*) a_{23}, a_{31}, a_{12} are components of a contravariant vector u^i of weight 1;
(*b*) b^{23}, b^{31}, b^{12} are components of a covariant vector v_i of weight −1:

$$[u^i = \tfrac{1}{2}\, \epsilon^{ijk}\, a_{jk}; \qquad v_i = \tfrac{1}{2}\, \epsilon_{ijk}\, b^{jk}].$$

23. If $T_{ij\cdots h}$, $S^{ij\cdots h}$ are absolute n-vectors (§ 170) in n-space, prove that $T_{12\cdots n}$ and $S^{12\cdots n}$ are relative scalars of weight 1 and −1.

$$[n!\, T_{12\cdots n} = \epsilon^{ij\cdots h}\, T_{ij\cdots h}.]$$

24. If v_i, u_j, a_{ij} are absolute tensors, prove that the bordered determinant in Prob. 13 is a relative scalar of weight 2.

25. Prove the "quotient law": If the set of functions $v_i T^{ij\cdots k}_{ab\cdots c}$ are tensor components of the type indicated by the indices for all absolute vectors v_i, then $T^{ij\cdots k}_{ab\cdots c}$ is a tensor of the same weight.

26. If u_k is a covariant vector, prove that the total differential equation $u_k\, dx^k = 0$ has the same form in all coordinate systems.

This equation is said to be *integrable* when there exists a function λ such that $\lambda u_k = \partial\varphi/\partial x^k$; for the equation is then equivalent to $d\varphi = 0$ and $\varphi =$ const is an integral. Show that

$$\epsilon^{ijk}\, u_i \mathrm{D}_j u_k = 0$$

is a necessary condition for the integrability of $u_k\, dx^k = 0$ and that the form of this condition is the same in all coordinate systems.

27. Let u, v be quantities (scalars, base vectors, tensors) whose "product" uv is distributive with respect to addition but not necessarily commutative. Show that the differential operator $\mathrm{D}_{ij} = \mathrm{D}_i\mathrm{D}_j - \mathrm{D}_j\mathrm{D}_i$ has the property,

$$\mathrm{D}_{ij}(uv) = (\mathrm{D}_{ij}u)v + u(\mathrm{D}_{ij}v).$$

28. Show by direct calculation that the operator D_{ij} transforms like a covariant dyadic:

$$\overline{\mathrm{D}}_{ij} = \frac{\partial x^a}{\partial \bar{x}^i} \frac{\partial x^b}{\partial \bar{x}^j}\, \mathrm{D}_{ab}.$$

29. Deduce (177.3) from (177.2) by applying the operator D_{ij} to $\mathbf{e}_h \cdot \mathbf{e}^k = \delta_h^k$. [Cf. Prob. 27.]

30. If **T** is a tensor (complete with base vectors) of weight N and valence m, prove that $\mathbf{e}^i\mathbf{e}^j\, D_{ij}\mathbf{T}$ is a tensor of weight N and valence $m + 2$.

[Since $E^{-N}\mathbf{T}$ is an absolute tensor, $E^{-N}\mathbf{T} = \bar{E}^{-N}\bar{\mathbf{T}}$; hence, from Prob. 28,

$$\bar{\mathbf{e}}^i\bar{\mathbf{e}}^j\, \bar{D}_{ij}(\bar{E}^{-N}\bar{\mathbf{T}}) = \bar{\mathbf{e}}^i\bar{\mathbf{e}}^j \frac{\partial x^a}{\partial \bar{x}^i}\frac{\partial x^b}{\partial \bar{x}^j}\, D_{ab}(E^{-N}\mathbf{T}) = \mathbf{e}^a\mathbf{e}^b\, D_{ab}(E^{-N}\mathbf{T}).$$

Thus (Prob. 26),

$$\mathbf{e}^i\mathbf{e}^j\, D_{ij}(E^{-N}\mathbf{T}) = E^{-N}\,\mathbf{e}^i\mathbf{e}^j\, D_{ij}\mathbf{T}$$

is an absolute tensor, which multiplied by E^N yields the relative tensor $\mathbf{e}^i\mathbf{e}^j\, D_{ij}\mathbf{T}$ of weight N.]

31. In Riemannian n-space with the metric tensor g_{ij} show that the gradient of an absolute scalar $\varphi(x^1, \cdots, x^n)$ is given by

$$\nabla\varphi = \mathbf{e}^i D_i\varphi = \mathbf{e}_j g^{ij}\, D_i\varphi = \frac{1}{g}\,(D_i\varphi)\mathbf{e}_j G^{ij},$$

or, in view of Prob. 13, by the bordered n-rowed determinant

$$(1) \qquad \nabla\varphi = \frac{(-1)^{n+1}}{g}\begin{vmatrix} \mathbf{e}_i & 0 \\ g_{ij} & D_j\varphi \end{vmatrix}.$$

Show that the scalar product of (1) by $\nabla\varphi = \mathbf{e}^k D_k\varphi$ gives

$$(2) \qquad \nabla\varphi \cdot \nabla\varphi = \frac{(-1)^{n+1}}{g}\begin{vmatrix} D_i\varphi & 0 \\ g_{ij} & D_j\varphi \end{vmatrix}.$$

32. In Riemannian n-space with the metric tensor g_{ij} show that the divergence of an absolute vector **v** is given by

$$\operatorname{div}\mathbf{v} = \frac{1}{\sqrt{g}}\, D_i(\sqrt{g}\, v^i) = \frac{1}{\sqrt{g}}\, D_i(\sqrt{g}\, g^{ij}v_j) = \frac{1}{\sqrt{g}}\, D_i\left(\frac{v_j}{\sqrt{g}}\, G^{ij}\right),$$

or by the bordered n-rowed determinant,

$$(1) \qquad \operatorname{div}\mathbf{v} = \frac{(-1)^{n+1}}{\sqrt{g}}\begin{vmatrix} D_i & 0 \\ g_{ij} & \dfrac{1}{\sqrt{g}}\, v_j \end{vmatrix}$$

where the determinant is to be expanded according to the elements of the first row and the operators D_i applied to differentiate their cofactors.

In particular, if $\mathbf{v} = \nabla\varphi$, $v_j = D_j\varphi$, we obtain the Laplacian,

$$\nabla^2\varphi = \frac{(-1)^{n+1}}{\sqrt{g}}\begin{vmatrix} D_i & 0 \\ g_{ij} & \dfrac{1}{\sqrt{g}}\, D_j\varphi \end{vmatrix}.$$

33. The equations of a surface in 3-space are $x^i = x^i(u^1, u^2)$. Show that

$$x_\alpha^i = \frac{\partial u^i}{\partial u^\alpha} \qquad (i = 1, 2, 3;\ \alpha = 1, 2)$$

is a contravariant vector in 3-space ($i = 1, 2, 3$; α fixed) and a covariant vector in the 2-space ($\alpha = 1, 2$; i fixed) formed by the surface.

Prove that

$$p_i = \tfrac{1}{2}\epsilon^{\alpha\beta}\,\epsilon_{ijk}\,x^j_\alpha\,x^k_\beta$$

is an absolute covariant space vector normal to the surface. Write out its three components in full.

34. In *special relativity* it is customary to use the independent variables,

$$x^1 = x,\quad x^2 = y,\quad x^3 = z,\quad x^4 = ct$$

where c is the speed of light. The *interval* ds between two events in space time is defined by

$$ds^2 = c^2\,dt^2 - dx^2 - dy^2 - dz^2 = g_{ij}\,dx^i\,dx^j. \tag{1}$$

Hence the metric tensor is given by

$$g_{11} = g_{22} = g_{33} = -1,\quad g_{44} = 1,\quad g_{ij} = 0\ (i \neq j). \tag{2}$$

If v is the speed of a particle relative to a frame $\mathfrak{F}$, we have $(ds/dt)^2 = c^2 - v^2$ from (1), and hence

$$\frac{ds}{dt} = \frac{c}{\gamma}\quad \text{where}\quad \gamma = \left(1 - \frac{v^2}{c^2}\right)^{-1/2}. \tag{3}$$

In a *rest frame* $\mathfrak{F}_0$ attached to the particle a clock registers the proper time τ. Putting $v = 0$, $t = \tau$ in (3) we have

$$\frac{ds}{d\tau} = c\quad\text{and}\quad \frac{dt}{d\tau} = \gamma. \tag{4}$$

Corresponding to the position vector (x, y, z) in space we now have the *event vector* $x^i = (x, y, z, ct)$ in space time. The velocity and acceleration *four-vectors* are now defined as

$$u^i = \frac{dx^i}{d\tau},\qquad w^i = \frac{du^i}{d\tau}. \tag{5}$$

Denoting t derivatives with dots, prove that

$$u^i = \gamma(\dot{x}, \dot{y}, \dot{z}, c),\quad u_i = \gamma(-\dot{x}, -\dot{y}, -\dot{z}, c);\qquad u^iu_i = c^2;$$

$$w^i = \gamma(\dot{\gamma}\dot{x} + \gamma\ddot{x}, \dot{\gamma}\dot{y} + \gamma\ddot{y}, \dot{\gamma}\dot{z} + \gamma\ddot{z}, \dot{\gamma}c);\qquad w^iu_i = 0.$$

Thus the velocity four-vector has the constant magnitude c and is always perpendicular to acceleration four-vector.

Show also that in the rest frame $\mathfrak{F}_0$ ($t = \tau$),

$$u_0^i = (0, 0, 0, c),\qquad w_0^i = (\ddot{x}, \ddot{y}, \ddot{z}, 0).$$

Covariant Differentiation.

35. Prove the product rule (§ 168, 1) for covariant differentiation without resort to geodesic coordinates.

36. If T^{ij} is an absolute bivector, show that

$$\operatorname{div} T = \nabla_j T^{ij} = \frac{1}{\sqrt{g}} \mathrm{D}_j(\sqrt{g}\, T^{ij}).$$

37. If $\mathbf{u}$ and $\mathbf{v}$ are absolute vectors, prove that

$$\mathrm{D}_h(g_{ij}\, u^i v^j) = u_i \nabla_h v^i + v_i \nabla_h u^h.$$

If $|\mathbf{u}|$ is the length of $\mathbf{u}$, prove that

$$\mathrm{D}_h |\mathbf{u}| = \frac{u_i \nabla_h u^i}{|\mathbf{u}|}.$$

38. Prove that

$$\nabla_{ij} v_k + \nabla_{jk} v_i + \nabla_{ki} v_j = 0. \qquad \text{[Cf. (177.5).}$$

39. Prove that

$$R_{ijk}^{\cdot\cdot\cdot h} + R_{jki}^{\cdot\cdot\cdot h} + R_{kij}^{\cdot\cdot\cdot h} = 0.$$

By contracting this identity with $h = j$, show that the Ricci tensor S_{ik} (§ 176, ex. 2) is symmetric.

40. In the Bianchi identity (177.6) raise the index h and contract with $j = h$, $k = m$ to obtain

$$\nabla_i R_{jk}^{\cdot\cdot jk} + \nabla_j R_{ki}^{\cdot\cdot jk} + \nabla_k R_{ij}^{\cdot\cdot jk} = 0.$$

On introducing the *mixed Ricci tensor*,

$$R_i^{\cdot j} = R_{ri}^{\cdot\cdot jr} \quad \text{and} \quad R = R_i^{\cdot i} = R_{ri}^{\cdot\cdot ir}$$

(its first scalar), show that this equation implies that the divergence of the tensor $R_i^{\cdot j} - \frac{1}{2}\delta_i^j R$ is zero.

41. Check the results of § 164, ex. 1, by differentiating $\mathbf{r} = \rho\mathbf{R}(\varphi) + z\mathbf{k}$ twice with respect to the time. [Apply (44.2) and (44.3).]

42. The generalized Kronecker delta $\delta_{ab\cdots d}^{ij\cdots h}$ has k subscripts and k superscripts ($1 \leqq k \leqq n$), each ranging from 1 to n. Its value is defined as follows: If both upper and lower indices consist of the same set of *distinct* numbers, chosen from $1, 2, \cdots, n$, the delta is 1 or -1 according to the upper indices from an even or an odd permutation of the lower; in all other cases the delta is zero.

Prove that if $\delta_{ab\cdots d}^{ij\cdots h}$ has $2k$ indices,

$$(n - k)!\, \delta_{ab\cdots d}^{ij\cdots h} = \epsilon^{ij\cdots hrs\cdots t}\, \epsilon_{ab\cdots drs\cdots t}$$

where the epsilons necessarily have n indices. Hence show that the delta is an absolute tensor.

CHAPTER X

QUATERNIONS

181. Quaternion Algebra. The problem of extending 3-dimensional vector algebra to include multiplication and division was first solved by Sir William Rowan Hamilton in 1843. He found that it was necessary to invent an algebra for *quadruples of numbers, or quaternions,* before a serviceable algebra for number triples, or vectors, was possible. Without attempting to motivate Hamilton's invention, we proceed to a brief account of quaternion algebra.

A real quaternion is a quadruple of real numbers written in a definite order. We shall designate quaternions by single letters, p, q, r; thus $q = (d, a, b, c)$, $q' = d', a', b', c')$. The fundamental definitions are the following.

Equality: $q = q'$ when and only when $d = d'$, $a = a'$, $b = b'$, $c = c'$.

Addition:

(1) $$q + q' = (d + d', a + a', b + b', c + c').$$

Multiplication by a Scalar λ:

(2) $$\lambda q = (\lambda d, \lambda a, \lambda b, \lambda c).$$

Negative: $-q = (-1)q$.

Subtraction: $q - q' = q + (-q')$. Hence

$$q - q' = (d - d', a - a', b - b', c - c').$$

The *zero quaternion* (0, 0, 0, 0) is denoted simply by 0.

From these definitions it is obvious that, as far as addition, subtraction, and multiplication by scalars are concerned, quaternions obey the rules of ordinary algebra:

(3) $$p + q = q + p, \qquad (p + q) + r = p + (q + r);$$

(4) $$\lambda q = q\lambda, \qquad (\lambda\mu)q = \lambda(\mu q);$$

(5) $$(\lambda + \mu)q = \lambda q + \mu q, \qquad \lambda(p + q) = \lambda p + \lambda q.$$

In order to define the product qq' of two quaternions in a convenient manner, we shall denote the four *quaternion units* as follows:

$$1 = (1, 0, 0, 0), \quad i = (0, 1, 0, 0), \quad j = (0, 0, 1, 0), \quad k = (0, 0, 0, 1).$$

Then, in view of the preceding definitions, we can write any quaternion in the form:

$$q = (d, a, b, c) = d1 + ai + bj + ck. \tag{6}$$

Definition of Multiplication: The quaternion product,

$$qq' = (d1 + ai + bj + ck)(d'1 + a'i + b'j + c'k),$$

is obtained by distributing the terms on the right as in ordinary algebra, except that *the order of the units must be preserved,* and then replacing each product of units by the quantity given in the following *multiplication table:*

(7) First factor

↗	1	i	j	k
1	1	i	j	k
i	i	-1	k	$-j$
j	j	$-k$	-1	i
k	k	j	$-i$	-1

Note that $i^2 = j^2 = k^2 = -1$, and the cyclic symmetry of the equations:

$$ij = k, \quad jk = i, \quad ki = j; \qquad ji = -k, \quad kj = -i, \quad ik = -j.$$

With this definition we find that

$$\begin{aligned} qq' &= dd' - aa' - bb' - cc' \\ &\quad + d(a'i + b'j + c'k) + d'(ai + bj + ck) \\ &\quad + \begin{vmatrix} i & j & k \\ a & b & c \\ a' & b' & c' \end{vmatrix}. \end{aligned} \tag{8}$$

If we form $q'q$ by interchanging primed and unprimed letters, the first two lines above remain unchanged, but the interchange of rows in the determinant is equivalent to changing its sign; hence

$q'q = qq'$ *only when the determinant is zero.* That $q'q$ and qq' differ in general was to be expected; for, from the table, $ij = k, ji = -k$.

The table shows that multiplying a unit by 1 leaves it unchanged; hence $(d1)q = q(d1) = dq$, $(d1)(d'1) = dd'1$; and, from (1), $d1 + d'1 = (d + d')1$. Quaternions of the form $d1$ therefore behave exactly like real scalars and may be identified with them. Henceforth we shall write $d1$ or $(d, 0, 0, 0)$ simply as d; in particular, 1 or $(1, 0, 0, 0)$ is regarded as the real unit 1.

We also may identify i, j, k with a dextral set of orthogonal unit vectors. For, if we make the orthogonal transformation,

$$\bar{i} = c_{11}i + c_{12}j + c_{13}k,$$
$$\bar{j} = c_{21}i + c_{22}j + c_{23}k,$$
$$\bar{k} = c_{31}i + c_{32}j + c_{33}k,$$

we have the relations $c_{ir}c_{jr} = \delta_{ij}$ (149.3) and

$$\bar{i}^2 = -c_{1r}\,c_{1r} = -1,$$

$$\bar{j}\bar{k} = -c_{2r}\,c_{3r} + \begin{vmatrix} i & j & k \\ c_{21} & c_{22} & c_{23} \\ c_{31} & c_{32} & c_{33} \end{vmatrix} = c_{11}i + c_{12}j + c_{13}k = \bar{i},$$

since the elements of det c_{ij} $(= 1)$ are their own cofactors (§ 149).

Thus every quaternion $q = d + ai + bj + ck$ is the sum of a *scalar* d and vector $v = ai + bj + ck$. With Hamilton we use the symbols Sq and Vq to denote the scalar and vector parts of q; thus

$$(9) \qquad Sq = d, \qquad Vq = ai + bj + ck, \qquad q = Sq + Vq.$$

The operations S and V are evidently distributive with respect to addition.

We are now in position to prove the fundamental

THEOREM. *Quaternion multiplication is associative and distributive with respect to addition; but the commutative law $pq = qp$ holds only when one factor is a scalar, or the vector parts of both factors are proportional.* In symbols:

$$(10) \qquad (pq)r = p(qr);$$

$$(11) \qquad p(q + r) = pq + pr, \qquad (p + q)r = pr + qr;$$

(12) $pq = qp$ only when $Vp = 0$, or $Vq = 0$, or $Vp = \lambda Vq$.

Proof of (10). It will suffice to verify (10) for all possible combinations of the units i, j, k. Since the multiplication table is unchanged under a cyclical permution of i, j, k, we need examine (10) only for those products whose left factor is i; thus we find

$$
\begin{aligned}
(ij)k &= k^2 = -1, & i(jk) &= i^2 = -1; \\
(ik)j &= -j^2 = 1, & i(kj) &= -i^2 = 1; \\
(ij)i &= ki = j, & i(ji) &= -ik = j; \\
(ik)i &= -ji = k, & i(ki) &= ij = k; \\
(ii)j &= (-1)j = -j, & i(ij) &= ik = -j; \\
(ii)k &= (-1)k = -k, & i(ik) &= -ij = -k; \\
(ii)i &= (-1)i = -i, & i(ii) &= i(-1) = -i.
\end{aligned}
$$

Proof of (11). If we form $p(q + r)$ and $pq + pr$ by formal expansion and without use of the multiplication table, the two expressions will agree term for term; hence they will agree also after the table is used.

Proof of (12). We have seen that $q'q = qq'$ only when the determinant in (8) vanishes. This occurs only in the three cases:

$$a = b = c = 0; \qquad a' = b' = c' = 0; \qquad \frac{a}{a'} = \frac{b}{b'} = \frac{c}{c'};$$

that is, either q or q' must be scalar, or Vq and Vq' must be proportional.

From (8), we have

$$S(qq') = S(q'q), \tag{13}$$

for both scalars equal $dd' - aa' - bb' - cc'$. From (13), we may prove that, *the value of the scalar part of a quaternion product is not changed by a cyclical permutation of its factors.* We have, for example,

$$S(p \cdot qr) = S(qr \cdot p) = S(q \cdot rp) = S(rp \cdot q),$$

and hence

$$S(pqr) = S(qrp) = S(rpq). \tag{14}$$

182. Conjugate and Norm. The *conjugate* of a quaternion q, written Kq, is defined as

$$Kq = Sq - Vq. \tag{1}$$

The conjugate of a sum of quaternions is evidently the sum of their conjugates:

$$(2) \qquad K(q + q') = Kq + Kq'.$$

Since the vector parts of q and Kq differ only in sign, $q(Kq) = (Kq)q$. This product is known as the *norm* of q and is written Nq. If

$$q = d + ai + bj + ck, \qquad Kq = d - ai - bj - ck,$$

we have, from (181.8),

$$(3) \qquad Nq = q(Kq) = (Kq)q = d^2 + a^2 + b^2 + c^2.$$

Therefore Nq is a scalar; and $Nq = 0$ implies that $a = b = c = d = 0$, that is, $q = 0$. If $Nq = 1$, q is called a *unit quaternion.*

If $v = ai + bj + ck$, we have, from (181.8),

$$(4) \qquad vv' = -(aa' + bb' + cc') + \begin{vmatrix} i & j & k \\ a & b & c \\ a' & b' & c' \end{vmatrix}$$

Since changing the sign of the determinant is equivalent to interchanging the second and third rows,

$$(5) \qquad K(vv') = v'v.$$

On taking the conjugate of every term in

$$qq' = (d + v)(d' + v') = dd' + dv' + d'v + vv',$$

we see that

$$(6) \quad K(qq') = dd' - dv' - d'v + v'v = (d' - v')(d - v) = (Kq')(Kq).$$

Therefore *the conjugate of the product of two quaternions is equal to the product of their conjugates taken in reverse order.* Since $Kv = -v$, (5) is a special case of (6).

We now use this property to compute the norm of a product. From (3), we have

$$N(pq) = pq \cdot K(pq) = pq \cdot Kq \cdot Kp = p \cdot Nq \cdot Kp = pKp \cdot Nq,$$

since Nq is a scalar and therefore commutes with Kp; hence

$$(7) \qquad N(pq) = Np \cdot Nq.$$

The norm of the product of two quaternions is equal to the product of their norms.

By mathematical induction we immediately may extend (6) and (7) to products of n quaternion factors:

(8) $$K(q_1q_2 \cdots q_n) = Kq_n \cdot Kq_{n-1} \cdots Kq_1,$$

(9) $$N(q_1q_2 \cdots q_n) = Nq_1 \cdot Nq_2 \cdots Nq_n.\text{*}$$

From (6), we conclude that *the product of two quaternions is zero only when one factor is zero.* Thus, if $pq = 0$, $Np \cdot Nq = 0$; and, since the norms are scalars, $Np = 0$ or $Nq = 0$, whence $p = 0$ or $q = 0$.

We now can appreciate Hamilton's exceptionally happy choice of a multiplication table for the quaternion units.† For quaternion algebra has a unique place among the algebras of hypernumbers,

$$x = x_1e_1 + x_2e_2 + \cdots + x_ne_n,$$

linear in the *units* e_i, with coefficients x_i in the field of real numbers, and for which the associative law of multiplication holds good. For it can be shown ‡ that *the most general linear associative algebra over the field of reals, in which a product is zero only when one factor is zero, is the algebra of real quaternions.*

Quaternions include the *real numbers* $(x, 0, 0, 0)$ with a single unit 1, and the complex numbers $(x, y, 0, 0)$ with two units $1, i$. Both real and complex numbers form a *field*, that is, a set of numbers in which the sum, difference, product, and quotient (the divisor not being zero) of two numbers of the set must be definite numbers belonging to the set. Moreover, quaternions include vectors $(0, x, y, z)$ in space of three dimensions. But (4) shows that *the product of two vectors is not in general a vector, but a quater-*

* This theorem applied to qq' gives Euler's famous identity:

$$(d^2 + a^2 + b^2 + c^2)(d'^2 + a'^2 + b'^2 + c'^2)$$
$$= (dd' - aa' - bb' - cc')^2 + (ad' + da' + bc' - cb')^2 + (bd' + db' + ca' - ac')^2 + (cd' + dc' + ab' - ba')^2.$$

† As to the genesis of quaternions, Hamilton himself has written:

"They started into life, or light, full grown, on the 16th of October, 1843, as I was walking with Lady Hamilton to Dublin, and came up to Brougham Bridge, which my boys have since called the Quaternion Bridge. That is to say, I then and there felt the galvanic circuit of thought close; and the sparks which fell from it were the fundamental equations in i, j, k; exactly such as I have used them ever since."

‡ Dickson, *Linear Algebras*, London, 1914, p. 10.

nion. Unlike addition—the sum of two vectors is always a vector—the multiplication of vectors leads outside of their domain. *Vector multiplication is not closed,* and, consequently, a *pure* vector algebra having all the desirable properties of quaternion algebra is not possible.

183. Division of Quaternions. If q is not zero, Nq is a non-zero scalar; and we may write the defining equation for the norm (182.3) as

$$q\frac{Kq}{Nq} = 1, \quad \text{or} \quad \frac{Kq}{Nq}q = 1.$$

We therefore define Kq/Nq as the *reciprocal* of q and write

$$(1) \qquad q^{-1} = \frac{Kq}{Nq}; \qquad \text{then} \quad qq^{-1} = q^{-1}q = 1.$$

The last equations show that $Nq \cdot Nq^{-1} = 1$, or

$$(2) \qquad Nq^{-1} = \frac{1}{Nq}.$$

In order to divide p by q ($\neq 0$), we must solve the equation

$$(3)(4) \qquad rq = p \quad \text{or} \quad qr = p$$

for r. This is easily done by multiplying by q^{-1} on the right in (3), on the left in (4). We thus obtain the *two* solutions,

$$(5)(6) \qquad r_1 = pq^{-1}, \qquad r_2 = q^{-1}p,$$

which are in general different. For this reason the symmetrical notation p/q will not be used. The notations (5), (6) are unambiguous: r_1 satisfies (3) and may be called the left-hand quotient of p by q; and r_2, the right-hand quotient, satisfies (4). These solutions are unique; if, for example,

$$rq = r_1q, \qquad (r - r_1)q = 0,$$

and, since $q \neq 0$, $r - r_1 = 0$, or $r = r_1$.

On taking norms in (5) and (6), we have

$$(7) \qquad Nr_1 = Nr_2 = \frac{Np}{Nq}.$$

The norm of either quotient of two quaternions is equal to the quotient of their norms.

From (182.8) and (182.9),

$$(8) \qquad (q_1 q_2 \cdots q_n)^{-1} = \frac{K(q_1 q_2 \cdots q_n)}{N(q_1 q_2 \cdots q_n)} = q_n^{-1} \cdot q_{n-1}^{-1} \cdots q_1^{-1}.$$

The reciprocal of the product of n quaternions is equal to the product of their reciprocals taken in reverse order.

The definition (1) shows that the reciprocal of a unit quaternion is its conjugate; the reciprocal of a unit vector is its negative.

Example. Solve the equations (3) and (4) when

$$p = 1 + 3i - j + k, \qquad q = 2 - i - 2k.$$

From (1), we have $q^{-1} = (2 + i + 2k)/9$; hence

$$r_1 = pq^{-1} = \tfrac{1}{9}(1 + 3i - j + k)(2 + i + 2k) = \tfrac{1}{9}(-3 + 5i - 7j + 5k);$$

$$r_2 = q^{-1}p = \tfrac{1}{9}(2 + i + 2k)(1 + 3i - j + k) = \tfrac{1}{9}(-3 + 9i + 3j + 3k).$$

Note that $Sr_1 = Sr_2$, $Nr_1 = Nr_2$ in conformity with (181.13) and (182.7).

184. Product of Vectors. The product of two vectors vv' is the quaternion (182.4) whose scalar and vector parts are

$$(1) \qquad S(vv') = -(aa' + bb' + cc'),$$

$$(2) \qquad V(vv') = \begin{vmatrix} i & j & k \\ a & b & c \\ a' & b' & c' \end{vmatrix}.$$

In order to find their geometric meaning we adopt a special basis i, j, k. Choose i as the unit vector along v, j as the unit vector perpendicular to i in the plane of v, v' and so directed that the angle (j, v') is not greater than 90° (Fig. 184). Then k is the unit vector which completes the *right-handed* orthogonal basis i, j, k. If the angle $(v, v') = \theta$, we have

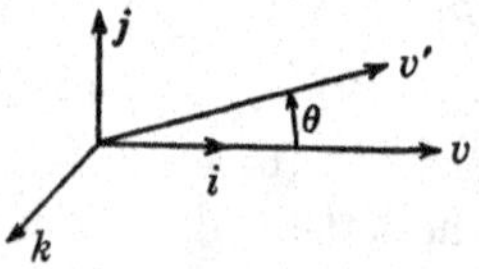

Fig. 184

$$v = |v|i, \qquad v' = |v'|(i \cos\theta + j \sin\theta),$$

$$vv' = |v||v'|(-\cos\theta + k \sin\theta);$$

$$(3) \qquad S(vv') = -|v||v'|\cos\theta,$$

$$(4) \qquad V(vv') = |v||v'|\sin\theta\, k.$$

Noting that k is the unit vector perpendicular to the plane of v and v' and directed so that v, v', k form a right-handed set, we see that (3) and (4) are geometric expressions for the scalar and vector parts of vv', entirely independent of the basis i, j, k. They form the cornerstones of the vector algebra of J. Willard Gibbs, who defined the scalar and vector product of two vectors **u**, **v** as

$$(5)(6) \qquad \mathbf{u} \cdot \mathbf{v} = -S(uv), \qquad \mathbf{u} \times \mathbf{v} = V(uv),$$

using the dot and cross to distinguish between these two types of "multiplication." Henceforth we shall use Suv, Vuv, Kuv to denote $S(uv)$, $V(uv)$, $K(uv)$; similarly $Suvw = S(uvw)$, etc.

A change in the order of the vectors in (3) has no effect, but in (4) reverses the direction of k; hence

$$(7)(8) \qquad Svu = Suv, \qquad Vvu = -Vuv.$$

From $u(v + w) = uv + uw$ and the distributive character of S and V,

$$(9)(10) \quad Su(v + w) = Suv + Suw, \qquad Vu(v + w) = Vuv + Vuw.$$

Since $Kuv = (-v)(-u) = vu$,

$$uv = Suv + Vuv, \qquad vu = Suv - Vuv;$$

$$(11)(12) \quad Suv = \tfrac{1}{2}(uv + vu), \qquad Vuv = \tfrac{1}{2}(uv - vu).$$

Turning now to products of three vectors, we have

$$(13) \qquad Suvw = Svwu = Swuv \qquad (181.14)$$

Since $Kuvw = (-w)(-v)(-u) = -wvu$ (§ 182),

$$(14) \qquad Suvw = SKuvw = -Swvu$$

$$(15) \qquad Vuvw = -VKuvw = Vwvu.$$

Also from $uvw = u(Svw + Vvw)$,

$$(16) \qquad Suvw = SuVvw,$$

$$(17) \qquad Vuvw = uSvw + VuVvw.$$

We now compute $Vuvw$ in another way:

$$\begin{aligned} 2Vuvw &= uvw - Kuvw \\ &= uvw + wvu \\ &= uvw + vuw - vuw - vwu + vwu + wvu \\ &= (uv + vu)w - v(uw + wu) + (vw + wv)u; \end{aligned}$$

hence, from (11),

(18) $$Vuvw = uSvw + wSuv - vSuw.$$

Comparison with (17) now gives the important formula:

(19) $$VuVvw = wSuv - vSuw.$$

Finally, we express these results in the dot and cross notation of Gibbs:

(5)(6) $\mathbf{u} \cdot \mathbf{v} = |\mathbf{u}||\mathbf{v}| \cos(\mathbf{u}, \mathbf{v}),\quad \mathbf{u} \times \mathbf{v} = |\mathbf{u}||\mathbf{v}| \sin(\mathbf{u}, \mathbf{v})\, \mathbf{k};$

(7)(8) $\mathbf{v} \cdot \mathbf{u} = \mathbf{u} \cdot \mathbf{v},\qquad \mathbf{v} \times \mathbf{u} = -\mathbf{u} \times \mathbf{v};$

(9) $$\mathbf{u} \cdot (\mathbf{v} + \mathbf{w}) = \mathbf{u} \cdot \mathbf{v} + \mathbf{u} \cdot \mathbf{w},$$

(10) $$\mathbf{u} \times (\mathbf{v} + \mathbf{w}) = \mathbf{u} \times \mathbf{v} + \mathbf{u} \times \mathbf{w};$$

(13) $$\mathbf{u} \cdot \mathbf{v} \times \mathbf{w} = \mathbf{v} \cdot \mathbf{w} \times \mathbf{u} = \mathbf{w} \cdot \mathbf{u} \times \mathbf{v},$$

(14) $$\mathbf{u} \cdot \mathbf{v} \times \mathbf{w} = -\mathbf{w} \cdot \mathbf{v} \times \mathbf{u};$$

(19) $$\mathbf{u} \times (\mathbf{v} \times \mathbf{w}) = \mathbf{v}(\mathbf{u} \cdot \mathbf{w}) - \mathbf{w}(\mathbf{u} \cdot \mathbf{v}).$$

From (5) and (6) we have also

(20) $$uv = -\mathbf{u} \cdot \mathbf{v} + \mathbf{u} \times \mathbf{v}$$

as the connecting link between quaternion and vector algebra.

185. Roots of a Quaternion.† Every quaternion

$$q = d + a\mathbf{i} + b\mathbf{j} + c\mathbf{k}$$

with real coefficients may be written as a real multiple of a unit quaternion:

(1) $$q = h(\cos\theta + \mathbf{e}\sin\theta), \qquad 0 \le \theta < 2\pi.$$

Here

$$h = \sqrt{d^2 + a^2 + b^2 + c^2},$$

$$\cos\theta = d/h, \qquad \sin\theta = \pm\sqrt{a^2 + b^2 + c^2}/h$$

and, when $a^2 + b^2 + c^2 \neq 0$, $\mathbf{e}$ is the unit vector:

(2) $$\mathbf{e} = \pm\frac{a\mathbf{i} + b\mathbf{j} + c\mathbf{k}}{\sqrt{a^2 + b^2 + c^2}}.$$

† In this and following articles we again denote vectors with bold-face letters; but **ab** denotes the quaternion (not dyadic) product.

When q is a real number, $\sin\theta = 0$, and $\mathbf{e}$ may be chosen at pleasure. Since $\mathbf{e}^2 = -1$, we have, by DeMoivre's Theorem,

$$q^n = h^n(\cos n\theta + \mathbf{e}\sin n\theta). \tag{3}$$

We now may find the nth roots of a real quaternion

$$Q = H(\cos\varphi + \mathbf{e}\sin\varphi), \qquad 0 \le \varphi \le \pi; \tag{4}$$

the angle φ always may be taken in the interval from 0 to π by choosing the appropriate $\mathbf{e}$ in (2). In solving $q^n = Q$, we consider two cases:

1. $\sin\varphi \ne 0$; we choose the $\mathbf{e}$ in q the same as in Q. Then

$$h^n = H, \qquad \cos n\theta = \cos\varphi, \qquad \sin n\theta = \sin\varphi,$$

and n nth roots of Q are given by (1), provided

$$h = H^{1/n}, \quad \text{the positive root,} \tag{5}$$

$$\theta = (\varphi + 2\pi m)/n \qquad (m = 0, 1, \cdots, n-1). \tag{6}$$

These n values of θ comprise all values in the interval $0 \leqq \theta < 2\pi$ which satisfy the preceding equations.

2. $\sin\varphi = 0$: the $\mathbf{e}$ in q is then an arbitrary unit vector.

$$\text{If } Q > 0: \quad \varphi = 0, \quad \theta = 2m\pi/n \qquad (m = 0, 1, \cdots, n-1).$$

When $n = 2$, the values $\theta = 0, \pi$ give just two roots $\pm\sqrt{Q}$, both real. When $n > 2$, some values of θ ($\ne 0$ or π) give non-real roots q with which any $\mathbf{e}$ may be associated.

$$\text{If } Q < 0: \quad \varphi = \pi, \quad \theta = (2m+1)\pi/n \qquad (n = 0, 1, \cdots, n-1).$$

In every case some values of θ ($\ne \pi$) give non-real roots q with which any $\mathbf{e}$ may be associated.

We summarize these results in the

THEOREM. *A quaternion with real coefficients, but not a real number, has exactly n nth roots. If Q is a positive real number, it has just two square roots $\pm\sqrt{Q}$; in all other cases a real number has infinitely many quaternion roots with real coefficients.*

In all cases the roots may be computed from (5) and (6). For example, if $Q = 1 + \mathbf{i} + \mathbf{j} + \mathbf{k}$, we write

$$Q = 2(\cos 60° + \mathbf{e}\sin 60°), \qquad \mathbf{e} = (\mathbf{i} + \mathbf{j} + \mathbf{k})/\sqrt{3}.$$

The cube roots of Q are then

$$q = \sqrt[3]{2}(\cos\theta + \mathbf{e}\sin\theta), \qquad \theta = 20°, 140°, 260°.$$

186. Great Circle Arcs. Every *unit* quaternion

$$q = d + a\mathbf{i} + b\mathbf{j} + c\mathbf{k} \qquad (Nq = 1)$$

can be expressed in the form

(1) $$q = \cos\theta + \mathbf{e}\sin\theta.$$

Here $\mathbf{e}$ is given by (185.2); and θ satisfies

(2) $$\cos\theta = d, \qquad \sin\theta = \pm\sqrt{a^2 + b^2 + c^2}.$$

If we choose the plus sign in these formulas, $0 \leqq \theta \leqq \pi$. In particular, if $q = 1, -1, \mathbf{e}$, the angle $\theta = 0, \pi, \frac{1}{2}\pi$, respectively.

THEOREM. *The unit quaternion* $\cos\theta + \mathbf{e}\sin\theta$ *may be expressed as the quotient* $\mathbf{ba}^{-1}$ *of any two vectors which satisfy the conditions:*

(i) $|\mathbf{a}| = |\mathbf{b}|$,
(ii) *Angle* $(\mathbf{a}, \mathbf{b}) = \theta$,
(iii) *Plane* $\mathbf{a}, \mathbf{b}$ *is perpendicular to* $\mathbf{e}$,
(iv) $\mathbf{a}, \mathbf{b}, \mathbf{e}$ *form a dextral set.*

Proof. In view of (i), we may write

$$\cos\theta + \mathbf{e}\sin\theta = \frac{|\mathbf{a}||\mathbf{b}|\cos\theta + |\mathbf{a}||\mathbf{b}|\sin\theta\,\mathbf{e}}{|\mathbf{a}|^2};$$

hence, if we choose the vectors $\mathbf{a}$ and $\mathbf{b}$ so that conditions (ii), (iii) and (iv) are fulfilled,

$$\cos\theta + \mathbf{e}\sin\theta = \frac{-S\mathbf{ab} + V\mathbf{ab}}{N\mathbf{a}} = \frac{-S\mathbf{ba} - V\mathbf{ba}}{N\mathbf{a}} = -\frac{\mathbf{ba}}{N\mathbf{a}} = \mathbf{b}\frac{K\mathbf{a}}{N\mathbf{a}}$$

or, in view of (183.1),

(3) $$\cos\theta + \mathbf{e}\sin\theta = \mathbf{ba}^{-1}.$$

From Fig. 186 we see that, when $\mathbf{a}, \mathbf{b}, \mathbf{e}$ form a right-handed set, the angle $(\mathbf{a}, \mathbf{b})$, when less than π, is counterclockwise, viewed from the tip of $\mathbf{e}$. We shall describe the sense of $(\mathbf{a}, \mathbf{b})$ as *positive relative to* $\mathbf{e}$. When the angle $(\mathbf{a}, \mathbf{b}) = 0$ or π, q is 1 or -1, respectively, and $\mathbf{e}$ is entirely arbitrary.

To every unit quaternion, $q = \cos\theta + \mathbf{e}\sin\theta$ corresponds to a great circle arc AB of a sphere centered at O, provided $\overrightarrow{OA} = \mathbf{a}$ and $\overrightarrow{OB} = \mathbf{b}$ satisfy the preceding conditions (i) through (iv). Thus q corresponds to an arc of a great circle whose plane is normal to $\mathbf{e}$ and whose central angle θ has the positive sense relative to $\mathbf{e}$. All such arcs of this great circle are equally valid representations. If the arc AB represents q, *AB is free to move about in its great circle, provided its length and sense remain unaltered.*

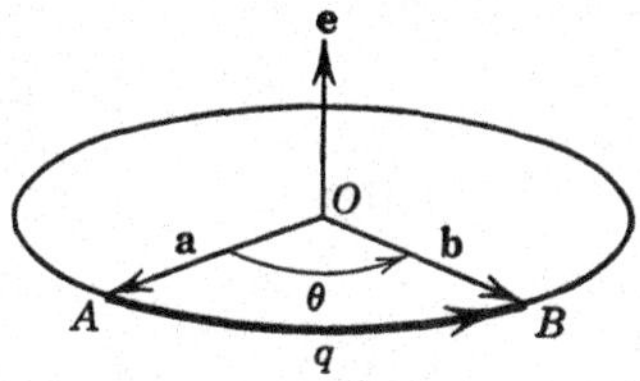

FIG. 186

The cases $q = 1$ $(\theta = 0)$ and $q = -1$ $(\theta = \pi)$, in which $\mathbf{e}$ is *arbitrary*, are exceptional. Any point of the sphere represents $q = 1$; and any great semicircle represents $q = -1$. A unit vector $q = \mathbf{e}$ $(\theta = \frac{1}{2}\pi)$ corresponds to a quadrantal arc in the plane through O normal to $\mathbf{e}$.

If $q = \cos\theta + \mathbf{e}\sin\theta$ corresponds to the arc AB,

$$(4) \qquad q^{-1} = Kq = \cos\theta - \mathbf{e}\sin\theta$$

corresponds to an arc having the same plane and angle, but whose sense is positive relative to $-\mathbf{e}$. Hence q^{-1} corresponds to the arc BA. Moreover,

$$(5) \qquad -q = -\cos\theta - \mathbf{e}\sin\theta = \cos(\pi - \theta) - \mathbf{e}\sin(\pi - \theta);$$

hence, if q corresponds to the arc AB, and AOA' is a diameter, $-q$ corresponds to arc $A'B$, the supplementary arc reversed in sense.

Using the sign $\sim$ to denote correspondence, we sum up our findings as follows:

$$1 \sim \text{point}, \quad -1 \sim \text{semicircle}, \quad \mathbf{e} \sim \text{quartercircle};$$

$$q \sim \text{arc } AB, \quad q^{-1} \sim \text{arc } BA, \quad -q \sim \text{arc } A'B.$$

The utility of this representation is due to a simple analytical method of adding great circle arcs "vectorially." *To add two arcs, shift them along their great circles until the terminal point of the first (AB) coincides with the initial point of the second (BC); then the great circle arc AC is defined as their vector sum.* If $\mathbf{a} = \overrightarrow{OA}$,

$\mathbf{b} = \overrightarrow{OB}$, $\mathbf{c} = \overrightarrow{OC}$, the preceding theorem shows that the arcs AB, BC, AC are represented by the quaternions $\mathbf{ba}^{-1}$, $\mathbf{cb}^{-1}$, $\mathbf{ca}^{-1}$, respectively. Write

$$p = \mathbf{ba}^{-1}, \quad q = \mathbf{cb}^{-1}; \qquad \text{then} \quad \mathbf{ca}^{-1} = \mathbf{cb}^{-1}\mathbf{ba}^{-1} = qp;$$

and the equation,

$$\text{arc } AB + \text{arc } BC = \text{arc } AC,$$

may be written

(6) $$\text{arc } p + \text{arc } q = \text{arc } qp.$$

For three arcs,

(7) $$\text{arc } p + \text{arc } q + \text{arc } r = \text{arc } qp + \text{arc } r = \text{arc } rqp;$$

and, in general, *the vector sum of any number of great circle arcs is given by the arc corresponding to the product of their representative quaternions taken in reverse order.*

In interpreting such arc-quaternion equations, remember that arc $q = 0$ means that $q = 1$; and, if arc q is any great semicircle, $q = -1$. For example, if arc p, arc q, arc r form the sides of a spherical triangle taken in circuital order,

$$\text{arc } p + \text{arc } q + \text{arc } r = 0, \qquad \text{arc } rpq = 0, \qquad rpq = 1.$$

In general, *the arcs representing the quaternions* $q_1, q_2, \cdots, q_n$, *taken in this circuital order, will form a closed spherical polygon when and only when*

(8) $$q_n q_{n-1} \cdots q_2 q_1 = 1.$$

Example. Spherical Trigonometry. Consider again the spherical triangle ABC of § 22. With the notation of this article

$$\text{arc } BC \sim \mathbf{cb}^{-1} = \cos\alpha + \mathbf{a}' \sin\alpha,$$

$$\text{arc } CA \sim \mathbf{ac}^{-1} = \cos\beta + \mathbf{b}' \sin\beta,$$

$$\text{arc } AB \sim \mathbf{ba}^{-1} = \cos\gamma + \mathbf{c}' \sin\gamma.$$

The "vector" equation

$$\text{arc } CA + \text{arc } AB = \text{arc } CB$$

corresponds to the quaternion equation $(\mathbf{ba}^{-1})(\mathbf{ac}^{-1}) = \mathbf{bc}^{-1}$, or

(i) $$(\cos\gamma + \mathbf{c}' \sin\gamma)(\cos\beta + \mathbf{b}' \sin\beta) = \cos\alpha - \mathbf{a}' \sin\alpha.$$

If we expand the left member, put

$$\mathbf{c}'\mathbf{b}' = -\mathbf{b}' \cdot \mathbf{c}' - \mathbf{b}' \times \mathbf{c}' = -\cos \alpha' - \mathbf{a} \sin \alpha';$$

and equate scalar parts in both members, we have

(ii) $$\cos \beta \cos \gamma - \sin \beta \sin \gamma \cos \alpha' = \cos \alpha.$$

This is the *cosine law* (22.6) of spherical trigonometry. On equating the vector parts in both members of (i) we have,

(iii) $$\mathbf{a}' \sin \alpha + \mathbf{b}' \sin \beta \cos \gamma + \mathbf{c}' \cos \beta \sin \gamma = \mathbf{a} \sin \alpha' \sin \beta \sin \gamma$$

and hence, on multiplication by $\mathbf{a} \cdot$,

$$\sin \alpha' \sin \beta \sin \gamma = \mathbf{a} \cdot \mathbf{a}' \sin \alpha = \mathbf{a} \cdot \mathbf{b} \times \mathbf{c},$$

or

$$\frac{\sin \alpha'}{\sin \alpha} = \frac{\mathbf{a} \cdot \mathbf{b} \times \mathbf{c}}{\sin \alpha \sin \beta \sin \gamma}.$$

Since the right member is unchanged by a cyclical permutation, we have

(iv) $$\frac{\sin \alpha'}{\sin \alpha} = \frac{\sin \beta'}{\sin \beta} = \frac{\sin \gamma'}{\sin \gamma},$$

the *sine law* (22.5) of spherical trigonometry.

187. Rotations. With the aid of quaternion algebra, finite rotations in space may be dealt with in a simple and elegant manner. This application depends upon the fundamental

THEOREM. *If q and r are any non-scalar quaternions, then*

(1) $$r' = qrq^{-1}$$

is a quaternion whose norm and scalar are the same as for r. The vector Vr' is obtained by revolving Vr conically about Vq through twice the angle of q. Thus if

$$q = \sqrt{Nq}(\cos \theta + \mathbf{i} \sin \theta),$$

Vr' is obtained by revolving Vr conically about $\mathbf{i}$ *through an angle 2θ.*

Proof. The norm and scalar of r' are

(2) $$N(qrq^{-1}) = Nq \cdot Nr \cdot Nq^{-1} = Nr \qquad (182.9),$$

(3) $$S(qrq^{-1}) = S(q^{-1}qr) = Sr. \qquad (181.14).$$

Moreover on writing $r = Sr + Vr$,

$$qrq^{-1} = Sr + q(Vr)q^{-1};$$

Now, from (3), $q(Vr)q^{-1}$ has the same scalar as Vr and is therefore a vector; hence

$$V(qrq^{-1}) = q(Vr)q^{-1}. \tag{4}$$

Let us now write

$$r = \sqrt{Nr}(\cos\varphi + \mathbf{e}\sin\varphi);$$

then from (4),

$$Vr' = \sqrt{Nr}\sin\varphi\,\mathbf{e}' \quad \text{where } \mathbf{e}' = q\mathbf{e}q^{-1}.$$

If we choose $\mathbf{j}$ in the plane of $\mathbf{e}$ and $\mathbf{i}$ (Fig. 187a) and $\mathbf{k}$ to complete the dextral set $\mathbf{i}, \mathbf{j}, \mathbf{k}$, then

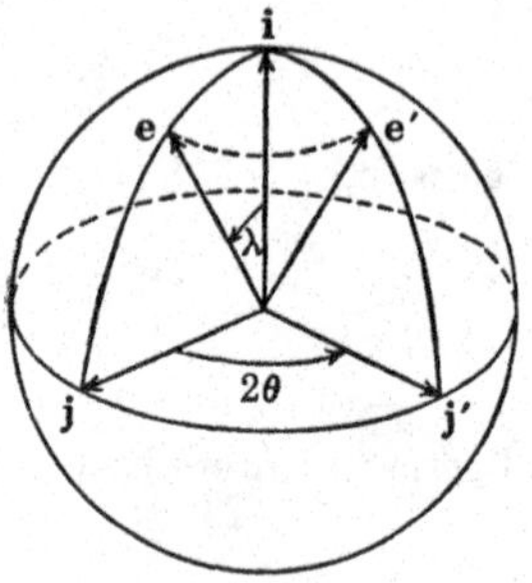

FIG. 187a

$$\mathbf{e} = \mathbf{i}\cos\lambda + \mathbf{j}\sin\lambda$$

$$\mathbf{e}' = (q\mathbf{i}q^{-1})\cos\lambda + (q\mathbf{j}q^{-1})\sin\lambda.$$

Since Vq is parallel to $\mathbf{i}$, $q\mathbf{i} = \mathbf{i}q$ and

$$q\mathbf{i}q^{-1} = \mathbf{i}qq^{-1} = \mathbf{i}.$$

Moreover,

$$\begin{aligned} q\mathbf{j}q^{-1} &= (\cos\theta + \mathbf{i}\sin\theta)\mathbf{j}(\cos\theta - \mathbf{i}\sin\theta) \\ &= (\mathbf{j}\cos\theta + \mathbf{k}\sin\theta)(\cos\theta - \mathbf{i}\sin\theta) \\ &= \mathbf{j}(\cos^2\theta - \sin^2\theta) + \mathbf{k}(2\sin\theta\cos\theta); \end{aligned}$$

hence $\mathbf{j}$ goes into

$$\mathbf{j}' = \mathbf{j}\cos 2\theta + \mathbf{k}\sin 2\theta,$$

a vector obtained by revolving $\mathbf{j}$ about $\mathbf{i}$ through an angle 2θ in the positive sense. Consequently

$$\mathbf{e} = \mathbf{i}\cos\lambda + \mathbf{j}\sin\lambda \rightarrow \mathbf{e}' = \mathbf{i}\cos\lambda + \mathbf{j}'\sin\lambda$$

and also $Vr \rightarrow Vr'$ by a conical revolution about $\mathbf{i}$ of the same amount. This completes the proof.

We note that vectors transform into vectors. In particular, if $q = \mathbf{a}$, a unit vector, $\theta = 90°$, and

$$\mathbf{e}' = \mathbf{a}\mathbf{e}\mathbf{a}^{-1} = -\mathbf{a}\mathbf{e}\mathbf{a}$$

is obtained by revolving $\mathbf{e}$ conically through 180° about $\mathbf{a}$. The transformation $-\mathbf{a}(\ )\mathbf{a}$ thus gives vectors a *half-turn* about $\mathbf{a}$.

The transformation $\mathbf{a}(\ )\mathbf{a}$ may be regarded as a *half-turn* followed by a *reversal*; a vector thus transformed is simply reflected in a plane normal to $\mathbf{a}$ (Fig. 187b). Thus $\mathbf{aea}$ *is the reflection of* $\mathbf{e}$ *in the plane normal to* $\mathbf{a}$.

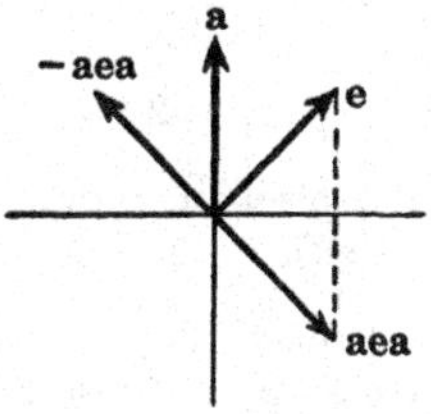

FIG. 187b

A rotation through an angle α about $\mathbf{a}$ implies that the angle turned has the positive sense relative to $\mathbf{a}$. If

$$p = \cos \tfrac{1}{2}\alpha + \mathbf{a} \sin \tfrac{1}{2}\alpha,$$

$$q = \cos \tfrac{1}{2}\beta + \mathbf{b} \sin \tfrac{1}{2}\beta$$

are unit quaternions, the operators $p(\ )p^{-1}$ and $q(\ )q^{-1}$ effect rotations of α about $\mathbf{a}$ and β about $\mathbf{b}$; for brevity we call these the rotations p and q. The succession of rotations p, q corresponds to the operator,

$$qp(\ )p^{-1}q^{-1} = qp(\ )(qp)^{-1};$$

since qp is also a unit quaternion, say

$$qp = \cos \tfrac{1}{2}\gamma + \mathbf{c} \sin \tfrac{1}{2}\gamma,$$

the resultant is equivalent to the single rotation qp, that is, a rotation through the angle γ about $\mathbf{c}$. Similarly, the resultant of the rotations q and p corresponds to the operator,

$$pq(\ )q^{-1}p^{-1} = pq(\ )(pq)^{-1}.$$

Since $pq = qp$ only when Vp and Vq are parallel (the values $p, q = \pm 1$ are excluded), the composition of rotations is non-commutative except when they have the same axis.

The rotation p followed by the rotation q is equivalent to the single rotation qp. More generally, *the succession of rotations* $q_1, q_2, \cdots, q_n$ *is equivalent to the single rotation* $q_n q_{n-1} \cdots q_2 q_1$.

Since $(-q)^{-1} = -q^{-1}$, the rotations $q(\ )q^{-1}$ and $(-q)(\ )(-q)^{-1}$ are the same. If $q = \cos\theta + \mathbf{e}\sin\theta$,

$$-q = \cos(\pi - \theta) + (-\mathbf{e})\sin(\pi - \theta);$$

thus the rotation $-q$ is a rotation through $2\pi - 2\theta$ about $-\mathbf{e}$; this produces the same result as the rotation q, namely 2θ about $\mathbf{e}$.

Since $q^{-1}q(\ )q^{-1}q = 1(\ )1$, the rotation $q^{-1}(\ )q$ is the reverse of $q(\ )q^{-1}$; this is also evident from $q^{-1} = \cos\theta - \mathbf{e}\sin\theta$.

Example 1. The rotation of 90° about $\mathbf{j}$ followed by a rotation of 90° about $\mathbf{i}$ is represented by the quaternion product

$$(\cos 45° + \mathbf{i} \sin 45°)(\cos 45° + \mathbf{j} \sin 45°) = \tfrac{1}{2}(1 + \mathbf{i} + \mathbf{j} + \mathbf{k});$$

that is, by

$$\frac{1}{2} + \frac{\mathbf{i} + \mathbf{j} + \mathbf{k}}{\sqrt{3}} \frac{\sqrt{3}}{2} = \cos 60° + \frac{\mathbf{i} + \mathbf{j} + \mathbf{k}}{\sqrt{3}} \sin 60°.$$

The resultant rotation is therefore a rotation of 120° about an axis equally inclined to the (positive) axes of x, y and z.

Example 2. The resultant of two reflections in planes normal to the unit vectors $\mathbf{a}, \mathbf{b}$ corresponds to the operator

$$\mathbf{ba}(\ )\mathbf{ab} = \mathbf{ba}(\ )(\mathbf{ba})^{-1}.$$

But if $\mathbf{a} \cdot \mathbf{b} = \cos \theta$, $\mathbf{a} \times \mathbf{b} = \mathbf{e} \sin \theta$, we have

$$\mathbf{ba} = -\mathbf{a} \cdot \mathbf{b} - \mathbf{a} \times \mathbf{b} = -(\cos \theta + \mathbf{e} \sin \theta) \qquad (184.20).$$

Now the rotation $q(\ )q^{-1} = (-q)(\ )(-q)^{-1}$; hence successive reflections in two plane mirrors is equivalent to a rotation about their line of intersection of double the angle between them.

Example 3. From (181.13) we know that $S(qp) = S(pq)$; moreover from (4)

$$V(qp) = V(qpqq^{-1}) = qV(pq)q^{-1}.$$

Hence $V(qp)$ is obtained by revolving $V(pq)$ about Vq through double the angle of q. Thus, if $\mathbf{u}$ is a vector, $V(\mathbf{u}p)$ is obtained by revolving $V(p\mathbf{u})$ 180° about $\mathbf{u}$; that is, *the vector* $\mathbf{u}$ *bisects the angle between* $V(p\mathbf{u})$ *and* $V(\mathbf{u}p)$.

Now let $\mathbf{a}, \mathbf{b}, \mathbf{c}$ be three radial vectors from the center of a sphere to its surface. Then $\mathbf{a}, \mathbf{b}, \mathbf{c}$ bisect the angles between $V\mathbf{abc}$, $V\mathbf{bca}$; $V\mathbf{bca}$, $V\mathbf{cab}$; $V\mathbf{cab}$, $V\mathbf{abc}$ respectively. In other words, if we form a spherical triangle whose vertices are $V\mathbf{abc}$, $V\mathbf{bca}$, $V\mathbf{cab}$, the middle points of the sides opposite lie on the vectors $\mathbf{b}, \mathbf{c}, \mathbf{a}$ respectively.

Example 4. If the sides of a spherical polygon are represented by the quaternions $q_1, q_2, \cdots, q_n$ taken in this circuital order, $q_n q_{n-1} \cdots q_2 q_1 = 1$ (186.8). Hence the succession of rotations,

$$q_n q_{n-1} \cdots q_2 q_1(\ ) q_1 q_2 \cdots q_{n-1} q_n = 1(\ )1,$$

about axes through a point O will restore a body to its original position. We state this result for the case of a triangle as follows:

Theorem (*Hamilton and Donkin*). *If ABC is any spherical triangle, three successive rotations represented by the directed arcs $2\,BC$, $2\,CA$, $2\,AB$ (about their polar axes) will restore a body to its original position.*

This same theorem applies to the polar triangle $A'B'C'$. Since the side $B'C' = \alpha' = \pi - A$ (§ 22) and has OA as polar axis, successive rotations of $2\pi - 2A$, $2\pi - 2B$, $2\pi - 2C$ about OA, OB, OC will restore a body to its original position. Since the rotations $2\pi - 2A$ and $-2A$ about OA give the same displacement, we may state the

THEOREM (*Hamilton*). *If ABC is any spherical triangle on a sphere centered at O, three successive rotations about OA, OB, OC through the angles 2A, 2B, 2C in the sense of CBA will restore a body to its original position.*

188. Plane Vector Analysis. The three-term quaternion $c + ai + bj$ has given rise to two types of vector analysis in the plane. The one interprets $c + ai$ as a vector w in the complex plane; then the product $w_1 w_2$ is always a complex vector. The other interprets $a\mathbf{i} + b\mathbf{j}$ as a "real" vector $\mathbf{w}$ and decomposes the quaternion product $w_1 w_2$ into its scalar and vector parts, which are used separately as "products."

To indicate the interpretation used, we denote complex and real vectors by italic and bold-face letters, respectively. Thus the plane vector whose components are u, v may be written as

$$w = u + iv, \quad \text{or} \quad \mathbf{w} = u\mathbf{i} + v\mathbf{j}.$$

In the first case,

(1) $$w_1 w_2 = (u_1 u_2 - v_1 v_2) + (u_1 v_2 + u_2 v_1)i;$$

in the second,

(2) $$\mathbf{w}_1 \mathbf{w}_2 = -(u_1 u_2 + v_1 v_2) + (u_1 v_2 - u_2 v_1)\mathbf{k}.$$

In Gibbs's notation,

$$u_1 u_2 + v_1 v_2 = \mathbf{w}_1 \cdot \mathbf{w}_2, \qquad u_1 v_2 - u_2 v_1 = \mathbf{k} \cdot \mathbf{w}_1 \times \mathbf{w}_2.$$

Let $\bar{w} = u - vi$ denote the conjugate of w; then, from (1) and (2),

$$\bar{w}_1 w_2 = \mathbf{w}_1 \cdot \mathbf{w}_2 + i\,\mathbf{k} \cdot \mathbf{w}_1 \times \mathbf{w}_2,$$

$$w_1 \bar{w}_2 = \mathbf{w}_1 \cdot \mathbf{w}_2 - i\,\mathbf{k} \cdot \mathbf{w}_1 \times \mathbf{w}_2,$$

and hence

(3) $$\mathbf{w}_1 \cdot \mathbf{w}_2 = \tfrac{1}{2}(w_1 \bar{w}_2 + \bar{w}_1 w_2),$$

(4) $$\mathbf{k} \cdot \mathbf{w}_1 \times \mathbf{w}_2 = \tfrac{1}{2}i(w_1 \bar{w}_2 - \bar{w}_1 w_2).$$

The conditions (128.6) and (128.7) for perpendicular and parallel vectors may be read from these equations.

We next consider corresponding differential invariants. The operator,

$$\nabla = \mathbf{i}\frac{\partial}{\partial x} + \mathbf{j}\frac{\partial}{\partial y} \sim \frac{\partial}{\partial x} + i\frac{\partial}{\partial y}$$

(read $\sim$ as *corresponds to*). If we introduce the conjugate variables,

$$z = x + iy, \qquad \bar{z} = x - iy,$$

$$\frac{\partial}{\partial x} + i\frac{\partial}{\partial y} = \left(\frac{\partial z}{\partial x} + i\frac{\partial z}{\partial y}\right)\frac{\partial}{\partial z} + \left(\frac{\partial \bar{z}}{\partial x} + i\frac{\partial \bar{z}}{\partial y}\right)\frac{\partial}{\partial \bar{z}}$$

$$= (1 + i^2)\frac{\partial}{\partial z} + (1 - i^2)\frac{\partial}{\partial \bar{z}},$$

or

$$(5) \qquad \frac{\partial}{\partial x} + i\frac{\partial}{\partial y} = 2\frac{\partial}{\partial \bar{z}}; \qquad \frac{\partial}{\partial x} - i\frac{\partial}{\partial y} = 2\frac{\partial}{\partial z}$$

follows in the same way.

Corresponding to the gradient $\nabla\varphi$ of a real function $\varphi(x, y)$, we have $2\,\partial\varphi/\partial\bar{z}$ in the complex plane, in which the variables x, y in φ are replaced by the values,

$$x = \tfrac{1}{2}(\bar{z} + z), \qquad y = \tfrac{1}{2}i(\bar{z} - z).$$

For example, $x^2 + y^2 = z\bar{z}$, and hence $\nabla(x^2 + y^2) = 2z$ in the complex plane.

The unit vector,

$$(6) \qquad \mathbf{e} = \mathbf{i}\cos\theta + \mathbf{j}\sin\theta \sim e^{i\theta} = \cos\theta + i\sin\theta.$$

In view of (3), the operator for differentiation in the direction $\mathbf{e}$, namely

$$(7) \qquad \mathbf{e}\cdot\nabla \sim e^{i\theta}\frac{\partial}{\partial z} + e^{-i\theta}\frac{\partial}{\partial \bar{z}}.$$

Therefore

$$(8) \qquad \frac{d\mathbf{w}}{ds} = \mathbf{e}\cdot\nabla\mathbf{w} \sim e^{i\theta}\frac{\partial w}{\partial z} + e^{-i\theta}\frac{\partial w}{\partial \bar{z}},$$

and hence $d\mathbf{z}/ds \sim e^{i\theta}$. If w is a complex function, dw/dz in the direction θ corresponds to the ratio of $d\mathbf{w}/ds$ to $d\mathbf{z}/ds$; hence

$$(9) \qquad \frac{dw}{dz} = \frac{\partial w}{\partial z} + e^{-2i\theta}\frac{\partial w}{\partial \bar{z}} \quad \text{in the direction } \theta.$$

When dw/dz is independent of θ, w is said to be an *analytic* function of z; for this, it is necessary and sufficient that

$$\frac{\partial w}{\partial \bar{z}} = 0, \quad \text{or} \quad \left(\frac{\partial}{\partial x} + i\frac{\partial}{\partial y}\right)(u + iv) = 0,$$

in view of (5). The last condition is equivalent to the familiar *Cauchy–Riemann* Equations:

$$(10)\qquad \frac{\partial u}{\partial x} - \frac{\partial v}{\partial y} = 0, \qquad \frac{\partial u}{\partial y} + \frac{\partial v}{\partial x} = 0.$$

We next find the correspondents for div $\mathbf{w} = \nabla \cdot \mathbf{w}$ and $\mathbf{k} \cdot \operatorname{rot} \mathbf{w} = \mathbf{k} \cdot \nabla \times \mathbf{w}$ by making use of (3), (4) and (5):

$$(11)\qquad \operatorname{div} \mathbf{w} \sim \frac{\partial \bar{w}}{\partial \bar{z}} + \frac{\partial w}{\partial z},$$

$$(12)\qquad \mathbf{k} \cdot \operatorname{rot} \mathbf{w} \sim i\left(\frac{\partial \bar{w}}{\partial \bar{z}} - \frac{\partial w}{\partial z}\right).$$

Moreover, for the Laplacian $\nabla^2 = \nabla \cdot \nabla$, we have

$$(13)\qquad \nabla^2 = 2\frac{\partial^2}{\partial \bar{z}\,\partial z} + 2\frac{\partial^2}{\partial z\,\partial \bar{z}} = 4\frac{\partial^2}{\partial z\,\partial \bar{z}}.$$

Thus a real function $\varphi(z, \bar{z})$ is a harmonic when $\partial^2\varphi/\partial z\,\partial \bar{z} = 0$; for example,

$$\log |z| = \tfrac{1}{2} \log z\bar{z} = \tfrac{1}{2}(\log z + \log \bar{z})$$

is harmonic.

If the plane vector $\mathbf{w} \sim w(z, \bar{z})$, the condition,

$$(14)\qquad \operatorname{rot} \mathbf{w} = 0 \sim \frac{\partial w}{\partial z} - \frac{\partial \bar{w}}{\partial \bar{z}} = 0.$$

When rot $\mathbf{w} = 0$,

$$\mathbf{w} = \nabla\lambda, \quad \text{where} \quad \lambda = \int_{\mathbf{r}_0}^{\mathbf{r}} \mathbf{w} \cdot d\mathbf{r};$$

hence, when $w(z, \bar{z})$ is irrotational,

$$(15)\qquad w = \frac{\partial \varphi}{\partial \bar{z}}, \quad \text{where} \quad \varphi = \int_{z_0}^{z} (\bar{w}\, dz + w\, d\bar{z})$$

is the real function $2\lambda(z, \bar{z})$. The field lines cut the curves $\varphi =$ const at right angles.

If the vector $\mathbf{w}$ is plane, the condition,

$$(16)\qquad \operatorname{div} \mathbf{w} = 0 \sim \frac{\partial w}{\partial z} + \frac{\partial \bar{w}}{\partial \bar{z}} = 0.$$

From (85.6), rot ($\mathbf{k} \times \mathbf{w}$) = $\mathbf{k}$ div $\mathbf{w}$; hence, *when* $\mathbf{w}$ *is solenoidal,* $\mathbf{k} \times \mathbf{w}$ *is irrotational.* Since $\mathbf{k} \times \mathbf{w}$ is $\mathbf{w}$ revolved through $+\pi/2$, $\mathbf{k} \times \mathbf{w} \sim iw$. Thus when $w(z, \bar{z})$ is solenoidal, (15) applies when w is replaced by iw (and $\bar{w}$ by $-i\bar{w}$). From this result, we conclude that

$$(17) \qquad w = i\frac{\partial \varphi}{\partial \bar{z}}, \quad \text{where} \quad \varphi = i\int_{z_0}^{z} (\bar{w}\,dz - w\,d\bar{z})$$

is a real function. The field lines are the curves $\varphi = $ const.

From these results we have

THEOREM 1. *If $\varphi(z, \bar{z})$ is a real function with continuous partial derivatives, $\partial\varphi/\partial\bar{z}$ and $i\,\partial\varphi/\partial\bar{z}$ give, respectively, an irrotational field with lines orthogonal to $\varphi = $ const, and a solenoidal field with lines $\varphi = $ const.*

The preceding results give a simple method for decomposing a plane vector function into an irrotational and a solenoidal part. For, if

$$\mathbf{w} = \mathbf{w}_1 + \mathbf{w}_2, \quad \text{rot}\,\mathbf{w}_1 = 0, \quad \text{div}\,\mathbf{w}_2 = 0,$$

there exist real functions φ, ψ such that

$$w = \frac{\partial \varphi}{\partial \bar{z}} + i\frac{\partial \psi}{\partial \bar{z}} = \frac{\partial}{\partial \bar{z}}(\varphi + i\psi);$$

$$\varphi + i\psi = \int w(z, \bar{z})\,d\bar{z} \quad (z \text{ const})$$

is determined to an arbitrary additive $f(z)$ and $w_1 = \partial\varphi/\partial\bar{z}$, $w_2 = i\,\partial\psi/\partial\bar{z}$.

If the vector field $w(z, \bar{z})$ is both irrotational and solenoidal, rot $\mathbf{w} = 0$ implies that

$$w = \frac{\partial \varphi}{\partial \bar{z}}; \qquad \text{then } \bar{w} = \frac{\partial \varphi}{\partial z}, \quad \text{div}\,w = 2\frac{\partial^2 \varphi}{dz\,d\bar{z}} = 0.$$

Hence φ is a real harmonic function, and $\bar{w}$ is an analytic function of z. Conversely, if $\bar{w}$ is an analytic function of z, $\partial\bar{w}/\partial\bar{z} = 0$; hence $\partial w/\partial z = 0$, and (11) and (12) show that w is solenoidal and irrotational.

THEOREM 2. *In order that the complex vector $w(z, \bar{z})$ be irrotational and solenoidal, it is necessary and sufficient that its conjugate be an analytic function of z.*

Example 1. The vector

$$w = \bar{z}^2 = (x^2 - y^2) - 2xy\, i,$$

is both irrotational and solenoidal; for its conjugate z^2 is an analytic function of z.

Example 2. To decompose the vector,

$$w = z^2 = (x^2 - y^2) + 2xy\, i,$$

into its irrotational and solenoidal parts, we may take

$$\varphi + i\psi = \int z^2\, d\bar{z} = z^2\bar{z}.$$

Since $\varphi - i\psi = \bar{z}^2 z$,

$$\varphi = \tfrac{1}{2} z\bar{z}(\bar{z} + z), \qquad \psi = \tfrac{1}{2} i\, z\bar{z}(\bar{z} - z);$$

$$w_1 = \frac{\partial \varphi}{\partial \bar{z}} = z\bar{z} + \tfrac{1}{2} z^2 = \tfrac{1}{2}(3x^2 + y^2) + xy\, i,$$

$$w_2 = i\frac{\partial \psi}{\partial \bar{z}} = -z\bar{z} + \tfrac{1}{2} z^2 = -\tfrac{1}{2}(x^2 + 3y^2) + xy\, i.$$

Stokes' Theorem in the plane,

$$\int \mathbf{k} \cdot \operatorname{rot} \mathbf{w}\, dA = \oint \mathbf{w} \cdot d\mathbf{r},$$

corresponds to

$$i \int \left(\frac{\partial \bar{w}}{\partial \bar{z}} - \frac{\partial w}{\partial z} \right) dA = \tfrac{1}{2} \oint (w\, d\bar{z} + \bar{w}\, dz),$$

by virtue of (12) and (3). If we replace w by $-iw$ (and $\bar{w}$ by $i\bar{w}$), we obtain, after canceling i,

$$i \int \left(\frac{\partial \bar{w}}{\partial \bar{z}} + \frac{\partial w}{\partial z} \right) dA = \tfrac{1}{2} \oint (-w\, d\bar{z} + \bar{w}\, dz).$$

On adding these equations and then replacing $\bar{w}$ by w, we have

$$\oint w\, dz = 2i \int \frac{\partial w}{\partial \bar{z}}\, dA. \tag{18}$$

When w is analytic in the region within the circuit, $\partial w / \partial \bar{z} = 0$, and (18) reduces to *Cauchy's Integral Theorem:* $\oint w\, dz = 0$.

Example 3. When $w = \bar{z}$ in (18),

$$2iA = \oint \bar{z}\, dz = \oint (x - iy)(dx + i\, dy) = i \oint (x\, dy - y\, dx).$$

This gives the well-known circuit integral for a plane area.

With $w = z\bar{z}$, we obtain the static moments of a plane area about the axes, expressed as circuit integrals.

189. Summary: Quaternion Algebra is a linear, four-unit $(1, i, j, k)$ associative algebra over the field of reals. The unit 1 has the properties of the real *one;* and

$$i^2 = j^2 = k^2 = -1, \qquad ij = k, \qquad ji = -k,$$

the last equations admitting cyclical permutations. Quaternions $q = d + ai + bj + ck$ include real and complex numbers $(d, d + ai)$. Since i, j, k may be interpreted as dextral set of orthogonal unit vectors, quaternions also include vectors $v = ai + bj + ck$. Thus $q = d + v$, a scalar plus a vector.

Quaternion multiplication is associative and distributive, but not in general commutative; in fact $pq = qp$ holds only when p or q is a scalar or when the vector parts of p and q are proportional.

The product vv' of two vectors is the quaternion:

$$vv' = -(aa' + bb' + cc') + \begin{vmatrix} i & j & k \\ a & b & c \\ a' & b' & c' \end{vmatrix},$$

$$= -\mathbf{v} \cdot \mathbf{v}' + \mathbf{v} \times \mathbf{v}' \quad \text{in Gibbs' notation.}$$

The quaternion $q = d + v$ has the *conjugate* $Kq = d - v$. The conjugate of the product qq' is $K(qq') = (Kq')(Kq)$.

The *norm* of a non-zero quaternion q is the positive real number,

$$Nq = q(Kq) = d^2 + a^2 + b^2 + c^2;$$

and $N(qq') = (Nq)(Nq')$. The equation $q = 0$ implies $Nq = 0$, and conversely; hence, if $qq' = 0$, either $q = 0$ or $q' = 0$.

A unit quaternion q $(Nq = 1)$ may be put in the form

$$q = \cos\theta + \mathbf{e} \sin\theta \qquad (|\mathbf{e}| = 1),$$

and associated with the great circle arc of angle θ and pole $\mathbf{e}$ on the unit sphere. On a fixed great circle, all arcs of the same length and sense correspond to the same q and are denoted by arc q. The scalars 1 and -1 correspond, respectively, to any point and to any great semicircle of the sphere.

Great circle arcs may be added vectorially; and

$$\text{arc } p + \text{arc } q = \text{arc } qp.$$

If three arcs form a spherical triangle, say

$$\text{arc } p + \text{arc } q + \text{arc } r = 0, \quad \text{then} \quad rpq = 1.$$

If $q = \sqrt{Nq}(\cos\theta + \mathbf{e}\sin\theta)$, the operator $q(\ )q^{-1}$ effects a conical revolution of 2θ about $\mathbf{e}$ on the vector of the operand; thus if $r' = qrq^{-1}$, then

$$Nr' = Nr,\ Sr' = Sr,\ Vr' = Vr \text{ revolved about } \mathbf{e} \text{ through } 2\theta.$$

When $q = \mathbf{e}$, a unit vector, $q^{-1} = -\mathbf{e}$; the operator $-\mathbf{e}(\ )\mathbf{e}$ gives vectors a *half-turn* about the axis $\mathbf{e}$.

The operator $\mathbf{e}(\ )\mathbf{e}$ *reflects* vectors in the plane normal to $\mathbf{e}$.

PROBLEMS

1. Solve the quaternion equations, $rq = p$, $qs = p$, for r and s when

$$q = 2 - i - 2k, \qquad p = 1 + 3i - j + k.$$

Verify that $Nr = Ns$.

2. Every quaternion q satisfies the quadratic,

$$q^2 - 2qSq + Nq = 0,$$

known as its *principal equation*. The conjugate Kq satisfies the same equation.

3. Show that $2 + 5i$ and $2 + 3j + 4k$ have the same principal equation; hence factor its left member in two different ways.

4. Show that if we identify the quaternion units with the 2×2 matrices

$$1 = \begin{pmatrix} 1 & 0 \\ 0 & 1 \end{pmatrix}, \quad i = \begin{pmatrix} -\iota & 0 \\ 0 & \iota \end{pmatrix}, \quad j = \begin{pmatrix} 0 & 1 \\ -1 & 0 \end{pmatrix}, \quad k = \begin{pmatrix} 0 & -\iota \\ -\iota & 0 \end{pmatrix},$$

where ι (iota) is the complex unit, $\iota^2 = -1$, these matrices satisfy the multiplication table (181.7).

5. If $\mathbf{u}, \mathbf{v}, \mathbf{w}$ are vectors prove the following identities:

(a) $\qquad \mathbf{u}^2 = -N\mathbf{u}; \qquad (\mathbf{u} - \mathbf{v})(\mathbf{u} + \mathbf{v}) = \mathbf{u}^2 + 2V\mathbf{uv} - \mathbf{v}^2;$

(b) $\qquad 2\,S\mathbf{uvw} = \mathbf{uvw} - \mathbf{wvu};$

(c) $\qquad 2\,V\mathbf{uvw} = \mathbf{uvw} + \mathbf{wvu};$

(d) $\qquad S(\mathbf{u} + \mathbf{v})(\mathbf{v} + \mathbf{w})(\mathbf{w} + \mathbf{u}) = 2\,S\mathbf{uvw}.$

6. Show that the multiplication table of the units i, j, k is completely given by

$$i^2 = j^2 = k^2 = ijk = -1.$$

7. Interpret the equations, $ij = k$, $jk = i$, $ki = j$ and $kji = 1$, in terms of "arc vectors."

8. Solve the equation $aq + qb = c$ for the quaternion q if a, b, c are known quaternions and $Na \neq Nb$. [$q = (ac - cb)/(Nb - Na)$.]

9. Solve the equation $qa + bq = q^2$. [Reduce to $aq^{-1} + q^{-1}b = 1$.]

10. If $\mathbf{a}, \mathbf{b}, \mathbf{c}$ are vectors for which $V\mathbf{abc} = 0$, prove that $\mathbf{a}, \mathbf{b}, \mathbf{c}$ are mutually orthogonal.

[$V\mathbf{abc} = \mathbf{a}S\mathbf{bc} + V\mathbf{a}V\mathbf{bc}$ and $V\mathbf{a}V\mathbf{bc} \perp \mathbf{a}$; hence $S\mathbf{bc} = 0$, $V\mathbf{a}V\mathbf{bc} = 0$.]

11. If $\mathbf{a}, \mathbf{b}, \mathbf{c}, \mathbf{d}$ are vectors for which $V\mathbf{abcd} = 0$, prove that

(*a*) $\mathbf{bcd}$ is a vector parallel to $\mathbf{a}$;

(*b*) $V\mathbf{bcda} = V\mathbf{cdab} = V\mathbf{dabc} = 0$;

(*c*) the vectors $\mathbf{a}, \mathbf{b}, \mathbf{c}, \mathbf{d}$ are coplanar;

(*d*) $\mathbf{cda}$ is a vector parallel to $\mathbf{b}$; and thus cyclically.

12. A body is revolved through 90 degrees about two axes $\mathbf{e}_1, \mathbf{e}_2$ which intersect at an angle θ ($\cos\theta = \mathbf{e}_1 \cdot \mathbf{e}_2$). Show that the equivalent single rotation is through an angle $2\cos^{-1}\frac{1}{2}(1 - \cos\theta)$ and about an axis parallel to $\mathbf{e}_1 + \mathbf{e}_2 - \mathbf{e}_1 \times \mathbf{e}_2$.

13. If $\mathbf{a}$ and $\mathbf{b}$ are unit vectors along intersecting axes, show that the half-turns $-\mathbf{a}(\ )\mathbf{a}$, $-\mathbf{b}(\ )\mathbf{b}$ in this order are equivalent to a rotation of twice the angle *from* $\mathbf{a}$ *to* $\mathbf{b}$ about their common perpendicular.

14. The quaternion $q = \mathbf{a}_1\mathbf{a}_2 \cdots \mathbf{a}_{n-1}\mathbf{a}_n$ is the product of n unit vectors. Show that if $Sq = 0$, q is the vector

$$\mathbf{q} = \pm\mathbf{a}_n\mathbf{a}_{n-1} \cdots \mathbf{a}_2\mathbf{a}_1 \ (+ \text{ when } n \text{ is odd}, - \text{ when } n \text{ is even}).$$

Hence show that the successive reflections $\mathbf{a}_i(\ )\mathbf{a}_i$ in the order $i = 1, 2, 3, \cdots, n$ reduce to the single reflection $\mathbf{q}(\ )\mathbf{q}$ when n is odd, and to a half-turn $-\mathbf{q}(\ )\mathbf{q}$ when n is even.

15. If the successive reflections $\mathbf{a}_i(\ )\mathbf{a}_i$ in the order $i = 1, 2, 3, \cdots, n$ reduce to a single reflection or half-turn, show that this is true for any cyclical permutation of this order.

16. Show that the succession of reflections in three coaxial planes reduces to a single reflection.

17. Prove the famous theorem of Euler (1776): *Any displacement of a rigid body which leaves the point O fixed is equivalent to a rotation about an axis through O.* [Let the displacement move the trihedral $\mathbf{ijk}$ in the body to the new position $\mathbf{i}_1\mathbf{j}_1\mathbf{k}_1$. This may be accomplished by two reflections: the first takes $\mathbf{i} \to \mathbf{i}_1$, $\mathbf{j} \to \mathbf{j}'$; the second takes $\mathbf{j}' \to \mathbf{j}_1$, leaving $\mathbf{i}_1$ undisturbed.]

18. If $q = \cos\theta + \mathbf{e}\sin\theta$ (not scalar), show that the rotation $q(\ )q^{-1}$ followed by the reflection $\mathbf{a}(\ )\mathbf{a}$ reduces to a single reflection $\mathbf{b}(\ )\mathbf{b}$ when and only when $\mathbf{a} \cdot \mathbf{e} = 0$; moreover,

$$\mathbf{b} = \mathbf{a}q = \mathbf{a}\cos\theta + \mathbf{a} \times \mathbf{e}\sin\theta.$$

19. If we define q^n for arbitrary real n by (185.3), show that any quaternion may be expressed as the power of a vector. When $q = \cos\theta + \mathbf{e}\sin\theta$, prove that $q = e^{2\theta/\pi}$; and that the rotation of angle φ about the axis $\mathbf{e}$ is represented by the operator $\mathbf{e}^{\varphi/\pi}(\ )\mathbf{e}^{-\phi/\pi}$.

20. Show that successive half-turns about three mutually orthogonal and intersecting axes will restore a body to its original position.

21. ABC is a spherical triangle and P, Q are the middle points of the sides AB, BC, respectively. Prove that two successive rotations represented by the arcs AB, BC are equivalent to the rotation represented by twice the great circle arc PQ.

22. Prove the theorem: Three successive rotations represented by the arcs AB, BC, CA of a spherical triangle ABC are equivalent to a rotation about OA through an angle equal to the spherical excess $(A + B + C - \pi)$ of ABC.

23. Prove *Rodriques' Construction* for the composition of two rotations through the angles α, β about the axes OA, OB:

Draw great circle arcs AC and BC such that

$$\text{angle } (AB, AC) = -\tfrac{1}{2}\alpha, \quad \text{angle } (BA, BC) = \tfrac{1}{2}\beta,$$

determining the spherical triangle ABC; then rotation α about OA followed by β about OB is equivalent to the rotation

$$\gamma = 2(CA, CB) \quad \text{about } OC.$$

[Use Hamilton's Theorem given in §187, Ex. 4.]

Give the construction when rotation β about OB is followed by rotation α about OA. Draw a figure to illustrate both constructions.

24. Prove that successive rotations through angles of φ, $\pi/2$, φ about the axes of x, y and z respectively are equivalent to a rotation of $\pi/2$ about the y-axis.

25. If $r = xi + yj + zk$ and $r' = x'i + y'j + z'k$, are position vectors and q is an arbitrary quaternion, show that the transformation $r' = qr(Kq)$ represents the most general rotation and expansion of 3-dimensional space. What is the ratio of expansion?

If $q = d + ai + bj + ck$, express x', y', z' in terms of x, y, z.

INDEX

The numbers refer to pages. A starred number locates the definition of the term in question. The letter f after a number means "and following pages."

Terms under a *noun* (key word) are to be read in *before* this word, unless preceded by a preposition such as "of" or "for." Terms under an *adjective* (key word) are to be read in *after* this word; the repeated adjective is indicated by dash.

A CATALOG OF SELECTED
DOVER BOOKS
IN SCIENCE AND MATHEMATICS

Mathematics–Bestsellers

HANDBOOK OF MATHEMATICAL FUNCTIONS: with Formulas, Graphs, and Mathematical Tables, Edited by Milton Abramowitz and Irene A. Stegun. A classic resource for working with special functions, standard trig, and exponential logarithmic definitions and extensions, it features 29 sets of tables, some to as high as 20 places. 1046pp. 8 x 10 1/2. 0-486-61272-4

ABSTRACT AND CONCRETE CATEGORIES: The Joy of Cats, Jiri Adamek, Horst Herrlich, and George E. Strecker. This up-to-date introductory treatment employs category theory to explore the theory of structures. Its unique approach stresses concrete categories and presents a systematic view of factorization structures. Numerous examples. 1990 edition, updated 2004. 528pp. 6 1/8 x 9 1/4. 0-486-46934-4

MATHEMATICS: Its Content, Methods and Meaning, A. D. Aleksandrov, A. N. Kolmogorov, and M. A. Lavrent'ev. Major survey offers comprehensive, coherent discussions of analytic geometry, algebra, differential equations, calculus of variations, functions of a complex variable, prime numbers, linear and non-Euclidean geometry, topology, functional analysis, more. 1963 edition. 1120pp. 5 3/8 x 8 1/2. 0-486-40916-3

INTRODUCTION TO VECTORS AND TENSORS: Second Edition--Two Volumes Bound as One, Ray M. Bowen and C.-C. Wang. Convenient single-volume compilation of two texts offers both introduction and in-depth survey. Geared toward engineering and science students rather than mathematicians, it focuses on physics and engineering applications. 1976 edition. 560pp. 6 1/2 x 9 1/4. 0-486-46914-X

AN INTRODUCTION TO ORTHOGONAL POLYNOMIALS, Theodore S. Chihara. Concise introduction covers general elementary theory, including the representation theorem and distribution functions, continued fractions and chain sequences, the recurrence formula, special functions, and some specific systems. 1978 edition. 272pp. 5 3/8 x 8 1/2. 0-486-47929-3

ADVANCED MATHEMATICS FOR ENGINEERS AND SCIENTISTS, Paul DuChateau. This primary text and supplemental reference focuses on linear algebra, calculus, and ordinary differential equations. Additional topics include partial differential equations and approximation methods. Includes solved problems. 1992 edition. 400pp. 7 1/2 x 9 1/4. 0-486-47930-7

PARTIAL DIFFERENTIAL EQUATIONS FOR SCIENTISTS AND ENGINEERS, Stanley J. Farlow. Practical text shows how to formulate and solve partial differential equations. Coverage of diffusion-type problems, hyperbolic-type problems, elliptic-type problems, numerical and approximate methods. Solution guide available upon request. 1982 edition. 414pp. 6 1/8 x 9 1/4. 0-486-67620-X

VARIATIONAL PRINCIPLES AND FREE-BOUNDARY PROBLEMS, Avner Friedman. Advanced graduate-level text examines variational methods in partial differential equations and illustrates their applications to free-boundary problems. Features detailed statements of standard theory of elliptic and parabolic operators. 1982 edition. 720pp. 6 1/8 x 9 1/4. 0-486-47853-X

LINEAR ANALYSIS AND REPRESENTATION THEORY, Steven A. Gaal. Unified treatment covers topics from the theory of operators and operator algebras on Hilbert spaces; integration and representation theory for topological groups; and the theory of Lie algebras, Lie groups, and transform groups. 1973 edition. 704pp. 6 1/8 x 9 1/4. 0-486-47851-3

Mathematics–Logic and Problem Solving

PERPLEXING PUZZLES AND TANTALIZING TEASERS, Martin Gardner. Ninety-three riddles, mazes, illusions, tricky questions, word and picture puzzles, and other challenges offer hours of entertainment for youngsters. Filled with rib-tickling drawings. Solutions. 224pp. 5 3/8 x 8 1/2. 0-486-25637-5

MY BEST MATHEMATICAL AND LOGIC PUZZLES, Martin Gardner. The noted expert selects 70 of his favorite "short" puzzles. Includes The Returning Explorer, The Mutilated Chessboard, Scrambled Box Tops, and dozens more. Complete solutions included. 96pp. 5 3/8 x 8 1/2. 0-486-28152-3

THE LADY OR THE TIGER?: and Other Logic Puzzles, Raymond M. Smullyan. Created by a renowned puzzle master, these whimsically themed challenges involve paradoxes about probability, time, and change; metapuzzles; and self-referentiality. Nineteen chapters advance in difficulty from relatively simple to highly complex. 1982 edition. 240pp. 5 3/8 x 8 1/2. 0-486-47027-X

SATAN, CANTOR AND INFINITY: Mind-Boggling Puzzles, Raymond M. Smullyan. A renowned mathematician tells stories of knights and knaves in an entertaining look at the logical precepts behind infinity, probability, time, and change. Requires a strong background in mathematics. Complete solutions. 288pp. 5 3/8 x 8 1/2. 0-486-47036-9

THE RED BOOK OF MATHEMATICAL PROBLEMS, Kenneth S. Williams and Kenneth Hardy. Handy compilation of 100 practice problems, hints and solutions indispensable for students preparing for the William Lowell Putnam and other mathematical competitions. Preface to the First Edition. Sources. 1988 edition. 192pp. 5 3/8 x 8 1/2. 0-486-69415-1

KING ARTHUR IN SEARCH OF HIS DOG AND OTHER CURIOUS PUZZLES, Raymond M. Smullyan. This fanciful, original collection for readers of all ages features arithmetic puzzles, logic problems related to crime detection, and logic and arithmetic puzzles involving King Arthur and his Dogs of the Round Table. 160pp. 5 3/8 x 8 1/2. 0-486-47435-6

UNDECIDABLE THEORIES: Studies in Logic and the Foundation of Mathematics, Alfred Tarski in collaboration with Andrzej Mostowski and Raphael M. Robinson. This well-known book by the famed logician consists of three treatises: "A General Method in Proofs of Undecidability," "Undecidability and Essential Undecidability in Mathematics," and "Undecidability of the Elementary Theory of Groups." 1953 edition. 112pp. 5 3/8 x 8 1/2. 0-486-47703-7

LOGIC FOR MATHEMATICIANS, J. Barkley Rosser. Examination of essential topics and theorems assumes no background in logic. "Undoubtedly a major addition to the literature of mathematical logic." – *Bulletin of the American Mathematical Society.* 1978 edition. 592pp. 6 1/8 x 9 1/4. 0-486-46898-4

INTRODUCTION TO PROOF IN ABSTRACT MATHEMATICS, Andrew Wohlgemuth. This undergraduate text teaches students what constitutes an acceptable proof, and it develops their ability to do proofs of routine problems as well as those requiring creative insights. 1990 edition. 384pp. 6 1/2 x 9 1/4. 0-486-47854-8

FIRST COURSE IN MATHEMATICAL LOGIC, Patrick Suppes and Shirley Hill. Rigorous introduction is simple enough in presentation and context for wide range of students. Symbolizing sentences; logical inference; truth and validity; truth tables; terms, predicates, universal quantifiers; universal specification and laws of identity; more. 288pp. 5 3/8 x 8 1/2. 0-486-42259-3

Mathematics–Algebra and Calculus

Mathematics–Probability and Statistics

BASIC PROBABILITY THEORY, Robert B. Ash. This text emphasizes the probabilistic way of thinking, rather than measure-theoretic concepts. Geared toward advanced undergraduates and graduate students, it features solutions to some of the problems. 1970 edition. 352pp. 5 3/8 x 8 1/2. 0-486-46628-0

PRINCIPLES OF STATISTICS, M. G. Bulmer. Concise description of classical statistics, from basic dice probabilities to modern regression analysis. Equal stress on theory and applications. Moderate difficulty; only basic calculus required. Includes problems with answers. 252pp. 5 5/8 x 8 1/4. 0-486-63760-3

OUTLINE OF BASIC STATISTICS: Dictionary and Formulas, John E. Freund and Frank J. Williams. Handy guide includes a 70-page outline of essential statistical formulas covering grouped and ungrouped data, finite populations, probability, and more, plus over 1,000 clear, concise definitions of statistical terms. 1966 edition. 208pp. 5 3/8 x 8 1/2. 0-486-47769-X

GOOD THINKING: The Foundations of Probability and Its Applications, Irving J. Good. This in-depth treatment of probability theory by a famous British statistician explores Keynesian principles and surveys such topics as Bayesian rationality, corroboration, hypothesis testing, and mathematical tools for induction and simplicity. 1983 edition. 352pp. 5 3/8 x 8 1/2. 0-486-47438-0

INTRODUCTION TO PROBABILITY THEORY WITH CONTEMPORARY APPLICATIONS, Lester L. Helms. Extensive discussions and clear examples, written in plain language, expose students to the rules and methods of probability. Exercises foster problem-solving skills, and all problems feature step-by-step solutions. 1997 edition. 368pp. 6 1/2 x 9 1/4. 0-486-47418-6

CHANCE, LUCK, AND STATISTICS, Horace C. Levinson. In simple, non-technical language, this volume explores the fundamentals governing chance and applies them to sports, government, and business. "Clear and lively ... remarkably accurate." – *Scientific Monthly*. 384pp. 5 3/8 x 8 1/2. 0-486-41997-5

FIFTY CHALLENGING PROBLEMS IN PROBABILITY WITH SOLUTIONS, Frederick Mosteller. Remarkable puzzlers, graded in difficulty, illustrate elementary and advanced aspects of probability. These problems were selected for originality, general interest, or because they demonstrate valuable techniques. Also includes detailed solutions. 88pp. 5 3/8 x 8 1/2. 0-486-65355-2

EXPERIMENTAL STATISTICS, Mary Gibbons Natrella. A handbook for those seeking engineering information and quantitative data for designing, developing, constructing, and testing equipment. Covers the planning of experiments, the analyzing of extreme-value data; and more. 1966 edition. Index. Includes 52 figures and 76 tables. 560pp. 8 3/8 x 11. 0-486-43937-2

STOCHASTIC MODELING: Analysis and Simulation, Barry L. Nelson. Coherent introduction to techniques also offers a guide to the mathematical, numerical, and simulation tools of systems analysis. Includes formulation of models, analysis, and interpretation of results. 1995 edition. 336pp. 6 1/8 x 9 1/4. 0-486-47770-3

INTRODUCTION TO BIOSTATISTICS: Second Edition, Robert R. Sokal and F. James Rohlf. Suitable for undergraduates with a minimal background in mathematics, this introduction ranges from descriptive statistics to fundamental distributions and the testing of hypotheses. Includes numerous worked-out problems and examples. 1987 edition. 384pp. 6 1/8 x 9 1/4. 0-486-46961-1

Mathematics-Geometry and Topology

PROBLEMS AND SOLUTIONS IN EUCLIDEAN GEOMETRY, M. N. Aref and William Wernick. Based on classical principles, this book is intended for a second course in Euclidean geometry and can be used as a refresher. More than 200 problems include hints and solutions. 1968 edition. 272pp. 5 3/8 x 8 1/2. 0-486-47720-7

TOPOLOGY OF 3-MANIFOLDS AND RELATED TOPICS, Edited by M. K. Fort, Jr. With a New Introduction by Daniel Silver. Summaries and full reports from a 1961 conference discuss decompositions and subsets of 3-space; n-manifolds; knot theory; the Poincaré conjecture; and periodic maps and isotopies. Familiarity with algebraic topology required. 1962 edition. 272pp. 6 1/8 x 9 1/4. 0-486-47753-3

POINT SET TOPOLOGY, Steven A. Gaal. Suitable for a complete course in topology, this text also functions as a self-contained treatment for independent study. Additional enrichment materials make it equally valuable as a reference. 1964 edition. 336pp. 5 3/8 x 8 1/2. 0-486-47222-1

INVITATION TO GEOMETRY, Z. A. Melzak. Intended for students of many different backgrounds with only a modest knowledge of mathematics, this text features self-contained chapters that can be adapted to several types of geometry courses. 1983 edition. 240pp. 5 3/8 x 8 1/2. 0-486-46626-4

TOPOLOGY AND GEOMETRY FOR PHYSICISTS, Charles Nash and Siddhartha Sen. Written by physicists for physics students, this text assumes no detailed background in topology or geometry. Topics include differential forms, homotopy, homology, cohomology, fiber bundles, connection and covariant derivatives, and Morse theory. 1983 edition. 320pp. 5 3/8 x 8 1/2. 0-486-47852-1

BEYOND GEOMETRY: Classic Papers from Riemann to Einstein, Edited with an Introduction and Notes by Peter Pesic. This is the only English-language collection of these 8 accessible essays. They trace seminal ideas about the foundations of geometry that led to Einstein's general theory of relativity. 224pp. 6 1/8 x 9 1/4. 0-486-45350-2

GEOMETRY FROM EUCLID TO KNOTS, Saul Stahl. This text provides a historical perspective on plane geometry and covers non-neutral Euclidean geometry, circles and regular polygons, projective geometry, symmetries, inversions, informal topology, and more. Includes 1,000 practice problems. Solutions available. 2003 edition. 480pp. 6 1/8 x 9 1/4. 0-486-47459-3

TOPOLOGICAL VECTOR SPACES, DISTRIBUTIONS AND KERNELS, François Trèves. Extending beyond the boundaries of Hilbert and Banach space theory, this text focuses on key aspects of functional analysis, particularly in regard to solving partial differential equations. 1967 edition. 592pp. 5 3/8 x 8 1/2. 0-486-45352-9

INTRODUCTION TO PROJECTIVE GEOMETRY, C. R. Wylie, Jr. This introductory volume offers strong reinforcement for its teachings, with detailed examples and numerous theorems, proofs, and exercises, plus complete answers to all odd-numbered end-of-chapter problems. 1970 edition. 576pp. 6 1/8 x 9 1/4. 0-486-46895-X

FOUNDATIONS OF GEOMETRY, C. R. Wylie, Jr. Geared toward students preparing to teach high school mathematics, this text explores the principles of Euclidean and non-Euclidean geometry and covers both generalities and specifics of the axiomatic method. 1964 edition. 352pp. 6 x 9. 0-486-47214-0